Bioanalytical Tools in Water Quality Assessment

Bioanalytical Tools in Water Quality Assessment

Second Edition

Beate Escher, Peta Neale and
Frederic Leusch

IWA
PUBLISHING

Published by **IWA Publishing**
Republic – Export Building, 1st Floor
2 Clove Crescent
London E14 2BE, UK
Telephone: +44 (0)20 7654 5500
Fax: +44 (0)20 7654 5555
Email: publications@iwap.co.uk
Web: www.iwapublishing.com

First published 2021

British Library Cataloguing in Publication Data
A CIP catalogue record for this book is available from the British Library

ISBN: 9781789061970 (paperback)
ISBN: 9781789061987 (eBook)

This eBook was made Open Access in June 2021.

Contents

About the Authors ... xv

Preface ... xvii

Foreword ... xxi

Acknowledgements (First Edition 2012) ... xxiii

Acknowledgements (Second Edition 2021) ... xxv

Chapter 1
Introduction to bioanalytical tools in water quality assessment ... *1*

1.1 Background ... 1
1.2 Organic micropollutants ... 2
1.2.1 Defining the issue ... 2
1.2.2 Transformation products ... 4
1.2.3 Low concentrations and mixtures ... 6
1.3 Environmental toxicology ... 6
1.4 Environmental risk assessment ... 8
1.5 Bioanalytical tools ... 9
1.5.1 *In vivo* and *in vitro* bioassays ... 9
1.5.2 Cell-based bioassays ... 12
1.5.3 Modes of action ... 13

1.6 Bioassay selection and design of test batteries 14
1.6.1 Design of test batteries 16
1.6.2 Protection-goal-motivated test battery design 16
1.6.3 Chemical-group-motivated test battery design 17
1.7 Chemical analysis and bioanalytical tools are complementary monitoring tools 18
1.8 Applications 22
1.9 Conclusion 22

Chapter 2
Risk assessment of chemicals* *25

2.1 Introduction 25
2.2 Current risk assessment of chemicals 26
2.2.1 Hazard identification 27
2.2.2 Effect assessment 28
2.2.3 Exposure assessment 30
2.2.4 Risk characterisation 30
2.2.5 Uncertainty analysis 32
2.2.6 Risk management 32
2.3 Application of bioanalytical tools in chemical risk assessment 33

Chapter 3
Water quality assessment and whole effluent toxicity testing* *35

3.1 Background 35
3.2 Derivation of guideline values 36
3.3 Human use of water 37
3.3.1 Drinking water 38
3.3.2 Recycled water, stormwater and managed aquifer recharge 39
3.3.3 Dealing with unregulated chemicals in water 40
3.4 Aquatic ecosystems 40
3.5 Comparison of environmental and drinking water guideline values 42
3.6 Whole effluent toxicity 42
3.6.1 Test systems in aquatic ecotoxicology commonly applied to WET testing 43
3.6.2 *In situ* WET testing 46
3.6.3 Biomarkers in WET testing 46
3.6.4 'WET testing' using bioanalytical tools 47
3.7 Conclusions 48

Chapter 4
Modes of action and toxicity pathways ***51***

4.1 Introduction 51
4.2 Toxicokinetics 52
4.2.1 Uptake, distribution and elimination 52
4.2.2 Xenobiotic metabolism 53
4.2.3 Toxicokinetic indicators of chemical exposure 53
4.3 Toxicodynamic processes: toxicity pathways 55
4.4 Mode of action classification 58
4.4.1 Non-specific toxicity 59
4.4.2 Specific modes of action 62
4.4.3 Reactive toxicity 64
4.5 Keeping the right balance: adaptive stress response pathways 67
4.6 Conclusions 70

Chapter 5
Toxicity pathways of chemicals in humans ***73***

5.1 Introduction 73
5.2 Route of exposure 74
5.3 Basal cytoxicity 77
5.4 Target organ toxicity 77
5.4.1 Hepatotoxicity 77
5.4.2 Nephrotoxicity 78
5.4.3 Cardiovascular toxicity 79
5.5 Non-organ-directed toxicity 81
5.5.1 Carcinogenicity 81
5.5.2 Developmental toxicology 82
5.6 System toxicity 83
5.6.1 Haematotoxicity 83
5.6.2 Immunotoxicity 84
5.6.3 Neurotoxicity 85
5.6.4 Endocrine toxicity 87
5.6.5 Reproductive toxicity 89
5.7 Conclusions 90

Chapter 6
Adverse outcome pathways of chemicals in aquatic organisms ***91***

6.1 Introduction 91
6.2 From the cellular level to the ecosystem 92
6.3 Adverse outcome pathways for aquatic organisms 93

6.3.1 Adverse outcome pathways for algae . . . 94
6.3.2 Adverse outcome pathways for invertebrates . . . 95
6.3.3 Adverse outcome pathways for fish . . . 97
6.4 Using *in vitro* assays to understand toxicity pathways in aquatic life . . . 98
6.5 Conclusions . . . 99

Chapter 7
Dose–response assessment . . . ***101***

7.1 Introduction . . . 101
7.2 Dose–response assessment . . . 102
7.2.1 Dose–response curves . . . 102
7.2.2 Dose benchmark values . . . 103
7.2.3 Continuum of toxicity . . . 104
7.3 Concentration–response assessment . . . 105
7.3.1 'Concentration' versus 'dose' . . . 105
7.3.2 'Response' can mean toxicity or effect . . . 105
7.3.3 Concentration–response modelling . . . 105
7.3.4 Concentration benchmark values . . . 107
7.3.5 Simultaneous effect and cytotoxicity in a cell-based assay . . . 109
7.3.6 Evaluating the linear portion of concentration–effect curves . . . 110
7.3.7 Antagonistic effects . . . 112
7.4 Concentration-response curves of water samples . . . 113
7.5 Bioanalytical equivalency concept . . . 115
7.5.1 Relative effect potency . . . 115
7.5.2 Toxic units and toxic equivalent concentration . . . 116
7.5.3 Effect units and bioanalytical equivalent concentration . . . 117
7.6 Conclusions . . . 118

Chapter 8
Mixtures . . . ***119***

8.1 Introduction . . . 119
8.2 Toxicity/effects of defined mixtures . . . 121
8.2.1 Independent action . . . 121
8.2.2 Concentration or dose addition . . . 122
8.2.3 Synergistic and antagonistic effects . . . 123
8.2.4 Grouping of chemicals . . . 125
8.2.5 Something from nothing? . . . 126

8.3 Assessment of concentration-additive effects using the toxic equivalency concept 129
8.4 Mixtures in risk assessment 130
8.4.1 Concepts 130
8.4.2 Do we have account for mixture effects in risk assessment? 131
8.4.3 Mixtures in chemicals regulations 133
8.5 Mixtures and water quality 135
8.5.1 What type of mixture effects occur in water samples? 135
8.5.2 How much of the measured effects in water sample can be explained by known and detected chemicals? 138
8.5.3 Mixture effects at very low effect levels (<10%) 139
8.5.4 Component-based prediction of mixture toxicity in water 140
8.6 Conclusion 142

Chapter 9
In vitro assays for the risk assessment of chemicals* *143

9.1 Introduction 143
9.2 Application of new approach methods in regulation 144
9.2.1 Alternatives to animal testing methods 144
9.2.2 Integrated testing strategy in the European Union 145
9.2.3 Toxicity testing in the 21st century (Tox21) strategy in the United States 147
9.3 Application of *in vitro* assays in risk assessment 148
9.3.1 A paradigm shift in human health risk assessment 148
9.3.2 Quantitative adverse outcome pathways 149
9.3.3 Quantitative *in vitro* to *in vivo* extrapolation 150
9.3.4 Next-generation risk assessment 151
9.3.5 Applications of new approach methods for environmental risk assessment 154
9.4 Exposure in *in vitro* bioassays 157
9.4.1 Dose-metrics in cell assays 158
9.4.2 Serum-mediated passive dosing 161
9.4.3 Metabolism in cell-based bioassays 163
9.5 Baseline toxicity and specificity of response 163
9.6 Practical considerations for dosing of chemicals 166
9.7 Conclusions 168

Chapter 10
Current bioanalytical tools for water quality assessment *169*

10.1 Introduction 169
10.2 Principles of cell-based bioassays 170
10.3 Bioassays indicative of xenobiotic metabolism 173
10.3.1 Aryl hydrocarbon receptor 173
10.3.2 Peroxisome proliferator-activated receptor γ 175
10.3.3 Pregnane X receptor 178
10.4 Bioassays indicative of hormone receptor-mediated effects . . 179
10.4.1 Estrogen receptor 179
10.4.2 Androgen receptor 182
10.4.3 Glucocorticoid receptor 188
10.4.4 Progesterone receptor 190
10.4.5 Thyroid receptor 194
10.4.6 Mineralocorticoid receptor 195
10.4.7 Retinoic acid receptor and retinoid X receptor 197
10.5 Bioassays indicative of other receptor-mediated Effects 199
10.5.1 Phytotoxicity 199
10.5.2 Neurotoxicity 200
10.5.3 Other assays 202
10.6 Bioassays indicative of reactive toxicity 202
10.6.1 Genotoxicity 203
10.6.2 Mutagenicity 207
10.6.3 Non-genotoxic electrophilic mechanisms 209
10.6.4 Oxidative stress 211
10.7 Bioassays indicative of adaptive stress responses 212
10.7.1 Oxidative stress response 212
10.7.2 p53 response 213
10.7.3 NF-κB response 213
10.8 Bioassays indicative of apical effects 215
10.8.1 Cytotoxicity 215
10.8.2 Algal growth inhibition 221
10.8.3 Fish embryo toxicity 222
10.9 Conclusions 222

Chapter 11
Quality assurance and quality control (QA/QC) *225*

11.1 Introduction 225
11.2 Method validation 226
11.2.1 Accuracy 226

11.2.2 Precision 226
11.2.3 Robustness 227
11.2.4 Quality 228
11.2.5 Matrix interference 228
11.2.6 Sensitivity 229
11.3 QA/QC in the laboratory 231
11.3.1 Practical considerations 232
11.3.2 Replication 233
11.3.3 Quality control samples 235
11.3.4 Control charts and fixed control criteria 240
11.3.5 Standardisation and documentation 242
11.3.6 Guidelines 242
11.3.7 High-throughput screening 243
11.4 Conclusions 243

Chapter 12
Sampling, sample preparation and dosing* *245

12.1 Introduction 245
12.2 Water sampling strategies 246
12.3 Sample pre-treatment options 249
12.3.1 Water sample preservation and storage 249
12.3.2 Water sample filtration 250
12.4 Extraction of water samples 252
12.4.1 Extraction versus testing the entire water sample 252
12.4.2 Solid-phase extraction 253
12.4.3 Passive sampling 254
12.4.4 Liquid–liquid extraction 254
12.4.5 Capturing volatile chemicals 255
12.5 Solid-phase extraction 255
12.5.1 Solid-phase extraction sorbents 255
12.5.2 Solid-phase extraction procedure 256
12.5.3 Effect recovery by solid-phase extraction 257
12.6 Sample collection and sample processing flow chart 260
12.7 Dosing into bioassays 261
12.8 Conclusions 263

Chapter 13
Design of test batteries and interpretation of bioassay results* *265

13.1 Introduction 265
13.2 Test batteries 267
13.2.1 Test battery design 267

13.2.2 Multiplex bioassays serving as test batteries 269
13.2.3 Routine test batteries for monitoring applications 269
13.3 Linking bioassay results with chemical analysis: iceberg modelling 271
13.3.1 Iceberg modelling 271
13.3.2 Effect-directed analysis 275
13.4 Category 1 and category 2 bioassays 278
13.5 Effect-based trigger values 280
13.5.1 Approaches to derive effect-based trigger values for category 1 bioassays 280
13.5.2 Approaches to derive effect-based trigger (EBT) values for category 2 bioassays 286
13.5.3 Approaches to derive effect-based trigger (EBT) values from read-across of *in vivo* data 291
13.6 What to do if a water extract exceeds the effect-based trigger value? 294
13.7 Conclusions 296

Chapter 14
Case studies* *299

14.1 Introduction 299
14.2 Case Study 1: treatment of drinking water 305
14.3 Case Study 2: quality of recycled water 307
14.4 Case Study 3: wastewater treatment 311
14.5 Case Study 4: surface water impacted by wastewater treatment plant effluent 314
14.6 Case Study 5: benchmarking surface water quality across the USA 318
14.7 Case Study 6: benchmarking surface water quality in small streams during rain events 320
14.8 Conclusions 324

Chapter 15
Application of bioanalytical tools beyond water: Sediment and biota* *325

15.1 Introduction 325
15.2 Suspended particulate matter, sediment and soil 326
15.2.1 Suspended particulate matter 326
15.2.2 Sediments 327
15.2.3 Soil 333
15.3 Particles in air and dust 333

15.4 Biota ... 334
15.4.1 Blood ... 334
15.4.2 Tissue ... 335
15.5 Human biomonitoring ... 336
15.6 Conclusion ... 336

Chapter 16
A promising future for bioanalytical tools ... 339

16.1 Introduction ... 339
16.2 Achievements so far ... 339
16.2.1 A sound guidance for selection of bioassays based on the conceptual framework of toxicity pathways ... 340
16.2.2 A more comprehensive measure of the realm of organic pollutants ... 340
16.2.3 Effect-based trigger values ... 341
16.3 Challenges ... 342
16.3.1 Matrix effects and extraction methods ... 342
16.3.2 Dosing into cell-based bioassays ... 343
16.3.3 Linking bioanalysis with chemical analysis ... 344
16.3.4 Linking bioanalysis with whole-animal testing ... 345
16.3.5 Bioassays that require further development ... 346
16.4 Future opportunities ... 347
16.4.1 The 'omics' revolution ... 347
16.4.2 Three-dimensional cell models and organ- and animal-on-a-chip systems to better model whole organism response ... 347
16.4.3 Moving from offline to online monitoring ... 348
16.4.4 Towards ultra-high-throughput testing, multiplex assays and artificial intelligence-assisted bioinformatics ... 349
16.5 The road to regulatory acceptance ... 350
16.6 Conclusions ... 352

Glossary ... 355

References ... 379

Index ... 439

About the Authors

Beate Escher is internationally recognised for her work on chemical pollution in the environment. She pioneered the field of water quality assessment by addressing complex mixtures of chemical pollutants using *in vitro* bioassays. Beate Escher obtained her PhD in 1995 and her Habilitation in 2002 at the Swiss Federal Institute of Technology ETH, Zürich, Switzerland and is head of the Department of Cell Toxicology at the Helmholtz Centre for Environmental Research in Leipzig, Germany and professor at the Eberhard Karls University Tübingen, Germany. She is also lecturer at ETHZ, Switzerland, holds an honorary professorship at the University of Queensland and an adjunct professorship at Griffith University, Australia. She was Associate Editor for Environmental Science and Technology from 2012 to 2020 and is member of the Board of Reviewing Editors at Science. In 2020, she was among the 'Highly Cited Researchers'. From 2011 to 2014 she held an Australian Research Council Future Fellowship. In 2013 she won the Australian Water Association AWA National Research Innovation Award for her work on cell-based bioassays in water quality assessment.

Peta Neale is a research fellow at Griffith University and the central theme of her research is to understand the fate and effect of emerging contaminants in the aquatic environment and engineered systems. She completed her PhD at The University of Edinburgh, United Kingdom, in 2009. She is currently an Associate Editor for Water Research. She was named as Australia's leading environmental sciences researcher by The Australian's 2018 research magazine. Her research covers a range of areas relevant to environmental toxicology and environmental chemistry and includes the application of bioanalytical tools to water, bioassay validation, iceberg mixture modelling and chemical fate and partitioning.

Frederic Leusch is professor and Deputy Head (Research) in the School of Environment and Science at Griffith University, Australia, where he teaches biology and environmental toxicology. Fred also leads the Toxicology Research Group (ARI-TOX) at the Australian Rivers Institute on the Gold Coast, which focuses on assessing the impact of environmental contaminants on humans and aquatic ecosystems. He completed his PhD at Lincoln University, New Zealand, in 2005. His current research focuses on endocrine disruption, developing novel bioassays for water quality assessment, validating ethical alternatives to animal toxicity testing, and the application of systems biology methods to assess exposure to environmental pollutants. He is Executive Editor for Environmental Science and Technology and serves on various national and international committees on issues related to trace organic pollutants and drinking and recycled water quality. He currently chairs the Water Quality Advisory Committee of the National Health and Medical Research Council (NHMRC), which provides advice to the NHMRC on the on-going revisions of the Australian Drinking Water Guidelines.

Preface

The last decades have seen an increase in research activities in the evaluation of organic chemicals that may pollute the aquatic environment and our drinking waters. While the majority of existing research has focused on identification and quantification of individual chemicals by chemical analysis techniques, effect-based methods have emerged in recent years to complement exposure-based measures of chemical contamination that are obtained by chemical analysis. These new effect-based methods include *in vitro* bioassays, and there are an ever-increasing number of bioanalytical tools that hold great promise for applications to water quality assessment.

The objective of this book is to summarise the scientific background underlying the application of bioanalytical tools in water quality assessment for both a specialist and non-specialist audience and to review the state-of-the-science. There is a focus on drinking water, but other water sources such as surface waters (both freshwater and marine), water from the urban water cycle (including wastewater and sewage), industrial effluents and storm water that may be available for beneficial reuse are also included, and we touch on applications of bioanalytical tools in other areas of research and monitoring.

Chapter 1 gives a general overview of the field and provides some background information on the type of chemicals that we are dealing with. The focus is on organic chemicals, such as pesticides, pharmaceuticals, personal care products,

Authors: Beate Escher, Peta Neale and Frederic Leusch
doi: 10.2166/9781789061987_xvii

consumer products and industrial chemicals, as well as their transformation products in the environment and in engineered systems.

Chapter 2 provides an introduction to risk assessment and international regulations of chemicals.

Chapter 3 introduces standards and guideline values defined for various types of water which are presented in a risk-based context. Applications of whole effluent toxicity (WET) assessment, also termed direct toxicity assessment (DTA), are also discussed.

The following chapters provide the scientific basis for bioanalytical tools. **Chapter 4** takes the reader to a cellular mechanistic level, introducing mode of action classification and toxicity pathways that are crucial for the design and application of bioanalytical tools. These cellular-level effects are the common root for effects on human health and ecosystems. **Chapter 5** summarises the potential human health effects from chemical exposure that are triggered by cellular-level effects and introduces related assessment endpoints. **Chapter 6** expands the idea of toxicity pathways, introduced in Chapter 4, to adverse outcome pathways that connect the dots to effects in environmental organisms, populations and ecosystems.

Chapter 7 describes dose–response assessments, data reporting and derivation of benchmark values. It also gives the mathematical background for calculating toxic units (TU) and bioanalytical equivalent concentrations (BEQ). **Chapter 8** provides an overview of mixture toxicity concepts, summarising the way chemicals can interact as mixtures and delving in depth into the concept of toxic equivalency, which is a means of reporting mixture toxicity using a simple metric.

Chapter 9 presents a brief overview of the application of high-throughput *in vitro* bioassay testing for chemical risk assessment. Huge databases of *in vitro* single chemical data have become available in the last 10 years and they serve well for inspiration on the selection of bioassays for water quality testing but also provides input data for the mixture models.

Chapter 10 provides a systematic review of bioassays currently applied for water quality assessment, demonstrating the breadth and depth of the types of endpoints that can now be accessed through effect-based monitoring.

The next chapters delve into practical aspects of bioassay application. Specific QA/QC steps required to ensure that bioassay data are reliable, repeatable and comparable across laboratories and demonstrate good assay practices are discussed in **Chapter 11**, while considerations on sampling, sample preparation and dosing are presented in **Chapter 12**.

Chapter 13 outlines how to develop a bioassay battery for water quality testing, which assays to include, and how to interpret bioassay results, introducing the concepts of iceberg mixture modelling and effect-based trigger (EBT) values.

Chapter 14 illustrates the benefits of including bioanalytical tools for water quality monitoring through selected case studies in surface water quality assessment, wastewater, advanced water treatment and drinking water.

Chapter 15 briefly introduces some applications of bioanalytical tools beyond water quality monitoring, providing some examples with sediment and biota assessment as well as the potential of these tools for human biomonitoring.

The final chapter, **Chapter 16**, provides a synthesis and an outlook to future developments in the field.

In addition, we have created a series of **online resources and tools** to apply some of the principles and data methods explained in this book. This supplementary information is available at www.ufz.de/bioanalytical-tools.

[illegible]

Foreword

The Global Water Research Coalition (GWRC) is an international organisation that is dedicated to the exchange and generation of knowledge to support sustainable development and management of the urban water cycle. The research agenda is developed by the member organisations of the GWRC and reflects their priorities and recognises global trends and drivers that affect the urban water cycle. The agenda of the GWRC always includes the monitoring contaminants of emerging concerns as one of the priority areas.

As we become more and more aware of the large number of pollutants (particularly organic micropollutants) in the aquatic environment, it is no longer possible to evaluate the elimination of the individual pollutants in water treatment plants or to guarantee the absence of their transformation products, including disinfection by-products. It is also difficult to evaluate which mixtures may induce adverse health effects at a later date, given that very low concentrations may already cause adverse effects, for example, endocrine disrupting effects. While these low concentrations are unlikely to pose a significant health concern, there is a scarcity of toxicity information on many of the chemicals currently in commercial use, and in most cases, it is impossible to conduct a proper risk assessment for all organic micropollutants.

The possible health impacts of these substances are of major interest to water operators and public consumers alike. As these concerns are widely debated today, they require a scientific, objective and rigorous assessment of consumer exposures and toxicity. There is an increased requirement to assess the level of risks to human health under high-throughput, cost-effective and predictive monitoring frameworks to better ensure that we limit our exposure to toxic chemicals and avoid early biological effects.

In 2019, the GWRC commenced a multinational research project on 'Effect Based Monitoring for Water Safety Planning'. The project builds upon the knowledge gained during the GWRC EDCI and EDCII Toolbox projects to develop new practices that support the application of bioanalytical tools within an internationally accepted water management framework, such as Water Safety Plans. The main added value of this project is to combine substance-based and effect-based monitoring tools to capture any adverse toxic chemicals missing from current conventional substance-based targeting and demonstrate application of this framework to assess the water quality profiles of different water from resource to tap using key case studies.

One critical barrier to wider implementation of effect-based monitoring methods is the lack of broader understanding of their usefulness in quantifying unknown pollutants and their rich potential applications. This is a key area that the GWRC project and this book aim to address. This book thoroughly lays the foundation behind the science and carefully guides the reader through the concepts needed to develop and successfully apply a battery of *in vitro* bioassays for enhanced water quality assessment. The GWRC is thus pleased to endorse this book and hopes that our joint effort and future reports will be useful to all who are active in the field of understanding and venturing into 'Effect Based Monitoring for Water Safety Planning'.

Stéphanie Rinck-Pfeiffer
December 2020

(Managing Director GWRC)

Acknowledgements (First Edition 2012)

This book was made possible thanks to the generous financial support of the Urban Water Security Research Alliance (UWSRA), a partnership between the Queensland Government, CSIRO's Water for a Healthy Country Flagship, Griffith University and The University of Queensland, Australia. The Alliance was formed to address South East Queensland's emerging urban water issues with a focus on water security and recycling. Particular thanks go to the Director of the Alliance, Don Begbie, for being a great advocate of bioanalytical tools and providing not only research opportunities but also his moral support and advice throughout the review process of this book.

We are indebted to the project reference panel of our UWSRA project 'Bioanalytical tools and risk communication' lead by Greg Jackson, who relentlessly motivated us and opened our eyes to the needs of the regulators and practitioners. The reference panel is composed of Michael Bartkow, Kelly Fielding, David Halliwell, Michael Lawrence, Richard Lim, Julia Playford, Annalie Roux, Louis Tremblay, Heather Uwins and Christine Yeates. We are especially grateful to Greg Jackson, Kelly Fielding, Annalie Roux and David Halliwell for reviewing various draft chapters of this book.

We would especially like to thank our co-authors: Heather Chapman for her significant contributions to Chapters 2 and 3, and Anita Poulsen for helping out with several chapters and scouring the literature for us. We are very grateful to have had the both of you working with us on this book!

We also thank the team of researchers at our home and associated institutions, namely A'edah Abu Bakar, Caroline Gaus, Eva Glenn, Marita Goodwin, Nat Ling Jin, Matti Lang, Miroslava Macova, Erin Maylin, Ben Mewburn, Jochen Mueller, Peta Neale, Janet Tang, Hanne Thoen and Wasa Wickramasinghe at the National Research Centre for Environmental Toxicology (Entox), Julien Reungoat, Maria-José Farré, Kristell LeCorré, Wolfgang Gernjak at the Advanced Water Management Centre of the University of Queensland, and Erik Prochazka, Vicky Ross and Phil Scott at the Smart Water Research Centre (Griffith University) for the enthusiasm they bring to their work every day and the impact they make on the development and application of bioanalytical tools in a both a strict and wider sense, and who not only helped with critical reviews of various draft chapters but also put up with sleep-deprived supervisors and colleagues during the book writing process. You are the ones who make our efforts worthwhile and we owe you!

The friends and colleagues who served as guinea pigs of our target audience (we are afraid that in this particular case *in vivo* testing could not be replaced by *in vitro* tools) are gratefully acknowledged: Rolf Altenburger, Janet Cumming, Meg Sedlak and Michael Warne.

Paul Knittel helped us with the graphic design. Maggie Smith of IWA was a great support during the publishing process.

There are many more people we want to thank as they have been important pieces of the puzzle and have supported and inspired us even if they have not laid a hand on the book itself. We won't list any more names, you know who you are and it is great to work and play with you.

Last but not least, we are grateful to our families for allowing us to vanish into this book for so many weekends and giving us, from time to time, a reality check that there is more to life than writing a book – although we had a lot of fun climbing the treacherous learning curves of the book writing process. We are looking forward to future editions as the field progresses.

Beate Escher
Frederic Leusch
August 2011

Acknowledgements (Second Edition 2021)

Ten years have passed. Our ‘little book’ proved to be much more useful than anticipated, for teaching, to train new students and staff, but also as a resource for this ever-growing field. What was initially a niche topic has exploded in the last 10 years: many barriers have been broken down and concepts have been steadily improved. Effect-based trigger values, a mere concept a decade ago, have been fine-tuned and now receive broader acceptance, although they are still only slowly being taken up by regulatory bodies. The first edition of the book has also been translated into Chinese, opening it up to a much wider audience than we had initially hoped.

Nevertheless, we had not planned on writing a second edition yet. Then came 2020 – the COVID pandemic turned the world upside down and Beate was stuck during lockdown in Brisbane. So close to Peta and Fred, and yet so far. What to do? Well … write a book!

We wish to thank the Global Water Research Coalition and especially Stéphanie Rinck-Pfeiffer for supporting this book by making material from the ongoing project ‘Effect Based Monitoring in Water Safety Planning’ available for this edition. We would like to acknowledge the project team (Geertje Pronk, Stefan Kools, Milou Dingemans, Jerome Enault, Jean-Francois Loret, Gaelle Meheut, Magali Dechesne, Charlotte Arnal) as well as the project technical advisory committee (David

Cunliffe, Dan Deere, Henner Hollert, Yves Levy, Alvine Mehinto, Leo Posthuma, Shane Snyder, Suzanne van Drunick, Etienne Vermeirssen and Viviane Yargeau).

We thank many of our colleagues for their inspiration and collaboration during the last 10 years – there are simply too many to name them all, but you know who you are (!).

We thank the Zelltox team at UFZ for 'testing' the chapters and providing many practical examples for the online resources and tools at www.ufz.de/bioanalytical-tools. We are particularly indebted to Rita Schlichting, Luise Henneberger, Maria König, Christin Kühnert and Lisa Glauch who were the driving forces of many of the innovations in our CITEPro facility at UFZ where we run *in vitro* bioassays now in a fairly high-throughput format. We also thank the ARI-TOX team at Griffith University for their never-ending curiosity, forcing us to constantly rethink our assumptions.

And, of course, we thank those who attended our many courses – starting in 2012 with a bioassay training course in California, becoming more formal within the UFZ Graduate School 'Higrade' and other venues, including courses in Australia, China, Germany and Switzerland – providing much useful input to the second edition.

A group of international students from Bachelor level to PhD level and some fellow researchers not only tested the effectiveness of the teaching material (slides and videos), questions and exercises on www.ufz.de/bioanalytical-tools but also helped to improve these supplementary resources. We thank Maricor Arlos, Kia Barrow, Victor Castaneda, Michelle Engelhard, Laura Fröhlich, Andrea Gärtner, Fritz Kramer, Yuyuan Liu, Yue Meng, Lili Niu, Audrey Ogefere, Pavel Sauer, Stephanie Spahr, Weiping Qin and Nelly Wang for their input and enthusiasm during online classes. Sandy Schöne helped with setting up the webpage www.ufz.de/bioanalytical-tool.

We thank Fabian Fischer for providing the photo of the cell lines on the cover page – they are MCF7 cells stained with Hoechst (33342) and Alexa (Flour 488) and photographed with a A Zeiss PALM CombiSystem enlargement 20× combining bright-field and fluorescence images.

The lake on the cover is the Haubitzer See, a former brown coal mine pit close to Leipzig, Germany, with the coal-fired Lippendorf power station in the background of the front cover and wind turbines on the back cover. From Lake Somerset dam in Southeast Queensland, Australia, on the cover of the first edition to the rehabilitated landscape of the German lake on the second edition – these pictures remind us of the global importance of water but also the threats to our water resources by human activities.

And last but not least, a huge thank you to our families, for putting up with our extended periods of work on the book, including on Christmas Day! They are a deep source of joy and a reminder that there is more to life than 'The Book'.

Beate Escher
Peta Neale
Frederic Leusch
December 2020

Chapter 1

Introduction to bioanalytical tools in water quality assessment

1.1 BACKGROUND

Chemical monitoring provides a quantitative assessment of individual organic contaminant concentrations in a water sample but does not account for the presence of unknown compounds such as transformation products, untargeted chemicals (*i.e.*, not previously known to be present) or for interactions among chemicals. Bioanalytical monitoring, also called effect-based monitoring (EBM), is complementary to chemical analysis and provides information on all bioactive micropollutants present in a sample ranked according to potency, that is, more toxic chemicals are weighted higher than less toxic chemicals.

Classical aquatic toxicity tests used in water quality assessment include *in vivo* assays with fish and aquatic invertebrates that measure endpoints such as mortality, growth, reproduction, behavioural and feeding responses. *In vitro* molecular and cell-based assays offer a sensitive, cost- and time-efficient ethical alternative to classical whole animal testing. The implementation of human and other living organism cell lines in water testing has facilitated high-throughput evaluation of toxicological endpoints relevant for assessing the potential for deleterious human and ecological health effects.

For the purpose of this book, we define 'bioanalytical tools' as cell-based *in vitro* and *in vivo* bioassays that can be run in well-plate formats and that are indicative of specific endpoints relevant for human and/or ecological health. These tools include whole cell assays and assays with genetically modified cells, where natural features have been over-expressed to enable more sensitive detection and/or where foreign

Authors: Beate Escher, Peta Neale and Frederic Leusch
doi: 10.2166/9781789061987_0001

features have been added for visualisation of effects. The cell membrane is an important barrier and target site, and cell-free assays (such as immunoassays and direct receptor binding assays) are generally excluded from this definition, with the exception of some key enzyme assays. Assays with unicellular organisms, such as algae, yeast or bacteria, and some high-throughput whole organisms such as crustacea and fish embryo assays are also included in our definition of 'bioanalytical tools'.

A major advantage of bioanalytical tools is the ability to detect the toxicity of mixtures of known and unknown compounds, whereas chemical analysis can only quantify the concentration of known, targeted chemicals irrespective of toxicity. By measuring the mixture toxicity of a water sample, the bioassay approach includes a risk perspective as it explicitly accounts for the differences in toxicity across different chemicals and for interactions among chemicals in a mixture. Many bioassays yield specific information on a given mode of action rather than merely answering whether or not the cells are dead or alive after exposure to the sample. This mechanistic information can be exploited by running a series of bioassays indicative of a range of different modes of action in parallel. In this way, a comprehensive bioanalytical test battery provides an integrated measure of the toxicity of the biologically active substances in a water sample. A bioassay can also be selected to target a specific protection goal such as the maintenance of hormone balance or photosynthesis.

This book aims to provide a comprehensive understanding of the key concepts and practical issues in the application of bioanalytical tools for water quality monitoring. The focus is exclusively on organic chemicals. It is also possible to target metals and inorganic pollutants with bioassays, however, sample treatment and data interpretation differs between organics and inorganics. In addition, while there are millions of organic chemicals, many of which may never be identified by chemical analysis, the limited number of inorganic elements allows comprehensive chemical analysis of metals and inorganics, reducing the need for effect-based analysis.

1.2 ORGANIC MICROPOLLUTANTS

1.2.1 Defining the issue

Organic micropollutants are a group of man-made chemicals such as pesticides, industrial chemicals, consumer products and pharmaceuticals (Schwarzenbach *et al.*, 2006) (Table 1.1). As the name implies, micropollutants occur in water and the environment in the microgram per litre concentration range ($1\ \mu g/L = 10^{-6}\ g/L = 0.000001\ g/L$) or even lower, in the nanogram to picogram range ($1\ ng/L = 0.000000001\ g/L$; $1\ pg/L = 0.000000000001\ g/L$).

Not all water pollutants are man-made, natural compounds such as human hormones and phytosterols can have adverse effects on aquatic life and natural

Table 1.1 Examples of ubiquitous organic water pollutants.

Origin/Usage	Class	Selected Examples
Industrial chemicals	Solvents	Tetrachloromethane
	Intermediates	Methyl-*t*-butylether
	Petrochemicals	BTEX (benzene, toluene, ethylbenzene, xylene)
	Plasticisers	Phthalates
	Lubricants	Polychlorinated biphenyls (PCB)
	Flame retardants	Polybrominated diphenylethers (PBDE), organophosphates
Consumer products	Detergents	Nonylphenol ethoxylates
	Pharmaceuticals	Antibiotics, painkillers
	Hormones	17α-Ethinylestradiol
	Personal care products	UV filters, hair dye, hydrotropes
Biocides	Pesticides	DDT, tributyltin, atrazine
	Non-agricultural biocides	Triclosan
Disinfectants	Bactericide, virucide	Isopropanol, alkyldimethyl-benzylammonium chloride
Natural chemicals	Taste and odour compounds	Geosmin, methylisoborneol
	Natural toxins	Microcystins, mycotoxins
	Human hormones	Estradiol, estrone, testosterone
	Phytosterols	Genistein, daidzein
Transformation products	Formed from all the above	Further details in Table 1.2

Adapted from Schwarzenbach *et al.* (2006).

toxins such as those produced by plants, cyanobacteria or fungi may be extremely harmful to aquatic life and human health.

In contrast to micropollutants, macropollutants are naturally occurring compounds that exist locally in excess concentration, for example, phosphate and nitrogen compounds, which can lead to eutrophication of surface waters (Schwarzenbach *et al.*, 2006). Macropollutants were the big environmental problem of the 1960s and 1970s. Today macropollutants are usually properly managed with the introduction of source controls and additional wastewater treatment requirements. As a result, attention has shifted to micropollutants, including inorganic and organic chemicals.

The widespread distribution of organic micropollutants in our waterways presents a hazard to aquatic life. Hazards to humans can occur through the consumption of food and drinking water and through other exposure routes such as inhalation and dermal contact. Micropollutants enter the aquatic environment via direct sources, such as industry and sewage effluent discharge, and via non-point sources such as urban runoff and agriculture. Due to the complex nature of the chemical mixtures occurring in domestic wastewater used for water recycling schemes, conventional treatment is not always sufficient to remove the entire contaminant load. Additional treatment steps such as ozonation and sorption to activated carbon have been introduced to wastewater and recycled water treatment to reduce more recalcitrant micropollutants. Membrane processes such as reverse osmosis can also reduce a wide range of micropollutants, but removal efficiency is related to the chemical structure and size, and some compounds may not be fully removed by reverse osmosis alone. Disinfection, such as chlorination and advanced oxidation processes, control human pathogens (microorganisms that cause disease). While conventional biological treatment and advanced treatment processes are very effective in eliminating most unwanted pathogens and many micropollutants, they also introduce other potentially harmful substances such as disinfection by-products and transformation products.

1.2.2 Transformation products

Transformation products are micropollutants that have undergone chemical reaction(s). It is currently unclear, which and how many transformation products are formed, in what quantities and what level of harm they may cause. Transformation products can arise from a variety of sources and can be formed in the environment as well as in engineered systems (Table 1.2). Pharmaceuticals are extensively metabolised in humans and animals and, hence, are typically not excreted in the same form as they were ingested but as a variety of metabolites (Lienert *et al.*, 2007). Many pharmaceuticals are activated inside the body to the pharmacologically active form, which may also be more potent than the precursor with respect to its adverse effect. Most pesticides and other micropollutants undergo biotic and abiotic transformation reactions in the environment. In surface water, for example, exposure to sunlight can cause direct photodegradation or indirect oxidation of micropollutants via formation of reactive oxygen species.

Biodegradation is extensive during biological wastewater treatment, yet full mineralisation (complete degradation to carbon dioxide and water) is incomplete for many chemicals, allowing biotransformation products to be formed. Hydrophobic micropollutants are also removed from water by adsorption to the sewage sludge without any transformation. Existing micropollutants in water can

Table 1.2 Transformation products of organic micropollutants and natural organic matter that are more problematic than the original chemical (for more details see Escher and Fenner, 2011).

Chemical Group	Transformation Process	Issues
Transformation products introduced into the water		
Pharmaceuticals	Metabolism of pharmaceuticals in humans and animals followed by excretion with urine and faeces to wastewater	Conjugates can later be cleaved back to the original compound Some pharmaceutical drugs are more potent after metabolic activation
Transformation products formed in the environment		
Pesticides	Abiotic and biotic reactions in the environment from direct runoff of pesticides	Pesticides (*e.g.*, organophosphates) can be activated by oxidation or other chemical reactions
All micropollutants	Direct and indirect photodegradation in surface water	Can lead to a product more persistent and toxic than the precursor (*e.g.*, photochemical condensation of triclosan to a dioxin-like structure)
Transformation products formed in engineered systems		
All micropollutants	Biodegradation during wastewater treatment	Some transformation products are more persistent than the precursor (*e.g.*, 4-nonylphenol as breakdown product of nonylphenol polyethoxylate)
All micropollutants and natural organic matter	Advanced oxidation and disinfection during water treatment	Disinfection and oxidation by-products (*e.g.*, trihalomethanes and haloacetic acids) from natural organic matter are putative carcinogens

be transformed during advanced oxidation and disinfection processes to more persistent and/or toxic disinfection by-products (Table 1.2).

Disinfection by-products are important contaminants of drinking water and are formed during disinfection from natural organic matter present even in the purest water used as a source for drinking water (Table 1.2). During chlorination of drinking water, a wide range of chlorinated chemicals are formed, for example, trihalomethanes and haloacetic acids. Nitrosamines are further formed during chloramination. Bromate can be formed after ozonation. As some disinfection by-products are known to cause cancer and other adverse effects (Hrudey and Fawell, 2015), their regulation and management is important while keeping in mind that protection from pathogens is of the high priority to safeguard health in the short term.

While most transformation products are less persistent, bioaccumulative and toxic than the original compounds (Boxall *et al.*, 2004), there are a number of prominent exceptions. Some transformation products are more persistent than their compounds of origin and thus accumulate in higher concentrations in the environment. Other transformation products are more toxic than the original chemicals (Escher and Fenner, 2011). An example is nonylphenol, which is a degradation product of the industrial surfactant nonylphenol polyethoxylate (NPE). Nonylphenol is highly persistent, bioaccumulative and in addition to being more toxic than NPE in terms of acute toxicity, it exhibits weak estrogenic effects (Fenner *et al.*, 2002).

1.2.3 Low concentrations and mixtures

Regulators are faced with vast numbers of largely unknown micropollutants and transformation products in water. Individual contaminants may be present at very low concentrations, most far below any concentration expected to cause adverse effects on their own but acting together in mixtures their biological activity may lead to detectable effects.

All chemical analysis is limited to the lowest level of resolution of each analytical method. In most analytical laboratories today, routine analysis is limited to the microgram per litre (µg/L) range, while specialised methods may resolve individual chemicals down to the nanogram (ng/L) or picogram per litre (pg/L) level. Chemical analysis can only identify the tip of the iceberg with no quantitative measure of the fraction that remains unaccounted for (Figure 1.1). Although bioanalytical tools do not quantify the individual components in the submerged part of the iceberg, they contribute a more complete picture of its total size, thereby improving our ability to predict the possible health significance of micropollutants. Bioassays indicative of specific modes of action, such as estrogenicity and genotoxicity, may help refine this picture further by pointing to specific groups of micropollutants with common modes of action, which often, although not always, comprise structurally similar chemicals.

1.3 ENVIRONMENTAL TOXICOLOGY

Environmental toxicology has evolved over the last few decades from an amalgamation of various scientific disciplines including biology, toxicology, environmental chemistry, biochemistry, pharmacology, medicine and ecology. The overall objective of environmental toxicology is to understand the impact of

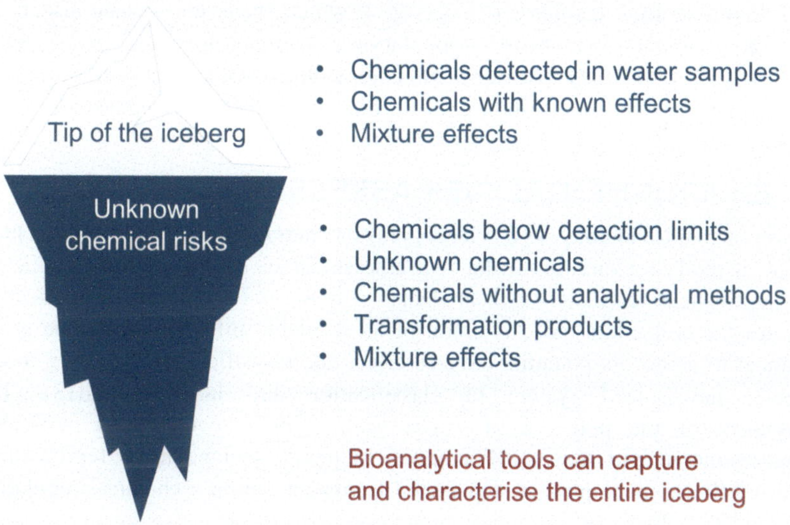

Figure 1.1 Organic micropollutants detected in water samples are only the tip of the iceberg – many other micropollutants may be present, including transformation products and known and unknown chemicals, which act together in mixtures and these mixture effects can be captured by bioanalytical tools.

environmental pollutants on humans and ecosystems, encompassing all levels of biological organisation. Effects range from biochemical interactions within organisms to whole animals, populations and ecosystems. In human toxicology, this sequence is paralleled starting from the toxicity pathways occurring at the cellular level to failure or malfunction at organ level to population effects and epidemiological studies (*e.g.*, cancer clusters).

Traditionally, environmental toxicology has been divided into ecological toxicology (or ecotoxicology) and human health toxicology. While the former discipline was generally associated with environmental sciences and biology, the latter is rooted in pharmacology and medicine. As these disciplines have become more focused at the molecular level, it has been recognised that mechanistic toxicity pathways have many common pathways and modes of action in all biota, and the fields have again grown closer.

Environmental toxicology comprises the following sub-disciplines:

- Environmental science, an interdisciplinary science that studies the earth, air, water, living environments and social components.
- Environmental chemistry and chemo-dynamics, the study of sources, reactions and fate and transport of chemicals in the environment.
- Classical toxicology, which aims to protect human health.
- Epidemiology to understand effects on human populations and communities.

- Ecotoxicology (ecology + toxicology), which seeks to evaluate effects on environmental organisms, populations, communities and ecosystems. Aquatic toxicology is a subset of ecotoxicology, where exposure occurs via aquatic ecosystems including saline, brackish and freshwater systems.

1.4 ENVIRONMENTAL RISK ASSESSMENT

Environmental toxicology, as a science, plays a central role in the development of robust methods for environmental risk assessment of chemicals. Environmental risk assessment consists of four steps (Figure 1.2). Hazard identification includes the collection and evaluation of all available information for the given chemical to assess its potential adverse effect and classification according to the globally harmonized system of classification and labelling (GHS). The assessment of the potential of a chemical to be categorised as persistent, bioaccumulative and toxic (PBT) or carcinogenic, mutagenic or reproduction toxic (CMR) is an important step in the European Union's chemical regulation to identify substances of very high concern. Effect assessment involves dose–response characterisation and extrapolation, which yield predicted 'derived no effect levels' (DNEL) and 'derived minimal effect levels' (DMEL) for humans and 'predicted no effect concentrations' (PNEC) for the environment. The exposure assessment involves evaluation of the expected exposure levels relevant for a given situation. For each exposure scenario, the risk is then characterised by comparing the expected exposure level with the DNEL/DMEL or PNEC, which should correspond to a safe dose over the entire lifetime of a human or environmental organism.

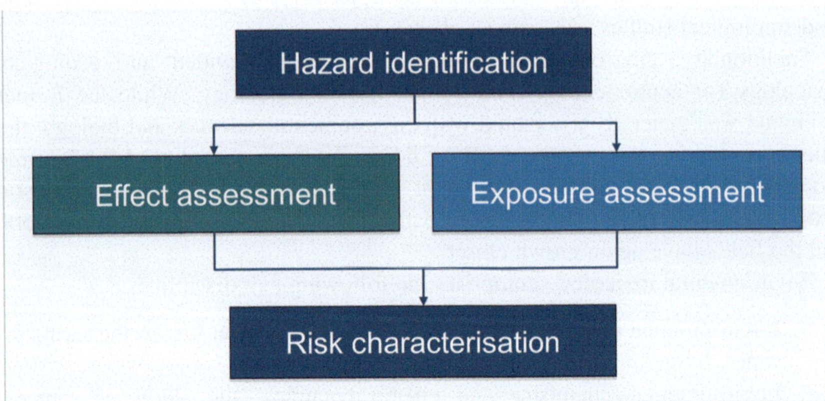

Figure 1.2 Environmental risk assessment of chemicals, simplified from the REACH Guidance Document for Chemical Safety Assessment (European Chemicals Agency, 2011).

In vitro assays are used in an early screening stage of hazard assessment as part of an integrated test strategy but if there is any indication that a chemical is of concern, risk assessment in the European Union legislation REACH needs to be based on *in vivo* information (EP&EC, 2006a). The U.S. risk assessment paradigm is different. The U.S. EPA is committed to reduce animal testing for risk assessment and to incorporate pathway-based *in vitro* toxicology in toxicity and risk assessment (NRC, 2017).

1.5 BIOANALYTICAL TOOLS

Bioanalytical tools are defined as *in vitro* cell-based and *in vivo* bioassays indicative of modes of action that are relevant for human and/or ecosystem health. These include whole cell and reporter gene assays, tests with unicellular and small organisms (*Daphnia*, fish embryo) as well as some enzyme and receptor-binding assays. Previous reviews had wider or narrower definition. Behnisch *et al.* (2001) included, for example, biomarkers and enzyme immunoassays, while Eggen and Segner (2003) only included assays describing a defined chemical–biological interaction excluding general cytotoxicity assays. The European technical report on aquatic effect-based tools under the Water Framework Directive compiled numerous *in vitro* and *in vivo* bioassays including biomarkers that have been applied for water quality monitoring (Wernersson *et al.*, 2015).

1.5.1 *In vivo* and *in vitro* bioassays

Toxicity testing can be performed at several levels of organisation. Epidemiological studies attempt to link observed clusters of disease with human exposure to chemicals. *In vivo* studies on individuals utilise historic case studies of human poisoning or perform animal tests (*e.g.*, using rodents) in order to obtain toxicological information at the whole organism or organ level (Figure 1.3). As with human toxicology, the *in vivo* scope of ecotoxicology may range across organisms, populations, ecosystems and model ecosystems.

In vivo assays are whole organism exposure tests used to determine the toxicity of a chemical, effluent or other mixture of interest to a target organism (Figure 1.4). In addition to survival and reproduction, sub-lethal and behavioural effects can be assessed. Sub-lethal effects are often quantified via biomarkers, which are molecular characteristics that are objectively measured as indicators of a normal biological process or in response to harm (Atkinson *et al.*, 2001).

In vitro technology merges human toxicology and ecotoxicology. *In vitro* assays are in a strict sense all assays that are performed in a controlled environment of a test tube or microtitre plate (Figure 1.4). In practice, the term is often used synonymously with 'alternative test methods' or 'new approach methods' (NAM) that do not make use of test animals. While most *in vitro* assays are cell-based, these also include isolated tissue (*e.g.*, metabolically active liver homogenate) and enzyme extracts. As cell lines (*e.g.*, mammalian, fish, yeast and bacteria) can be

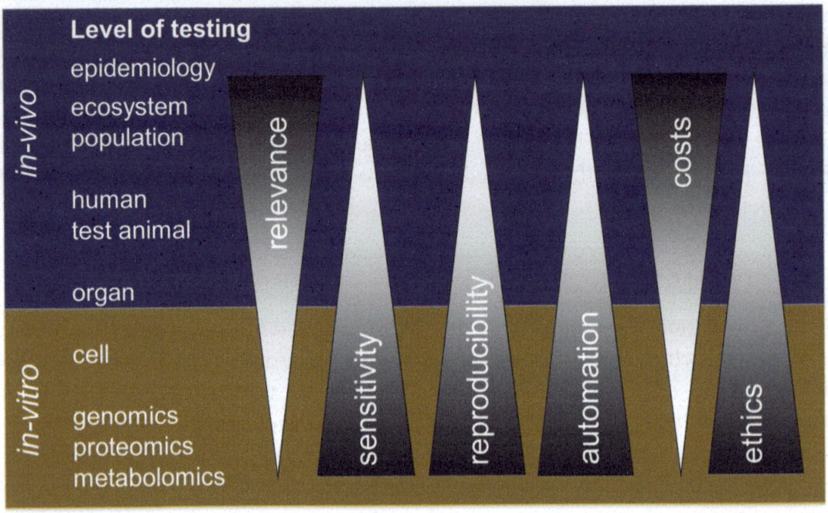

Figure 1.3 Different levels of toxicological testing in human toxicology and ecotoxicology and their agreed attributes.

obtained and grown without sacrificing test animals, molecular and cell-based assays have the advantage of being of low ethical impact compared to *in vivo* assays (Blaauboer, 2002; Hartung, 2010). Some mammalian cells cannot be maintained in culture for a long time and have to be isolated from tissue (primary cells), but other cell cultures, especially mammalian cancer cells and fish cells, are immortal, that is, they can be cultured and reproduced indefinitely.

In vitro assays generally require less space (lower volumes) and are often more practical for assessment of environmental samples with low levels of micropollutants, which need to be enriched prior to toxicity testing. Cell-based assays allow automation and high-throughput screening resulting in time- and cost-effectiveness. Increased sensitivity can also be achieved through genetically modified cell lines with amplified response (Figure 1.4).

Some *in vivo* assays share some of the advantages of *in vitro* assays. The fish embryo test (FET), for instance, is a recommended alternative to traditional ecotoxicological protocols. The FET is used in Germany to assess the quality of wastewater before introduction into environmental waters (Embry *et al.*, 2010). *In vivo* biomarkers such as vitellogenin, a marker for estrogenicity, are also sensitive and informative indicators of endocrine disruption (Purdom *et al.*, 1994) (Figure 1.4).

Yet, while *in vivo* bioassays are valuable for ecotoxicological assessment of pure chemicals, applications for monitoring of water quality are generally limited to whole effluent testing and low-complexity assays including those based on biomarker responses (Figure 1.4). Reproductive and developmental effects are

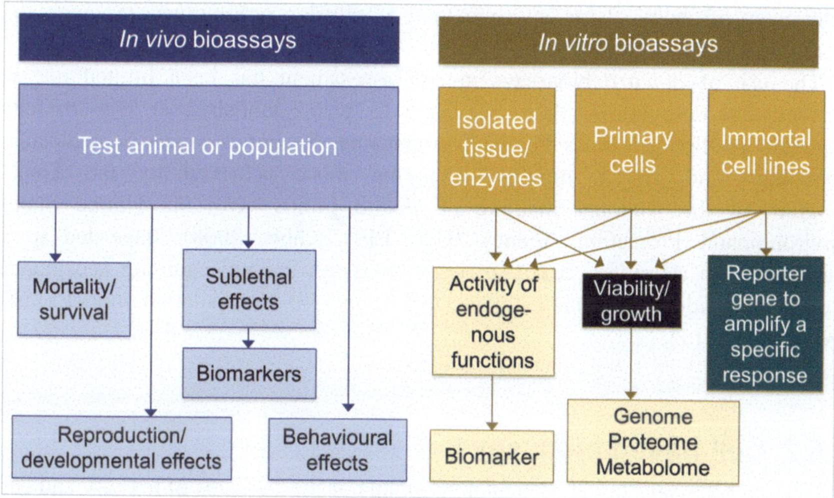

Figure 1.4 Principles of *in vivo* and *in vitro* bioassays used in water quality monitoring.

rarely assessed (with the exception of the FET). As *in vivo* bioassays are typically performed with whole water samples, either directly or in diluted form, only relatively polluted water can be tested using these methods. In contrast, water is typically extracted and enriched before administration to *in vitro* bioassays, thus allowing a much wider range of sample matrices (*e.g.*, from wastewater to drinking water) to be tested.

Additional very promising tools for hazard assessment of chemicals arise from the emerging field of toxicogenomics. Toxicogenomics is the science of applying genomic technologies to elucidate the toxicity pathways and modes of action triggered by a micropollutant (Nuwaysir *et al.*, 1999). Technologies applied in toxicogenomics include profiling at the gene (transcriptomics) and protein (proteomics) expression levels as well as profiling of the metabolic products arising from biological reactions (metabolomics). Ecotoxicogenomics takes this approach one step further by linking these cellular level effects with adverse outcomes for whole organisms, populations and ecosystems (Ankley *et al.*, 2006; Fedorenkova *et al.*, 2010). However, despite significant progress over the last decade in pathway analysis and data curation (including the Comparative Toxicogenomics Database), it is still difficult to link a specific gene, protein or metabolic change to organism, population and ecosystem health outcomes (Bahamonde *et al.*, 2016), a process rendered even more difficult when applied to mixtures of contaminants in environmental samples (Altenburger *et al.*, 2012). Thus, while omics profiling can offer novel insights during hazard identification, toxicogenomic techniques are yet to be validated for use in regulatory risk assessment of chemicals or environmental monitoring programmes (Sauer *et al.*, 2017).

The use of *in vitro* bioassays in risk assessment has been limited due to difficulty in determining their relevance to well-established *in vivo* toxicity tests and predicting effects in whole organisms. Recent advances in molecular toxicology and system biology, including those achieved by the Tox21 program of the National Institute of Health jointly with the United States Environmental Protection Agency (U.S. EPA, Gibb, 2008), have led to a paradigm shift (Hartung, 2010). *In vitro* bioassays are now gaining acceptance provided an (ideally mechanistic) *in vitro* to *in vivo* extrapolation model exists (Wetmore, 2015).

1.5.2 Cell-based bioassays

Cell-based bioassays target particular endpoints or mechanisms of toxicity and can be divided into two groups:

- Bioassays with native cells (primary cells and immortal cell lines)
- Bioassays with recombinant cell lines

1.5.2.1 Native cells

Native cells are cells that have not been genetically modified. Primary cells can be sourced directly from tissue samples but have limited life span *in vitro*. Immortal cell lines are mutated cell lines that can proliferate indefinitely. Immortal cells are preferable due to their high reproducibility and improved animal ethics and cost. In mammals, only cancer and stem cells are immortal. More recently, methods have become available to immortalise cells but so far, they have not been widely used in practical applications for water quality assessment. In fish, any cell type can theoretically be cultured and transformed from a primary to an immortal cell line (Schirmer, 2006).

Native cells typically respond to all bioactive substances in a given sample and are suitable for assessment of non-specific toxicity. Non-specific toxicity is typically measured in bioassays that quantify cell growth/viability (cytotoxicity) (Figure 1.5). Cytotoxicity assays can be more specific if cells are derived from particular tissues such as pulmonary epithelial cells or liver cells. Growth of neuronal cell lines can be used to assess not only cytotoxicity but also neurite development as a more specific endpoint. The differential toxicity between different cell types can further give an indication of the mode of action of the chemicals in the sample. Some cells react specifically to groups of chemicals with common modes of action by expressing a specific physiological response such as direct inhibition of photosynthesis in algae or the proliferation of breast cancer cells in the presence of estrogen.

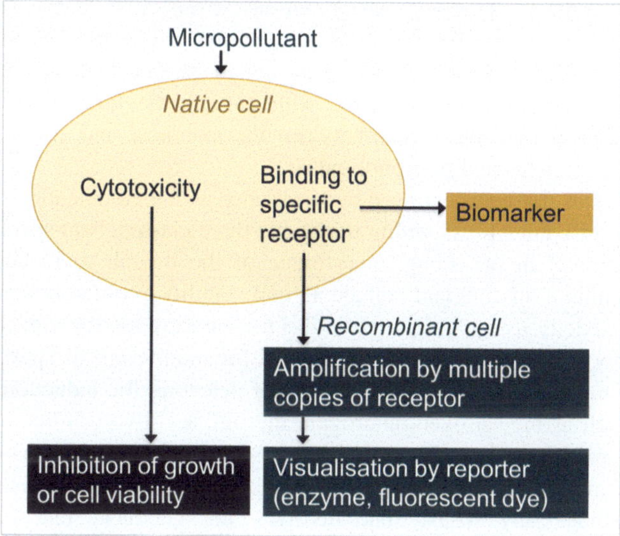

Figure 1.5 Design of cell-based bioassays. Receptor refers to a structural element in the cell to which a chemical can bind and, subsequently, induce processes that may cause toxicity. Reporters are gene products that are not naturally present in the cell but have been introduced through genetic modification to allow visualisation of the receptor binding.

1.5.2.2 Genetically modified cells

Recombinant cell bioassays use genetically modified cell lines and have emerged in the last decade to detect and amplify specific toxic responses (Figure 1.5). Examples include hormone-mimetic activity and induction of the arylhydrocarbon receptor (AhR). The general design of recombinant cell bioassays is the integration of a reporter plasmid into a cell (*e.g.*, human or mammalian immortal cell line). A plasmid is a circular DNA molecule, which carries a responsive element for the receptor of interest, followed by a reporter gene that encodes a measurable feature such as an enzyme (*e.g.*, β-galactosidase or luciferase) or an easily measured fluorescent protein. The amount of response quantified via the enzyme activity or the fluorescence intensity of the fluorescent protein is proportional to the amount of chemical bound to the receptor.

1.5.3 Modes of action

Modes of action (MOA) can be classified into three major groups: non-specific, specific and reactive toxicity (see Chapter 4 for more details). Non-specific toxicity refers to baseline toxicity and is the minimum toxicity that any compound exerts without evoking specific effect. This minimum toxicity occurs at the same critical membrane concentrations irrespective of the cell type or organism (Escher *et al.*, 2019). Cytotoxicity assays will mainly be indicative of non-specific toxicity. Specific toxicity refers to all mechanisms that involve binding to a receptor or interference with an enzyme function. For reactive toxicity, chemical reactions occur between the chemical and cell components, such as DNA, proteins and phospholipids.

Bioassays can also be categorised in accordance with their potential to detect and quantify the different MOAs within the above three classes. Non-specific toxicity assays are crucial in providing an estimate of the overall toxic burden of all chemicals within a mixture and include all cell viability/proliferation assays.

Specific toxicity assays target particular toxicant groups through detection of specific endpoints. Typical bioassays applied for monitoring of specific toxicity include recombinant cell bioassays capable of detecting the induction of nuclear receptors, such as the estrogen, androgen, thyroid, aryl-hydrocarbon and retinoic acid receptors. Bioanalytical capabilities are improving rapidly worldwide by covering an increasing number of receptor-mediated toxicity endpoints. Reactive toxicity includes any MOA that involves the chemical reaction between chemicals and biological molecules, including DNA damage (genotoxicity, mutagenicity), reactivity towards proteins, peptides and lipids, as well as oxidative stress. The main focus has been on detection of mutagenicity and genotoxicity using the classic Ames test (for mutagenicity) and assays indicative of DNA repair. Genotoxicity tests based on mammalian cell lines with detection of DNA damage using the Comet assay and/or the micronucleus assay have also been introduced to water quality testing. Test battery-type approaches combine a number of assays within and/or across the above categories enabling a more comprehensive characterisation of various aspects of toxicity.

1.6 BIOASSAY SELECTION AND DESIGN OF TEST BATTERIES

Cell-based assays have been applied for monitoring of water quality worldwide since the 1960s (Figure 1.6). Early work mainly focused on assays indicative of carcinogenicity (reactive toxicity) and general (non-specific) toxicity. The Ames test (Ames *et al.*, 1975), for example, has been used for water monitoring since the 1970s (Simmon and Tardiff, 1976) and is still widely used. The Microtox assay, which measures bioluminescence inhibition in the marine luminescent bacteria *Aliivibrio fischeri* (formerly named *Vibrio fischeri*) as an indicator of cytotoxicity, was first applied for water samples in the early 1980s (Chang *et al.*, 1981).

Researchers have applied single and multiple assays for water quality assessment for decades; however, since testing was expanded from contaminated sites and effluents to surface waters and highly treated waters, battery applications have dramatically increased particularly in the last decade (Figure 1.6). Sanchez *et al.*

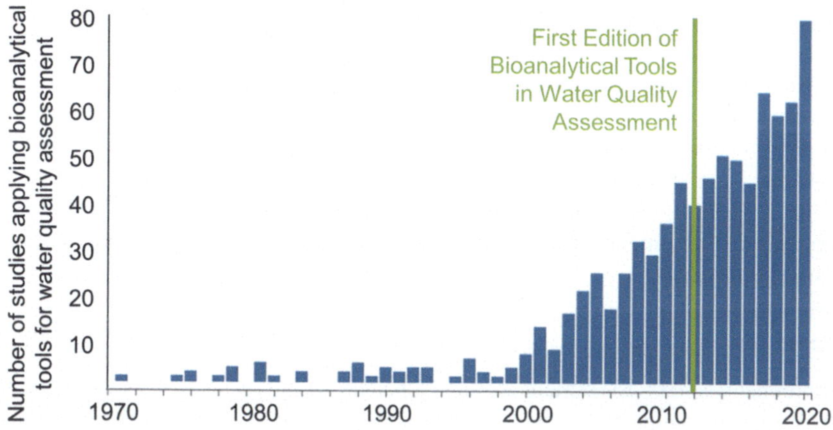

Figure 1.6 Increase of studies that have applied bioanalytical tools in water quality assessment since 1970. Search in Web of Science with the keywords '(*in vitro* or *vitro* or bioanalytical) and battery and bioassay* and water and quality' on 18 November 2020.

(1988) were among the first to employ an assay battery to evaluate the toxicity of industrial effluents. The battery included five acute toxicity assays (three bacterial, one *in vivo* and one molecular) and three mutagenicity assays (the Ames test, and the *Escherichia coli* and *Saccharomyces cerevisiae* (yeast) reverse mutagenicity assays).

Specific toxicity assays emerged in the 1990s to monitor endocrine disrupting compounds (EDCs), which raised much concern due to the potential adverse effects of these xenobiotics to wildlife. In particular, the observation of reduced fish reproduction at environmentally relevant exposure concentrations sparked much attention (Jobling *et al.*, 1998).

The field really started to explode in the early 2000s (Figure 1.6). Mammalian cell-based assays became more abundant in water quality testing as the focus expanded from ecosystem health targeting surface water quality and wastewater treatment to also include human health by considering advanced and drinking water treatment and associated water quality (*e.g.*, Brand *et al.*, 2013; Leusch *et al.*, 2014a, 2014b; Hebert *et al.*, 2018).

Most research has focused on surface waters and domestic and industrial wastewaters. Scattered studies on pulp and paper mill effluents as well as oil field-produced effluents are also found in the literature. In recent years, screening of wastewater and advanced water treatment processes, disinfected drinking water and recreational waters have emerged. Improved sample preparation and sample enrichment methods as well as the introduction of more sensitive bioassay endpoints have enabled progression from highly contaminated water samples to high-quality water such as purified recycled water and drinking water.

Apart from testing quality of different water types, an important application for bioassays has been the assessment of treatment efficacy of a certain technical or natural process, to evaluate trends in effect over time and to benchmark the quality of water from different origins. Hence, the effects are typically compared within a process, along a time axis or across different locations. This allows calculation of treatment efficacies and can help to evaluate natural and engineered treatment processes.

1.6.1 Design of test batteries

Comprehensive risk assessment requires a battery of bioassays to cover a range of MOAs and/or recipients relevant for the water sample to be tested. Two distinct approaches can be applied to design a test battery; one is driven by the protection goal, while the other is driven by the chemical groups of concern and their modes of action (Figure 1.7; see Chapter 13 for more details).

1.6.2 Protection-goal-motivated test battery design

A protection goal targets a health endpoint, organism or ecosystem process that it is desired to be protected. A goal could be to minimise cancer occurrences in humans or to ensure healthy fish reproduction in an aquatic ecosystem. Protection goals set the context for all chemical risk assessment legislation and are often translated into

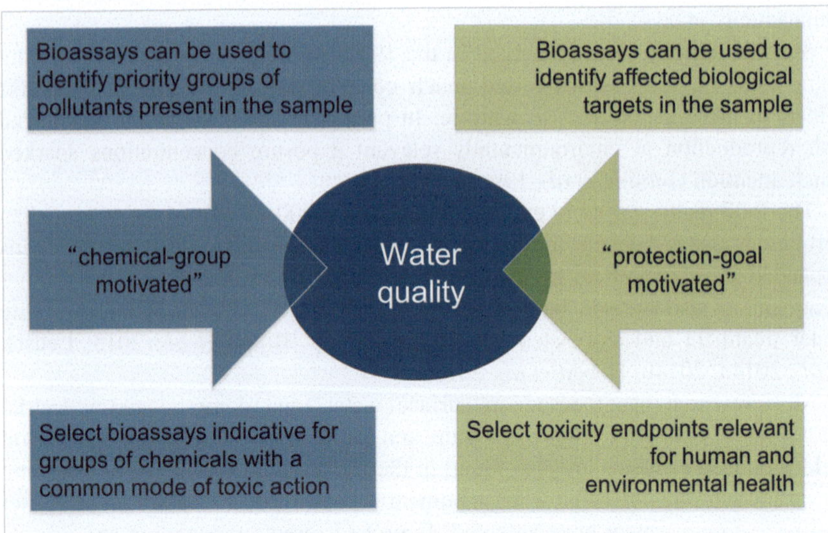

Figure 1.7 Design of test batteries for water quality assessment.

specific assessment endpoints. Depending on the protection goal, a test battery needs to include the relevant assessment endpoints. When selecting a battery of bioassays, it is important to consider what we seek to protect (*e.g.*, human health versus aquatic ecosystem health, marine species versus freshwater species) as well as the suite of tools necessary to conduct the assessment. The most appropriate exposure route (and recipient tissue) must also be carefully evaluated. If exposure to humans is via drinking water, for instance, the oral route is the most important exposure pathway. If, on the other hand, exposure is to recreational water via swimming, dermal contact is likely of higher significance than ingestion. When the relevant organism(s), exposure route(s) and potential risks posed to that organism(s) have been established, the relevant bioassays can be selected. Assays relevant for inclusion will be specific to the protection goals identified as part of hazard identification within a risk assessment framework.

1.6.3 Chemical-group-motivated test battery design

Chemicals with a common MOA and existing together in mixtures act according to the concept of concentration addition (see Chapter 8 for more details). Chemicals with a common MOA are also often (but not always) structurally similar. MOA-specific bioassays can thus be used to identify relevant toxicant groups present in a sample.

As all known and unknown chemicals with a common MOA will contribute to the mixture toxicity in an associated bioassay, application of a MOA-based test battery can help generate a more comprehensive picture of the toxic potency of a water sample than chemical analysis alone. If the sample is suspected to contain hormones (*e.g.*, wastewater), it is sensible to include a test indicative of endocrine disruption, particularly estrogenic and glucocorticoid activity. If herbicides are likely to be present (*e.g.*, agricultural runoff), a phytotoxicity assay will be suitable.

Effect-based batteries can be advanced further by including several assays for a given toxic MOA. Estrogenic effects, for example, can be activated by direct binding of the estrogen receptor (ER) by estrogenic compounds but also by indirect mechanisms such as activation of the AhR by polyaromatic hydrocarbons (PAHs) or inhibition of the cytochrome P450 19A aromatase that transforms testosterone to 17β-estradiol.

Application of broad test batteries covering both non-specific cytotoxicity and several specific endpoints allows the assessor to account for unexpected toxicant groups that may otherwise go undetected. In the chemical-oriented design, quantification of the risks posed by relevant groups of chemicals is prioritised. Bioassays of high sensitivity towards the toxicant group of interest may therefore be selected irrespective of their (lack of) direct relevance to the protection goal. In order to assess drinking water for the presence of herbicides, for example, it may be appropriate to include an algal assay, even if the water tested is destined for human consumption and the protection goal is to achieve good human health.

Photosynthetic organisms are, however, particularly sensitive to herbicide exposure and the test indicates exposure even in the absence of information on effects in humans.

Both test battery approaches may lead to comparable and often overlapping sets of bioanalytical tools as it is not possible to view chemicals independently of their MOA. When researchers design test batteries, they will often consider both approaches. It must further be noted that not all bioassays are fully selective and 100% indicative of a given MOA. All cell-based bioassays will be influenced by a combination of non-specific and specific toxicity. In a water sample, there may be a multitude of chemicals, of which only a fraction will respond specifically to the endpoint featured in the applied assay. Within a range of concentrations, a window typically exists where the specific effect sets in but is not yet suppressed by cytotoxicity. The wider this window is, the more useful a given bioassay is for application with complex water samples.

1.7 CHEMICAL ANALYSIS AND BIOANALYTICAL TOOLS ARE COMPLEMENTARY MONITORING TOOLS

Bioanalytical tools do not replace chemical analytical monitoring. Both approaches provide complementary information (Figure 1.8) and, if combined appropriately, will allow for comprehensive assessment of organic micropollutants in a water sample. What we can detect with chemical analysis is dependent on the sample preparation, the chromatographic method (typically high-performance liquid chromatography (HLPC) for waterborne pollutants but also gas chromatography (GC)) and the detection method (mostly mass spectrometry (MS)). Precise identification and quantification of chemicals in water samples can only be done with target analysis using standards but it is possible to extend the list of target analytes with suspect screening methods that apply a smaller range of standards but can quantify larger sets of chemicals by a combination of retention times and exact mass filtering (Krauss *et al.*, 2010; Schymanski *et al.*, 2014; Alyzakis *et al.*, 2018). Non-target analysis, that is, the identification of chemicals by high-resolution MS, also shows enormous promises (Hollender *et al.*, 2017) and has already been applied to better understand pollution patterns in rivers (Albergamo *et al.*, 2019; Beckers *et al.*, 2020). While there are clear promises of non-target screening, it does not allow accurate quantification of micropollutants, and does not provide any indication of the toxicity of the newly detected pollutants.

Typical monitoring programs routinely assess from 10 to 100 individual chemicals, with more recent research publications including as much as 500 individual chemicals (Malaj *et al.*, 2014; Bradley *et al.*, 2017; Kandie *et al.*, 2020). Despite these amazing advances in analytical chemistry, it will never be possible to quantify all chemicals. 175 million organic and inorganic chemical substances were registered in the CAS (Chemical Abstract Services) Registry at the end of 2020. Several million of these chemicals are commercially available with more

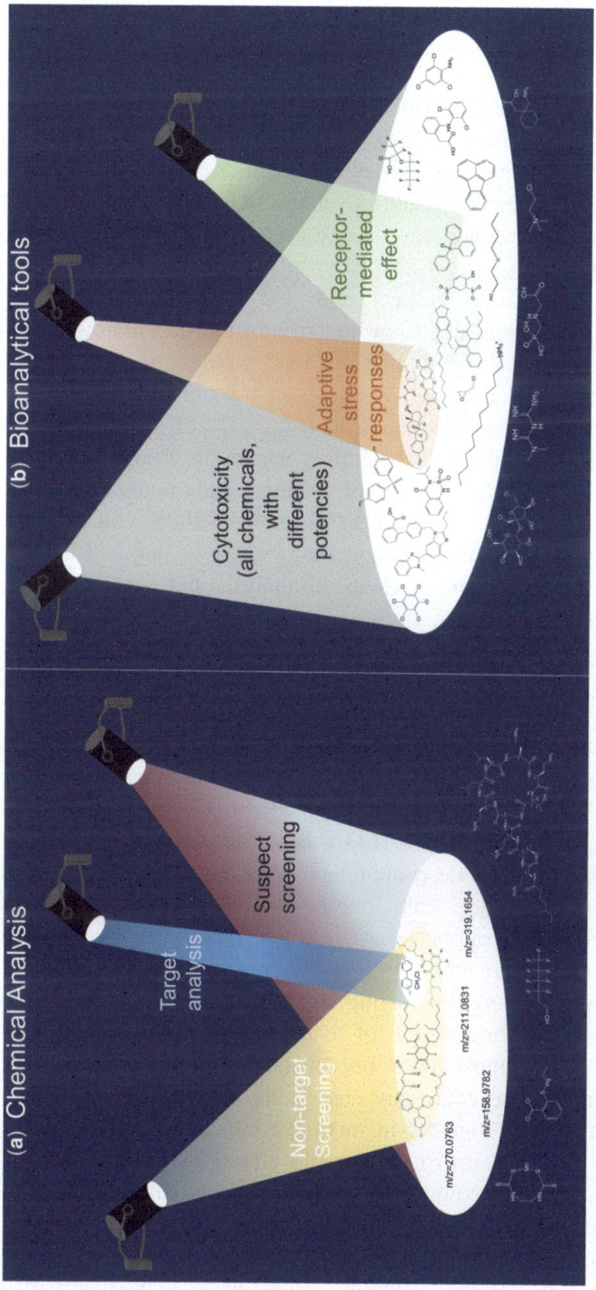

Figure 1.8 The universe of organic micropollutants and the targets for chemical analysis (a) and bioanalytical tools (b). The spotlights are turned onto a small fraction of all organic micropollutants present in a water sample. Adapted from Escher *et al.* (2020d).

than 350,000 estimated to be in commercial use presently (Wang *et al.*, 2020), not counting transformation products. There are more than 1000 registered active ingredients of pesticides in the USA alone, and up to 4000 for pharmaceutical use. In the European Union >100,000 chemicals have been registered in REACH, the European Union industrial chemicals' regulation as of 2020.

Thus, only a small fraction of micropollutants potentially present in water can be monitored by chemical methods. Whether or not the monitored chemicals are relevant from a risk perspective, and cover the majority of the overall toxicant burden, can only be ascertained via mixture toxicity assessment using bioassays. As Figure 1.8a illustrates, target analysis can shed light only on a very small fraction of chemicals. With suspect and non-target screening, many more chemicals can be identified (albeit not quantified), but an unknown fraction still lurks in the dark. With apical endpoints such as cytotoxicity, we can capture and quantify the toxicity caused by all chemicals in a water sample acting together, but we cannot identify the causative agents (Figure 1.8b). By applying bioassays with specific modes of action, we can further narrow in groups of chemicals that act according to the same mode of action, and better characterise the toxicological profile. The results from the different methods can be connected in a quantitative way through mixture modelling, which is outlined in Chapter 13. The capabilities and limitations of chemical analysis and bioanalytical tools are compared in Table 1.3.

Bioassays provide a comprehensive picture of biologically active chemicals present in a sample. It is, however, not possible to elucidate which chemical(s) is (are) responsible for the observed toxicity. To address this, a sample can be fractionated, and the individual fractions tested for biological activity. The active fractions may need to be fractionated further before the chemical(s) that caused the effect can be identified. This so-called bioassay-directed fractionation technique or effect directed analysis (EDA, Brack *et al.*, 2008) is very promising for water samples, where single contaminants dominate the overall toxicity such as, for example, after an accidental release or illegal dumping of a chemical. Generally, and for most applications in water quality monitoring, including wastewater treatment, water recycling and drinking water treatment, there will be no individual component(s) dominating the toxicity. The observed effect will more likely reflect the combination of a large number of chemicals and their transformation products, most likely present at individual concentrations below the thresholds necessary to cause any observable single-chemical effect. Indeed, when toxic chemicals act together in mixtures, concentrations below individual effect thresholds may add up to measurable effects (Silva *et al.*, 2002). Such mixture effects cannot be accounted for using chemical analysis alone.

Bioassays that are selective for specific endpoints, such as binding to ER, will respond to subgroups of chemicals that exhibit common MOAs and act together in mixtures via concentration addition. Non-selective bioassays that detect non-specific indicators, such as cytotoxicity or growth inhibition, are true sum

Table 1.3 Comparison of capabilities and limitations of chemical analysis and bioanalytical tools.

Chemical Analysis		Bioanalytical Tools	
+	Capable of quantifying individual chemicals (typically 10–500).	–	Cannot resolve individual compounds.
		+	Selective for specific endpoints.
		+	Sum parameter for chemicals with the same MOA.
–	Interactive mixture effects cannot be assessed.	+	Detects mixtures of known and unknown bioactive chemicals because all will respond to various degree (weighted by potency).
		±	Mixture toxicity measured but interactive effects cannot be differentiated from simple additive effects.
–	Full extent of chemical burden is unknown.	+	Non-specific endpoints account for the sum of bioactive chemicals present in a sample.
–	In 'clean' water samples, individual components fall below the limit of detection although they may still contribute to cumulative mixture toxicity.	±	Bioassays are often less sensitive than chemical analysis for individual chemicals, in complex mixtures their detection limit is much better.
–	Transformation products need to be identified before they can be quantified.	+	Transformation products are accounted for by measurement of mixture toxicity, but individual contributions cannot be resolved.

parameters of the entire burden of micropollutants in a given water sample (Figure 1.8). Effect-based sum parameters are weighted according to the toxic potency of each individual mixture component and are superior to chemical sum parameters, such as dissolved organic matter, where each component is weighted according to its amount contribution to the mixture regardless of differing individual toxicity.

Biodegradation and advanced oxidation processes will lead to substantial formation of transformation products. Chemical analysis can only quantify transformation products that are known beforehand and/or are present in relatively high concentrations. The identification of unknown transformation products is only possible with highly sophisticated approaches and instrumentation (Kern *et al.*, 2009). As transformation products will contribute to the mixture toxicity, their effect can be quantified with bioanalytical tools although their quantitative contribution cannot be resolved.

1.8 APPLICATIONS

Bioanalytical tools have been applied widely to assess treatment efficiency of technical processes, particularly primary and secondary wastewater treatment (Prasse *et al.*, 2015) (see Chapters 10 and 14 for more details). Advanced treatment using flocculation and oxidation by ozone followed by biological treatment has been shown to reduce specific toxicity (*e.g.*, estrogenicity) to below the limit of quantification, and to substantially reduce non-specific toxicity (Kim *et al.*, 2007; Tsuno *et al.*, 2008; Escher *et al.*, 2009; Stalter *et al.*, 2011). In 2019, 46 comprehensive studies were available that applied bioassays to assess the treatment efficacy of ozonation and activated carbon treatment (Volker *et al.*, 2019). Despite the observed reduction in toxicity, micropollutant concentrations were still found to be sufficiently high to elicit distinct responses in some of the selected bioassays. In this way, the bioassays enabled evaluation of the different steps of the treatment process. In a comparison of ozonation, ultrafiltration and reverse osmosis (RO), Cao *et al.* (2009) found RO to be the most efficient technology for removing genotoxicity, mortality of the water flea *Daphnia magna*, and effects caused by binding to the retinoic acid receptor (RAR). In another study, the final steps of treatment following RO removed residual effects even further, although many endpoints at this advanced stage of treatment fell below detection limits (Escher *et al.*, 2011).

The other main application of bioanalytical tools is to benchmark water quality. This can be done by comparison between sites and time of sampling. In the last years, effect-based trigger (EBTs) values have been developed that serve to differentiate between acceptable and poor water quality (see Chapter 13 for more details). Although EBTs have not yet been implemented in regulation, they are widely used for research purposes and various methods have been developed for their derivation. Most of the EBTs for drinking water have been developed by read across from drinking water guideline values, in some cases using toxicokinetic corrections. EBTs for surface water are also mainly read-across methods from guideline values and environmental quality standards but several methods specifically account for mixture effects.

1.9 CONCLUSION

Water samples contain an innumerable variety of contaminants from human activities, such as pharmaceuticals, personal care products, industrial compounds, pesticides and others, as well as their many transformation products. It is thus impossible to analyse all contaminants in a water sample using chemical analysis techniques. Bioanalytical tools can detect both known and non-target chemicals, and also provide a measure of the potency of these chemicals to interact with biological targets, and ultimately produce an adverse effect in exposed organisms. Bioanalytical tools allow us to view the whole iceberg of micropollutants, not just the top part that we see through a conventional chemical analysis prism.

Chemicals can affect living organisms via a range of modes of action. Some affect cause non-specific effects, which leads to cytotoxicity. Others will cause specific toxic effects, such as binding to a receptor or interference with enzyme function. Others yet can induce reactive toxicity, by interacting with cell components such as DNA, proteins and phospholipids. It is therefore important to consider and include multiple modes of action when designing a test battery. A test battery design can be motivated either by a specific protection goal, or focused on a specific mode of action, and there are a range of bioassays to choose from, from native cells to genetically engineered platforms.

Yet bioanalytical tools also have their own limitations, including the fact that they do not resolve individual compounds present in the sample. Bioanalytical tools complement chemical analytical monitoring, greatly improving the overall assessment of quality of any water type, including wastewater, surface, drinking and reclaimed water.

The aim of this book is to provide the reader with an in-depth perspective on the concepts and ideas in the application of bioanalytical tools to water quality monitoring, including practical advice on sampling, analysis and interpretation of effect-based monitoring. Ultimately, we want to enable the reader to apply bioanalytical tools for water quality monitoring.

Chapter 2

Risk assessment of chemicals

2.1 INTRODUCTION

There exist more than 175 million chemicals and over 350,000 chemicals and their mixtures have been registered in chemical inventories of 19 countries worldwide (Wang *et al.*, 2020). We know still too little about the risk they may pose to humans and ecosystems.

Risk in the context of chemicals is the probability of an adverse effect on humans or the environment occurring from exposure to chemicals. Risk assessment consists of an objective evaluation of risk, in which assumptions and uncertainties are clearly considered and presented. All activities, processes and products have some degree of risk. The ultimate aim of chemical risk assessment is to provide the scientific, social and practical information so that decisions can be made on the best way to manage chemicals. The use of quantitative risk assessment in decision making is becoming increasingly important as situations cannot be judged simply binomially as 'safe' or 'unsafe'. Risk assessment of chemicals is based on scientific evidence, while risk management explores regulatory options by weighing risk assessment with political and socio-economic factors.

The terms hazard and risk are frequently misunderstood and often incorrectly used interchangeably. A hazard is a substance or event that has the potential to cause harm. Risk is the probability or likelihood that this harm will occur. If exposure is low or absent, then the risk is correspondingly low or absent, irrespective of its potential to cause harm. In addition, if exposure is likely but the effects are low or absent, the risk is low. The concentration of a chemical

Authors: Beate Escher, Peta Neale and Frederic Leusch
doi: 10.2166/9781789061987_0025

does not need to be nil necessarily, but it does need to be below a certain threshold level of toxicity. This is based on the premise that at very low doses, below that threshold, the chemical is considered to be safe over a lifetime of exposure. Carcinogenic chemicals are treated differently in risk assessment and are called 'non-threshold chemicals' because it is assumed that there is no safe concentration.

Most countries have traditionally introduced separate legislations for risk assessment of chemicals related to the environment – ecological (or environmental) risk assessment (ERA) – and those related to human health – human health risk assessment (HHRA). While terminology often differs, the essential steps are the same in ERA and HHRA. The borders between ERA and HHRA were broken down with the implementation of the European legislation, REACH (Registration, Evaluation, Authorisation and Restriction of Chemical substances), where a coherent assessment strategy applies for both fields (EP&EC, 2006a), and also the Australian Guidelines for Environmental Health Risk Assessment (enHealth, 2012). In keeping with this logical development, we have attempted to integrate ERA and HHRA in the following overview. There are differences, nevertheless. In HHRA we want to protect everyone, and especially the most vulnerable members of the human population, unborn children, mothers and the elderly from harm, while in ERA we want to protect most species and the ecosystem in its structure and functioning, but not each and every member of the ecosystem.

2.2 CURRENT RISK ASSESSMENT OF CHEMICALS

Risk assessment of chemicals encompasses the evaluation of impacts on human or environment health arising from exposure to those chemicals. In the estimation of risk, a number of steps are required, involving inputs from various disciplines. Regulatory risk assessment of chemicals in most jurisdictions follows the framework developed by the United States Environmental Protection Agency (U.S. EPA, 1976) in the late 1970s (Figure 2.1). This process was reaffirmed by the US National Research Council (NRC, 1983) and has since been adopted in many national regulations among them the European Union (European Chemicals Agency, 2011) and Australia (enHealth, 2012).

Hazard identification sets the scene and is followed by parallel effect and exposure assessment (Figure 2.1). In the risk characterisation step, the probability of exposure is evaluated against the severity of effect and conclusions are drawn that inform risk management. The process is not linear, and includes feedback loops that engage stakeholders, risk assessors, scientists, risk communicators and communities (Figure 2.1).

The various published versions of the four-step framework for risk assessment can be focused on environmental or on human health impacts, either from direct exposure (*e.g.*, food and water consumption) or indirect exposure (*e.g.*, air toxins, recreational exposure). Slight variations in the methods and terminology from this

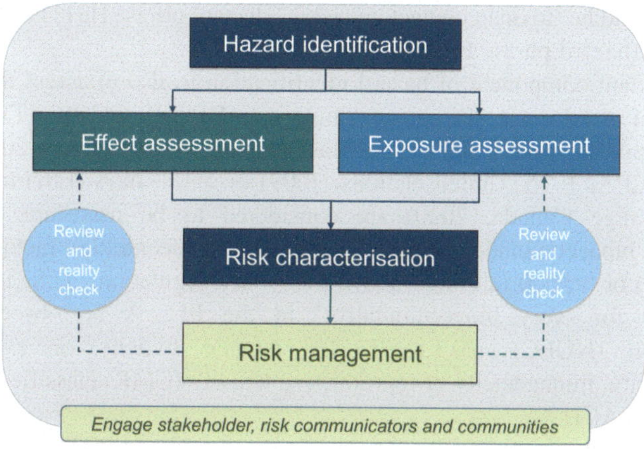

Figure 2.1 Generic risk assessment framework including feedback loops. Adapted from enHealth (2012).

basic framework have developed in different documents (*e.g.*, 'hazard identification' is sometimes called 'issue identification', 'effect assessment' is sometimes called 'hazard assessment' and 'risk assessment' is sometimes called 'safety assessment') but the principles remain similar.

2.2.1 Hazard identification

Hazard identification involves an issue identification step to define why a risk assessment is required and the concerns that the assessment is to address. Then, information on the inherent potential of chemicals to cause adverse effects is collected, including physicochemical properties, information on modes or mechanisms of toxic action, and human health and ecosystem effects.

A crucial outcome of hazard identification is classification and labelling of chemicals and products. In terms of international trade, it is vital to implement an internationally accepted way of labelling. The GHS – 'Globally Harmonized System of Classification and Labelling' (United Nations, 2019) – is widely accepted and has been implemented in many national legislations, for example in 2009 in the EU in the form of the Directive for Classification, Labelling and Packaging as an important complement to the European Chemicals policy REACH. In the GHS, criteria developed to assess the physical, health and environmental hazard of chemicals or products lead to a GHS label in the form of a pictogram (a stylised picture), a signal word and a hazard statement. A pictogram is printed on the packaging and provides an indication of the type of hazard the contents pose, for example, a dead fish with a dead tree indicates 'dangerous for the environment'. Examples of signal words are 'danger' for severe hazard categories or 'warning' for less severe hazard. A standard hazard statement would be 'toxic in contact with skin' (hazard phrase H311) or 'harmful to aquatic life' (hazard phrase H402).

An important component of hazard identification is also to assess the inherent properties of persistence (P), bioaccumulation (B) and toxicity (T) (EP&EC, 2006b). Chemicals with a half-life in water of >40 days (EP&EC, 2006b), >60 days (U.S. EPA, 1976; United Nations, 2009) or >180 days (Environment and Climate Change Canada, 2020) are considered to be persistent. The main criterion for bioaccumulation is that the aquatic bioconcentration factor (BCF) of >2000 (EU) or >5000 (all other above-mentioned regulations, U.S. 1000–5000) and ≥5000 for 'very bioaccumulative' in the EU. A no observed effect concentration (NOEC) <0.1 mg/L for aquatic toxicity or evidence of carcinogenicity, mutagenicity or reproductive toxicity (CMR) classifies chemicals as toxic in REACH.

If all three criteria are fulfilled, chemicals are considered PBT chemicals and a full chemical safety assessment (EU term for risk assessment) must be conducted (European Chemicals Agency, 2011). PBT and CMR chemicals are also added to the candidate list of substances of very high concern (SVHC) in the EU. SVHCs are intended to be phased out eventually, either by removing them from the market entirely (restriction) or allowing them for specific uses only (authorisation).

PBT assessment is also essential for the international Stockholm Convention (United Nations, 2009), which has the goal of protecting humans and the environment from persistent organic pollutants (POP). In addition to being persistent, bioaccumulative and toxic, a POP must also have long-range transport potential, which means it can be found far away from its source. POPs such as organochlorines have been found in the polar regions due to their combination of longevity and physicochemical properties that makes them semi-volatile and hydrophobic (Wania, 2003).

2.2.2 Effect assessment

Dose–response assessment is the term normally used in HHRA and characterises safe levels of exposure to a variety of populations including children and the elderly. To define a safe level, experimental (animal) toxicity data are collected and extrapolated to a 'derived no-effect level' (DNEL) or 'derived minimal effect level' (DMEL) for cancer. Ecological risk assessment on the other hand seeks to protect a percentage (usually 95%) of species from adverse effects by deriving a 'predicted no effect concentration' (PNEC) from experimental ecotoxicity data on selected species that are representative for the ecosystem.

DNELs are derived from the lowest effect level identified in a large set of experimental acute and chronic animal toxicity studies (*i.e.*, the lowest 'no observed adverse effect level' (NOAEL) or 'benchmark dose' (BMD)) and applying relevant uncertainty factors (also called extrapolation, safety or assessment factors, Figure 2.2). The uncertainty factors can range from 10 to

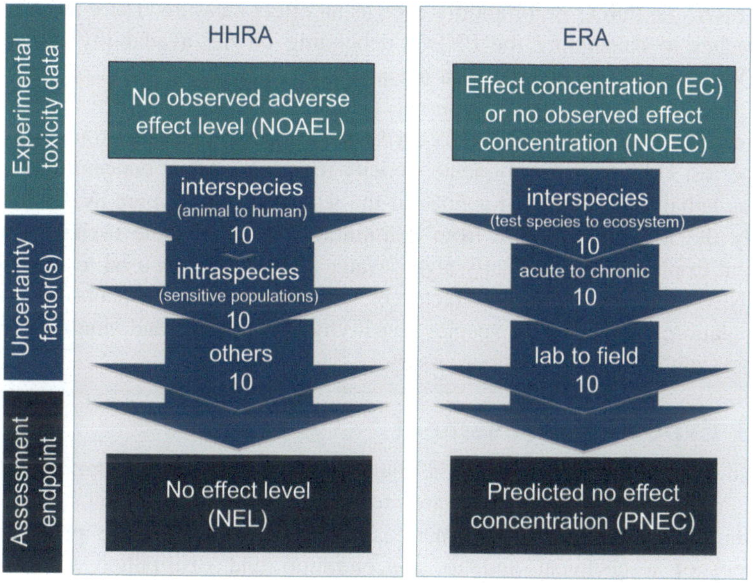

Figure 2.2 Derivation of no-effect levels (NEL) in HHRA and PNEC in ERA.

10,000 and account for extrapolation from animal studies to humans, from individual people to a whole population, for differences in exposure duration, quality and comprehensiveness of the database and for all other uncertainties related to extrapolation from a model system to a human population (Ritter *et al.*, 2007). In HHRA, DNELs, also often referred to as 'acceptable daily intake' (ADI), 'tolerable daily intake' (TDI) or 'reference dose' (RfD), are defined as intakes of chemicals in mg per kg body weight over a set window of time (usually a day, but sometimes a week in the case of tolerable weekly intake (TWI)) that do not pose an appreciable risk over the lifetime of a person (typically 70 years).

For the so-called 'non-threshold effects' (*e.g.*, caused by carcinogens), the effect expected is based on the assumption that the risk is proportional to the dose at all low-dose levels. In other words, the sum of a number of small exposures has the same effect as one larger exposure. While the threshold model assumes very minor exposures are likely to have a negligible effect, it is believed that there is no safe level for carcinogens. A linear extrapolation from the BMD_{10} indicative of 10% tumour incidences to zero is therefore used to calculate a 'cancer slope factor' (CSF) for non-threshold effects, such as chemically induced carcinogenesis. Cancer risk is usually reported as an additional number of people affected out of a million people per year, and 'acceptable risk' is determined by regulation as a likelihood of a deleterious health outcome of 10^{-6} (one in a million) (Australia, Europe for consumers) or 10^{-5} (one in hundred thousand) (WHO, U.S., Europe for workers).

In ERA, the PNEC is commonly used as an effect measure. There are different approaches to calculating the PNEC, depending on the availability of toxicity data. When sufficient toxicity data are available (*i.e.*, toxicity data of preferably more than 15 species from different taxonomic groups), 'species sensitivity distributions' (SSD) can be used to derive the concentration that protects 95% of the species. However, if only acute toxicity data (*e.g.*, effect concentration (EC) such as lethal concentration for 50% of the test species LC_{50}) are available, then usually the lowest EC value from a minimum of three acute toxicity tests at different trophic levels (typically algae, water flea and fish) is used to extrapolate the PNEC with an uncertainty factor of 1000 to account for acute to chronic extrapolation, differences in species sensitivity, lab to field and single organism to ecosystem extrapolation, as necessary.

2.2.3 Exposure assessment

Exposure assessment determines the magnitude, frequency, character, extent and duration of exposure to a hazard for an exposed population. An initial requirement for exposure assessment is an understanding of the presence (or absence) of a chemical and its concentration and distribution in different environmental compartments (air, water, soil, sediment). In the absence of actual exposure data, mathematical models can be used to predict exposure. In these models the emissions are typically estimated from production volumes and knowledge on application and use of the chemicals, and multimedia fate models are used to evaluate the partitioning among different environmental compartments and the degradation processes in each compartment. From these models predicted environmental concentrations (PEC) are derived for multiple compartments.

For HHRA, the uptake of chemicals from various sources (air, water, food) and via various exposure routes (inhalation, ingestion, dermal uptake) are combined to derive a total daily intake.

2.2.4 Risk characterisation

Risk characterisation is the final step in risk assessment and determines whether adverse health effects could occur at a particular exposure concentration. Risk quotients (RQ) as defined in Equation (2.1) are often used and are expressed as the ratio of the exposure to an acceptable effect level. In ERA the exposure level is the PEC and the acceptable effect level is the PNEC.

$$RQ = \frac{\text{exposure level}}{\text{acceptable effect level}} \tag{2.1}$$

Different terminologies for RQ are used in different legislation (*e.g.*, hazard quotient (HQ) is used by some), but their meaning is essentially the same and the only requirement is that both terms used in their calculation, that is, exposure level and acceptable effect level, must have the same units, for example, aqueous concentrations for risk to aquatic organisms or oral dose for HHRA. If RQ <1 no risk is expected. The margin of safety (MOS) or margin of exposure (MOE) is calculated as the inverse of the RQ if RQ <1. It gives a measure of how much difference there is between exposure and effect and is typically applied in HHRA: the larger the MOS, the less concern there is that a chemical might exceed the acceptable effect level.

An RQ ≥1 indicates a possibility of harm. Because exposure is likely to exceed the acceptable effect levels, there is a requirement for a more in-depth risk assessment and/or implementation of risk reduction measures. A chemical can have multiple RQs for different protection goals (*e.g.*, human health or occupational health) and for different environmental compartment (*e.g.*, water, air, soil, sediment).

More recently, probabilistic methods have been introduced to better describe variability and uncertainty of the many factors influencing exposure and effects. Series of measured environmental concentrations or a Monte Carlo simulation of various predicted environmental concentrations from exposure models can be used to construct distributions of exposure levels. Distributions of effect data such as ECs and NOECs capture variability in species sensitivity. If the distributions of exposure and effect data overlap, there is a risk (Figure 2.3a). Conversely, if the upper fifth percentile of the distribution of exposure concentrations and lower fifth percentile of distribution of effect data do not overlap, the chemical should be safe by the indicated MOS (Figure 2.3b).

The process of risk characterisation integrates information from effect assessment and exposure assessment, provides an overview of the quality of the process and describes the risks to individuals, communities and populations. This is the information that is communicated to the risk managers. The summary should

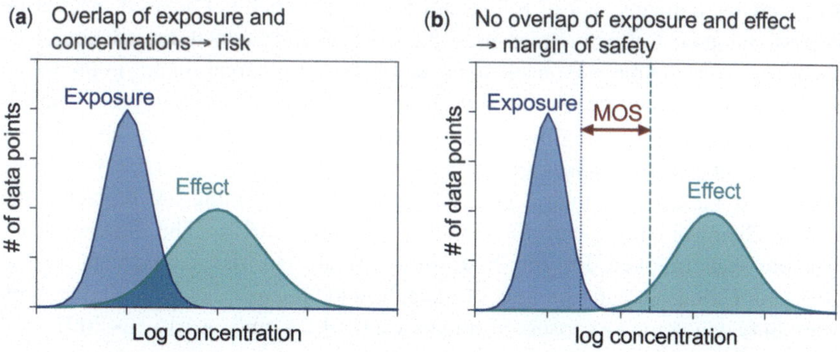

Figure 2.3 Probabilistic risk assessment. (a) Distributions of exposure and effect concentrations overlap, indicating risk; (b) upper 5% percentile of distribution of exposure and lower fifth percentile of distribution of effect data do not overlap and a MOS can be derived. MOS = margin of safety.

include a description of the key issues, and the overall strengths and limitations (including uncertainties) of the conclusions. This may well result in a requirement for additional information to improve the risk characterisation or may determine that no further actions are necessary with the available information. Thus, risk assessment is an iterative process where screening information is used to derive a precautionary initial assessment. If this step identifies a problem, a more refined assessment is carried out to reduce the uncertainty.

2.2.5 Uncertainty analysis

Given that all steps leading to risk characterisation apply simplified assumptions and generalisation, an uncertainty analysis is vital for risk assessment. Uncertainty may relate to lack of or limited knowledge of the true value of any parameters or relationships among parameters. Uncertainty can be caused by indeterminacy, when the true value of a parameter is not known, and variability, when the parameters cover a range, such as temperature, system homogeneity and species' and organisms' sensitivity. In response to the need for uncertainty analysis, the European Directive REACH has implemented a specific guidance document for uncertainty analysis (European Chemicals Agency, 2012). This is especially important when considering that the current international risk assessment have an asymmetric perspective because they minimise the likelihood of positive results by only continuing the process of the RQ ≥ 1 but stopping the process if RQ <1. Hence there could be a false-negative outcome and the reliability of the statement 'no risk' remains unknown.

An answer to uncertainty is the Precautionary Principle, which in its form as Principle 15 of the Rio Declaration (United Nations, 1992) reads as follows: 'In order to protect the environment, the precautionary approach shall be widely applied by States according to their capabilities. Where there are threats of serious or irreversible damage, lack of full scientific certainty shall not be used as a reason for postponing cost-effective measures to prevent environmental degradation'. The Precautionary Principle goes back as far as 1974 as 'Vorsorgeprinzip' in the Clean Water Act of West Germany and was declared as basis of the European Union's environmental policy by the Maastricht Treaty in 1992. History has shown how damaging misuse or neglect of the Precautionary Principle can be when one chemical is banned but many others that are similarly acting are replacing the banned chemical (European Environment Agency, 2013). The Precautionary Principle was invoked for the first time formally in risk assessment in the European Union to ban the use of pentabromodiphenylether due to the high uncertainty concerning exposure of infants via mothers' milk as early as 2001.

2.2.6 Risk management

Risk management is a broader evaluation of the results of the risk assessment and takes into account not only the scientific data but also social, economic and

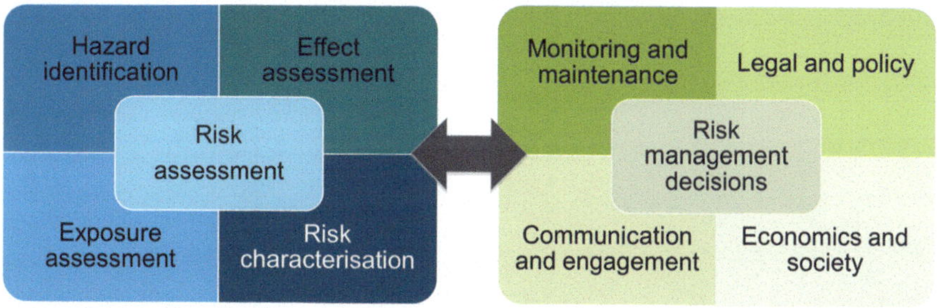

Figure 2.4 Risk assessment/risk management paradigm. Risk assessment and risk management are informed by each other and work towards ongoing improvement in the process.

political considerations (Figure 2.4). Risk reduction measures can aim to replace a chemical or implement control strategies to minimise exposure. Safety standards are set by regulatory action and define safe levels of certain chemicals in various environmental compartments (see Chapter 3).

Risk communication should be seen as a process to enable all stakeholders to make an informed judgement about a risk and its management. There are different perspectives to risk including actual risk, estimated risk and perceived risk. Thorough risk assessment and risk communication minimise the mismatch between the different perspectives. Risk management also monitors and evaluates the effectiveness of the actions.

2.3 APPLICATION OF BIOANALYTICAL TOOLS IN CHEMICAL RISK ASSESSMENT

In vitro methods can support risk assessment at all three main steps. During hazard identification, *in vitro* tools can provide an initial screening of potential toxic effects and classification of modes of toxic action.

During effect assessment, *in vitro* methods can provide additional evidence to (1) identify mechanisms of chemically induced biological activity, (2) prioritise chemicals for more extensive toxicological evaluation and (3) develop predictive models of *in vivo* biological response (Shukla *et al.*, 2010).

During exposure assessment, *in vitro* methods can serve as markers of exposure, similar to biomarkers, if the relationship between chemical exposure and magnitude of effect in an *in vitro* assay is established. Under certain circumstances, *in vitro* methods may also serve as surrogates of chemical analysis, especially for highly specific bioassays that are affected by relatively limited numbers of contaminants.

More recently, there have been first attempts to base the screening-level risk assessment solely on *in vitro* data. This is discussed in more detail in Chapter 9.

Chapter 3

Water quality assessment and whole effluent toxicity testing

3.1 BACKGROUND

Water quality refers to the physical, chemical and biological characteristics of water. For the purpose of this book the focus is on chemical water quality although it should be emphasised that protection from pathogens is a key concern in drinking water quality guidelines. Chemical quality encompasses salts, metals and organic compounds, and our focus is on organic micropollutants.

The classical approach used for chemical water quality monitoring is to compare detected individual chemical concentrations measured by targeted chemical analysis to chemical guideline values (GVs) or standards. Safety standards are defined to protect humans and the environment from unwanted chemicals and are the foundation of water quality-based pollution control. There are several levels of control (van Leeuwen and Vermeire, 2007):

- **Water quality criteria** are based on data, scientific judgement, environmental and human health effects and provide guidance for regulators when they are setting the standards, but they are not laid down in any legislation. Despite this they provide a valuable tool in the management of water pollution.
- **Water quality guidelines** provide recommendations on safe levels, but they are not legally enforceable. They provide targets but exceeding them does not necessarily result in clean up or enforcement actions.
- **Water quality standards (QS)** are upper exposure limits that are enshrined in legislation. They are based on water quality guidelines or derived from

Authors: Beate Escher, Peta Neale and Frederic Leusch
doi: 10.2166/9781789061987_0035

scientifically based water quality criteria both by applying safety factors and by political decision making.

The U.S. and European Countries have defined national standards for drinking water quality: the National Primary and Secondary Drinking Water Regulations as part of the Safe Drinking Water Act in the U.S. and the Drinking Water Directive 2020/2184 of the European Parliament and of the Council (EP&EC, 2020). Australia (NHMRC, 2011) and Canada (Health Canada, 2020) on the other hand rely on the guideline approach at the national level, which some states/provinces have adopted, upon which they become legally binding standards. The World Health Organisation has also defined drinking water guidelines (WHO, 2017b). While these are evidently not legally binding standards, they are meant to assist policy makers in the development of national standards.

There are guidelines for recycled water in some parts of the world. For example, Australia has guidelines for potable water reclaimed from sewage (NRMMC/EPHC/NHMRC, 2008), stormwater harvested for reuse (NRMMC/EPHC/NHMRC, 2009b) and managed aquifer recharge (NRMMC/EPHC/NHMRC, 2009a). The WHO has the 'Potable Reuse: Guidance for Producing Safe Drinking-water' (WHO, 2017c). Some U.S. states also have their own refined guidance documents, for example, the 'Water quality control policy for recycled water' for the State of California (State Water Resources Control Board, 2019).

Surface water guidelines and/or standards are intended to protect aquatic ecosystems. They can have the character of standards, such as in the Water Framework Directive (WFD) of the European Union (EP&EC, 2000) or guidelines, such as the Australia and New Zealand Guidelines for Fresh and Marine Water Quality (Australian Government, 2018a).

These documents provide frameworks for managing water quality, including by setting chemical guideline/standard values for a range of chemicals. The WFD contains environmental quality standards (EQS) for (groups of) 45 priority substances (EP&EC, 2013), while the Australian Guidelines for Water Recycling for Augmentation of Drinking Water Supplies (NRMMC/EPHC/NHMRC, 2008) provide guidelines for over 200 chemicals. Chemical guidelines cannot possibly capture all chemicals potentially present in water, including contaminants of emerging concern. Consequently, the recent revision of the EU Drinking Water Directive allows risk-based monitoring approaches, provided that they ensure full protection of public health (EP&EC, 2020). This revision allows monitoring programmes to focus on chemicals that are relevant for a specific water system.

3.2 DERIVATION OF GUIDELINE VALUES

There is similarity in the approaches for drinking water and surface water despite the difference in protection goals (Figure 3.1). Drinking water GVs are typically derived

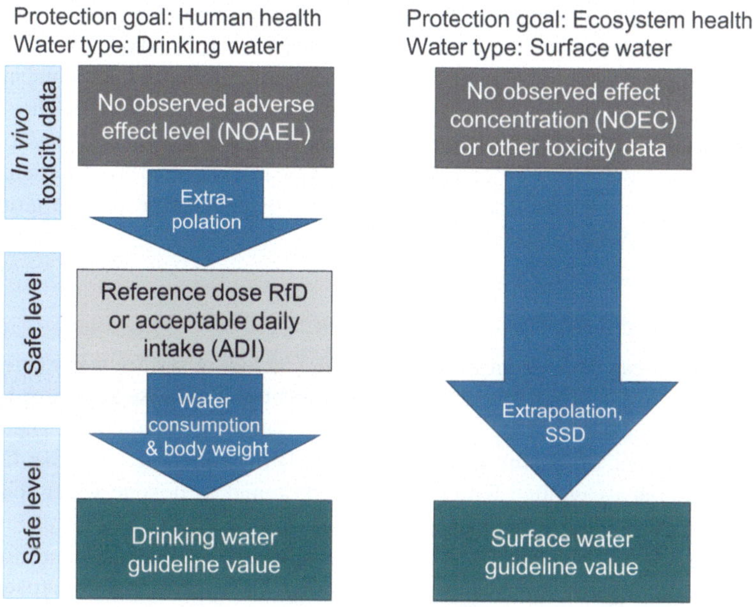

Figure 3.1 Guideline values for single chemicals for drinking water and surface water and how they are derived from *in vivo* toxicity data. NOAEL = no observed adverse effect level, NOEC = no observed effect concentration, SSD = species sensitivity distribution.

from an acceptable daily intake (ADI–also sometimes referred to as tolerable daily intake (TDI) or reference dose (RfD)), which itself is usually derived from a no observed adverse effect level (NOAEL) established from animal toxicity testing and extrapolated to a human context (WHO, 2017b, Baken *et al.*, 2018). The ADI represents a daily intake that can be ingested over a lifetime without adverse health effects. The ADI is multiplied by the average human body weight (typically 60–70 kg) and divided by water consumption (typically 2 L/day) to derive a safe concentration in water. This safe concentration is also often multiplied by a relative source contribution factor to account for other sources of exposure (not shown in Figure 3.1).

Surface water GVs are derived from no observed effect concentrations (NOECs) using species sensitivity distributions (SSD) or extrapolation methods (European Commission, 2011; Australian Government, 2018a) (Figure 3.1).

3.3 HUMAN USE OF WATER

Many chemical guidelines have been developed from risk assessment processes based on typical exposure scenarios. For humans, the exposure is usually based on a daily consumption of 2 L of water per day over a lifetime, typically 70 years. This information is used on the assumption that there will be no toxicological effects despite continuous exposure over the entire lifetime. Human health standards in food, water and air are based on the fact that for the vast majority of chemicals there is a safe level of exposure, below which no adverse health effects occur. Chemicals can sometimes be grouped with a typical compound used to represent the whole group, such as, for example, benzo[a]pyrene for polycyclic aromatic hydrocarbons (PAHs).

3.3.1 Drinking water

Drinking water can come from a range of sources including surface water (streams, rivers and/or lakes), groundwater or rainwater directly. It can also be sourced from seawater or wastewater treated and purified for human consumption. There are growing numbers of chemical GVs and standards emerging, but the list of regulated chemicals is often based on chemicals found in pristine water sources, which are not applicable to less conventional water sources. Not all chemicals have standards or guidelines if they are reasonably expected not to occur in a drinking water source. Increasing global human population now encroach on water catchments, with impacts from urban development, agriculture and forestry, and seepage from landfill and runoff from mining to name just a few. In many low- and middle-income countries there is simply insufficient access to safe water sources causing local human populations to use water of impaired quality. It is important to note that some chemicals do not have water quality criteria because we do not yet have sufficient toxicological data to establish them. The current World Health Organisation guidelines (WHO, 2017b) emphasise preventative management of drinking water quality and the use of multiple treatment barriers.

The Safe Drinking Water Act implemented by the United States Environmental Protection Agency (U.S. EPA) sets legal limits on certain contaminants in drinking water. They define the drinking water equivalent level (DWEL) as the concentration in drinking water that over a lifetime exposure is protective of human health, at least for threshold chemicals (*i.e.*, non-carcinogenic chemicals). The DWEL reflect the best available technology at the time and are subject to ongoing review. In addition to the legal limits, the U.S. EPA determines water testing schedules and methods that water providers must follow. Updated drinking water standards and health advisories have recently been published by the U.S. EPA (2018).

In Europe, EU member states must comply with the EU Council Directive on the quality of water intended for human consumption (EP&EC, 2020) but they can have separate national regulations, provided they comply with the overarching Directive. The Directive requires a regular monitoring programme using the analytical methods specified therein, or equivalent methods. In the previous version, only few organic contaminants were specifically regulated in the EU. The broad standard for individual pesticides was initially set in 1998 at 0.1 μg/L with all pesticides combined not exceeding 0.5 μg/L (EP&EC, 1998). The 2020 revision of the Directive laid down the essential QS at EU level (EP&EC, 2020) that were derived from the WHO drinking water guidelines requiring monitoring and regular testing of 48 microbiological, chemical and indicator parameters.

The Australian Drinking Water Guidelines (ADWG) are built around a 12-point framework for the management of drinking water quality (NHMRC, 2011) and provide the Australian community and water industry with guidance on the provision of safe drinking water. The ADWG are part of the National Water Quality Management Strategy (NWQMS, Australian Government, 2018b), a nationally coordinated framework to facilitate consistent water quality management across different types of waters (fresh water, marine water, groundwater, estuarine water and recycled water) intended for a variety of uses (for drinking, the environment, primary industry, recreation, industry and cultural and spiritual values). The ADWG are subject to a rolling revision process with regular amendment, the latest being from 2018, and the GVs have been risk-based all along (NHMRC, 2011). The ADWG are intended to provide a framework for good management of drinking water supplies that, if implemented, will assure safety at the point of use. The ADWG does not provide mandatory standards but gives guidance to agencies that have responsibilities associated with the supply of drinking water, including catchment and water resource managers, water regulators and health authorities in the states and territories of Australia.

3.3.2 Recycled water, stormwater and managed aquifer recharge

Similar to the WHO and ADWG, the Australian Guidelines for Water Recycling (AGWR, phases 1 and 2, NRMMC/EPHC/AHMC 2006; NRMMC/EPHC/NHMRC 2008, 2009a, 2009b) are based on a preventative approach to water safety management. Phase 1 of the AGWR provides a generic framework for management of recycled water quality that applies to all combinations of recycled water and end uses. These guidelines provide specific advice on the reuse of treated sewage and grey water for purposes other than drinking and environmental flows. Phase 2 extends the guidance in phase 1 on the planned use of reclaimed water from sewage and stormwater to augment drinking water supplies. The document focuses on the source of the water, initial treatment processes and the blending of the water with drinking water sources (NRMMC/EPHC/NHMRC, 2008). There is an increasing emphasis on the use of a multi-barrier approach in preventing water quality incidents, rather than a response when one occurs.

The AGWR phase 2 (NRMMC/EPHC/NHMRC, 2008) provide significantly more GVs than the existing ADWG. This is because the source waters, in this case sewage and stormwater, are expected to contain a broader range of chemical contaminants than conventional drinking water sources such as protected surface water catchments. The process for setting these guidelines is a hierarchical decision tree involving a number of steps including determining a list of chemicals of interest, if there is an existing guideline, if the chemical is a pharmaceutical and if there is health and toxicological information on which to base the setting of a guideline. In the absence of these data it is then determined if the chemical is likely to cause cancer, in which case it would be classified as having no threshold of effect. If it is not causing cancer, then a threshold of toxicological concern can be calculated as a conservative estimate of safe concentrations and used together with exposure assessment data as a basis for risk characterisation. This is a precautionary approach and protective of public health.

3.3.3 Dealing with unregulated chemicals in water

A serious drawback of most drinking water regulations is that they cannot react promptly to contaminants of emerging concern. The AGWR (NRMMC/EPHC/NHMRC, 2008) and Schriks *et al.* (2010a) have proposed an pragmatic approach to derive provisional drinking water GVs for unregulated chemicals once they have been detected in drinking water in the absence of statutory GVs and this work was updated recently (Baken *et al.*, 2018). The approach prioritises available ADI, RfD or TDI values, and if those are not available, ADIs are calculated from available toxicity data or, if needed, thresholds of toxicological concern (TTC). When comparing GVs derived via this approach with measured water concentrations of emerging pollutants, there is often a substantial margin of safety between the measured water concentration and the provisional GVs, so that there is no immediate action necessary (Schriks *et al.*, 2010a, Baken *et al.*, 2018). As mixtures become more and more complex and as derivation of health-based GV entail a risk-based approach, Dingemans *et al.* (2019) also suggested that effect-based methods might be implemented in future drinking water legislations.

3.4 AQUATIC ECOSYSTEMS

The overarching goal of water quality risk assessment for ecosystems is to protect biodiversity. A very important component of this is a thorough characterisation of the ecosystem at risk and the environmental value placed on that ecosystem. Environmental values are defined as values of the environment used for a healthy ecosystem or for public benefit, welfare and safety and that require protection. Important environmental values include aquatic ecosystems, primary industries, recreation and aesthetics, and cultural and spiritual values (Australian Government, 2018a). All water resources are subject to at least one environmental value and in most cases several apply.

In the U.S. the Federal Water Pollution Control Act (U.S. EPA, 1976) from 1948 with amendments through to 1987 (now the Clean Water Act CWA) employs a variety of regulatory and non-regulatory tools to reduce direct pollutant discharges into waterways and to manage pollutant runoff. The intention is to provide a range of tools to achieve the broader goal of restoring and maintaining the chemical, physical and biological integrity of waters that can support aquatic life. In the earlier years of the legislation there was a focus on regulating discharges from point sources such as municipal sewage treatment plants and industrial facilities. Since the 1980s efforts to reduce non-point source pollution (*e.g.*, from runoff) have been introduced including cost sharing with landowners as a key tool. Under the CWA, the U.S. EPA has implemented control programmes such as setting water QS (U.S. EPA, 2018). This includes numeric and narrative water quality criteria, for example, 'waters shall be free from toxic pollutants in toxic amounts'. Whole effluent toxicity (WET) testing, which is described in more detail in Section 3.6, is an important component of the National Pollutant Discharge Elimination System. The most recent update (2015) includes 50 chemical CMCs (criterion maximum concentration) for the protection of aquatic life.

In Europe, the WFD (EP&EC, 2000) has set a goal to achieve 'good ecological status' and 'good chemical status'. Good chemical status equates to achieving the QS established for chemical substances at the European level. 45 priority substances have been assigned EQS (EP&EC, 2013), which have been derived according to a technical guidance document for the derivation of EQS (European Commission, 2011). In addition, there is a surface water 'watch list' of potential water pollutants that are monitored to determine the risk they pose to the aquatic environment and whether they should be included in the priority list. The EQS are linked with emission limit values and discharge permits to ensure compliance with the WFD. In the technical guidance document, QS are defined for the three environmental compartments water, sediment and biota, considering the various receptors at risk (humans, benthic biota, pelagic biota and top predators (birds and mammals)). QS for biota refer the consumption of fish by humans or secondary poisoning of aquatic organisms. Not all combinations of compartment and receptor require the definition of QS for a given chemical in relation to the physicochemical properties that define its environmental fate. However, if several combinations are relevant, for example, for a hydrophobic and bioaccumulative chemical, the QS are derived for all compartments and QS_{biota} and $QS_{sediment}$ are translated to water concentrations. The lowest of these values is adopted as the overall EQS. The effect assessment conducted in REACH (EP&EC, 2006a) and the approach to estimate the QS share many principles of derivation. There are two types of EQS defined in the WFD:

- the annual average EQS (AA-EQS) refer to the annual average concentrations and are derived from chronic toxicity data, and
- the maximum acceptable concentrations EQS (MAC-EQS) refer to the maximum concentration measured and are derived from acute toxicity data.

Von der Ohe *et al.* (2011) evaluated and prioritised 500 existing and emerging micropollutants with this method and found in a monitoring study covering four

European river basins that 44 of these 500 micropollutants exceeded the tentative EQS, most of them pesticides. Monitoring of 223 pollutants at 4000 sites confirmed that European freshwater is under pressure and that poor chemical status was associated with poor ecological status (Malaj *et al.*, 2014).

In Australia, all water quality is managed within the NWQMS (Australian Government, 2018b), as noted above. The Australian and New Zealand Guidelines for Fresh and Marine Water Quality (Australian Government, 2018a) provide specific guidance to manage fresh and marine water quality for a variety of uses, including the protection of aquatic ecosystems. There are default GV for various stressors but site-specific GV are recommended that are relevant to local conditions. Where possible, default GVs are derived using the SSD approach; for chemicals with insufficient toxicity data for an SSD approach, GVs are derived from predicted no effect concentration (PNEC) using an assessment factor approach (ANZECC/ARMCANZ, 2000). The national guidelines are not mandatory. Their enforcement is a state or territory responsibility through their legislation. This extends to the requirement for WET testing (referred to locally as direct toxicity assessment (DTA); see Section 3.6), which has now been incorporated into discharge licences in various parts of Australia under state and territory legislation for discharge to aquatic ecosystems. The most recent update (2018) includes 138 default GVs for chemicals or chemical groups.

3.5 COMPARISON OF ENVIRONMENTAL AND DRINKING WATER GUIDELINE VALUES

It might come as a surprise but for many chemicals the drinking water GV are higher than their equivalent GV for the protection of aquatic ecosystems. For example, Figure 3.2 compares WHO drinking water guidelines (WHO, 2017a, 2017b, 2017c) with the relevant European EQS values. This difference can be rationalised by the fact that environmental GVs need to protect the most sensitive aquatic species and that aquatic organisms are continuously exposed to the water in which they live, while drinking water GVs protect just one species (humans) only intermittently exposed to the water when they drink (an average of 2 L/day).

3.6 WHOLE EFFLUENT TOXICITY

Whole effluent toxicity (WET), whole effluent assessment (WEA) and direct toxicity assessment (DTA) all refer to the assessment of the combined toxicity of the mixture of all micropollutants in an effluent sample to aquatic organisms using a suite of standardised aquatic toxicology assays (Gruiz *et al.*, 2016). WET testing has become an important component of the municipal and industrial National Pollutant Discharge Elimination System (NPDES) in the U.S. (Grothe *et al.*, 1995) and as WEA in the European Union. The Australian and New Zealand Guidelines for Fresh and Marine Water Quality (Australian Government,

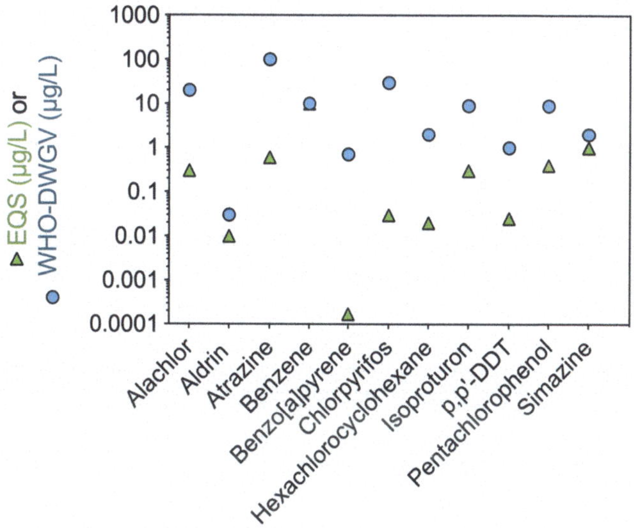

Figure 3.2 Comparison of environmental quality standards (EQS) from the EU's Water Framework Directive or Swiss law with drinking water guideline values (DWGV; WHO 2017b). Data compiled in the Supplementary Information of Escher *et al.* (2018).

2018a) recommend the use of DTA both to monitor the impact of environmental discharges as well to derive site-specific guidelines together with measurements of single chemicals and biological monitoring (van Dam and Chapman, 2001).

The primary purpose of WET testing is to confirm that effluents discharged into receiving waters do not adversely affect aquatic life. An advantage of testing whole effluents is that it integrates the effect of all of the constituents in discharge water.

Typical aquatic toxicology tests are presented below. These test systems are applied both for environmental risk assessment of chemicals (Chapter 2) and for derivation of water quality criteria/standards, as well as for WET testing using effluents or complex mixtures, but the focus of the sections below is on their application in WET.

3.6.1 Test systems in aquatic ecotoxicology commonly applied to WET testing

Acute and chronic WET testing methods had their beginnings in the 1950s and 1980s, respectively (Grothe *et al.*, 1995) and were used to estimate the toxicity of wastewaters. WET testing evaluates the adverse effects or toxicity to a population of aquatic organisms determined experimentally in the laboratory with surrogate organisms believed to be representative of those in the environment exposed to the effluent discharge. This enables a situation-specific assessment. The method can be used, for example, to derive guidance on the amount of dilution required to safely discharge an effluent to an aquatic environment or monitoring the effectiveness of an effluent discharge management programme. The method can also be used as a monitoring tool, testing ambient water that has or is suspected of receiving a chemical pollutant discharge. More recently there had been a move towards application of alternative test methods to WET to reduce animal testing (Norberg-King *et al.*, 2018).

Acute toxicity of effluents is generally measured using the original sample and a minimum of five dilution concentrations. The tests are designed to produce concentration–effect data expressed as a per cent dilution that is lethal (or causes the defined effect) to 50% of the test organisms within a specified time interval (24–96 h) or the highest concentration that is not statistically different from the control (NOEC). A negative result in a single acute test does not preclude the possibility of chronic toxicity or the possibility of temporal variability in an effluent discharge. It also does not preclude the possibility of effects with some taxonomic groups (*e.g.*, plants) but not others (*e.g.*, fish or crustaceans).

If toxicity tests are performed with single species, they should be representative of the different trophic levels, that is, the position in the aquatic food chain. The three most common taxa considered in aquatic toxicology are green algae as representatives of primary producers, aquatic invertebrates such as water fleas as primary consumers, and fish as aquatic vertebrates and secondary consumers (Figure 3.3).

Typical test species in the U.S. EPA 'Methods for Measuring Acute Toxicity to Freshwater and Marine Organisms' (U.S. EPA, 2002a) include freshwater species such as water fleas (*Ceriodaphnia dubia and Daphnia* ssp.) and fish species including fathead minnow (*Pimephales promelas*) and rainbow trout

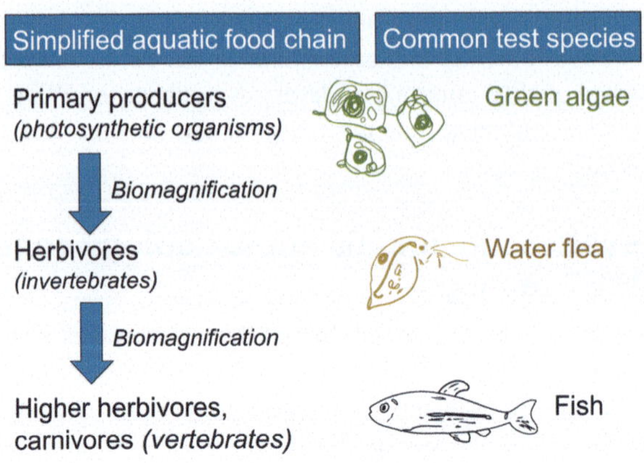

Figure 3.3 Representative test organisms in aquatic toxicology, representing different trophic levels: algae, water flea (*Daphnia)* and fish also simulate a simplified food chain.

(*Oncorhynchus mykiss*). *Marine* species include fish such as the sheepshead minnow (*Cyprinodon variegates*) and *silverside* (*Menidia spp.*), and the mysid shrimp (*Americamysis bahia*). Chronic toxicity testing is also conducted using similar species but also including the freshwater algae (*Selanastrum capricornutum*) for growth, and the sea urchin (*Arbacia punctulata*) fertilisation test (U.S. EPA, 2002b). Correct endpoint selection is critical to estimate the sample effect concentration and to enable management strategies to be developed. The use of standardised species and testing methods is advantageous in that the results can enable comparison between effluents from different industries and will lead to sound scientific data, however, they may be less relevant to specific sites of interest in that endemic species may differ in sensitivity. The methods do, however, allow for prediction of a safe concentration for similar species so is still useful as a screening approach.

There are several standardised guidelines for toxicity testing of aquatic species (OECD, 2006) and the International Organisation for Standardization (ISO) has provided a number of guideline documents on water quality testing that have been either directly adopted or further adapted by national bodies (*e.g.*, DIN in Germany, ASTM International in the U.S.). For algae, an exposure period of 72 h followed by assessment of the inhibition of growth rate or biomass yields an EC_{50} value (effect concentration for effect on 50% of the population; for definition see Chapter 7) that is considered to represent the acute toxicity towards algae, and the NOEC of the same concentration–effect curve is considered as the chronic endpoint (ISO8692, 2004; OECD, 2011). For the water flea *Daphnia magna*, the most popular invertebrate in ecotoxicity testing, the EC_{50} for immobilisation after exposure of 24 hours to the chemical is taken as indicator of acute toxicity (ISO6341, 1996), and the NOEC for the reproductive success (number of progeny per adult) after 21 days of exposure is considered representative of chronic toxicity (ISO10706, 2000).

The acute toxicity test for adult fish generally lasts 96 hours and allows the derivation of the LC_{50}, that is the concentration that is lethal for 50% of the test fish (OECD, 1992; ISO7346-3, 1996). Warm-adapted fish species are often used, such as fathead minnow or guppy, but many national regulations prefer the use of more representative native species. Toxicity testing with fish should ideally be performed in flow-through aquaria to ensure that the chemical exposure is constant during the entire experiment. Since the life cycle of a fish may be several years, chronic toxicity testing is limited only to sensitive life stages, usually early life stages. Ethical issues with vertebrate testing have put more pressure on replacing standard *in vivo* fish tests with alternative test methods. The 'early life stage test' evaluates the embryo and egg yolk larvae stage for mortality, growth and deformation (ISO12890, 1999). The early part of this test, the 'fish embryo test' (FET), which uses fish embryo up to hatching, is considered an *in vitro* method in most legislations (OECD, 2013). This is discussed further in Section 3.6.4.

A WET testing study of effluent from various wastewater treatment plants (WWTPs) in Sydney, Australia, serves as an illustrative example (Bailey *et al.*, 2005). Effluent samples were tested with the acute toxicity test over 48 hours with the water flea *C. dubia*. An effluent sample collected from one WWTP exhibited acute toxicity to *C. dubia* as part of a routine screening programme. The effluent sample tested was a composite of surface water grab samples collected from the discharge stream. The 48 h-LC_{50} for *C. dubia* test was 31.9% effluent, thus the pure effluent would be almost fully lethal for this species. A toxicity identification evaluation (TIE) identified chlorfenvinphos, an organophosphorous insecticide from a pet grooming business, as the source of the high toxicity (Bailey *et al.*, 2005).

3.6.2 *In situ* WET testing

Standardised laboratory tests combined with chemical characterisation of effluents can be used to predict safe discharge concentrations to a receiving environment based on the amount of dilution expected to occur on discharge. Field validation of laboratory results is required to gain confidence in the ability of the laboratory methods to extrapolate to field effects. For a controlled discharge of effluents, the amount of dilution required can be calculated from the laboratory WET testing combined with field studies. This can include the use of caging animals in the field (Lazaro-Cote *et al.*, 2018), and would typically involve having a series of cages from the point of discharge to a predicted safe distance downstream in a river for example, or beyond the mixing zone in the case of lakes and ocean outfalls, to test the accuracy of the dilution prediction for the concentration of contaminants to become acceptable. After an appropriate period of exposure (determined by site and species) the animals are brought back to the laboratory and examined for effects and biomarkers.

Meso- and macrocosms, enclosed experimental environments that replicate larger ecosystems, are viable at least over one growth period (6–8 months), although they rarely include higher vertebrates such as fish and reptiles. Finally, outdoor bypass systems and field studies encompass the interactions in the community and indirect effects such as predation, but the drawbacks are the low number of possible replicates and high costs.

3.6.3 Biomarkers in WET testing

WET testing is based on standard testing methods that can consist of measuring a range of endpoints including biomarkers of exposure and of effect (or sometimes both) such as vitellogenin (egg protein) induction in male fish (Sumpter and Jobling, 1995), or biochemical markers in liver and kidneys in fish (Petala *et al.*, 2009). This may require sacrificing test animals or taking of fluid samples such as blood, however, this is seen as more acceptable than conducting experiments on live animals. Biomarkers are a popular means of quantifying exposure to chemicals. For example, the induction of the egg yolk protein vitellogenin in male fish is an indicator of the presence of estrogens and xenobiotic estrogenic compounds in water (Jobling and Tyler, 2003).

3.6.4 'WET testing' using bioanalytical tools

In vitro bioassays normally require sample extraction and enrichment (*e.g.*, solid-phase or liquid–liquid extraction, see Chapter 12) to compensate for the dosing factor in the assay medium. This has drawn some criticism because some compounds (*e.g.*, inorganics and metals) are lost during the extraction step. As discussed earlier, bioanalytical tools are particularly useful to assess organic chemicals in water because of the sheer number of chemicals that may be present. Other inorganic water pollutants, such as metals, are more limited in number and can be analysed by already exquisitely sensitive methods. Nevertheless, several bioanalytical tools can and have been used in WET testing.

A small number of cell-based bioassays are classified as whole organism tests (*e.g.*, the Microtox assay based on bacteria, and chlorophyll fluorescence assays based on green algae), and several studies have applied those alongside with more conventional WET tests to a variety of whole effluents (Chang *et al.*, 1981; Dizer *et al.*, 2002; Latif and Licek, 2004; Zurita *et al.*, 2019).

A few studies have even adapted other bioanalytical tools to a WET format, although some minimal sample preparation is often still required, for example, filtration, pH adjustment, addition of powdered medium (Wagner and Oehlmann, 2009; Zegura *et al.*, 2009; Niss *et al.*, 2018). The introduction of whole effluent into *in vitro* assays can, however, result in unpredictable effects that are not necessarily associated with actual toxicity, but side effects caused by the matrix. This requires thorough testing and validation of robustness to matrix interference.

The FET is a short-term toxicity test on embryo and sac fry stages of fish that may serve as an ethical alternative in WET testing (Norberg-King *et al.*, 2018). In this test the embryos of the zebrafish (*Danio rerio*) or another fish species are exposed in 24-well plates to a range of dilutions of wastewater. A variety of parameters are observed frequently during the course of the exposure. These parameters include survival/mortality at the different stages, time to hatching, length, morphological, physiological (*e.g.*, heart rate) and behavioural abnormalities. For wastewater, a shortened 48 h exposure standard test on the egg stage alone has been established by the ISO (ISO15088, 2007). The zebrafish FET has been mandatory in Germany for testing wastewater discharges since 2005 and has fully replaced the 96-h acute toxicity test on adult fish.

Lahnsteiner (2008) applied the zebrafish FET to screen wastewater quality and compared the obtained results with acute toxicity testing with adult fish. Six types of wastewater were sampled from Austrian factories involved in industrial processes from the internal sewage collection point in each factory and from the receiving environment from the sewage treatment plant. For dilution of the wastewater samples, groundwater was used. Acute toxicity tests for the fish eggs were conducted using exposure for 48 hours and the results reported as EC_{50} values. Embryos were defined as 'dead' when they showed no heartbeat, no somites were differentiated, egg yolk material was coagulated, or the tail was not detached from the yolk sac. Most investigated wastewaters did not affect zebrafish embryo viability. Only undiluted or marginally diluted wastewater from hide tanning and galvanising metal industries induced effects in the FET. The FET and acute adult fish toxicity agreed fairly well in this study. Gartiser *et al.* (2009) applied the FET with zebrafish to a wider range of industrial effluents and compared with other *in vivo* tests such as toxicity towards algae and water flea. Algae turned out to be the most sensitive endpoint but as they cover a different spectrum of pollutant and effects by colour cannot be excluded, the authors recommended the use of a comprehensive test battery.

The FET has also been applied to investigate the success of advanced water treatment. Cao *et al.* (2009) investigated the WET of secondary effluents treated with chlorination, ozonation and UV irradiation using the FET with Japanese medaka. While the controls and reverse osmosis permeate had >90% hatching success, this was reduced to less than 40% in the secondary effluent. All oxidative treatment steps reduced the toxicity towards embryos with a hatching success increasing from 45 to 65% after treatment. Parallel to the decrease hatching success, the percentage of dead and abnormal embryos was increased as compared to the controls. Another study applied the FET with rainbow trout (*Oncorhynchus mykiss*) eggs on a full-scale wastewater treatment plant with an additional ozonation and sand filtration step (Stalter *et al.*, 2010a, 2010b). All waters had to be filtered as the raw water caused severe effects due to microbial contamination. The membrane-filtered water did cause a slight time delay in hatching, especially for ozonated wastewater but still between 70 and 80% of larvae hatched as compared to 90% in the control. Only after the larvae transitioned to the juvenile stage and started to feed were more significant effects after ozonation observed, although those effects disappeared again after the subsequent sand filtration. In both of these studies, water concentrated by solid-phase extraction was tested with cell-based assays in parallel to the FET assay. *In vitro* tools and the FET gave complementary information on groups of chemicals being reduced (or not) with the different treatment and the overall effect of the treated water, which can also be caused by mobilised organic matter and transiently formed polar and reactive metabolites.

3.7 CONCLUSIONS

It is likely that chemical-by-chemical risk assessment for new and emerging chemicals using whole animals will continue for some time yet for registration purposes and where there are a single or limited number of chemicals being discharged from a specific point source. Likewise, chemical GV will continue to be an important regulatory tool to assure good water quality for different uses. However, as we realise that we are not exposed to single chemicals but rather complex mixtures in real life and as alternatives to animal testing gain momentum, this is likely to change (Norberg-King *et al.*, 2018). Already, some guidelines recommend the use of both *in vivo* and *in vitro* bioassay methods for monitoring purposes. The concept of WET testing relies on testing non-concentrated water and can be applied to a variety of test systems. It is the test media that separates WET from other methods rather than the actual endpoints themselves. WET testing has some parallels with bioanalytical methods in that it can measure the aggregate effect of a range of chemicals in a mixture. The limitation, however, is that when chemicals occur at trace concentrations (*e.g.*, pg/L or ng/L) whole-organism tests may not be sensitive enough to detect those minor changes in water quality, which is dominated by bulk properties such as salinity, pH or organic matter. A combination of WET testing methods with the bioanalytical methods presented in this book may provide a very powerful approach as neither one can replace the other but they provide complementary information.

Chapter 4

Modes of action and toxicity pathways

4.1 INTRODUCTION

When humans or wildlife are exposed to chemicals, several barriers must be overcome before a chemical can elicit an adverse effect. The processes that occur between exposure to that chemical and the adverse cellular effect can be broken down into two phases: the toxicokinetic and toxicodynamic phases (Figure 4.1).

The toxicokinetic phase describes all processes that link the external exposure (*e.g.*, via drinking water) to the biologically effective concentration within the cell. Toxicokinetics encompass absorption and excretion, and internal distribution and metabolism of a chemical within the whole body and within cells.

The toxicodynamic phase describes the cellular toxicity pathways taking place inside the cell starting with the initial molecular interaction of the chemical and its biological target. These interactions can induce cellular defence mechanisms and other cellular responses that ultimately lead to observable toxic effect(s).

For the application of bioanalytical tools to be meaningful, the selected assays must cover not only well-defined toxic mechanisms but also relevant toxicokinetic steps. Cells can be thought of as simple models of organisms that simulate many crucial processes. The lipid bilayer of the cell membrane is a major barrier to chemical exposure. This is the main reason for advocating for the use of whole-cell bioassays for the assessment of environmental samples and for advising against molecular-based cell-free bioassays such as enzyme or receptor-binding assays, which do not include a toxicokinetic component.

Authors: Beate Escher, Peta Neale and Frederic Leusch
doi: 10.2166/9781789061987_0051

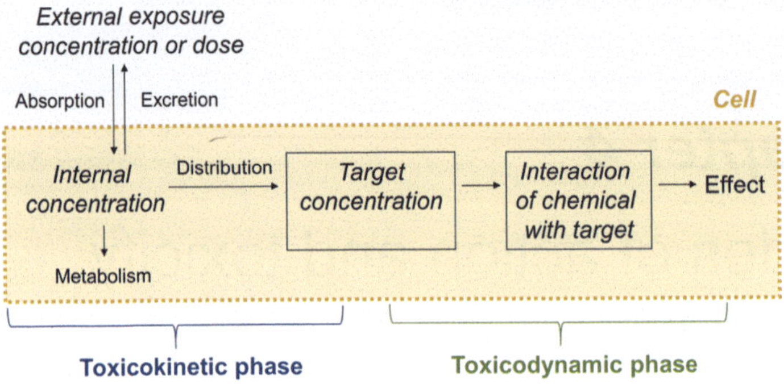

Figure 4.1 Pathway from exposure to effect. Adapted from Escher and Hermens (2002).

Cell-based bioassays can yield information on both the general toxicity to cells (cytotoxicity) and on specific modes of action (MOA). This is important because groups of chemicals with common MOA act together in mixtures by concentration addition (Chapter 8). Using a suite of bioassays that covers various MOA enables the generation of mechanistic information relevant for predicting adverse health outcomes.

In this chapter, the structuring principles of toxicity pathways are summarised to provide a better understanding of the processes that occur in cells and to introduce a mode of action classification that serves as a basis for the selection of bioassays discussed in Chapter 10.

4.2 TOXICOKINETICS

4.2.1 Uptake, distribution and elimination

Uptake and elimination can be a passive or an active process. Passive uptake is the concentration-dependent diffusion of chemicals over cellular barriers (*e.g.*, epithelial cells or biological membranes) and depends on the physicochemical properties of the chemical. Hydrophobic chemicals accumulate in biota to a higher extent but via slower uptake kinetics than more hydrophilic chemicals. Active transport processes require energy and are capable of moving chemicals even against a concentration gradient. Active transport is generally more important for metals than for organics but one group, the ATP-binding cassette (ABC) family of drug transporters, is also of importance for organic chemicals.

Uptake and elimination steps in cell-based bioassays are governed by the same processes as in the whole body. There are, however, quantitative differences and the most important step for *in vitro* to *in vivo* extrapolation is to account for the higher complexity of uptake, distribution and elimination processes that occur in a whole organism.

Once taken up by the organism, chemicals are distributed via the lymph and blood stream to organs and various tissues ultimately reaching the target cells. Distribution is also relevant on the cellular level as it determines if a chemical can reach its target site. As a hydrophobic and reactive chemical will more likely be accumulated in biological membranes, for example, it is prevented from reaching and reacting with DNA.

4.2.2 Xenobiotic metabolism

Xenobiotics are chemicals that are not native to an organism's normal biochemistry. When cells absorb xenobiotic chemicals, they are metabolised, a process called biotransformation. This typically proceeds in three phases (Figure 4.2). Phase I enzymes, such as the cytochrome P450 monooxygenases (CYP), one of the most important families of metabolic enzymes, oxidise chemicals by adding functional groups such as hydroxides to the molecules. In phase II reactions these functional groups can conjugate with molecular entities such as sulphate and glucuronic acid to yield larger and highly water-soluble metabolites, which are more easily excreted from the body (Omiecinski *et al.*, 2011). Phase III refers to the active transport of chemicals across cell membranes by the ABC transporters mentioned above. Phase III processes are not strictly metabolic; however, they do contribute to increased elimination of chemicals from the cell and, therefore, are often presented alongside phase I and II metabolic processes.

While the role of metabolism is primarily to detoxify chemicals, it can in some cases produce more toxic metabolites, particularly in the oxidation reactions of phase I. One prominent example is the bioactivation (*i.e.*, oxidation) of polycyclic aromatic hydrocarbons (PAH) to reactive epoxides, which may cause DNA damage and subsequently carcinogenesis.

4.2.3 Toxicokinetic indicators of chemical exposure

Many xenobiotic chemicals trigger metabolic pathways, whereby they activate and/or increase the metabolic activity within a given cell. Most cell types exhibit

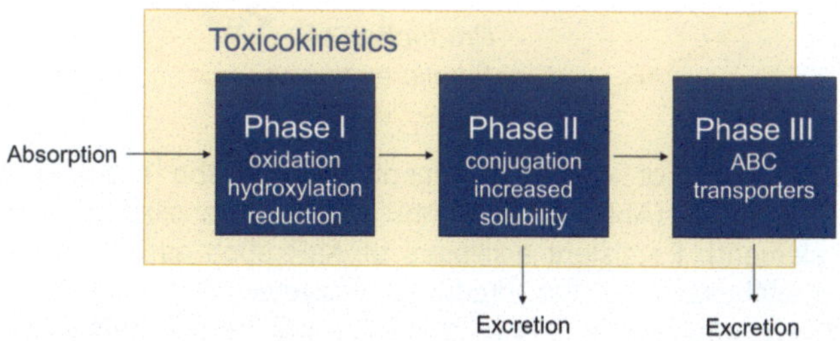

Figure 4.2 Three phases of xenobiotic metabolism in a cell in relation to the other toxicokinetic processes of absorption and excretion.

some degree of metabolic capacity. Liver cells (hepatocytes) have a particularly large capacity for biotransformation.

Metabolic pathways themselves can be used to indicate the presence of chemicals. Cellular pathways related to metabolism are regulated by the so-called xenobiotic receptors (Omiecinski *et al.*, 2011). The most prominent member of this family of nuclear receptors is the aryl hydrocarbon receptor (AhR), a nuclear receptor that is responsive to dioxin-like chemicals and other ligands.

Specific chemicals bind to these xenobiotic receptors and induce the transcription of genes that encode metabolic enzymes. The binding of a nuclear receptor to its nuclear binding site is not a toxic process in itself; however, it indicates the presence of xenobiotic chemicals. Furthermore, the metabolic machinery set off by binding to the receptor will change the structure of the molecule.

All xenobiotic nuclear receptors function in a similar way. In principle, a chemical (or ligand) binds to the receptor (*e.g.*, dioxin binding to AhR), which causes bound proteins (*e.g.*, heat-shock protein and other subunits in the case of AhR) to dissociate from the receptor. The ligand–receptor complex can then translocate into the nucleus, where in the case of AhR it associates with AhR nuclear translocator (ARNT) to facilitate binding to a receptor-specific response element on the DNA (*e.g.*, dioxin response element DRE in the case of the AhR), thus triggering the expression of the associated gene (*e.g.*, CYP1A1 in the case of the AhR) and the production of associated metabolic enzymes (Figure 4.3).

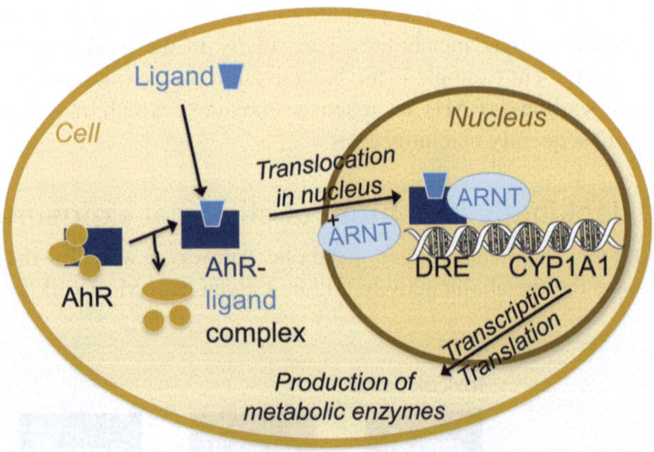

Figure 4.3 Activation of xenobiotic receptors using the example of the aryl hydrocarbon receptor (AhR; ARNT = AhR nuclear translocator, DRE = dioxin-responsive element). Persistent activators of AhR such as dioxin-like chemicals cannot be metabolised by the produced enzymes and lead to a range of AhR-related toxic effects (Denison *et al.*, 2011), while those AhR ligands that can be metabolised after activating the AhR, such as polycyclic aromatic hydrocarbons (PAH), cause a different spectrum of effects.

Table 4.1 Functions of the currently known nuclear xenobiotic receptors related to metabolism and examples of chemicals that induce them.

Nuclear Receptor	Function	Inducing Chemicals
Pregnane X receptor (PXR)	Induction of various phase I enzymes (CYP)	Steroids
Constitutive androstane receptor (CAR)	Protective role against toxicity induced by bile acid, regulation of physiological functions	Indirectly activated by phenobarbital, various pharmaceuticals
Peroxisome proliferator receptor (PPAR)	Glucose, lipid and fatty acid metabolism	Phthalates, fibrate pharmaceuticals
Aryl hydrocarbon receptor (AhR)	Induction of cytochrome P450 (CYP1A1)	PAH, PCDD

CYP = cytochrome P450 monooxygenase; PAH = polycyclic aromatic hydrocarbons; PCDD = polychlorinated dibenzodioxins.

Table 4.1 lists the currently known nuclear xenobiotic receptors that are involved in regulation of metabolism and which are all relevant for water quality testing. Each receptor has several functions taking part in various metabolic processes and in cell homeostasis. The AhR is the most pertinent receptor for toxicological investigations. While the full physiological role of the AhR remains unclear, activation of this receptor contributes to carcinogenicity via the CYP enzymes, which can convert many of its ligands to reactive intermediates, consequently causing DNA damage. Persistent AhR activators, such as dioxin-like chemicals, cannot be metabolised by the produced enzymes and lead to a range of AhR-related toxic effects (Denison *et al.*, 2011), while those AhR ligands that can be metabolised after activating the AhR, such as PAH, cause a different spectrum of effects.

4.3 TOXICODYNAMIC PROCESSES: TOXICITY PATHWAYS

Toxicity pathways are defined as the cellular response pathways induced after chemical exposure that are expected to result in adverse health effects (Collins *et al.*, 2008) (Figure 4.4). The starting point is the molecular interaction between the xenobiotic chemical and the receptor or other biomolecules. This is called the molecular initiating event (MIE). The chemical–biomolecule interaction triggers a cellular response (*e.g.*, translocation of the complex from the cytoplasm to the nucleus, activation of genes, production or depletion of proteins or altered protein signalling) that ultimately leads to observable endpoints or disease. We can capture either critical steps of these cellular responses, the so-called key events

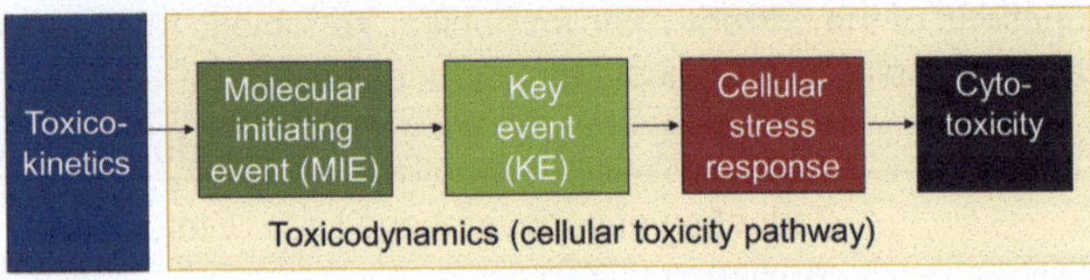

Figure 4.4 Principle of cellular toxicity pathways. Adapted from Collins *et al.* (2008); Ankley *et al.* (2010).

(KE), or the more general cellular stress responses, also called adaptive stress responses.

As cellular responses occur via multiple steps, there are many points of cross-over and branching both within and between toxicity pathways. Some of the pathways induced by chemicals are natural endogenous pathways, whereby the xenobiotic chemical simply replaces a natural ligand. Some authors thus advocate the use of the term biological pathway instead of toxicity pathway. Biological pathways may not directly cause an adverse effect but changed levels of activity are still indicative of the presence of xenobiotic chemicals.

In ecotoxicology, the concept of toxicity pathways has been expanded to the so-called 'adverse outcome pathways' (AOP, Ankley *et al.*, 2010). An AOP links the toxicity pathway (at the cellular level) with the response at the organ level, followed by the response of the organism and finally the effect on the population (Figure 4.5). Organ-level responses include altered physiology of the organ, disruption of homeostasis, altered tissue development and/or disruption of organ function. On the organism level, these effects translate to impaired development, reproduction and/or death. These responses may then be observed across a population and with potential implications for population and ecosystem health. Organ- and organism-level responses are discussed in relation to human health in Chapter 5. The AOP principles integrate human health and environmental/ecological risk assessment. Representative test organisms and population-level endpoints typically applied in environmental risk assessment are discussed in more detail in Chapter 6.

In vitro cell-based assays can be used to indicate toxicity pathways at the cellular level. Cellular responses do not necessarily imply higher-level effects in an

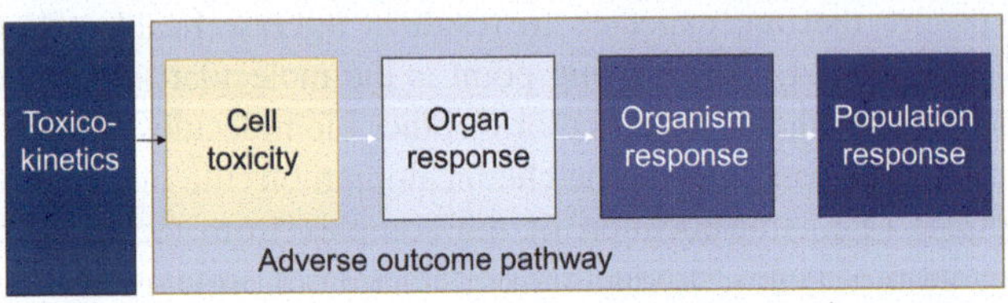

Figure 4.5 Principle of adverse outcome pathways (AOP). Adapted from Ankley *et al.* (2010).

organism, but they are a prerequisite. Species-specific factors and the inherent sensitivity of individuals due to genetic polymorphism will further modulate the causal chain. In addition, environmental factors have the potential to affect individual and population health (Gohlke and Portier, 2007).

Examples of the processes taking place in a cell exposed to xenobiotic chemicals are depicted in Figure 4.6. Chemicals can disturb membrane integrity and thus membrane function by non-specific partitioning into the membranes of cells and organelles. Further, xenobiotic chemicals can bind non-specifically and specifically to proteins. Non-specific interaction with proteins can lead to protein depletion, which ultimately causes oxidative stress. Specific binding to proteins (*e.g.*, receptors and enzymes) can result in inhibition or stimulation of endogenous processes. Most often, binding to enzymes causes a blockage of the active site thus inhibiting enzyme activity. Receptor binding can induce endogenous processes. The (weak) binding of nonylphenol to the estrogen receptor is one example of such agonistic effect on a receptor. Xenobiotic chemicals can also block access of the endogenous agonist to the receptor, hence decreasing normal activity – this is referred to as antagonistic activity. Finally, the interaction (intercalation or covalent binding) of a chemical with DNA can result in errors during replication and transcription. Repair and defence mechanisms are in place to protect the cell from DNA damage up to a certain threshold, above which, the damage becomes permanent.

Direct measurement of the interactions between a chemical and its cellular target is difficult. The associated cell responses (*e.g.*, gene activation after a

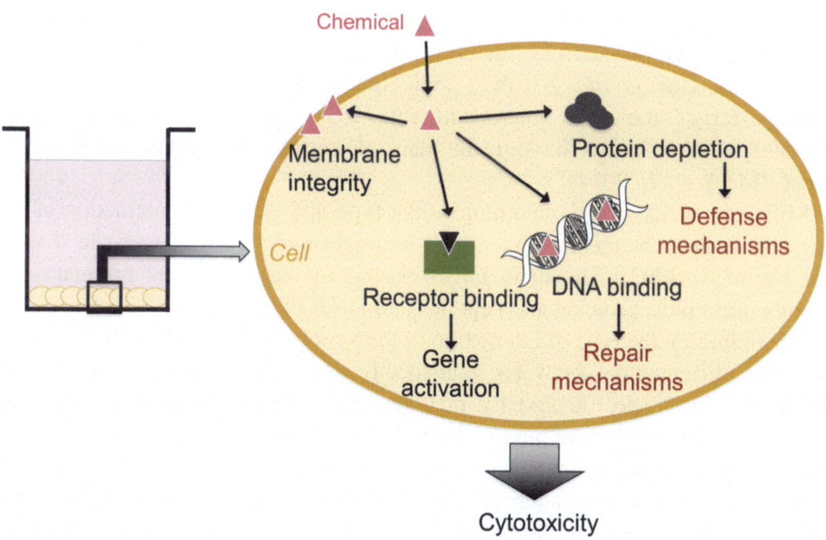

Figure 4.6 Possible toxicity pathways in a cell-based *in vitro* bioassay, ▲ = chemical.

receptor–ligand has been formed or the induction of DNA repair) are, however, useful surrogate measures of these primary interactions (Figure 4.6). When the capacity of repair and defence mechanisms is exceeded, all mechanisms ultimately lead to cytotoxicity, that is, cell death.

There are two types of cell death; (i) apoptosis or programmed cell death and (ii) necrosis, which occurs following irreversible inhibition of vital cell function. Apoptosis is initiated to remove damaged cells and plays an important role in the elimination of pre-cancerous cells. Both types of cell death can occur in relation to non-specific and specific toxicity.

4.4 MODE OF ACTION CLASSIFICATION

With only a limited number of biomolecule types and a myriad of different receptors and molecule–receptor permutations possible, some structuring principles can help classify chemicals in groups that act together in a similar way.

A mode of action (MOA) is a common set of physiological and behavioural signs that characterise a particular type of adverse biological response (Rand, 1995). These responses can be caused by a range of molecular (toxic) mechanisms. Molecular (toxic) mechanisms represent the crucial biochemical processes and/or xenobiotic–biological interactions underlying a given mode of action. It must be noted that the MOA is not a universal property of a chemical but is related to the target organism, organ and/or tissue. As a given chemical can exhibit multiple mechanisms of toxicity, the MOAs displayed may vary with exposure duration (acute vs. chronic) and organism (a human will respond differently to a shrimp) (Escher and Hermens, 2002).

The terminology for MOA and mechanism of action is not consistent in the literature. For the purposes of this book and to be consistent with the framework of AOP (Ankley *et al.*, 2010), a toxic mechanism is defined as the initial chemical–target molecule interaction (the MIE) and the resulting cellular response. MOA refers to 'biologically plausible series of key events leading to an effect' (Meek *et al.*, 2014).

MIEs can be classified according to the type and degree of interaction taking place between a chemical pollutant and its target molecule or target site (Escher and Hermens, 2002). The main target classes for environmental pollutants are (membrane) lipids, proteins and peptides and DNA (Table 4.2).

Depending on the type of interaction of the xenobiotic chemical with the target, one can differentiate between non-specific, specific and reactive toxicity (Table 4.3). Non-specific toxicity involves partitioning to the target site only, whereas specific effects are the results of three-dimensional interactions including specific H-donor/acceptor interactions and ionic interactions between the chemical and target molecules. MOAs are classified as reactive when covalent bonds are formed between the chemical and its target or when chemical reactions are involved (*e.g.*, oxidative stress) (Escher and Hermens, 2002). This generic

Table 4.2 Target molecules and target sites where the molecular initiating event (MIE) takes place.

Target Class	Target Molecules	Target Site
Lipid	Phospholipid	Biological membrane (phospholipid bilayer)
Lipid	Triolein and other glycerols	Storage lipid
Protein	Structural protein, *e.g.*, collagen	Tissue
Protein	Enzyme	All cell types
Protein	Nuclear receptor	All cell types
DNA	DNA bases (nucleic acids)	Nucleus
DNA	DNA backbone	Nucleus

classification scheme can be further refined by differentiation between more specialised target sites such as specific enzymes and receptors. Particularly prominent is the nuclear receptor super family, a class of proteins that sense hormones and regulate gene expression. Evaluation of hormone-induced responses is more complicated than evaluating those of enzyme inhibition because receptor binding effects can lead to complex feedback loops making causal relationships difficult to elucidate. The same holds true for DNA damage, for which numerous repair mechanisms have evolved. The induction of repair processes is, however, still a valuable surrogate for the damage that has occurred.

The MOA classification is universal for all species and provides a common link between human health and ecological risk assessment. Features exist, however, that are unique to certain species, organ and tissue types. Only plants, algae and certain bacteria have the capability to perform photosynthesis. The effect of herbicides that specifically bind to and block the photosystem will therefore only be observable in photosynthetically active cells. Chemically induced immunosuppression will only be relevant for species that have developed an immune system. Yet, there are many highly conserved features in all cells, and many are similar between eukaryotic and prokaryotic (bacterial) cells.

The three major classes of MOAs are reviewed further in the following subchapters to provide a basic understanding of the concept. For a detailed treatise the reader is referred to Timbrell's Principles of Biochemical Toxicology (Timbrell, 2009).

4.4.1 Non-specific toxicity

Non-specific toxicity encompasses all cytotoxic responses that are not mediated by specific or reactive mechanisms. Non-specific toxicity is often termed 'narcosis' or 'baseline toxicity' in ecotoxicology and 'basal toxicity' in human toxicology. Chemicals disrupt membrane function by merely accumulating in biological

Table 4.3 Mode of action classification scheme

MOA Class	Target Molecules or Site(s)	Molecular Mechanism(s)	Mode of Action
Non-specific	All membranes	Non-specific disturbance of membrane structure and functioning	Baseline toxicity
Specific	Energy transducing membranes (mitochondria)	Ionophoric shuttle mechanisms	Uncoupling/depletion of ATP
	Energy transducing membranes (mitochondria)	Blocking of quinone and other binding sites, etc.	Inhibition of the electron transport chain
	Energy transducing membranes (mitochondria)	Blocking of proton channels and other transport channels	Inhibition of ATP synthesis/depletion of ATP
	Photosynthetic membranes	Blocking of photosynthetic electron transport	Photosynthesis inhibition
	Nerve cell membranes	Interference with signal transduction	Neurotoxicity
	Specific enzymes	Binding to enzymes	Enzyme inhibition, *e.g.*, AChE
	(Nuclear) receptors	Binding to (nuclear) receptors	Inhibition or induction of (nuclear) receptors, *e.g.*, AhR, ER
Reactive	DNA, RNA	Base modification and damage: electrophilic (alkylation) and oxidative damage, bulky adducts	Direct genotoxicity (frameshift, cross-links, strand breaks, deletion, etc.)
	Enzymes and receptors	Non-covalent or covalent binding to enzymes of the nucleic acid metabolism, effect on replication or repair	Indirect genotoxicity (DNA repair, recombination, regulation)
	All proteins, peptides	Electrophilic reactivity, alkylation and oxidation of proteins and GSH	Damage and depletion of biomolecules
	All membranes	Formation of reactive intermediates (*e.g.*, ROS) causing peroxidation of membrane lipids and membrane proteins	Degradation of membrane lipids and membrane proteins

This classification is illustrative although not comprehensive.
AChE = acetylcholinesterase, AhR = arylhydrocarbon receptor, ATP = adenosine-5'-triphosphate, DNA = deoxyribonucleic acid, ER = estrogen receptor, GSH = glutathione, RNA = ribonucleic acid, ROS = reactive oxygen species.
Adapted from Escher and Hermens (2002).

membranes and at the interfaces of membrane proteins (van Wezel and Opperhuizen, 1995). As cells lose their integrity, ion and proton gradients cannot be maintained across membranes and ATP is depleted, consequently impairing active transport and other ATP-dependent processes.

The AOP for baseline toxicity encompasses three possible MIEs: narcosis, direct mitochondrial inhibition and decompartmentalisation (Vinken and Blaauboer, 2017). All lead to the key event of mitochondrial dysfunction followed by cell death, both due to apoptosis and necrosis (Figure 4.7). In this scheme 'narcosis' refers to the above-mentioned intercalation of chemicals in biological membranes and 'decompartmentalisation' to the disturbance of cellular organelles' structure and functioning. Mitochondria can be shut down by the specific mechanisms of uncoupling, inhibition of the electron transport chain and the ATP synthase but partitioning of chemicals into energy-transducing membranes also leads to mitochondrial dysfunction (Escher *et al.*, 2002).

Baseline toxicity is the minimum toxicity any organic chemical can exhibit. All chemicals cause baseline toxicity with the same intrinsic potency, which means that there are constant critical membrane concentrations independent of chemical structure. All chemicals are equipotent when the effect is related to concentrations in the biological membranes, the target concentration, but due to differences in cellular uptake and makeup of the bioassay the nominal effect concentrations for baseline toxicity in cell assays differ between chemicals and assays (Escher *et al.*, 2019). Even specifically acting and reactive toxicants induce baseline toxicity as the underlying toxic mechanism, but the concentration necessary to induce baseline toxicity is typically much higher than that required to induce specific effects, so baseline toxicity does not play a role. For reporter gene assays, the cytotoxicity is often caused by baseline toxicity and can be recorded independently from the reporter gene activation. More details in Chapter 9.

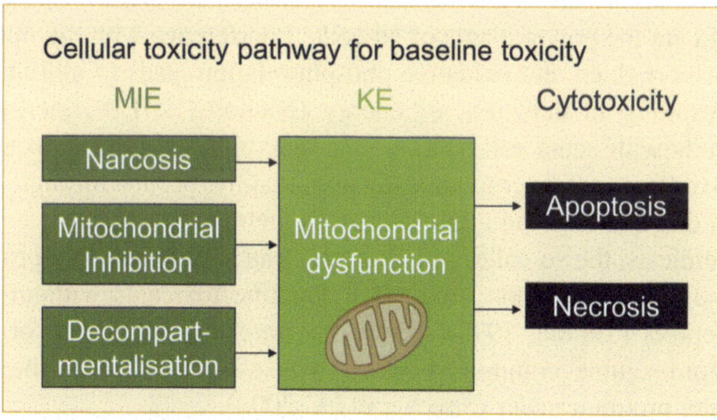

Figure 4.7 Cellular toxicity pathway for baseline or basal toxicity. Adapted from Vinken and Blaauboer (2017).

4.4.2 Specific modes of action

The common mechanistic basis for most specific MOAs is the selective binding of a chemical to a protein (enzyme or receptor) target site (Table 4.3). The following sub-sections discuss the molecular and cellular basis for a selection of relevant MOAs for toxicological assessment. Integrative effects on the human organ and organ system level are detailed in Chapter 5 and ecotoxicological endpoints are reviewed in Chapter 6.

4.4.2.1 Enzyme inhibition

The most direct pathway for inactivation of enzymes is binding of the chemical to the active site of the enzyme. Organophosphates, for instance, are inhibitors of acetylcholinesterase (further discussed in Section 4.4.2.3).

Enzyme function can also be adversely affected via indirect toxicity pathways. Haloacetic acids have been used as pesticides in the past and are also disinfection by-products formed during chlorination in drinking water treatment. Haloacetic acids are capable of replacing acetate in the mitochondrial tricarboxylic acid (TCA) cycle, which is important for cell energy metabolism (Landis and Yu, 2004). Fluoroacetate is formed from haloacetic acids in this process and undergoes all steps of the TCA cycle leading to formation of fluorocitrate, which is a potent inhibitor of the aconitase, the enzyme that converts citrate to isocitrate. In this example, the toxicity is caused by the metabolite rather than by the haloacetic acid itself.

Many enzymes require cofactors (various metal ions, *e.g.*, Fe^{3+}, Ca^{2+}) or organic coenzymes (*e.g.*, nicotinamide adenine dinucleotide phosphate (NADPH)) for their catalytic function. Chemicals that destroy or deplete these cofactors will also degrade the enzyme's catalytic function. For example, fluoride complexes with the cofactors Ca^{2+} and Mg^{2+} and inhibits the activity of important enzymes that require these cofactors.

4.4.2.2 Disturbance of energy production

Mitochondria are the power plants of all cells. Interference with the mitochondrial electron transport chain and oxidative phosphorylation leads to inhibition of ATP synthesis, resulting in depletion of energy (Nicholls, 2013). Energy depletion affects all cells with acute cell death as the outcome. In addition to non-specific toxicity, disturbance of energy transduction mainly occurs through binding to proteins and disruption of ion gradients across membranes.

Some chemicals, the so-called 'uncouplers', can shuttle ions and protons across membranes and are thus more toxic than baseline toxicants without binding to specific receptors (Terada, 1990). Uncouplers are typically weak organic acids that form lipid-soluble conjugated bases, whose diffusion over the membrane results in a net proton transfer (Spycher *et al.*, 2008).

Chemicals such as cyanide, strobins and rotenone bind to the quinone binding sites in the mitochondrial electron transfer chain and inhibit electron transport and thus ultimately energy production, while organotin compounds (*e.g.*, tributyltin) and N,N′-dicyclohexylcarbodiimide directly inhibit ATP synthase.

In analogy, for photosynthetic organisms, energy production through photosynthesis is inhibited by chemicals that block the photosystem or the electron transport chain of chloroplasts (Moreland, 1980). Many herbicides such as triazines (*e.g.*, atrazine) or phenylureas (*e.g.*, diuron) are direct inhibitors of photosystem II. While herbicides exhibit low toxicity to mammals and most vertebrates, some are suspected of possessing additional MOAs, for example, atrazine is considered to be a modulator of aromatase (see Section 4.4.2.4).

4.4.2.3 Neurotoxicity

Many insecticides are neurotoxicants that act through interference with electrical signal transduction or by inhibition of chemical signal transduction at the synapse. At the molecular level, natural and synthetic pyrethroids (*e.g.*, pyrethrin, permethrin) inhibit sodium channels, which are responsible for transmission of electrical signals through cells. By slowing down the re-closure of the sodium channels, pyrethroids cause over-excitation. Organophosphate pesticides bind to the enzyme acetylcholinesterase, inhibiting the cleavage of acetylcholine and hence interfering with chemical signal transduction. Similarly, the neonicotinoid imidacloprid acts as an antagonist on the nicotinic acetylcholine receptor.

The γ-aminobutyric acid (GABA) receptor is another target in nerve cells. The GABA receptor acts as a gate for chloride channels and as an inhibitory neurotransmitter by reducing the flow of chloride ions across chloride channels. Some pesticides such as dieldrin, lindane (γ-hexachlorocyclohexane) and avermectins are GABA agonists.

Insecticides have a lower toxicity to humans than to insects for a variety of reasons. The organophosphates, for example, are better detoxified (metabolised) by mammals than by insects. For some insecticides, the relevant receptors simply play a different role in mammals and insects. The GABA receptor is important for the peripheral nervous system of invertebrates, in which agonistic activity will lead to paralysis. Conversely in mammals, the GABA receptors are only important for the central nervous system and as many of the GABA agonistic insecticides such as the macrocyclic lactones are incapable of crossing the blood–brain barrier, mammals are not affected. This example demonstrates that even highly conserved molecular targets can lead to very different adverse outcomes depending on the organism of interest. This needs to be considered when using bioassays as a tool for tracking specific group of chemicals.

4.4.2.4 Modulation of endocrine functions

Hormones are chemical signalling agents. When hormones bind to receptors, the receptor–ligand complex triggers a series of effects through cell surface or internal (cytosolic) receptors. The level of hormones is modulated by a negative

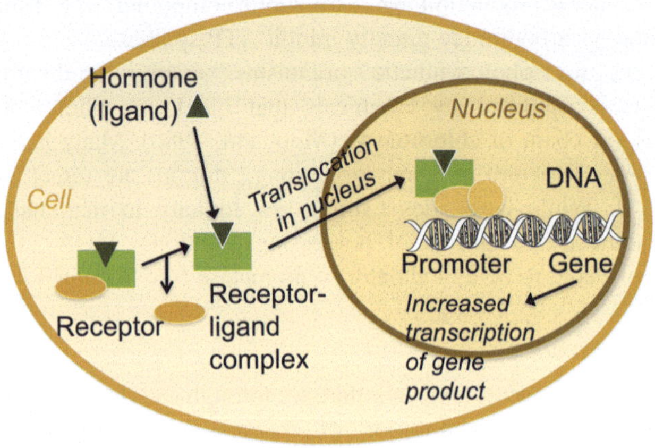

Figure 4.8 Genomic pathway of hormone action.

feedback system. In the case of cytosolic receptors, the receptor–ligand complex crosses the nuclear envelope into the nucleus, where the complex binds to a specific promoter region on the DNA triggering the transcription and translation of specific gene products (genomic pathway of hormone action, Figure 4.8). Various components of the endocrine system are modulated by various micropollutants as further described in Chapter 5.

Chemicals can interfere with hormone receptors by agonism and antagonism. Agonists mimic the natural function of hormones, whereas antagonists block the hormone receptor (Figure 4.9). Both functions can result in adverse outcomes.

Endocrine disruption is relevant for both humans and wildlife. A well-known example of endocrine disruption in wildlife is the observed feminisation of male fish caused by natural and synthetic human hormones as well as some industrial chemicals in wastewater (Sumpter, 2002).

Chemicals can also interfere with the endocrine system via non-receptor-mediated pathways such as inhibition of enzymes relevant for hormone production. Hydroxylated polychlorinated biphenyls (OH-PCBs), for example, inhibit the estrogen sulphotransferase causing increased estrogen levels in blood. The enzyme aromatase is responsible for transforming testosterone to estradiol. This important process can be induced by, for example, atrazine, or inhibited by, for example, triorganotin compounds.

4.4.3 Reactive toxicity

The toxicity of reactive chemicals is caused by their reaction with endogenous molecules. Examples of biological molecules (nucleophiles) that are attacked by reactive chemicals (electrophiles) are the amino acid cysteine in peptides and proteins, the bases in DNA, and the double bonds in phospholipids.

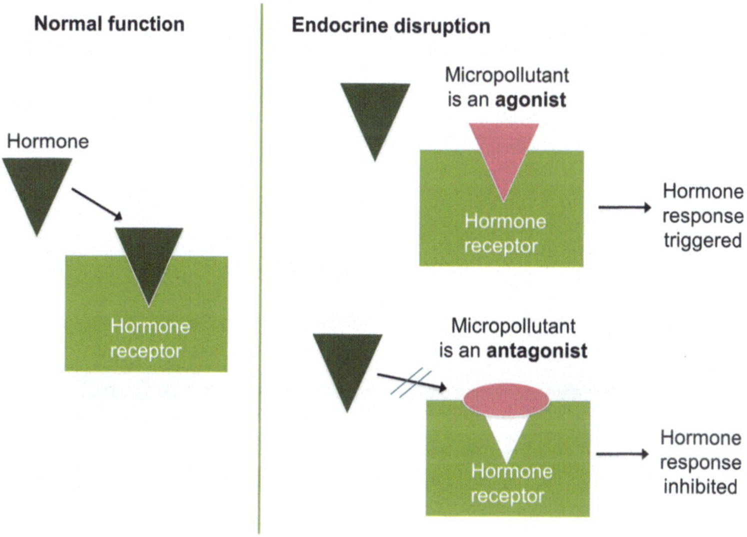

Figure 4.9 Agonistic and antagonistic effects of chemicals.

4.4.3.1 Direct genotoxicity

DNA can be damaged by direct reaction with chemicals, reactive oxygen species (ROS) or stressors such as ultraviolet light (UV). Alkylating agents, for example, fluorouracil or methyl iodide, can covalently bind to DNA creating methyl adducts particularly via the nitrogen atoms of guanine and adenine bases in DNA (Figure 4.10). Electrophilic reactions of larger multifunctional molecules can produce cross-links within or between DNA strands and large adducts can create errors in translation or replication. Furthermore, large planar molecules can intercalate into the DNA thus distorting its structure without directly reacting and modifying DNA. Such distortion can nevertheless lead to mutations and other errors during replication.

Enzymes are able to recognise damaged DNA and trigger repair mechanisms. Sensors of the p53 pathway (see Section 4.5), for example, can detect strand breaks and trigger DNA repair.

Methyl adducts that have been formed by alkylation of DNA can be de-methylated by the enzyme alkyl transferase. Small lesions are repaired by base excision repair and larger adducts are repaired by nucleotide-excision repair (Figure 4.10). Repair mechanisms are, however, prone to error. Failure to repair DNA generally triggers cell death via apoptosis.

DNA damage can (but does not necessarily) lead to loss of bases or strand breaks into which incorrect bases can be inserted, resulting in irreversible mutations. Mutations can cause errors in protein synthesis and are a major cause of cancer.

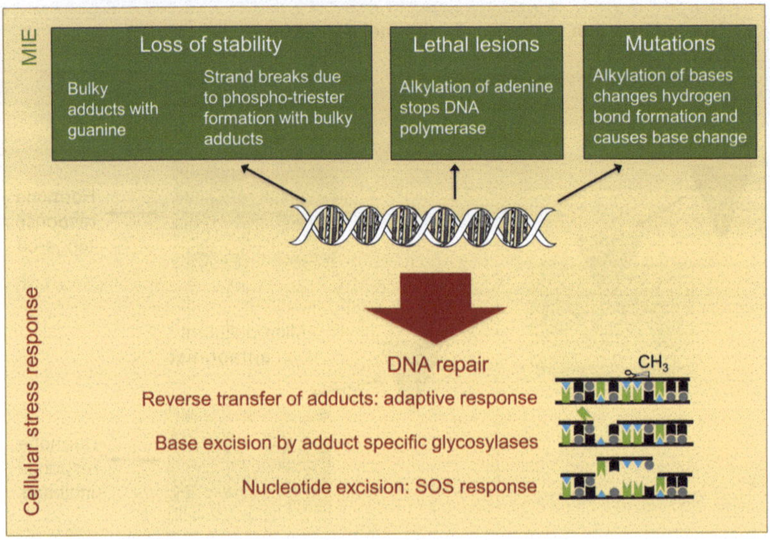

Figure 4.10 Reactive mechanisms leading to DNA damage and repair mechanisms.

4.4.3.2 Non-specific reactivity towards proteins

Biocides such as the antifoulant Sea-Nine, electrophilic chemicals (*e.g.*, acrylates) and dithiocarbamate pesticides can directly react with the thiol group in the amino acid cysteine. Heavy metals such as mercury (Hg^{2+}) and cadmium (Cd^{2+}) can also form complexes with thiol groups. These complexes can cause structural damage to proteins and if this damage affects an enzymatic site of a protein, non-specific enzyme inhibition may also occur.

Glutathione (GSH) is a peptide that contains cysteine and plays an important role in the defence against reactive chemicals and internal reactive oxygen species (ROS). Exposure to micropollutants and subsequent defence mechanisms can lead to GSH depletion, which can cause proteins to lose their protection against oxidative stress resulting in direct protein damage.

4.4.3.3 Oxidative stress

ROS such as the superoxide radical ($O_2 \cdot ^-$), hydrogen peroxide (H_2O_2) and the hydroxyl radical (OH·) are formed during normal cell processes, particularly in mitochondria during electron transport and NADPH-dependent enzyme processes (Figure 4.11). ROS can also be formed by certain radical chemicals (*e.g.*, paraquat) and redox cyclers (*e.g.*, quinones). Inhibition of the mitochondrial electron transfer chain will also lead to the formation of ROS. In the presence of divalent iron (Fe^{2+}), reactive hydroxyl radicals will be formed. ROS can cause lipid peroxidation, DNA damage and oxidation of proteins followed by loss of enzymatic activity.

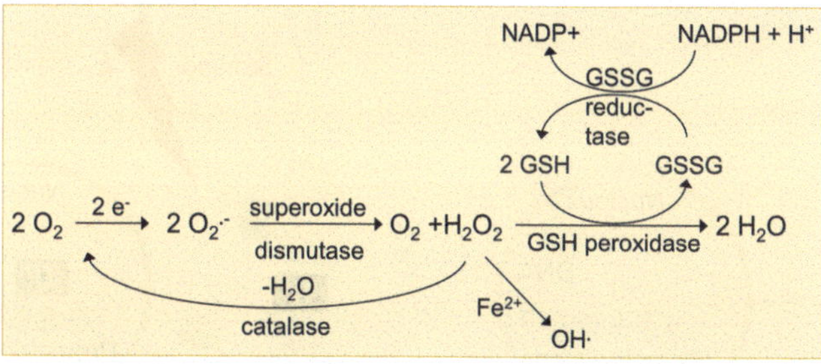

Figure 4.11 Formation and deactivation of reactive oxygen species (ROS). GSH = glutathione GSSG = glutathione disulphide, O_2 = molecular oxygen, $O_2\cdot^-$ = superoxide, H_2O_2 = hydrogen peroxide, OH· = hydroxyl radicals, $NADP^+$ = nicotinamide dinucleotide phosphate, NADPH = reduced $NADP^+$.

Cells have developed sophisticated systems to detoxify ROS and keep the redox balance in the cell stable. Chemical stressors can, however, put more pressure on the cellular redox balance overcoming the natural compensation mechanisms.

During detoxification of ROS, GSH is oxidised to the dimer glutathione disulphide (GSSG) (Figure 4.11). A change in the ratio of GSH to GSSG is an indicator of oxidative stress and ultimately leads to a disturbance of the cellular redox homeostasis. Such imbalance will also impact other redox systems in the cell. The hydrogen transferring coenzymes $NADP^+$/NADPH, for example, will be affected by a change in GSSG/GSH because NADPH is needed to reduce GSSG back to GSH. Oxidative stress can in this way reduce the amount of NADPH available for other vital functions, such as acting as coenzyme for the phase I metabolic enzyme cytochrome P450.

4.4.3.4 Lipid peroxidation

ROS can not only damage DNA and proteins but also play a role in lipid peroxidation. Polyunsaturated phospholipids are particularly vulnerable to this attack, which leads to a chain reaction breakdown of fatty acids, which are important components of membrane lipids. The degradation of fatty acids leads to a change in (un)saturation, which causes alteration in fluidity of membranes and structural damage of membranes. Lysosomes may lose their hydrolytic content and the function of membrane-bound enzymes in the mitochondria and endoplasmic reticulum can be disturbed.

4.5 KEEPING THE RIGHT BALANCE: ADAPTIVE STRESS RESPONSE PATHWAYS

Damage to cellular macromolecules and cellular structures including nucleus, mitochondria, endoplasmic reticulum and lysosomes triggers one or more cellular stress pathways crucial for maintaining the balance in the cell (cell homeostasis)

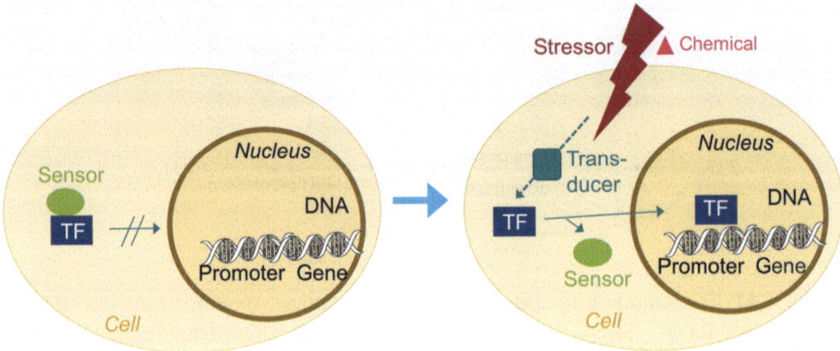

Figure 4.12 Activation of an adaptive stress response pathway. Adapted from Simmons *et al.* (2009). TF = transcription factor.

and/or for repairing damage by transcriptional activation of genes that protect the cells (Simmons *et al.*, 2009). These stress responses are only induced by chemicals or other stressors and thus referred to as adaptive. Adaptive stress response pathways are activated and measurable at much lower concentrations than cytotoxicity and can therefore serve as early warning signals of exposure to chemicals or other stressors.

The principle of an adaptive stress response pathway is depicted in Figure 4.12. On the left, the cell is shown under normal conditions. The transcription factor (TF), which is the protein responsible for triggering the adaptive response, is silenced by a sensor molecule. In this state, the sensor–TF complex cannot enter the nucleus. When cells are under stress, the transducers break the sensor–TF complex, setting the TF free. The TF then enters the nucleus, where it binds to specific sites on the DNA (response elements), which in turn trigger the expression of the associated genes.

This general adaptive stress response pathway shares similarities with both the xenobiotic metabolism and the hormone response pathways in that they all involve some mediating proteins, either nuclear receptors or TF. An important difference is that the adaptive stress pathways occur in all cells, while other toxicity pathways are specific to certain tissues and organs, for example, the liver or reproductive organs.

The heat-shock response was the first stress response pathway discovered and is important for the adaptation to hyperthermia (Table 4.4). The resulting gene products help prevent heat denaturation of proteins. Chemicals that denature proteins also trigger this protective pathway.

Exposure to metals and carbon monoxide can cause cell oxygen levels to be depleted, which activates the hypoxia stress response pathway, triggering for example transcription of proteins that increase transport of oxygen and iron (Table 4.4). The metal response pathway differs from the other stress response

Table 4.4 Relevant adaptive cellular stress response pathways (Simmons *et al.*, 2009).

Pathway	Sensor	Transcription Factor (TF)	Inducing Chemicals and Stressors
Heat-shock response	Hsp90	HSF-1	Temperature, metals
Hypoxia	VHL	HIF-1	Oxygen depletion
Metal stress	VHL	MTF-1	Metals
Endoplasmic reticulum stress	BiP	XBP-1	Norephedrine, diphenylcyclopropene[a]
Osmotic stress	None	MTF-1	High salt, glycol
Inflammation	IkB	NF-κB	Metals, PCBs, smoke, particles
Oxidative stress	Keap1	Nrf2	Chemicals that produce ROS
DNA damage	MDM2	p53	Electrophilic chemicals, UV radiation

[a]Hirota *et al.* (2010); Yang *et al.* (2011).

pathways in being constitutive (always expressed) as opposed to adaptive (only expressed after activation). Its activation induces increased synthesis of metallothionein proteins, which are small cysteine-rich proteins that chelate metals.

The endoplasmic reticulum plays a central role in lipid synthesis and folding and maturation of proteins. The endoplasmic reticulum stress response pathway induces genes that help refold proteins and remove damaged ones.

Osmotic stress triggers a pathway that ultimately leads to increased solute transport across membranes. The inflammatory stress response is mediated by the nuclear factor kappa B (NF-κB), which is closely related to immune responses and causes induction of cytokines, cytochromes P450 and regulators of apoptosis.

The mammalian cellular defence mechanism against oxidative stress is primarily mediated at the transcriptional level by Nrf2 (NF-E2-related factor 2), which is responsible for the induction of detoxification and antioxidant genes (Nguyen *et al.*, 2009) (Table 4.4). Nrf2 activates the transcription of sequences containing the antioxidant response element (ARE). ARE is found in the promoter region of genes encoding the major detoxification enzymes including glutathione S-transferase A2 (GSTA2) and NADPH:quinone oxidoreductase 1 (NQO1), which are the two major contributors to cellular protection. These enzymes serve to neutralise ROS and reactive chemicals, biosynthesise GSH, direct xenobiotic efflux and remove oxidised proteins. The net result is to limit oxidative damage and to detoxify cells.

The most important response to DNA damage is regulated by the p53 family of TFs. Under normal conditions, p53 is negatively regulated by the sensor MDM2. Upon DNA damage, the p53 is stabilised and triggers a series of DNA repair mechanisms. p53 is also referred to as 'the tumour suppressor gene' and regulates cell cycle arrest and apoptosis.

As discussed, stimulation of adaptive stress response pathways is not a direct indicator of toxicity but an early indicator of the presence of stressors. Since activation occurs at concentrations of micropollutants lower than those required to elicit an observable adverse effect, these pathways are useful early warning signs with potential for application in water quality assessment.

4.6 CONCLUSIONS

All the different toxicity pathways discussed in this chapter are highly interconnected. The complete picture is thus much more complex than presented through independent view of the individual processes. Figure 4.13 expands the simplistic picture drawn in Figure 4.6 and interconnects all these different processes. First, metabolism can lead to both toxification and detoxification, and the reactive metabolites of phase I in particular may cause direct reactive toxicity and oxidative stress. Second, GSH acts as scavenger of reactive intermediates but

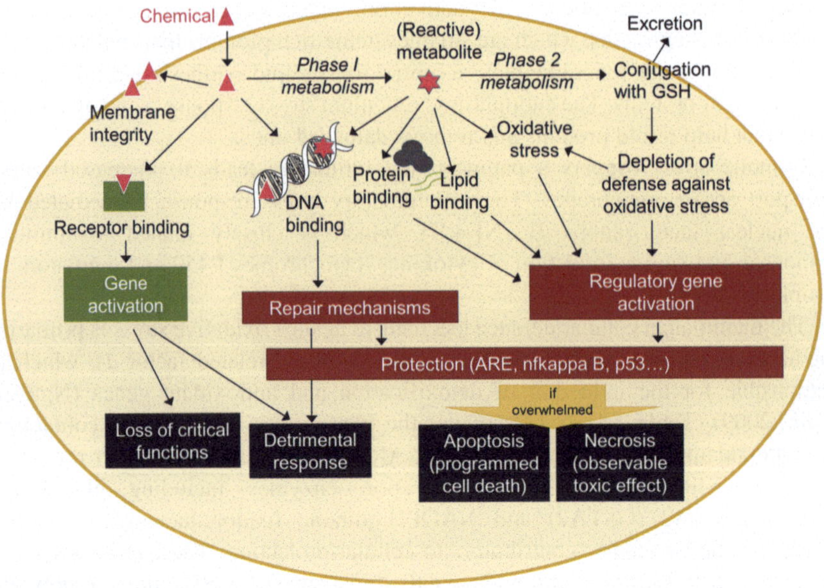

Figure 4.13 Interplay between various toxicity pathways and the effects induced if injury is beyond repair.

if it is depleted, it cannot continue its role in keeping the redox homeostasis. The cell then invokes a second line of damage control, represented by the adaptive stress response pathways discussed in Section 4.5.

The lower part of Figure 4.13 highlights how the defence and repair mechanisms initially serve to protect the cell but can become overwhelmed if the damage is too great. As a final resort, the cells can invoke programmed cell death (apoptosis), but if the damage is too severe, necrotic cell death will occur.

Chapter 5

Toxicity pathways of chemicals in humans

5.1 INTRODUCTION

This chapter explores the different types of toxicity that may be associated with chemical contaminants in drinking water, including toxicity to specific organs (*e.g.*, liver, kidneys, heart), organ systems (*e.g.*, blood formation, immune, nervous, endocrine systems) and integrated organism effects (*e.g.*, developmental and reproductive effects, carcinogenicity). Molecular and/or cellular level toxicity that is described in Chapter 4 can translate into effects at the tissue, organ, organ system and eventually at the organism and population level (Figure 5.1).

Within the framework of the adverse outcome pathway (AOP) (Chapter 4), once a toxicant has reached its target site (toxicokinetic phase) and affected its biological target (toxicodynamic phase), the resulting cellular-level effect(s) can lead to dysfunctions at higher levels depending on its severity and the capacity of repair and compensation mechanisms (Figure 5.2). It should be noted that on the organism scale the toxicokinetic processes are broader than at the cellular level, and are referred to as absorption, distribution, metabolism and excretion (ADME). These processes deliver the chemical to its cellular target site, where the AOP is initiated.

Exposure to environmental pollutants can result in a variety of effects in the exposed organism. In broad terms, toxicity is an adverse effect on the production, function and/or survival of cells. Some of these toxic effects are very general and can potentially affect all types of cells, while others are specific to certain tissues

Authors: Beate Escher, Peta Neale and Frederic Leusch
doi: 10.2166/9781789061987_0073

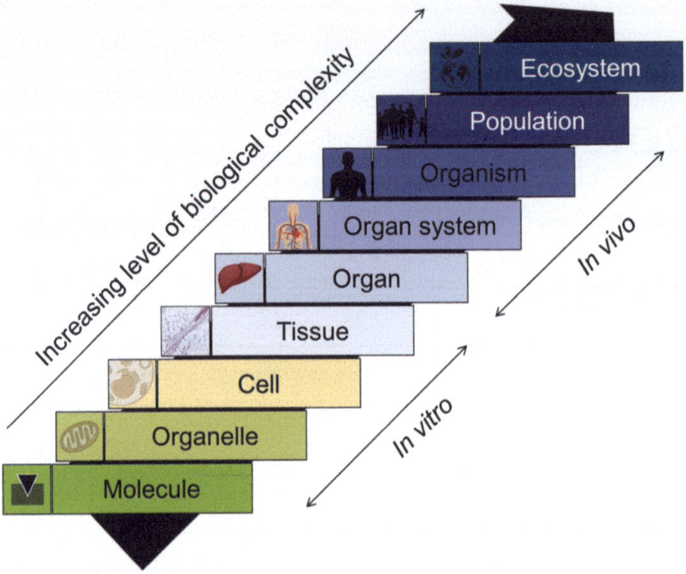

Figure 5.1 Simplified scheme of biological organisation illustrating, where *in vitro* and *in vivo* investigations fit in the sequence of effects that take place from the molecular to the ecosystem level.

due to their unique structure and/or function. Some biological functions are fulfilled by multiple organs (*e.g.*, the endocrine system), and toxicity to any of the organs involved may result in failure of the whole system. A thorough understanding of these concepts for each potential site of toxicity is critical when developing a comprehensive screening battery for risk assessment.

Unless otherwise indicated, the information in this chapter is based on classical toxicology textbooks, in particular, Ballantyne *et al.* (1995), Fox (1991) and Klaassen (2013).

5.2 ROUTE OF EXPOSURE

The route of exposure is significant because chemicals can reach different targets depending on the point of entry into an organism (Figure 5.3). For drinking water, oral ingestion is the main route of exposure and the digestive system will be the focal point of entry into the organism.

The majority of ingested toxicants will be absorbed in the small intestine, which has a very large, specialised surface area making it very efficient at absorbing nutrients but also unfortunately toxicants from food and water. Absorption can take place via active transporters but is usually a passive process, where toxicants traverse the epithelial barriers and reach blood capillaries by diffusing through

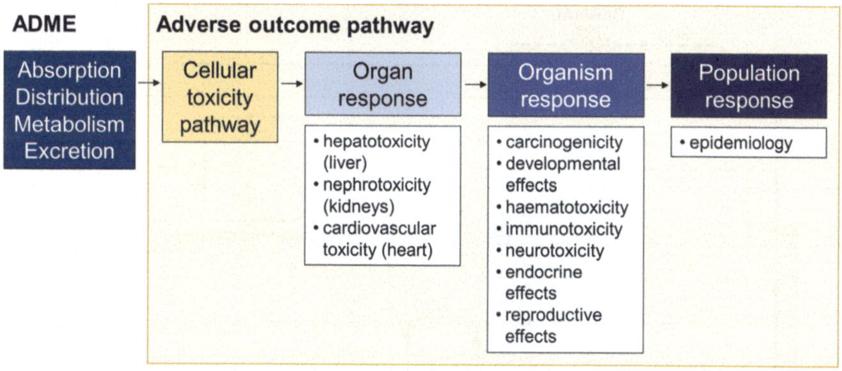

Figure 5.2 Toxicokinetic processes on the organism level (ADME – absorption, distribution, metabolism and excretion) connected to the adverse outcome pathway (AOP) that links cellular-level events to whole organism effects.

cell membranes. Lipid solubility is usually the most important property influencing absorption with lipophilic (fat-loving) chemicals more readily absorbed than hydrophilic (more water-soluble) substances, but size and charge also influence diffusion with small and uncharged substances more easily absorbed. Xenobiotics absorbed in the gastrointestinal tract are transported to the liver via the hepatic portal vein.

Once in the liver, the absorbed chemicals will undergo 'first-pass metabolism': biotransformation by cytochrome P450 enzymes and conjugation with large hydrophilic molecules (such as glucuronide or sulphate). The biological purpose of first-pass metabolism is to make lipophilic toxicants more water soluble, thus facilitating their excretion. After biotransformation, xenobiotics can travel via either of two routes: large water-soluble chemicals are excreted back into the small intestine via the bile duct and eventually excreted from the body in faeces. Those chemicals no longer pose a risk to other organs because no further contact will occur (unless they are susceptible to 'enterohepatic recycling', where the conjugate is cleaved off by microbial activity in the intestinal tract and the chemical can be reabsorbed and sent back to the liver). Alternatively, xenobiotics that are still sufficiently small and lipophilic after first-pass metabolism can instead enter the systemic blood circulation, where they can reach and affect any tissue perfused by blood – in other words, all tissues – particularly if they are lipid soluble. Hydrophilic toxicants will eventually be excreted into bile by the liver or into urine by the kidneys. Highly lipophilic chemicals that are resistant to biotransformation (such as polyhalogenated biphenyls and chlorinated hydrocarbons) are very hard to eliminate and tend to accumulate in the body upon repeated exposure – a process called 'bioaccumulation'.

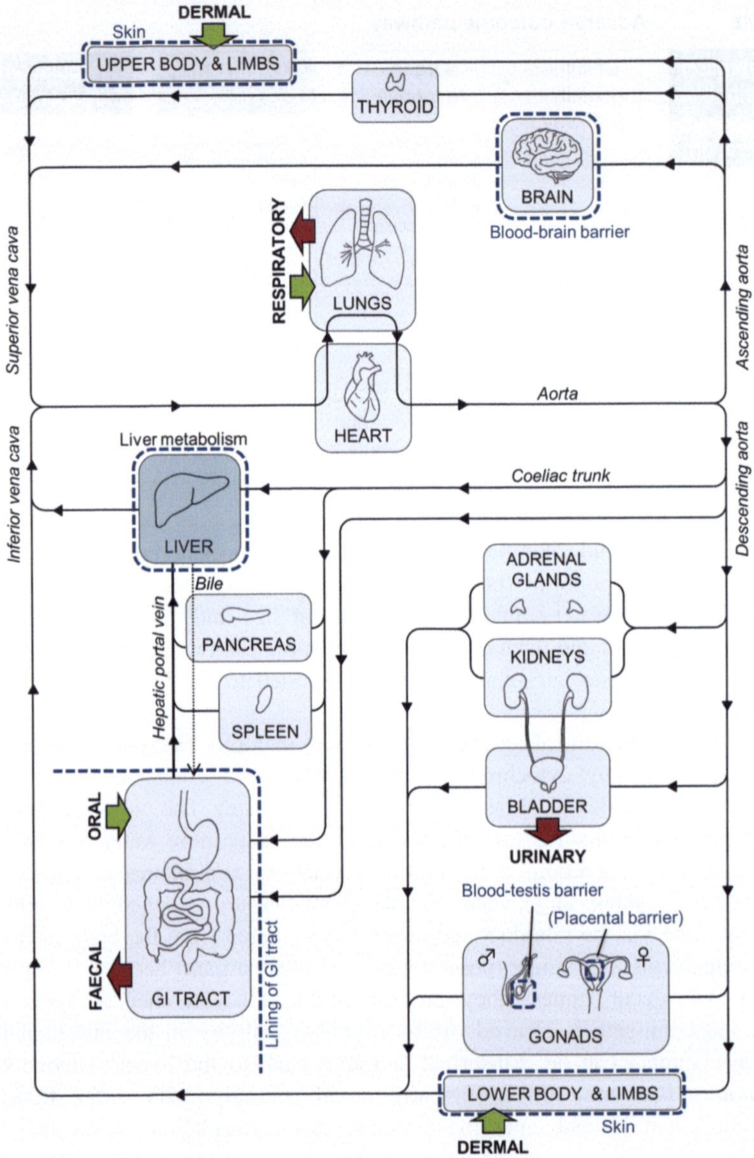

Figure 5.3 Simplified absorption, distribution, metabolism and excretion (ADME) diagram showing sites of intake (large green arrows: oral, respiratory and dermal intake), excretion (large red arrows: faecal, urinary and respiratory excretion) and internal barriers for xenobiotics (dashed blue lines: lining of the GI tract, skin, blood–brain barrier, blood–testis barrier in males, placental barrier in pregnant females and liver metabolism). Oral ingestion is the main route of exposure for contaminants in drinking water. GI = gastrointestinal.

5.3 BASAL CYTOXICITY

Basal cytotoxicity refers to the ability of a chemical substance to damage living cells and may be described as any general cellular-level effect that leads to dysregulation of on-going cellular activity (*e.g.*, disruption of ATP production, protein synthesis), impairment of cellular maintenance, dysregulation of gene expression and/or physical damage to cell structures (*e.g.*, proteins, plasma membranes) (Vinken and Blaauboer, 2017). Basal cytotoxicity is not cell-type specific – it can affect any tissue type and occurs at similar critical membrane concentrations in all cell types (Escher *et al.*, 2019). It is therefore also often called 'baseline toxicity' (see Chapter 4). This impaired cell structure and function can lead to cell and tissue death, either planned (apoptosis) or unplanned (necrosis) cell death.

5.4 TARGET ORGAN TOXICITY

In the context of drinking water, toxicity to three organs is particularly relevant: the liver (hepatotoxicity), the kidneys (nephrotoxicity) and the heart/blood system (cardiovascular toxicity). Gastrointestinal tract and bladder toxicity are of course also important, but the main mechanisms of toxicity to these targets are basal cytotoxicity (Section 5.3) and carcinogenicity (Section 5.5.1).

5.4.1 Hepatotoxicity

The liver is the main organ where exogenous chemicals are metabolised to make them more hydrophilic for excretion. Consequently, liver cells can be exposed to high concentrations of toxicants. Thankfully, a healthy liver has an immense capacity for self-repair and once the toxicant is removed recovery is usually possible.

After absorption by the small intestine, ingested nutrients, vitamins, metals, drugs and environmental toxicants are all distributed to the liver via the hepatic portal vein (Figure 5.3). Efficient scavenging or uptake processes extract these absorbed materials from the blood for catabolism, storage and/or excretion into bile. Hepatocytes (liver cells) are rich in mitochondria to provide for their high energy needs and cytochrome P450 enzymes, which conduct the liver's main function of metabolism and detoxification. Hepatocytes also have a significant role in protein synthesis by recycling all major plasma proteins, carbohydrate and lipid metabolism, cholesterol production, and bile secretion, which can be an important detoxification mechanism.

There are several key factors that modulate hepatotoxicity:

- Uptake and concentration: the liver is immediately 'downstream' of the gastrointestinal tract, and as such receives the highest concentration of lipophilic drugs and environmental pollutants from the oral route. Other toxins are rapidly extracted from the blood into hepatocytes via active transport mechanisms.

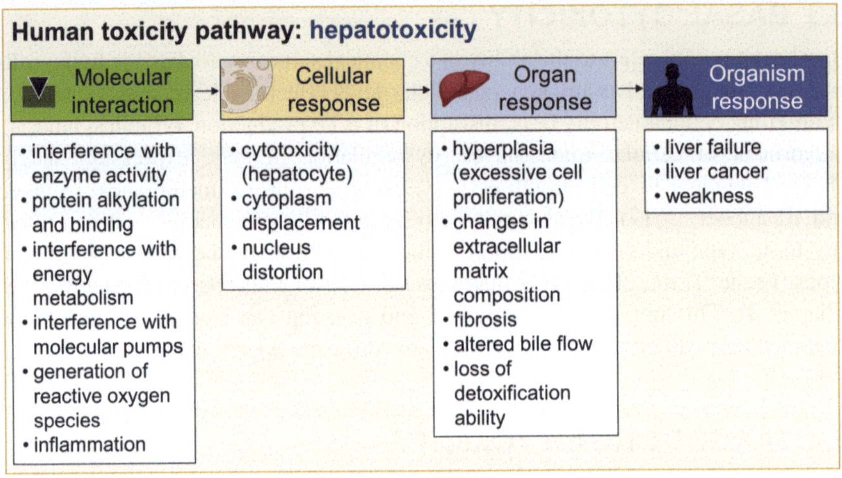

Figure 5.4 Spectrum of hepatotoxicity from the molecular to the organism level.

- Bioactivation and detoxification: one of the vital functions of the liver is to eliminate exogenous chemicals and toxic endogenous intermediates. Biotransformation, and in particular phase I metabolism, can, however, generate reactive metabolites, which can interact with proteins and other biological molecules.
- Regeneration: the liver has a high capacity to restore lost tissue and function by regeneration. Loss of hepatocytes triggers proliferation of mature hepatocytes to replace the lost tissue, which is initiated by cytokines and growth factors. Nevertheless, chemicals that can interfere with the cell cycle (*e.g.*, colchicine) can block that regenerative ability.

Hepatotoxicity results in impaired liver function and the potential build-up of toxic by-products of cellular metabolism (Figure 5.4). There are several known mechanisms of toxicity to liver cells, including direct cytotoxicity to hepatocytes (*e.g.*, acetaminophen, carbon tetrachloride, microcystin), damage to epithelial cells of liver capillaries (*e.g.*, after excessive dose of acetaminophen, endotoxin, microcystin), impaired bile excretion (usually from interference of bile salt export pumps by toxicants such as pharmaceuticals, hormones and metals) and excessive cell proliferation to replace dead cells (hyperplasia; *e.g.*, after chronic exposure to excess androgens, alcohol and aflatoxin).

5.4.2 Nephrotoxicity

The principal role of the kidneys is to filter blood and maintain total body homeostasis. The kidneys play a central role in excretion of metabolic wastes (such as urea) and in the regulation of extracellular fluid volume, electrolyte composition and pH of the blood. The kidneys also produce hormones (renin and erythropoietin) that regulate extracellular volume and red blood cell production and metabolise vitamin D3 to its active form. Similar to the liver, the kidneys are equipped with a variety of detoxification mechanisms and have considerable functional reserve and regenerative capacities.

The kidneys are particularly sensitive to blood-borne toxicants as they receive about a quarter of the cardiac output (Figure 5.3). The processes involved in the production of urine may also concentrate potential toxicants in the tubular fluid. A wide variety of pharmaceuticals (*e.g.*, antibiotics, analgesics, radiocontrast media, anti-cancer agents and angiotensin inhibitors and blockers), environmental chemicals and metals can cause nephrotoxicity via structural and/or functional damage. Proper kidney function is highly dependent on passive and active (ATP-driven) transport mechanisms. Toxicant-induced interruptions in energy production for any of these active transport mechanisms or interference with critical membrane-bound enzymes and/or transporters can thus seriously impact kidney function (Figure 5.5). The efficiency of the kidneys is also dependent on tight control of capillary pressure, and the kidneys are particularly sensitive to vasoactive substances, that is, substances that modulate blood pressure.

5.4.3 Cardiovascular toxicity

Cardiovascular toxicology focuses on adverse effects on the heart and the vascular system. Exposure to toxic chemicals can result in alterations of biochemical pathways, defects in cellular structure and function, and pathogenesis of the affected cardiovascular system.

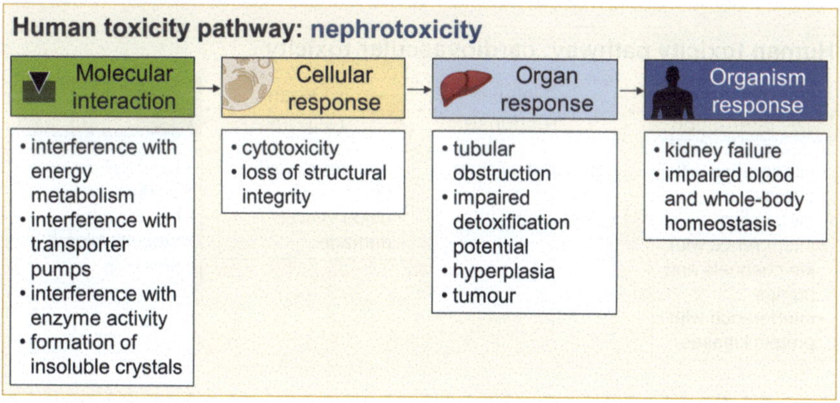

Figure 5.5 Spectrum of nephrotoxicity from the molecular to the organism level.

5.4.3.1 Cardiotoxicity

Heartbeat is controlled by specialised pacemaker cells, and cardiac electrophysiology and function are under neuro-hormonal regulation. The primary contractile unit in the heart is the cardiac muscle cell, or cardiomyocyte. Stimulation of cardiomyocytes through bioelectricity is due to carefully orchestrated transport of three positively charged ions: calcium, sodium and potassium. Each of the ions has specific channels and pumps on the membrane of cardiac myocytes.

The very high energy requirements of the heart muscle (continuous synthesis of ATP via mitochondrial oxidative phosphorylation is required for cardiomyocyte function) and heavy reliance on ion channels and pumps are particularly relevant for cardiotoxicity. Not surprisingly, many substances can cause cardiac toxic responses, mostly by affecting ion channels (*e.g.*, the anti-arrhythmic drugs verapamil and quinidine), calcium ion homeostasis (*e.g.*, pharmaceuticals such as ouabain, some antimicrobial and antiviral agents, aldehydes, halogenated alkanes and metals), and electrical excitability and action potential generation (*e.g.*, local anaesthetics like benzocaine or procainamide) (Figure 5.6).

5.4.3.2 Vascular toxicity

Toxic responses of the vascular system include changes in blood pressure and damage to blood vessels (Figure 5.6). The main function of the vascular system is to provide oxygen and nutrients and to remove carbon dioxide and metabolic products to/from organ systems. The vascular system also delivers hormones and cytokines to target organs (Figure 5.3). Dilation and constriction of blood vessels are controlled remotely by neurons and hormones, such as epinephrine, norepinephrine and angiotensin, and locally by oxygen supply and

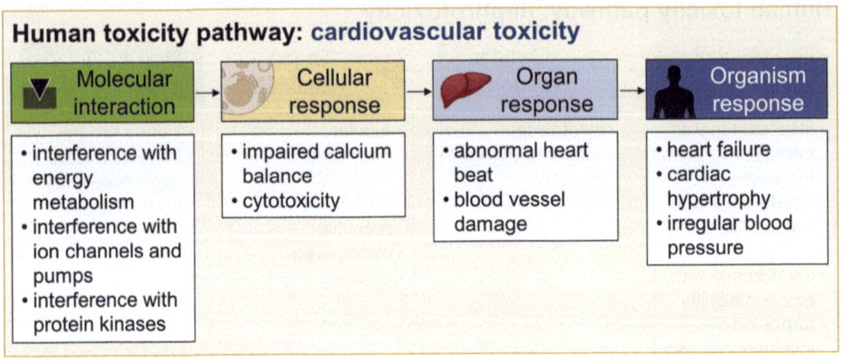

Figure 5.6 Spectrum of cardiovascular toxicity from the molecular to the organism level.

endothelium-derived relaxing factor (EDRF). Thus, neurotoxicity or endocrine disruption may also affect vascular function.

Blood vessels are mostly composed of epithelial cells enveloped by smooth muscle cells. After absorption, all chemicals come into contact with the vascular system. Specific vascular toxicity can occur from damage to either epithelial cells (*e.g.*, aspirin, endotoxins, carbon monoxide) or smooth muscle cells (*e.g.*, metals interfering with calcium homeostasis) or from exposure to vasoactive chemicals (*e.g.*, cocaine, nicotine, metals). It is not entirely clear how toxic responses of the vascular system affect physiological function and/or cause toxicity to other organs, but damage to vascular epithelial cells could produce reactive oxygen species (ROS) and subsequent oxidative injury.

5.5 NON-ORGAN-DIRECTED TOXICITY

Non-organ-directed toxicity includes carcinogenicity and developmental toxicity.

5.5.1 Carcinogenicity

Chemicals that induce cancer have been broadly classified in two categories: (i) genotoxic carcinogens (*e.g.*, PAHs) that interact physically with DNA to alter or damage its structure, and (ii) epigenetic carcinogens that impact DNA expression through DNA methylation, protein phosphorylation and receptor-mediated effects, without directly affecting DNA structure (Figure 5.7). Either case can eventually lead to aberrant cell cycle kinetics and unregulated cell growth.

Carcinogenesis develops over three stages:

- Initiation is the introduction of a 'mistake' (mutation) in the DNA sequence. Initiation can be caused by genotoxic carcinogens binding to DNA and causing errors during DNA synthesis. Initiation on its own is not sufficient to cause abnormal cell growth because DNA damage can sometimes be repaired or because the cell can lose its viability due to the mutation.
- Promotion is the selective expansion of initiated cells.

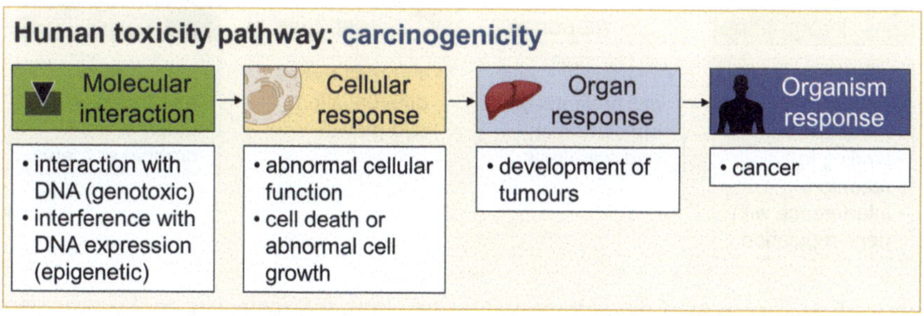

Figure 5.7 Spectrum of carcinogenicity from the molecular to the organism level.

- Progression involves the conversion of unstable promoted cells into stable malignant tumours. Due to the resulting increased DNA synthesis, additional genotoxic events may occur at this stage, resulting in additional DNA damage including chromosomal aberrations and translocations.

Complete carcinogens have the ability to function at all levels, that is, initiation, promotion and progression. Many carcinogens are not intrinsically carcinogenic but require metabolic activation to become carcinogenic. This may also result in tissue-specific effects, as different tissues can have different levels of enzyme expression.

5.5.2 Developmental toxicology

Developmental toxicity focuses on adverse effects on development caused by exposure to toxicants. Development is characterised by changes that are orchestrated by a cascade of factors regulating gene transcription (Figure 5.8). A particularity of developmental toxicology is that the sensitivity of the organism to toxicants can vary depending on its developmental stage.

Embryotoxic chemicals affect the conceptus prior to the foetal stage (usually up to 8 weeks in humans). Imprinting, implantation, gastrulation and organogenesis all occur during embryo development, and toxicants that interfere with cell proliferation, differentiation and/or apoptosis may lead to embryotoxicity (*e.g.*, cyclophosphamide).

Foetotoxic chemicals affect the conceptus from the foetal stage onwards (usually after 8 weeks in humans).

Teratogens are toxic chemicals that cause birth defects and can lead to pre- and postnatal mortality. Gastrulation and organogenesis during embryo development and the subsequent tissue differentiation and growth during foetal development are particularly sensitive to teratogens. Toxicants that can affect cell migration,

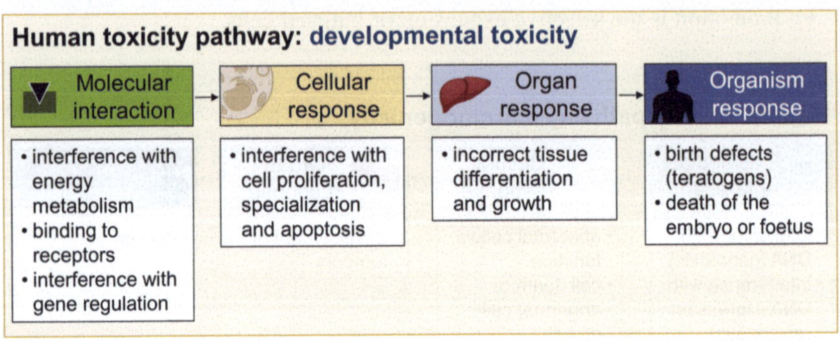

Figure 5.8 Spectrum of developmental toxicity from the molecular to the organism level.

cell–cell interactions, differentiation, morphogenesis and energy metabolism are often teratogens (Figure 5.8).

Exposure to developmental toxicants can result in death of the embryo, death of the foetus or teratogenesis. Endocrine-disrupting compounds (EDCs, *e.g.*, diethylstilbestrol) may also negatively affect development of the conceptus (see Section 5.6.4 for more details on endocrine disruption).

5.6 SYSTEM TOXICITY

Some biological functions are fulfilled by systems composed of multiple organs (*e.g.*, the immune system), and toxicity to any of the organs involved may result in failure of the whole system. This section discusses toxicity towards the blood system (haematotoxicity), immune system (immunotoxicity), nervous system (neurotoxicity), endocrine system (endocrine disruption) and reproduction (reproductive toxicity).

Some health outcomes of relevance for risk assessment of new chemicals have been excluded from this review because they were deemed unlikely to result from exposure to chemical contaminants in drinking water (although they may occur from different exposure routes to that same water, *e.g.*, showering). These include:

- Sensory organ toxicity, including ocular toxicity,
- Respiratory toxicity,
- Cutaneous toxicity,
- Musculoskeletal toxicity (*e.g.*, myotoxicity).

5.6.1 Haematotoxicity

The production of blood cells (haematopoiesis) is a highly regulated sequence of events by which blood cell precursors proliferate and differentiate to meet the body's relentless needs for oxygen transport, host defence and repair and blood homeostasis. The main organs involved in haematopoiesis are the bone marrow and the spleen. A haematotoxicant is a toxicant that either interferes with haematopoiesis or affects the viability of red blood cells, which can result in anaemia and hypoxia (lack of oxygen). Effects on white blood cell viability are covered in the section on immunotoxicity below.

Haematopoiesis requires carefully orchestrated cell maturation and differentiation and is particularly sensitive to cytoreductive or antimitotic drugs used for cancer treatment and toxicants that can interfere with differentiation and maturation of blood cell precursors.

The viability of red blood cells can be affected by oxidative damage, which can interfere with the oxygen-carrying capacity of haemoglobin, or by modification of cell surface proteins (*e.g.*, mefenamic acid), which can lead to loss of 'self' antigens (cell surface markers that identify the cell as part of the self, as opposed to foreign) and subsequent destruction by white blood cells (Figure 5.9).

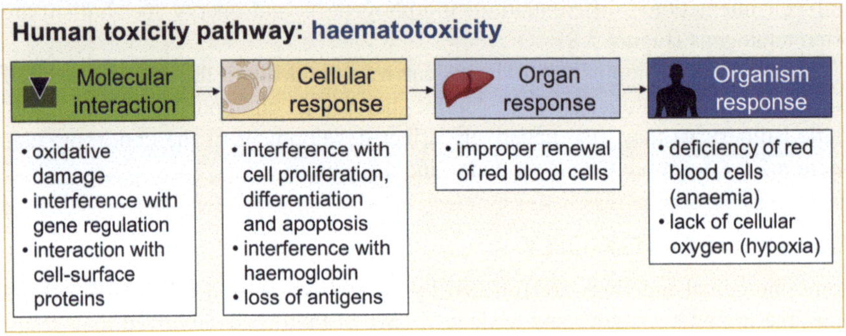

Figure 5.9 Spectrum of haematotoxicity from the molecular to the organism level.

5.6.2 Immunotoxicity

Broadly defined, immunotoxic agents adversely affect the immune system, which protects the organism against pathogens and tumours. The immune system comprises numerous lymphoid organs (*e.g.*, bone marrow, thymus, spleen, lymph nodes) and cell populations with a variety of functions. Antigen recognition is the cornerstone of the immune system. Antigens, usually protein or polysaccharide 'signatures' of foreign material, are recognised by specific antibodies, which subsequently initiates an immune response.

There are two types of immune response: innate and adaptive. The innate immune system is non-specific and is the body's primary defence mechanism. It relies on a variety of proteins (called the complement system) and involves several immune cells, such as natural killer cells, macrophages and neutrophils. Natural killer cells release cytokines and cytolytic chemicals that destroy the target cell. Macrophages and neutrophils are phagocytic cells that eliminate most microorganisms through the release of ROS.

The adaptive (or 'acquired') immune system is an antigen-specific response triggered by the innate immune system. In simple terms, immune cells learn to recognise the invading pathogen and deploy a more sophisticated set of specialised cells, such as helper T-cells and killer T-cells. Helper T-cells secrete cytokines and help direct the immune response depending on the nature of the threat. Killer T-cells bind to the target cell and release the content of cytolytic granules, which contain cytokines, perforins and other enzymes, on the target cell, a process called degranulation. Once degranulated, the killer T-cell releases the dying target cell and moves on to kill other target cells.

The immune system can also call upon other specialised cells when fighting inflammation, such as basophils and mastocytes. When stimulated, these cells degranulate to release histamine, proteoglycans, proteolytic enzymes, leukotrienes and cytokines. These chemicals attract other immune cells.

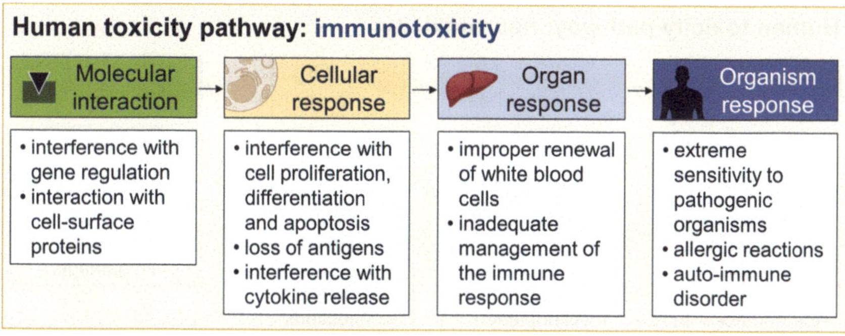

Figure 5.10 Spectrum of immunotoxicity from the molecular to the organism level.

The immune system must strike a delicate balance between excessive and insufficient immune response. Toxicant exposure can result in dysfunctions of the immune system:

- Immunosuppression results in reduced efficacy of the immune response (*i.e.*, impaired resistance), while immunostimulation stimulates the immune system, which may result in excessive immune response (Figure 5.10). A very wide range of toxicants have been shown to suppress or stimulate the immune system, including polychlorinated biphenyls (PCB), polycyclic aromatic hydrocarbons (PAH), pesticides, metals, solvents, hormones, pharmaceuticals and UV radiation. Some toxicants (*e.g.*, sulphamethoxazole) can stimulate immune cells directly by binding to their membrane receptors.
- Hypersensitivity reactions (allergies) result from the immune system responding in an exaggerated or inappropriate manner (*e.g.*, penicillin). Hypersensitivity has been linked with exposure to industrial chemicals, metals, solvents and pharmaceuticals.

Autoimmune disease occurs when the reactions of the immune system are directed at the body's own tissues. It is more difficult to establish a clear link between toxicant exposure and autoimmunity. Some chemicals have, however, been implicated in chemical-induced autoimmunity. These include some pharmaceuticals, plastic monomers (vinyl chloride), mercury and some pesticides (*e.g.*, hexachlorobenzene). Interaction between toxicants and endogenous proteins can sometimes result in the altered protein no longer being recognised as own tissue (*e.g.*, penicillin).

5.6.3 Neurotoxicity

A neurotoxicant is a toxic chemical that affects the development, function or viability of neurons and the nervous system (Figure 5.11). The nervous system coordinates numerous functions in the organism via neurons and

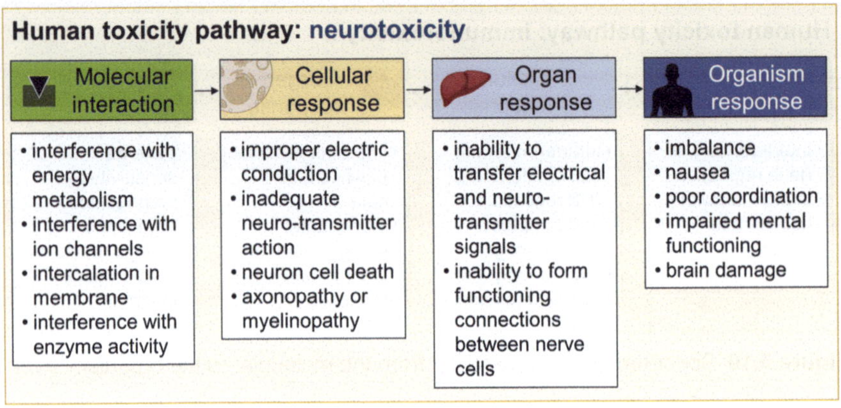

Figure 5.11 Spectrum of neurotoxicity from the molecular to the organism level.

neurotransmitters. There are two cell populations in nervous tissues: the neurons, which specialise in generation, reception and transfer of information (transmitted by neurotransmitters such as acetylcholine and epinephrine); and glial cells, which provide support and nutrition to neurons.

The four most common targets of neurotoxicants are the neuron, the axon (the neuron's projection towards other neurons), the myelinating cell and the neurotransmitter system.

- Neuronopathy: Although the neuron is similar to other types of cells in many respects, some features of the neuron are unique, and provide distinctive vulnerabilities. Some of those unique features are a high metabolic rate, a long cellular process supported by the cell body (the axon) and an excitable membrane that is rapidly depolarised and repolarised. A large number of chemicals are known to result in toxic neuronopathy, including metals (aluminium, arsenic, lead, manganese, mercury, methyl mercury), industrial chemicals (trimethyltin), pharmaceuticals and solvents.
- Axonopathy: Some toxicants can physically damage the axon, resulting in a degradation of neuron transmission. Many chemicals have been linked to axonopathy, including metals (gold and platinum), alkaloids, pharmaceuticals, industrial chemicals (acrylamide), solvents and pesticides.
- Myelinopathy: Myelin provides electrical insulation of neuronal processes, and its absence leads to a slowing and/or aberrant conduction of electrical impulses. Some toxicants can interfere with myelin maintenance or function.
- Neurotransmitter-associated toxicity: A wide range of naturally occurring toxins as well as pesticides and pharmaceuticals can inhibit normal neurotransmitter function. Organophosphate and carbamate pesticides, for example, inhibit the enzyme acetylcholinesterase, which is responsible for recycling the neurotransmitter acetylcholine.

There are several morphological idiosyncrasies of the nervous system. Some, such a mesh of endothelial cells called the 'blood–brain barrier' (Figure 5.3), provide an additional barrier to toxicants reaching the central nervous system. On the other hand, some make it more sensitive to toxicants. The unusual cell morphology of neurons, for example, which are very elongated rather than small and spherical, creates extraordinary demands on protein synthesis and transport of vesicles and organelles. The myelin sheet, which is rich in lipids and dependent on the proper function of a number of membrane-associated proteins, is also a sensitive target site for toxicants. Finally, the high energy requirements of neurons make them extremely sensitive to interruptions in the supply or oxygen or glucose that can be caused by toxicants such as cyanide and carbon monoxide as well as very sensitive to mitochondrial toxicants.

Astrocytes (a type of glial cell) play an important role in defence against neurotoxicants. They are activated by hypoxia and inflammation and have far greater antioxidant abilities than neurons, protecting axonal structure and processes. Intercellular communication between astrocytes and neurons also involves organelle exchange, including the transfer of healthy astrocytic mitochondria to adjacent neurons to restore degraded mitochondrial function, in exchange for damaged and defective neuronal mitochondria, which are then broken down in the astrocytes (a process called 'mitophagy').

5.6.4 Endocrine toxicity

Endocrine glands are collections of specialised cells that synthesise, store and release their secretions (hormones) directly into the bloodstream. As sensing and signalling devices capable of responding to changes in the internal and external environments, the hormone system coordinates a multiplicity of activities that maintain homeostasis. Disruption of normal endocrine function can thus have a wide range of effects, potentially affecting many different organ systems (Figure 5.12). The main endocrine glands of the body are the following:

The pituitary gland is a small protrusion off the hypothalamus at the base of the brain, which secretes hormones (including the so-called 'trophic hormones' to other endocrine glands) under stimulation of the hypothalamus. The pituitary releases hormones related to growth (growth hormone), lactation (prolactin), reproductive function (gonadotropins and corticotropic hormones) and thyroid activity (thyroid-stimulating hormone).

The adrenal glands are located above both kidneys and are mainly responsible for regulating the stress response through the synthesis of corticosteroids (cortisol and aldosterone) and catecholamines (epinephrine, norepinephrine and dopamine). They affect glucose metabolism (glucocorticoids) and reproduction (androgens, estrogens and progestins).

The pancreas produces digestive enzymes and hormones that regulate glucose metabolism (insulin, glucagon and somatostatin).

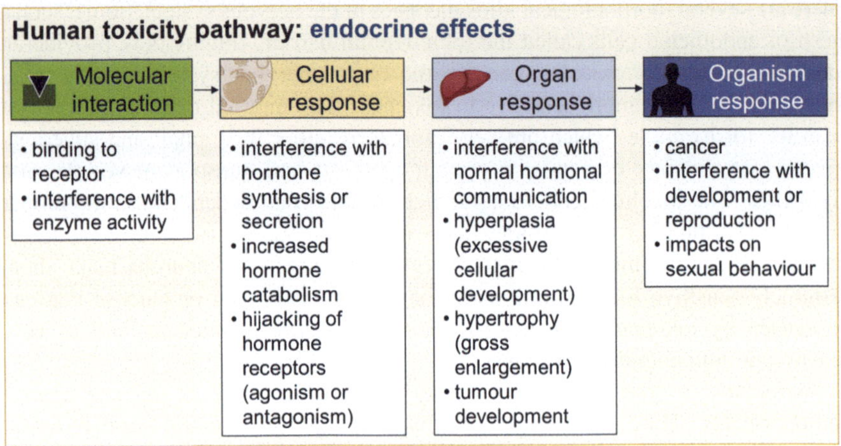

Figure 5.12 Spectrum of endocrine effects from the molecular to the organism level.

The thyroid gland secretes the thyroid hormones thyroxine (T4) and triiodothyronine (T3) under stimulation of the pituitary gland. Thyroid hormones increase metabolic rate and glucose availability, stimulate new protein synthesis, heart rate, cardiac output and blood flow, and increase neuronal development in young animals. The thyroid gland also produces calcitonin, which is involved in calcium homeostasis.

The parathyroid gland produces hormones involved in calcium homeostasis (parathyroid hormone, calcitonin and vitamin D) under stimulation of calcium-sensing receptors. This unique feedback system is sensitive to toxicants similar to calcium ions (*e.g.*, aluminium).

The gonads (testes in males, ovaries in females) produce sex hormones (androgens, estrogens) and progestogens (progesterone). Hormone production and gametogenesis in the gonads is under direct pituitary hormonal control. Gonads are sensitive to toxic substances because gametogenesis relies on rapidly dividing cells, which are often vulnerable to chemical destruction. The blood–testes barrier controls the entry of large molecules and toxicants into the seminiferous tubules, where gametogenesis occurs.

There are four main mechanisms of endocrine toxicity (Figure 5.12):

- Excessive stimulation can cause hyperplasia (excessive cellular development) and hypertrophy (gross enlargement) of individual endocrine organs, and eventually lead to tumour development.
- Interference with hormone synthesis or secretion. For example, some pharmaceuticals (*e.g.*, sulphonamides, 2,4-dihydroxybenzoic acid, aminotriazole, antipyrine) and pesticides (*e.g.*, amitrole) interfere with thyroid hormone synthesis.

- Increased hormone catabolism (destruction). Toxicants that induce liver enzymes (*e.g.*, phenobarbital, benzodiazepines, DDT, chlorinated hydrocarbons) can increase the rate of conjugation and excretion of hormones such as T3 and T4.
- Interference with hormone signalling (endocrine disruption). Hormones act by binding to highly specific hormone receptors (*e.g.*, estrogen receptor, androgen receptor, progesterone receptor, thyroid receptor), which results in a biochemical cascade that eventually triggers the intended effect. Toxicant-induced interference with hormone signalling can result in erroneous endocrine communication and/or interfere with the complex system of hormonal feedback loops. Some toxicants can mimic hormone activity (agonists) while others can inhibit normal hormonal function (antagonists). In some instances, this effect is intentional (*e.g.*, pharmaceuticals used for birth control such as ethinylestradiol or levonorgestrel), but toxicants can also be unintentional endocrine disruptors (*e.g.*, industrial chemicals like bisphenol A, phthalates). Interference or mimicry of sex steroids (estrogens and androgens) can also significantly affect reproduction. Most of the activity of the endocrine system has sensitive feedback loops, and toxicants that affect or mimic hormones often affect multiple endocrine glands.

5.6.5 Reproductive toxicity

The purpose of the reproductive system is to produce good quality gametes, capable of fertilisation and producing a viable offspring, which can in turn successfully reproduce. This requires a large number of complex processes, orchestrated in a precise order for optimal performance at different life stages. The fact that chemicals can adversely affect reproduction in males and females is not a new notion. One only has to look at the importance of drugs as contraceptives to realise how sensitive the reproductive system can be to external chemical influences. Endocrine communication is critical to proper reproductive function, and toxicants that can adversely affect endocrine glands also generally result in reproductive toxicity. A wide range of environmental chemicals are known to mimic or inhibit androgens (*e.g.*, trenbolone, vinclozolin, procymidone, linuron, p,p′-DDE, phthalates), estrogens (*e.g.*, methoxychlor metabolites, ethinylestradiol, bisphenol A, nonylphenol, DDT) or progestogens (*e.g.*, levonorgestrel, norethindrone). Exposure to these hormone mimics can adversely affect reproductive functions.

Sex hormones (androgens and estrogens) are particularly important in foetal reproductive organ development, puberty and sexual maturation. These stages are thus inherently susceptible to endocrine disruption. Toxicants such as PCBs, DDT/DDE, brominated flame retardants, dioxins, hexachlorobenzene, personal care products and heavy metals have been linked to reproductive abnormalities, although their exact mechanism is often unknown.

The female reproductive cycle relies on subtle hormonal communication between the pituitary and ovarian secretions of progesterone and estrogens. These hormones determine ovulation and prepare the female accessory organs to receive male sperm. Disruption of these hormonal cues can lead to infertility. Disruption of the luteinising hormone (LH) surge, for example, by the pesticides chlordimeform and *N*-methyldithiocarbamate prevents or delays ovulation, which has been shown to cause infertility in laboratory animals.

Male reproductive processes also rely on carefully orchestrated hormonal communication through the hypothalamic–pituitary–testes axis, and endocrine disruption can also affect male reproduction. The vast majority of male reproductive toxicants affect sperm production (spermatogenesis). Some of these toxicants act via indirect routes, such as nutrient disruption after exposure to zinc, or increased steroid clearance by the liver due to carbon tetrachloride exposure. Most toxicants affecting spermatogenesis, however, do so via a direct effect on testis or spermatogenesis itself by interfering with or damaging Sertoli cells (*e.g.*, phthalate ester metabolites, dibromoacetic acid, m-dinitrobenzene) or interfering with energy production in sperm cells (*e.g.*, chlorosugars, epichlorohydrin).

Reproductive toxicants can also affect fertilisation and implantation as successful pregnancy depends heavily on complex and subtle hormonal communication. These processes are also susceptible to endocrine disruption (*e.g.*, some pharmaceuticals used to terminate pregnancies interfere with progesterone synthesis).

5.7 CONCLUSIONS

The purpose of this chapter was to give the reader an appreciation of the normal function and significance of different organs and organ systems in the human body, and to describe toxic effects and define mechanisms of toxicity. Humans exposed to contaminated water can exhibit a wide variety of tissue-, organ- and organ system-level responses, many of which can be traced back to the effect of the toxicant at the molecular or cellular level, illustrating the concept of toxicity pathways introduced in the previous chapter. Monitoring those molecular or cellular events using *in vitro* bioassays may therefore provide a simple screening method to detect toxicants in water, and Chapter 10 reviews *in vitro* methods available to measure toxic effects discussed here.

Chapter 6

Adverse outcome pathways of chemicals in aquatic organisms

6.1 INTRODUCTION

Ecotoxicology is the study of the effects of toxic substances and other stressors on the structure and function of ecosystems. The discipline has evolved relatively independently of human toxicology but with the concept of adverse outcome pathways (AOPs) and the recognition that chemical stressors perturb cellular functions in a similar manner, the two fields have come closer together.

Aquatic organisms can bioaccumulate pollutants via direct uptake from water across skin or gills and from their diet. The level of bioaccumulation is dependent on the physicochemical properties of the chemicals with lipophilic chemicals accumulating to a greater extent than hydrophilic chemicals.

The aquatic component of the ecosystem is in perpetual contact with the other environmental compartments including those of air, sediment and soil, and the ecotoxicological principles for all compartments are similar.

A crucial driver of the discipline of ecotoxicology was Rachel Carson's 'Silent Spring', published in 1962 (Carson, 1962). It denounced the negative impacts of organochlorine pesticides such as DDT on humans and the environment. A comprehensive treatise on ecotoxicology can be found in various textbooks, with recommendations for Newman's 'Fundamentals of Ecotoxicology' (2019), Walker's 'Principles of Ecotoxicology' (2012) and Landis and Yu's 'Introduction to Environmental Toxicology: Impacts of Chemicals Upon Ecological Systems' (2004).

Authors: Beate Escher, Peta Neale and Frederic Leusch
doi: 10.2166/9781789061987_0091

In this chapter, we focus on some of the basics of ecotoxicology and mechanistic approaches to aquatic toxicology and ecosystem health. This chapter builds on the foundation laid in Chapter 4 and uses the AOP concept in a conceptual manner rather than detailing individual AOPs. The purpose of this chapter is to understand the toxicity pathways of chemicals in aquatic organisms sufficiently to select appropriate *in vitro* tools for water quality assessment. Since the first edition of this book, which appeared shortly after the introduction of the AOP concept to this field of research, more than 350 AOPs have been developed that contain >1000 key events and almost 2000 key event relationships (aopwiki.org, accessed on 20 December 2020). Many of the available AOPs describe adverse outcomes on fish health with only a few AOPs addressing algae and invertebrates.

6.2 FROM THE CELLULAR LEVEL TO THE ECOSYSTEM

To understand the effects of chemicals on ecosystems, we need a mechanistic understanding on how chemicals act on individual organisms as well as an understanding of the complex interactions occurring in ecosystems.

Comprehensive field studies in natural ecosystems are limited to cases of unintentional contamination because obviously planned exposure studies and the intentional contamination at the ecosystem level are ethically unacceptable. The majority of ecotoxicology research has therefore focused on laboratory studies with single species that are representative of different trophic levels. An overview of commonly used test species and endpoints is given in Chapter 3. While integrated assessment in ecosystems may be more relevant than laboratory studies, they are less controlled and can only detect overall apical responses that are difficult to interpret. Studies with simplified model ecosystems (*e.g.*, meso- and macrocosm) are viable alternatives to real ecosystem assessments because aspects of food chains and ecosystem interaction can be integrated (Chapter 3). Further, controlled model systems have the benefit of being able to include quality assurance/quality control (QA/QC) measures such as negative and positive controls as well as replicate experiments (Chapter 11). Despite these obvious advantages, model ecosystem studies are very expensive and typically only conducted in higher tiers of pesticide risk assessment.

If single organisms are tested, apical endpoints such as developmental dysfunction, reproductive failure and death are often measured (Chapter 3). Acute toxicity tests are performed over a timeframe of several hours to several days. To be protective of the entire ecosystem, one must extrapolate the information gained from acute tests to safe levels for the entire population and from short-term to long-term exposure (Figure 6.1, top panel). Such extrapolation can be accomplished by applying extrapolation and uncertainty factors (Chapter 2). This process must be completed for many aquatic species to obtain an idea about the species sensitivity distributions (SSD). Generally, the lower fifth percentile (HC5) of a distribution of chronic 'no observed effect concentrations' (NOEC) is

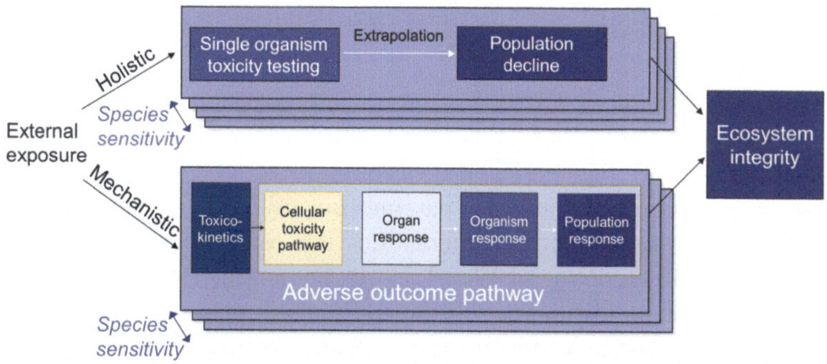

Figure 6.1 Comparison of the traditional holistic approach for aquatic toxicity assessment of chemicals and the novel mechanistic approach based on AOPs.

considered sufficiently low to protect ecosystem integrity and functioning (Posthuma *et al.*, 2002), although concentrations as low as one percentile may be needed for chemicals known to bioaccumulate.

In the AOP approach the cellular level response pathway, which is often highly conserved in different species, is used to obtain a mechanistic picture of the species sensitivity, which facilitates extrapolation from cellular responses to expected effects on ecosystems (Figure 6.1, bottom panel).

Although AOPs are central organising principles, the quantitative link to individual organism responses must still be established to allow the implementation of AOPs to population level models (Kramer *et al.*, 2011). Species sensitivity differences can be related to differences in toxicokinetics, specifically the metabolic capacity of different species, and to differences in the toxicity pathways. Species with shared ancestry will have similar AOPs due to the close homology of genes (Celander *et al.*, 2011). Read-across is thus possible for highly homologous species. Read-across from human drug targets to adverse effects in fish is even possible for evolutionarily well-conserved targets.

6.3 ADVERSE OUTCOME PATHWAYS FOR AQUATIC ORGANISMS

Chapter 3 introduced the most fundamental members of aquatic food chains – primary producers, invertebrates and vertebrates. Common apical endpoints measured for these organism groups are growth inhibition, immobilisation and mortality, respectively. In the following section, the concept of AOPs for aquatic organisms is illustrated using examples from these three taxonomic groups. For each organism group, the AOPs for non-specific toxicity are compared with one selected specific mode of action (MOA), using specific toxicity of herbicides, insecticides and estrogens as illustrative examples for algae, water fleas and fish, respectively.

6.3.1 Adverse outcome pathways for algae

6.3.1.1 Baseline toxicity

We have learnt in Chapter 4.4.1 that baseline toxicity is the minimal toxicity that any compound exhibits. This is of high relevance in the aquatic environment, where all chemicals present act together in a concentration additive manner, potentially resulting in an appreciable level of baseline toxicity. The molecular targets of baseline toxicants are biological membranes, which can become leaky, disrupting their structure and function (van Wezel and Opperhuizen, 1995) (Figure 6.2). In algae, such membrane disruption will indirectly affect photosynthesis efficiency, as the electron transfer chain of algal photosystems is embedded in the chloroplast membrane.

The biologically effective concentration of baseline toxicants is the same for all chemicals, that is, every chemical has the same baseline effect once it has crossed the membrane. The apparent differences in potency between different chemicals thus stem exclusively from differences in uptake and other toxicokinetic processes. The more hydrophobic a chemical is, the higher its degree of accumulation in algae and consequently the lower the dosed concentration required to trigger the baseline-toxic effect.

A feature of baseline toxicity is that it is a reversible MOA. Once the algae are transferred to clean water, the baseline toxicants can be depurated, enabling the algae to recover from the toxic stress, provided that no irreversible secondary effects have occurred. Furthermore, baseline toxicity-mediated photosynthesis inhibition occurs at much higher chemical concentrations than specific inhibition of photosynthesis. Nevertheless, at the organism level, reduced photosynthesis leads to lowered energy for cellular growth causing cells to stop dividing. At the population level, this effect manifests itself by a slower population growth rate or decline in population size (Figure 6.2).

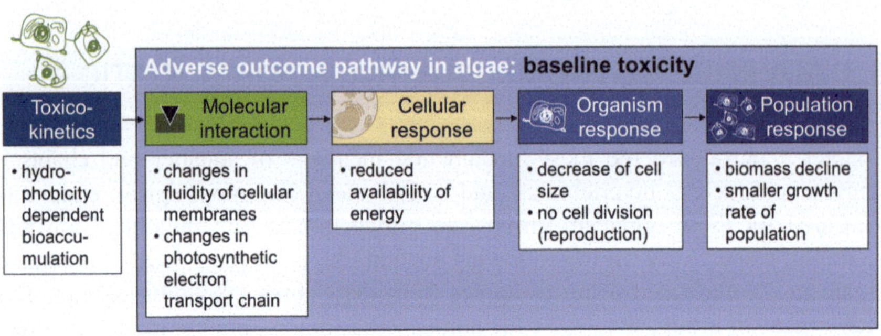

Figure 6.2 Adverse outcome pathway for baseline toxicants in algae. Adapted from Ankley *et al.* (2010).

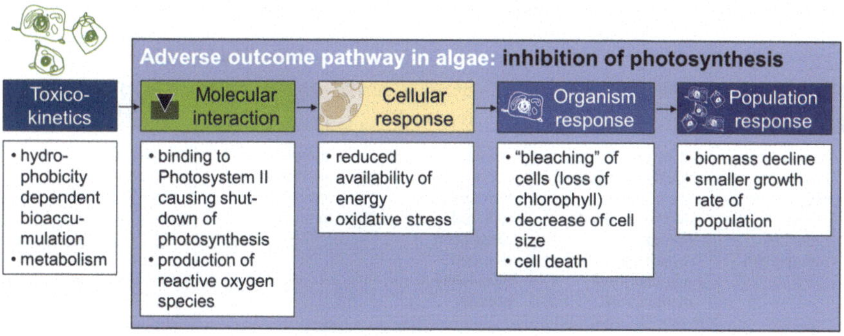

Figure 6.3 Adverse outcome pathway for herbicides in algae.

6.3.1.2 Inhibition of photosynthesis by herbicides

Herbicides are used for weed control but they also have the negative effect of inhibiting photosynthesis in non-target plants and green algae. Important classes of herbicides are the triazines, such as atrazine, simazine and irgarol, and phenylurea herbicides, such as diuron and isoproturon. Both groups of herbicides bind to a quinone-binding site on photosystem II, blocking photosynthetic electron transfer (Figure 6.3). As a result, no energy in form of ATP is produced, causing a loss of chlorophyll, slower growth, smaller cell size and ultimately death of algal cells. The ultimate observable effects are the same as for baseline toxicity, but these specific effects occur at chemical concentrations that are orders of magnitude lower than baseline toxicity.

Vogs and Altenburger (2016) developed a combined toxicokinetic/toxicodynamic model that could differentiate well between different AOPs with effect progression, that is, time to progress from molecular initiating event (MIE) to apical effect, being shortest for specific photosynthesis inhibitors followed by chemicals that cause oxidative stress and those which interfere with lipid synthesis.

6.3.2 Adverse outcome pathways for invertebrates

6.3.2.1 Baseline toxicity

Baseline toxicity occurs in aquatic invertebrates such as the water flea *Daphnia magna* at the same internal concentration as in algae. As water fleas are metabolically more active, the baseline toxicants can be better detoxified and thus the apparent toxicity in relation to the external exposure concentration may be lower. Apart from this one exception, the molecular interactions of baseline toxicants in cell membranes of water fleas and algae are the same (Figure 6.4). Mitochondrial dysfunction is also one of the characteristics of baseline toxicity (Vinken and Blaauboer, 2017). Despite this similarity in MIE, the apical

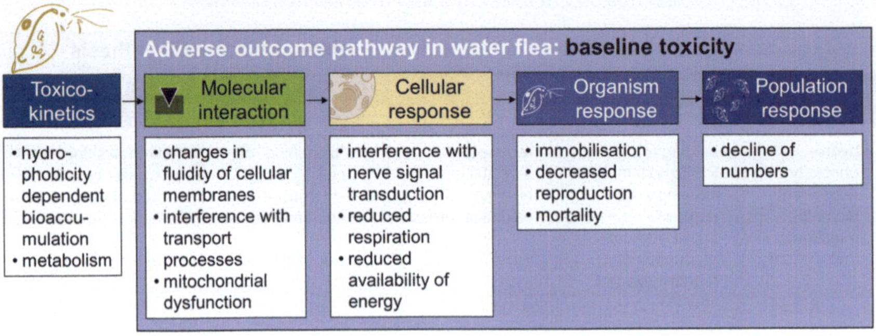

Figure 6.4 Adverse outcome pathway for baseline toxicants in the water flea *Daphnia magna*. Adapted from Ankley *et al.* (2010) and Vinken and Blaauboer (2017).

endpoints observed in water fleas differ from algae. The effects in water fleas will mainly manifest themselves as reduced ventilation/respiration rate, which equates to immobilisation and eventually death at the organism level.

6.3.2.2 Activity of insecticides

Insecticides such as organophosphates or carbamates inhibit the enzyme acetylcholinesterase (AChE). AChE is responsible for breaking down the neurotransmitter acetylcholine. Inhibiting this breakdown process induces a constant chemical signal firing across the synapse, leading to over-excitation of nerves. This over-excitation is initially expressed as rapid movement, consequently leading to excessive expense of energy before the ultimate responses of immobilisation and death occur. Invertebrates such as insects and, in our example, water fleas are particularly sensitive to organophosphates. First, in terms of toxicokinetics, organothiophosphates such as diazinon must be metabolically activated to their bioactive form, for example, diazoxon, in order to induce their MIE of binding to AChE (Figure 6.5). This metabolic activation is particularly efficient in water fleas with little detoxification but strong activation

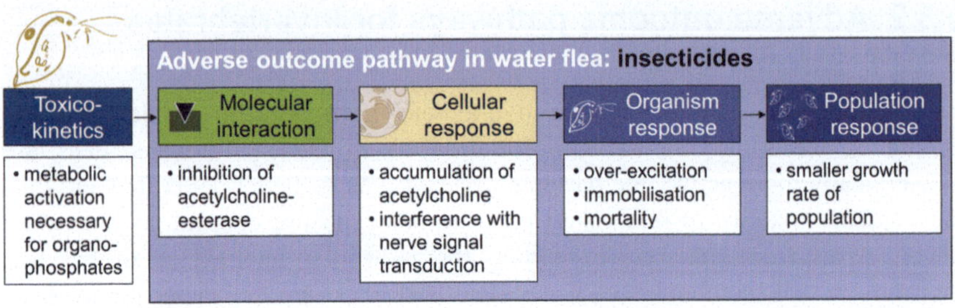

Figure 6.5 Adverse outcome pathway for the inhibition of acetylcholinesterase (AChE) in the water flea *Daphnia magna*. Adapted from Russom *et al.* (2014).

(Kretschmann *et al.*, 2011). Second, in terms of toxicodynamics, the AChE of *Daphnia magna* is particularly sensitive to organophosphates (Kretschmann *et al.*, 2012). Carbamates differ from organophosphates in that metabolic activation is not required for the binding, which is reversible (Jeon *et al.*, 2013). Organophosphates bind covalently to the enzyme, causing it to hydrolyse further, which makes the binding irreversible.

Russom *et al.* (2014) developed a formal AOP for inhibition of AChE leading to acute mortality for various taxa, recognising the highly conserved MIE but also species-specific differences in toxicokinetics and AChE binding domains.

6.3.3 Adverse outcome pathways for fish

6.3.3.1 Baseline toxicity and mitochondrial dysfunction

Baseline toxicity affects the same molecular and cellular mechanisms in fish as in the invertebrate and algal species discussed above, but the consequences are somewhat different (Figure 6.6). The initial effect of baseline toxicity is a loss of equilibrium. Fish that normally swim in fast flowing rivers against the current can no longer maintain their position. Non-targeted swimming may lead to increased vulnerability to predators and difficulty in mating. At higher concentrations, baseline toxicity leads to narcotic effects and eventually to death.

Mitochondrial dysfunction is also one of the key features of baseline toxicity but there are also specific effects on mitochondria that lead to the same symptoms at much lower concentrations of chemicals. Souders *et al.* (2018) demonstrated how oxidative respiration in fish embryos can be used to identify specific MIEs such as uncoupling of oxidative phosphorylation, specific inhibition of the electron transfer chain and inhibition of ATP synthesis.

6.3.3.2 Reproductive toxicity

Many pathways can lead to reduced reproductive success or even reproductive failure (Figure 6.7). Natural estrogens and xenoestrogens (*e.g.*, nonylphenol and

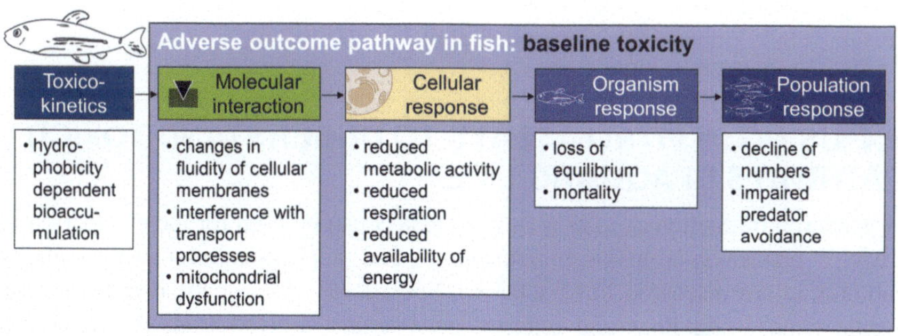

Figure 6.6 Adverse outcome pathway for baseline toxicants in fish.

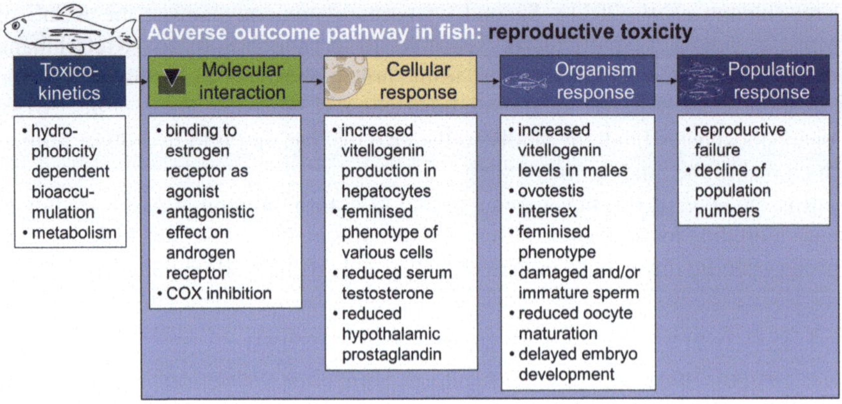

Figure 6.7 Adverse outcome pathway for reproductive effects in fish. COX = cyclooxygenase. Adapted from Ankley *et al.* (2010), Martinovic-Weigelt *et al.* (2017) and Martyniuk *et al.* (2020).

bisphenol A) occurring in environmental waters can bind to estrogen receptors in fish, triggering a number of cellular responses that would normally be triggered by endogenous hormones. Inducing such cellular responses at the wrong time can cause both structural and functional problems in the fish. One major problem of estrogenic effects is that they can induce feminisation of male fish. At the cellular level, this translates to feminisation of the phenotype and production of vitellogenin (an egg yolk protein precursor produced in liver cells) in both females and males. In male fish, estrogenicity can further induce notable changes in sexual development such as the formation of ovotestis (egg cells in the testis tissue) and intersex. Such changes evidently lead to reproductive failure (or reduced reproductive success) and can eventually wipe out entire populations (Kidd *et al.*, 2007). Anti-androgenic chemicals can similarly antagonistically impact the androgen receptor, reducing testosterone levels and cause reproductive failure due to damaged or immature sperm. Cyclooxygenase (COX) inhibitors disrupt prostaglandin synthesis, leading also ultimately to reproductive toxicity in female fish by preventing oocyte maturation.

6.4 USING *IN VITRO* ASSAYS TO UNDERSTAND TOXICITY PATHWAYS IN AQUATIC LIFE

A general introduction as to how *in vitro* assays can be used for the elucidation of toxicity pathways is given in Chapter 9 and applications for water quality monitoring are discussed in Chapter 10. Specifically, for aquatic organisms, one would choose cell lines and reporter gene assays derived from the tissues of aquatic biota such as fish cell lines (Schirmer, 2006).

However, given the high degree of conservation of cellular pathways, it is often acceptable to use a construct from different species. The yeast estrogen screen (Routledge and Sumpter, 1996), for example, is based on a yeast cell line transfected with a human estrogen receptor, but the assay is a widely applied bioanalytical tool for assessing estrogenicity in water, where fish are the primary target for EDCs. Likewise, reporter gene assays based on human and zebrafish nuclear receptors yielded similar responses of water samples (Neale *et al.*, 2020c).

6.5 CONCLUSIONS

The concept of AOPs bridges the initial cellular level effects of toxicants with the organism level outcome. The application of this concept to aquatic organisms demonstrates the ability of *in vitro* bioassays to detect subtle toxic effects in aquatic organisms, thereby highlighting the potential of *in vitro* bioassays to replace, reduce and/or refine whole organism testing.

Chapter 7

Dose–response assessment

7.1 INTRODUCTION

The central paradigm in toxicology is that 'the dose makes the poison'. In fact, all substances, even something as innocuous as table salt, are poisonous if taken in sufficient quantity. Understanding the dose–response relationship can therefore, quite literally, be a matter of life or death. A dose–response curve is the mathematical display of this paradigm. From a dose–response curve we can derive various descriptors of effect, for example, the lethal dose LD_{50} or the highest concentration where no effect was observed, the no observed effect concentration (NOEC).

A dose is the total quantity of a chemical delivered to a test animal or system. We know doses from pharmaceuticals we take, for example, one tablet or 200 mg of aspirin if we have a headache. In toxicity testing it makes more sense to report a normalised dose, for example, dose per kg of body weight of the test animal. For *in vitro* bioassays, the dose is difficult to quantify because it will depend on the number of cells and the medium volume. Therefore, concentrations are the typical dose-metrics in cell assays. A concentration is the mass or molar amount of a chemical divided by the volume of the test system.

This chapter introduces dose–response assessment in general terms before we focus on concentration–response curves (CRC) in *in vitro* assays. We differentiate between the activation of a certain response on the level of molecular initiating event or key event (Chapter 4) and cytotoxicity. We present various options for modelling CRCs and deriving benchmark doses (BMD) or

Authors: Beate Escher, Peta Neale and Frederic Leusch
doi: 10.2166/9781789061987_0101

benchmark concentrations (BMC). Effect concentrations of water samples are expressed in units of relative enrichment factors (REF) instead of concentrations because water samples contain undefined complex mixtures of chemicals but can be translated into bioanalytical equivalent concentrations (BEQ) for an easier analogy to the effects caused by single chemicals.

7.2 DOSE–RESPONSE ASSESSMENT

7.2.1 Dose–response curves

A dose–response curve plots the response (*e.g.*, death, or a more subtle sublethal effect) of a population under study (*e.g.*, rats, mice) against increasing dose of the test chemical applied to the system (Figure 7.1). The dose–response curve looks like a saturation curve on a linear dose scale but is typically sigmoidal on a logarithmic scale.

On the lower left of the log dose–response curve, at very low doses, there is no measurable response in the test system. It is important to note, however, that a close look at the lower end of the dose–response curve on a linear dose scale reveals that the relationship between dose and response is in fact linear at low doses when the dose is not in logarithmic scale (Figure 7.1).

Figure 7.1 shows that as the dose is increased, there is initially no detectable increase in the monitored response. Eventually, a detection threshold is reached above which the response quickly increases with increasing dose in a near linear manner (actually log-linear, as the x-axis is on a log-scale). The dose–response curves then levels to a maximum of 100%. This dose–response curve is common to all biological responses, although it can sometimes be affected when more than one type of effect co-occur in the test system as is discussed in more detail for cytotoxicity and activation of an effect.

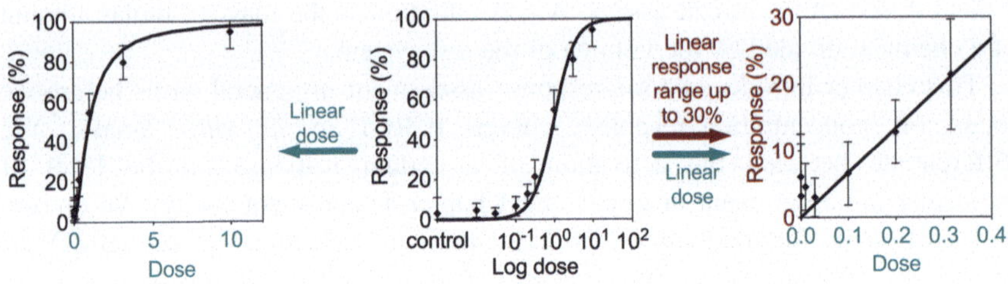

Figure 7.1 A dose–response curve with the dose on the x-axis and the response (in %) on the y-axis. If the dose is plotted on a logarithmic scale, the typical sigmoidal form is visible. The figure on the right depicts the low-response level linear portion of the dose–response curve.

7.2.2 Dose benchmark values

Several important parameters can be described by the dose–response curve. The most common descriptor is the LD_{50}, the dose that is lethal to 50% of the organisms in a test population (Figure 7.2). The highest dose tested that is not statistically different from the negative control is referred to as the 'no observed adverse effect level' (NOAEL), while the next dose above, that is, the lowest dose tested that is statistically different from the negative control, is called the 'lowest observed adverse effect level' (LOAEL). In general term, these dose values are called point of departure (POD) in risk assessment for extrapolation to safe levels for sensitive human populations.

A major issue with LOAEL and NOAEL as expressions of toxicity is that they are dependent on the experimental design (*i.e.*, dose spacing) and variability of the data, which could be affected by the number of replicates for each dose or the inherent variability of the test system. Therefore, parameters that are calculated from the whole dose–response relationship are preferred expressions of toxicity, such as the BMD. The BMD_{10} for example is the dose calculated from the dose–response curve required to produce 10% of the maximum response. Another common parameter is the benchmark dose level (BMDL), which is the lower confidence boundary of the BMD. The BMD is preferred over NOAEL and LOAEL because it does not depend on the experimental doses and takes into account the shape of the full dose–response curve (black line in Figure 7.2). A limited number of data points or high variability in the controls can cause the NOAEL to be much higher, as high as 40–50% of the response, while for the same situation the BMDL becomes smaller due to the higher uncertainty (Haber *et al.*, 2018), which is more precautious from a risk assessment perspective.

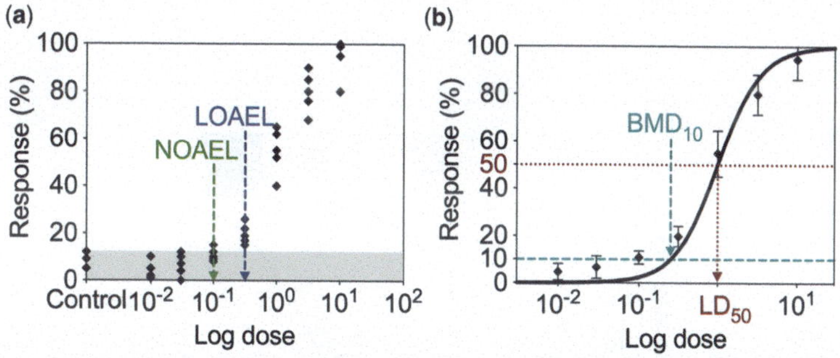

Figure 7.2 A typical dose–response curve with the logarithm of the dose on the x-axis and the response (in %) on the y-axis. (a) Individual response data points and derivation of no observed adverse effect level (NOAEL) and lowest observed adverse effect level (LOAEL). The grey area represents the variability of the negative control. (b) The same data plotted as averages and described with a dose–response model used to derive the lethal dose (LD_{50}) and the benchmark dose (BMD_{10}).

7.2.3 Continuum of toxicity

As described in Chapter 4, in the case of chemical-mediated toxicity, effects at low levels of biological complexity (*e.g.*, molecular and cellular) can translate into higher-order effects at the tissue, organ and eventually organism level, depending on the severity of the effect and the ability of compensation and repair mechanisms to cope with the damage. This continuum of toxicity can also be plotted on a dose–response graph, which shows that as the dose increases so does the level of biological complexity affected (Figure 7.3).

Let us delve into this continuum by using the example of a liver toxicant that interferes with a liver enzyme. At low dose, the toxicant has no noticeable effect on the liver enzyme. As the dose increases, increasing amounts of the enzyme is inactivated by the toxicant. Initially, this effect has no further consequence beyond the molecular level, but as more and more of the enzyme is affected, the increasing molecular dysfunction starts to affect cellular health and finally, cytotoxicity becomes evident. The organ, in this example the liver, can deal with a certain amount of cellular damage and initially the death of a few individual liver cells has no further impact. But above a certain dose, the cellular damage exceeds the compensatory capacity of the organ, and organ damage occurs. Eventually, sufficient organ damage occurs to negatively affect the whole organism, ultimately resulting in death.

The continuum of toxicity is one of the reasons it is logical to attempt to detect toxicants by their effect at the molecular and cellular level, as is done with *in vitro* bioassays. Indeed, the molecular and cellular effect will occur at much lower doses than those causing organ and/or organism-level effects detectable in *in vivo* tests. Using a well-designed battery of *in vitro* bioassays therefore allows a sensitive assessment of toxic potential. As previously mentioned, a response at the molecular and cellular level does not necessarily translate into higher-order

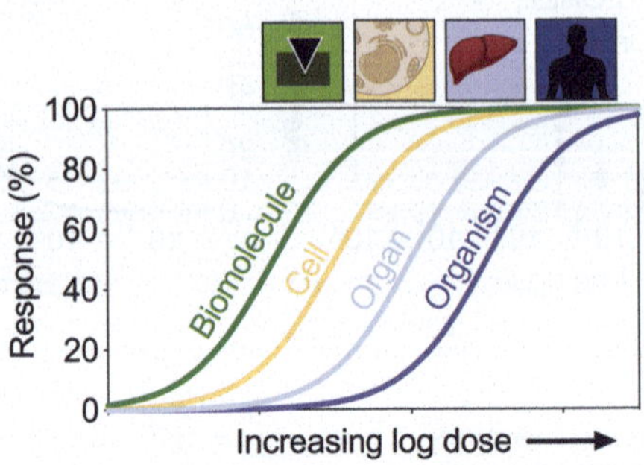

Figure 7.3 Continuum of toxicity. As the dose increases, so does the level of biological complexity that is affected by the toxic compound.

effects, but in chemical-mediated toxicity, higher-order effects cannot occur without preliminary molecular or cellular level effect.

7.3 CONCENTRATION–RESPONSE ASSESSMENT

7.3.1 'Concentration' versus 'dose'

In aquatic ecotoxicology and in cell-based bioassays, the exposure is usually expressed as the concentration of chemical in the medium (*e.g.*, ng test compound per litre of water) instead of the dose (*e.g.*, mg test compound per kg body weight). The cellular concentration is the biologically effective concentration (BEC), but we cannot easily measure this dose as is outlined in more detail in Chapter 9. The concentration in the medium surrounding the cells provides a better estimate of the BEC than any actual dose parameter (Kisitu *et al.*, 2019). The dose in an *in vitro* test would relate to the amount of chemical added to the well of a microtitre plate. How much of that dose reaches the cells depends on the number of cells and especially the volume of the medium. For the same number of cells and the same dose added, there will be threefold increase in concentration if the volume of the medium is reduced by a factor of 3. Therefore, 'concentration' is more accurate than 'dose' with aquatic test systems, and the 'dose–response curve' in cell-based bioassays is more commonly referred to as the 'concentration–response curve' (CRC). More details on dose-metrics are discussed in Chapter 9.4.1.

7.3.2 'Response' can mean toxicity or effect

In order to avoid confusion, we are referring to any kind of biological effect as a 'response'. All CRC can be fitted with the same mathematical models but, depending on the bioassay, the response can refer to toxicity (lethality, inhibition of growth, inhibition of cell viability) or to effects, which may include binding to nuclear receptors, induction or inhibition of metabolic enzymes or activation of adaptive stress responses. Effects do not necessary lead to toxicity and manifestation of an effect in an *in vitro* bioassay such as a reporter gene assay does not necessarily equate to effects on higher levels of biological organisation, and it is inaccurate to use the term 'toxicity' when describing sublethal responses. In the section below, we therefore describe the CRC equations in general terms before discussing benchmark values for toxicity and effects independently.

7.3.3 Concentration–response modelling

There are many different approaches to fit the CRC to the raw data. Many parametric models are available in the literature that have been successfully used to fit sigmoidal logarithmic CRCs of various forms (Scholze *et al.*, 2001). More recently, Bayesian approaches have become popular for CRC modelling. While they can fit even the noisiest data, such models might not be useful for the purpose of water quality monitoring. For water quality monitoring it is essential to be able to compare between samples and to compare samples with individual components' CRCs as well as process large numbers of CRCs.

Among the many parametric models to choose from, the log-logistic equation presented in Equation 7.1 (also called Verhulst or Hill equation) has the advantage that it provides the median response concentration RC_{50} directly and that the other parameters of the equation (min, max and slope) all have biological relevance. Figure 7.4 depicts a CRC for bacterial cytotoxicity caused by the pharmaceutical diclofenac. Note that because the response is inhibition of bioluminescence, the RC_{50} is termed inhibitory concentration IC_{50} in this example.

$$\text{Response (\%)} = \text{min} + \frac{\text{max} - \text{min}}{1 + 10^{\text{slope} \cdot (\log RC_{50} - \log C)}} \tag{7.1}$$

where min is the minimum response (0% inhibition in the example in Figure 7.4), max is the maximum response (100% inhibition in Figure 7.4), slope is the slope of the curve (2.4 in Figure 7.4), log RC_{50} is the logarithmic value of the RC_{50} ($\log IC_{50} = -3.9$ in this example, corresponding to an IC_{50} of 1.3×10^{-4} M in Figure 7.4), and log C is the logarithmic concentration.

If the response ranges from 0 to 100%, the equation simplifies to

$$\text{Response (\%)} = \frac{1}{1 + 10^{\text{slope} \cdot (\log RC_{50} - \log C)}} \tag{7.2}$$

The concentration associated with a y% response (RC_y) can be calculated by solving the equation for log RC_y, as shown in Equation 7.3.

$$\log RC_y = \log RC_{50} - \frac{1}{\text{slope}} \cdot \log\left(\frac{\text{max} - \text{min}}{y - \text{min}} - 1\right) \tag{7.3}$$

where $\log RC_y$ is the logarithm of the response concentration RC_y, $\log RC_{50}$ is the logarithmic RC_{50} value from the curve equation, max is the maximal response

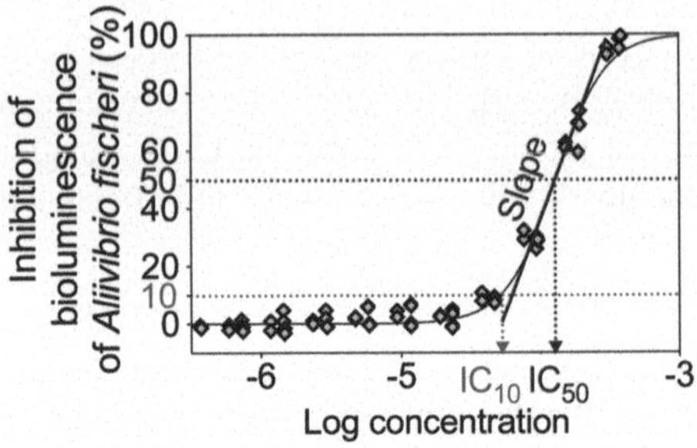

Figure 7.4 Typical concentration–response curve on the example of diclofenac in the bioluminescence inhibition assays with *Aliivibrio fischeri* (Baumer *et al.*, 2017).

(usually 100%), min is the minimal response (usually 0%), y is the % response for which you wish to determine the concentration (*e.g.*, for RC_{10} the response is 10%), and slope is the slope of the curve.

7.3.4 Concentration benchmark values

Concentration–response models allow the computation of the concentration associated with any specific response level, where RC_y is the effective concentration required to cause y% of the response. The most commonly used value is the RC_{50}, the concentrations required to cause 50% of the response. In assays for cytotoxicity, the response is usually referred to as lethal, and the concentration referred to as lethal concentration (LC_y). Likewise, inhibitory concentrations (IC_y) are used when the assay quantified an inhibitory response. In the example in Figure 7.4, the IC_{50} identifies the concentration that causes 50% inhibition of bioluminescence, while the IC_{10} is the concentration that causes 10% inhibition.

In parallel to the dose–response analysis (see Section 7.2.2), the 'lowest observed effect concentration' (LOEC) is the lowest concentration tested that produce a significant deviation from the control (unexposed bacteria in this example) and the 'no observed effect concentration' (NOEC) is the highest tested concentration that does not produce an effect significantly different from the control (Figure 7.5). Just like the LOAEL and NOAEL, the LOEC and NOEC are affected by the specifics of the experimental design (concentrations tested, number of replicates and variability of the test system). They are therefore falling out of favour, replaced by parameters calculated from the full CRC such as the concentration required to produce 10% of the response (RC_{10}) – commonly

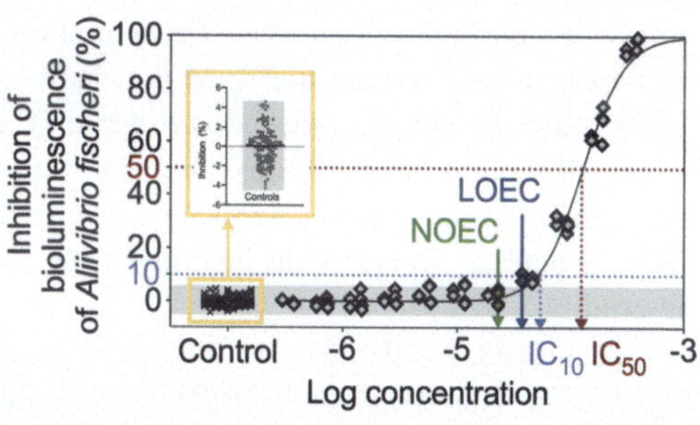

Figure 7.5 Typical concentration–toxicity curve with diclofenac in the bioluminescence inhibition assays with *Aliivibrio fischeri* as an example (Baumer *et al.*, 2017) with 'no observed effects concentration' (NOEC), 'lowest observed effects concentration' (LOEC), and inhibitory concentrations IC_{10} and IC_{50} indicated. The grey bar depicts the variability of the controls (unexposed cells) with a mean of 0.01 ± 1.9% inhibition, which is enlarged in the inset.

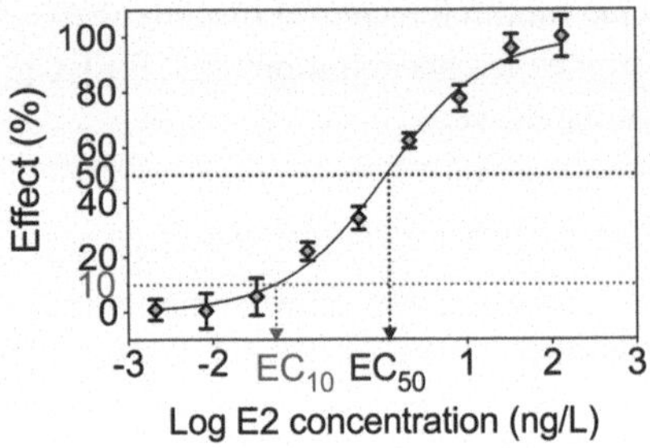

Figure 7.6 Typical concentration–response curve (Equation 7.1) with EC_{10} and EC_{50} values indicated. The depicted data are typical of average ± standard deviation of an experiment with the ER CALUX bioassay (Leusch, unpublished).

referred to as EC_{10}, LC_{10} or IC_{10} ('effective', 'lethal' and 'inhibitory' concentrations, respectively) or for reporter gene and transactivation assays PC_{50}, the concentration of a test chemical, at which the measured activity is 50% of the maximum activity induced by the positive control (PC). The big advantage of synthesising the whole concentration–response relationship into a few parameters is that the entire dataset is used rather than individual points (such as LOEC and NOEC). This provides more confidence in the reliability of the analysis.

Another example of a CRC with a reporter gene assay (Figure 7.6) identifies the concentration of 17β-estradiol concentration required to produce 10% and 50% of the maximal luciferase induction in the ER-CALUX reporter gene bioassay for estrogenicity, calculated as $EC_{50} = 1.2$ ng/L and $EC_{10} = 0.06$ ng/L using Equation 7.1 with EC replacing RC.

Tox21 employs a specific data evaluation pipeline in R called ToxCast Analysis Pipeline (tcpl) for high-throughput screening data, which includes three CRC models. One is simply a constant effect over concentration (slope = 0), the second is equivalent to Equation 7.1, and the third is a gain-loss model, which can describe CRC that after saturation comes down again (Filer *et al.*, 2016). This inversion is typically caused by cytotoxicity masking the effect and therefore, for the purpose of water quality assessment, we recommend using a cut-off for cytotoxicity instead as described in Section 7.3.5, rather than using a CRCs with a turning point.

The benchmark values used by Tox21 are indicated in Figure 7.7 using the same data as Figure 7.6. The AC50 and AC10 are equivalent to EC_{50} and EC_{10}, respectively. The activity concentration at cut-off (ACC) is set at a user-defined cut-off (usually 15%–20%). The activity concentration at baseline (ACB) uses the two lowest concentration to derive the baseline band (BMAD) accounting for the noise of the bioassay by using three times median absolute deviation (MAD) over all responses of the two lowest concentrations around the zero effect.

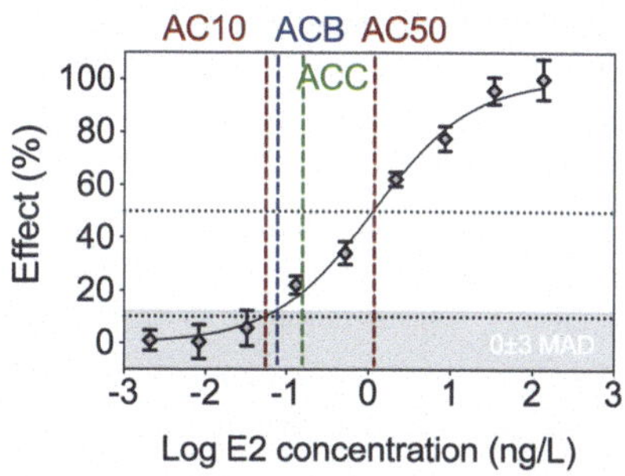

Figure 7.7 Benchmark values applied in the ToxCast Analysis Pipeline (tcpl). Modified on the example of the data from Figure 7.6 according to Filer *et al.* (2016). ACC set to 20% cut-off.

7.3.5 Simultaneous effect and cytotoxicity in a cell-based assay

In many cell-based assays there are two response types occurring at the same time – the actual effect of interest, and cytotoxicity. Enzyme induction in a liver cell line will, for example, increase with increasing concentration at first but decrease at higher doses. This is a typical example of cytotoxic interference, where measurable enzyme activity decreases at higher doses because the toxicant is becoming cytotoxic and starts destroying the liver cells (Figure 7.8). The same phenomenon can be observed in reporter gene assays and transactivation assays, where the activation of the reporter gene or the transactivation decreases when cells are dying. A related phenomenon is the cytotoxicity burst (Judson *et al.*, 2016), which occurs close to cell death and where all reporter genes can be non-specifically induced in a last-ditch attempt to rescue the cell.

In bioassays where a specific response is normalised to the number of cells, cytotoxicity (a decrease in cell number) will cause an increase in the response:cell

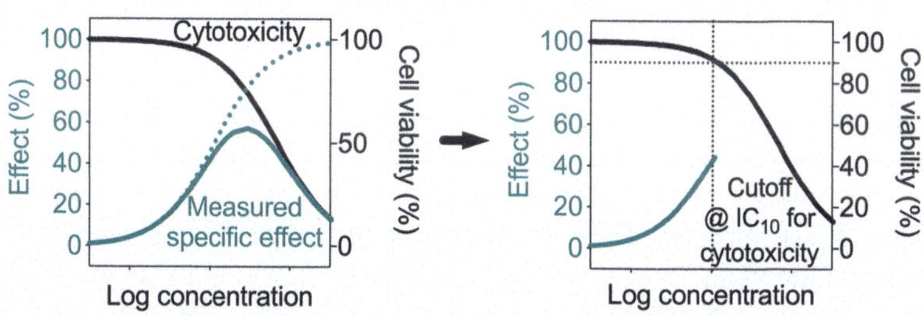

Figure 7.8 Example of a bioassay response with cytotoxicity interference. The dotted line shows the theoretical effect but due to cytotoxicity (black line is cell viability), the measured effect has an inverted U-shape. In this case, the effect can only be evaluated up to the IC_{10} for cytotoxicity.

ratio, thus in some cases wrongly suggesting an increase in the specific response. The CRC will in this case likely take an exponential shape because some of the cells found dead (decreased cell count) at the end of the exposure period will have produced a response in the earlier stages before succumbing to the cytotoxicity.

Figure 7.8 demonstrates how we can deal with CRCs that are compromised by cytotoxicity interference. The effect data are removed above concentrations that cause 10% cytotoxicity (IC_{10}). The downside of the need to omit any cytotoxic concentrations is that it often leaves an incomplete CRC, which means that the classical models cannot be applied because there are not enough values above 50% effect included for curve fitting. This problem can be alleviated by focusing on the linear low effect portion of the CRC.

7.3.6 Evaluating the linear portion of concentration–effect curves

The higher effect levels are often compromised due to cytotoxicity in reporter gene assays when environmental samples are tested. Therefore, it is sensible to evaluate only the low-level linear range of the CRC. In addition, many water types such as drinking water and recycled water as well as pristine surface water contain few micropollutants and high effect levels are rarely reached.

CRCs are typically linear up to 30% effect or cytotoxicity (Escher *et al.*, 2018) and of the form of Equation 7.4, where y is the % effect or % cytotoxicity. Variability of the negative controls in cell-based assays is usually much tighter than it is in whole organism assays, and 10% effect is typically above the limit of detection (LOD, see Chapter 11). It is thus possible to derive EC_{10} or IC_{10} values for the linear CRCs using Equation 7.5 for EC_y and Equation 7.6 for derivation of the standard error SE of EC_y.

$$y = \text{slope} \cdot \text{concentration} \tag{7.4}$$

$$EC_y = \frac{y}{\text{slope}} \tag{7.5}$$

$$SE\,(EC_y) = \frac{y}{\text{slope}^2} \cdot SE\,(\text{slope}) \tag{7.6}$$

In practice, it is recommended to first derive the IC_{10} for cytotoxicity from the linear portion of the CRC (% cytotoxicity = 100% – % cell viability). There are various possibilities to measure cell viability in parallel to effect endpoints as presented in Chapter 10. Then, only concentrations $< IC_{10}$ are used for the derivation of the EC_{10}. In addition, the CRC is only linear up to about 30% effect (Figure 7.1), and any concentrations that trigger >30% effect also have to be omitted for CRC modelling (Figure 7.9).

Certain reporter gene assays, for example, those indicative of adaptive stress responses or DNA repair, have no maximum for the CRC curves. For these types of assays, the signal can be converted to an induction ratio (IR) by dividing the signal by the signal of the negative control (Escher *et al.*, 2014). The IR of the

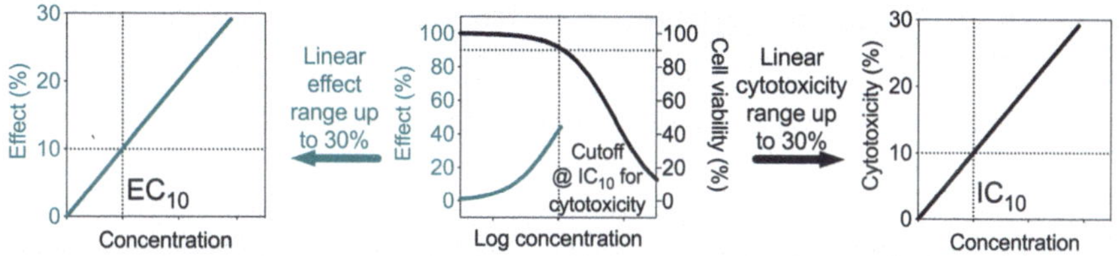

Figure 7.9 Derivation of inhibitory concentration causing 10% cytotoxicity IC_{10} and effect concentration causing 10% effect EC_{10} from the linear portion of the CRC. Modified from Escher *et al.* (2018). Reprinted with permission by John Wiley and Sons © 2018.

negative control is then by definition 1 and accordingly the linear portion of the CRC has an intercept of 1 (Equation 7.7). The benchmark concentration is EC_{IRz}, with an IR of z, often z is 1.5, that is, 50% over the control and the EC is called $EC_{IR1.5}$ (Equation 7.8 with standard error in Equation 7.9).

$$y = 1 + \text{slope} \cdot \text{concentration} \tag{7.7}$$

$$EC_{IR1.5} = \frac{0.5}{\text{slope}} \tag{7.8}$$

$$SE\,(EC_{IR1.5}) = \frac{0.5}{\text{slope}^2} \cdot SE\,(\text{slope}) \tag{7.9}$$

As for bioassays where a maximum can be reached, we still have to deal with cytotoxicity interferences as Figure 7.10 shows. Concentrations > IC_{10} for toxicity must be removed and the CRC is typically only linear up to an IR of approximately 3–4.

Depending on the bioassay, the maximum response can be at IR of 2 up to over 100. For a response where the maximum IR reaches 6, then the $EC_{IR1.5}$ is equivalent to the EC_{10}, but if the maximum IR is 2, then the $EC_{IR1.5}$ is equivalent to the EC_{50}. Hence, direct comparisons between EC values becomes cumbersome but the bioanalytical equivalency concept outlined in Section 7.5 can help us to overcome this difficulty.

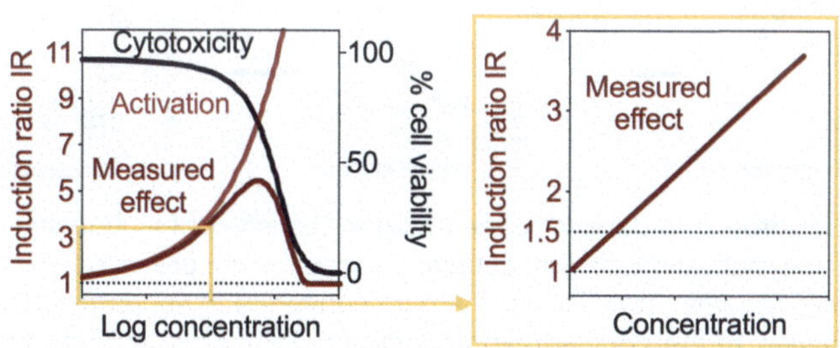

Figure 7.10 Derivation of effect concentration triggering an IR of 1.5 $EC_{IR1.5}$ from the linear portion of the CRC.

7.3.7 Antagonistic effects

Effects on hormone receptors elicited by environmental chemicals in water samples may not only be of activating nature (agonism) but also inhibitory (antagonism). As a mirror to estrogenic effects, we can often observe anti-androgenic effects in surface water impacted by wastewater treatment plant effluent. Antagonistic effects are quantified in the presence of a constant concentration of an agonist, with antagonists suppressing the signal of the agonist. Typically, a concentration of agonist (*e.g.*, estradiol) is chosen that has an effect of approximately 80% of the maximum effect (Figure 7.11a).

A suppression ratio SPR can be calculated from Equation 7.10 from the effect (*e.g.*, activation of the estrogen receptor) of the sample run in antagonist mode divided by the effect of the agonist alone. Note that previous work has abbreviated suppression ratio as 'SR', but we are using 'SPR' to avoid any confusion with the specificity ratio, which has the abbreviation SR (see Chapter 9).

$$\mathrm{SPR} = 1 - \frac{\mathrm{activation}_{\mathrm{sample}}}{\mathrm{activation}_{\mathrm{agonist}}} \tag{7.10}$$

The SPR, often expressed as percentage (1 = 100%), is then also plotted against the logarithm of the concentration of the antagonist (Figure 7.11b). As for agonism, the CRC is also linear up to a SPR of 30% (Figure 7.11c). However, it is typically not possible to derive an effect concentration for the SPR_{10} because this is often still within the variability of the agonist (red bar in Figure 7.11). Therefore, it is common practice to derive the $\mathrm{EC}_{\mathrm{SPR20}}$ from the linear portion of the CRC with Equation 7.11 (Escher *et al.*, 2014).

$$\mathrm{EC}_{\mathrm{SPR20}} = \frac{20\%}{\mathrm{slope}} \tag{7.11}$$

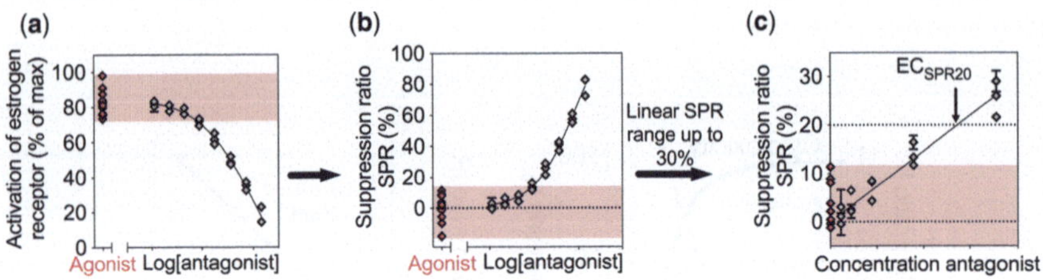

Figure 7.11 Approach to determine the antagonistic effect of a chemical or sample. (a) CRC of the activation of the estrogen receptors in presence of a constant concentration of agonist; (b) CRC for the data converted to suppression ratio (SPR) and (c) derivation of the concentration causing a SPR of 20%. Data from Escher *et al.* (2014).

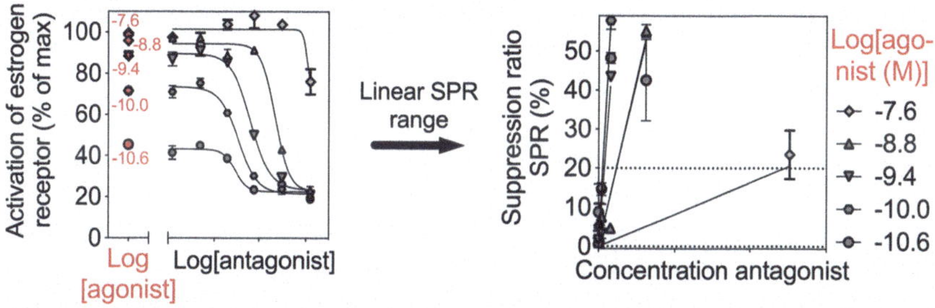

Figure 7.12 Example of the antagonistic effect of tamoxifen against different constant concentrations of 17β-estradiol as agonist. Data from Neale and Leusch (2015).

There are four major caveats when it comes to antagonism testing. First, the EC_{SPR20} is heavily dependent on the concentration of the competing agonist. This is illustrated in Figure 7.12 for 17β-estradiol as agonist at different starting concentration with varying concentration of the antagonist tamoxifen (Neale and Leusch, 2015). Most stable and reproducible CRCs of SPR are obtained with a concentration of agonist that produces about 80% of the maximum effect (Neale and Leusch, 2015).

Second, antagonism response curves are often more variable than their agonist equivalents. This is in part because it can be difficult to hit exactly the correct effect with the competing agonist, and even a small deviation from the recommended 80% can produce large deviations in the response.

Third, antagonism testing is even more susceptible to cytotoxicity interference because concomitant cytotoxicity produces the same suppression of the activation signal as antagonism does. Therefore, the cytotoxicity cut-off has to be even stricter than in the case of activation, where we recommended to exclude all concentrations $> IC_{10}$ from the evaluation of activation (see Section 7.3.6). In antagonist mode, 10% cytotoxicity would cause a suppression of more than 10%, hence it is recommended to use an even stricter cytotoxicity cut-off.

Finally, natural organic matter present in the sample after extraction can serve as a binding site for the competing agonist. This causes an apparent decrease in the activation signal that looks like antagonism but is really only an artefact due to loss of the agonist available for interaction with the reporter gene (Neale *et al.*, 2015a).

7.4 CONCENTRATION-RESPONSE CURVES OF WATER SAMPLES

In applying *in vitro* bioassays to water quality assessment, it is imperative to be able to express the bioassay results in quantitative terms. Being able to express a bioassay result as a discrete value allows, for example, quantitative comparisons and ranking among different water samples, or comparison between 'before' and 'after' samples to calculate treatment efficacy. Testing of water samples in *in vitro* bioassays is very similar to testing of model compounds. Once the water sample has been collected, extracted and concentrated, the extract is diluted to create a serial dilution series (*i.e.*, a stepwise dilution that results in a range of concentrations) and each dilution is tested in the assay, as one would test different concentrations of a test chemical. In the case of water samples, each dilution will not have a typical concentration unit (such as μg/L or mol/L). Instead, the 'concentration' can be expressed as a relative extraction factor (REF), which is unitless.

The REF is calculated from Equation 7.12 as the product of the extraction factor (EF) from the extraction and concentration step and the dosing factor of the sample in the bioassay. If an undiluted water sample is used, the EF is set to 1. The EF can be calculated with Equation 7.13 and the dosing factor with Equation 7.14.

$$\text{Relative extraction factor REF} = \frac{\text{mass or volume extracted}}{\text{final volume in bioassay}} = \text{EF} \cdot \text{DF}\left[\frac{\text{kg}_{\text{matrix}} \text{ or } \text{L}_{\text{water}}}{\text{L}_{\text{bioassay}}}\right] \tag{7.12}$$

$$\text{Extraction factor EF} = \frac{\text{mass or volume extracted}}{\text{final volume of extract}}\left[\frac{\text{kg}_{\text{matrix}} \text{ or } \text{L}_{\text{water}}}{\text{L}_{\text{extract}}}\right] \tag{7.13}$$

$$\text{Dosing factor DF} = \frac{\text{volume of extract dosed}}{\text{final volume in bioassay}}\left[\frac{\text{L}_{\text{extract}}}{\text{L}_{\text{bioassay}}}\right] \tag{7.14}$$

These equations work for any matrix, not just water. For water, which is often enriched by solid-phase extraction or liquid–liquid extraction, the EF is typically called 'enrichment factor'. Other matrices from contaminated sites such as sediment and soil may have higher loads of contaminants and therefore are not enriched by the extraction process. We have therefore used 'extraction factor' as the more general term. Practical aspects of extraction and dosing are discussed in more detail in Chapter 12.

The biological response can then be plotted against the REF of the sample for a range of dilutions to generate typical CRCs for diverse water samples with log REF as the concentration unit (Figure 7.13). Testing a water sample without enrichment would only really be worthwhile at the inlet of a wastewater treatment plant (WWTP), which generally produces a detectable response even at an REF 1 (log REF = 0). Even then, effects after wastewater treatment are typically below 10% (Figure 7.13a) and so it would be difficult to quantify treatment efficacy if water was not extracted and enriched. If we enrich by up to an REF of 100 (log REF = 2) and run full CRCs, then there is a very visible difference between untreated and treated wastewater (Figure 7.13a). Even field and lab blanks can show an effect (mostly due to impurities in extraction solvents), but normally only above an REF 30-100. For advanced treatment, drinking water and bottled water, it is even more important to enrich a lot: no effects would be visible in the

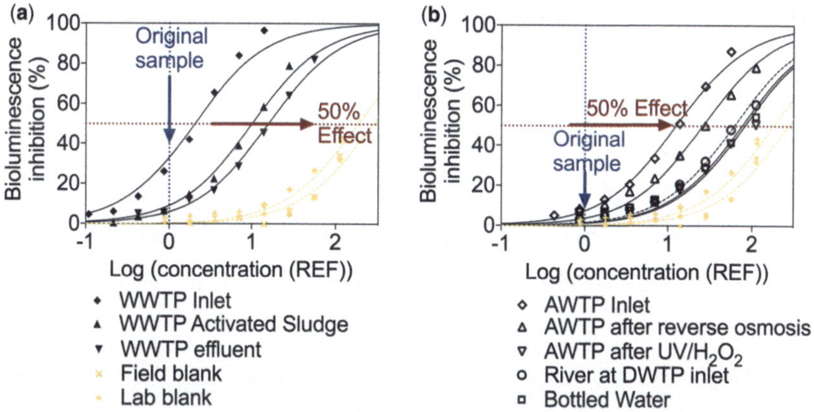

Figure 7.13 Concentration–inhibition curves of water samples from the Oxley Creek WWTP (a) and from the Bundamba AWTP and Brisbane River water (b) in the Microtox assay of bioluminescence inhibition in *Aliivibrio fischeri.* Concentrations are expressed as relative enrichment factor (REF, Equation 7.12). WWTP = wastewater treatment plant, AWTP = advanced water treatment plant, DWTP = drinking water treatment plant. Data from Macova *et al.* (2011).

original water sample, while extracted samples have a spread of EC_{50} values. It is important to check and assure that all CRCs show distinct differences from the field and lab blanks (Figure 7.13b).

7.5 BIOANALYTICAL EQUIVALENCY CONCEPT

7.5.1 Relative effect potency

Being able to express the entire CRC in mathematical terms using as little as one or two parameters ($\log EC_{50}$ and slope for sigmoidal CRCs and slope for linear CRCs) allows comparisons of the potency (*i.e.*, the strength) of different samples or test compounds. To express the potency of one chemical i relative to another (usually the reference compound used in the assay), the CRCs of both chemicals are compared (Figure 7.14).

The relative effect potency (REP_i) of the test compound i relative to the reference compound can then be calculated by Equation 7.15 (Villeneuve *et al.*, 2000) with the standard error $SE(REP_i)$) by Equation 7.16.

$$REP_i = \frac{EC_y(\text{reference})}{EC_y(\text{i})} \tag{7.15}$$

$$SE(REP_i) = \sqrt{\frac{1}{EC_y(\text{i})^2} \cdot SE(EC_y(\text{reference}))^2 + \frac{EC_y(\text{reference})^2}{EC_y(\text{i})^4} \cdot SE(EC_y(\text{i}))^2} \tag{7.16}$$

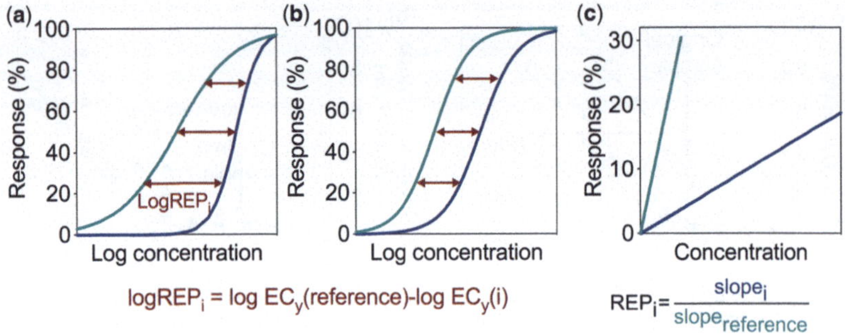

Figure 7.14 Concentration–effect curves of a reference and a test compound. REP_i is the relative effect potency of the test compound i compared with the reference compound. (a) REP_i varies with effect level if the logarithmic CRCs are not parallel. For logarithmic CRC with the same slope (b) and for linear CRCs (c), the REP_i are independent of the effect level. Reprinted with permission from Escher *et al.* (2018). The advantages of linear concentration–response curves for *in vitro* bioassays with environmental samples. *Environmental Toxicology and Chemistry*, 37(9): 2273–2280. © 2018. John Wiley and Sons.

As shown in Figure 7.14a and b, the REP_i is only independent of the effect level if the logarithmic CRCs are parallel to each other, that is, have the same slope. In reality this is rarely the case and resulting issues are discussed in more detail in Villeneuve *et al.* (2000). In contrast, if the linear CRCs are used to derive the REP_i (Figure 7.14c), the ratio is independent on the effect level and can also be calculated from the inverse ratio of the slopes of the linear CRC (Escher *et al.*, 2018).

The REPs derived from both *in vitro* and *in vivo* bioassays are used to determine the more comprehensive 'toxic equivalency factors' (TEFs). The TEF concept is used frequently in risk assessment, for example in the assessment of dioxin-like effects by the World Health Organization (Van den Berg *et al.*, 2006). The TEF concept will be expanded upon in Chapter 8.

7.5.2 Toxic units and toxic equivalent concentration

Since small values of LC_y or IC_y equate to high toxicity and high values of LC_y or IC_y to low toxicity, we often use the inverse of the LC_y or IC_y to visualise toxicity as toxic units (TU, Equation 7.17). If the TU is directly derived from a bioassay experiment, we refer to it as TU_{bio}. The TU can also be calculated from the chemical concentrations detected and their effect and are then termed TU_{chem}, which will be further discussed in Chapter 8 on mixtures.

$$TU_{bio} = \frac{1}{IC_y(\text{sample})} \tag{7.17}$$

In the same way that REP was used to quantify the relative effect of individual chemical compounds, the water sample can be quantified relative to the assay reference compound using toxic equivalent concentrations (TEQ). The TEQ is the concentration of the reference compound that would be required to produce the equivalent effect in an assay as a given sample concentration. As highlighted in the discussion on REP above, it is important to ensure equality of slopes if sigmoidal logarithmic CRCs are used. Linear CRCs and linearly derived LC_{10} and IC_{10} are preferable for determination of TEQ, where possible. The TEQ_{bio} is the ratio of the LC_y or IC_y of the reference compound and the LC_y or IC_y of the water sample (Equation 7.18), where the former is expressed either as a molar concentration (*e.g.*, 3 pmol/L) or as a mass-based concentration (*e.g.*, 0.88 ng/L), while the latter is expressed in REF (*i.e.*, unitless):

$$TEQ_{bio} = \frac{IC_y(\text{reference})}{IC_y(\text{sample})} \tag{7.18}$$

7.5.3 Effect units and bioanalytical equivalent concentration

Formerly, TU and TEQ were also used to express effect concentrations. This is misleading however, because, as discussed above, a measured effect in an *in vitro* assay is not necessarily a toxic effect – it might be an adverse effect or even a beneficial effect, such a defence mechanism. Therefore, we differentiate in this book strictly between 'toxicity' and 'effect', using 'response' as the general term. The analogous term to TU then is effect units (EU), and EU_{bio} can be calculated by Equation 7.19.

$$EU_{bio} = \frac{1}{EC_y(\text{sample})} \tag{7.19}$$

The BEQ is the ratio of the EC_y of the reference compound (expressed as a molar or mass-based concentration) and the EC_y of the water sample (expressed as REF, *i.e.*, unitless) (Equation 7.20) with the SE in Equation 7.21.

$$BEQ_{bio} = \frac{EC_y(\text{reference})}{EC_y(\text{sample})} \tag{7.20}$$

$$SE(BEQ_{bio}) = \sqrt{\frac{1}{EC_y(\text{sample})^2} \cdot SE(EC_y(\text{reference}))^2 + \frac{EC_y(\text{reference})^2}{EC_y(\text{sample})^4} \cdot SE(EC_y(\text{sample}))^2} \tag{7.21}$$

When using the BEQ concept, it is important to pay particular attention to the choice of the reference compound. The optimal reference compound should be (i) a chemical linked to the mode of action of the bioassay, (ii) one of the most potent compounds in the bioassay, and (iii) likely to be present in water samples. 17β-Estradiol is an example of a good reference compound for bioassays that detect estrogenic endocrine disrupting compounds: it is a natural hormone and the native ligand for the estrogen receptor (the keystone of the estrogenic response), it is one of the most potent estrogenic compounds (second only to synthetic estrogens such as ethinylestradiol) and is commonly found in sewage-impacted waters.

As an example, if the EC_{10} for estradiol in a bioassay for estrogenicity was 0.5 ng/L and a water sample had an EC_{10} of 0.1 REF, the BEQ_{bio} expressed as estradiol equivalent concentration (EEQ) for this sample would be calculated as $EEQ = 0.5 / 0.1 = 5$ ng/L.

There are many other BEQs – some commonly used BEQs are dihydrotestosterone equivalent (DHTEQ) used in bioassays for androgenic endocrine activity, diuron equivalent (DEQ) used in bioassays for photosynthesis inhibition, and chlorpyrifos equivalent (ChlEQ) used in bioassays for acetylcholinesterase inhibition. More examples are given in Chapter 10, where individual bioassays are discussed.

7.6 CONCLUSIONS

CRCs are the graphical representation of the response of a biological system to toxicants, and one of the cornerstones of toxicology. The concentration–response relationship can be expressed in simple mathematical terms using, for example, a log-logistic equation and at low effect levels even by a linear regression. This allows quantitative comparisons of different CRCs, determination of REPs of different toxicants and calculation of TEQ and BEQ for water samples.

In the supplementary information to this book on www.ufz.de/bioanalytical-tools we provide additional resources for concentration–response assessment including example spreadsheets for the methods discussed in this chapter.

Chapter 8

Mixtures

8.1 INTRODUCTION

Chemicals rarely occur alone in environmental water samples and wastewater, but instead occur as mixtures. It is the mixtures that threaten water quality (Kortenkamp *et al.*, 2019). While the concentrations of individual chemicals are often below any toxicity threshold and below the limit of detection (LOD) of chemical analysis, the effect of the mixture may still be a cause for concern.

A wealth of literature is available on mixture toxicity experiments with binary (two components) and multiple component (more than two components) mixtures (Kortenkamp *et al.*, 2009). Systematic investigations of complex mixtures comprising individual components at very low concentrations, such as found in wastewater and recycled water, are still not very common but we review in this chapter a number of emerging studies and introduce typical mixture designs.

Unfortunately, many of the available mixture studies provide only anecdotal evidence and lack an explicit mechanistic understanding. In the last decades, concepts from pharmacology have been adapted to toxicology (Kortenkamp *et al.*, 2009; Rider *et al.*, 2018). With these came a conceptual breakthrough in that mixture effects are now typically categorised in four classes (Table 8.1). Two of these classes, called independent action (IA) and concentration addition/dose addition (CA/DA), are more common and have underlying mathematical models. IA applies when chemicals act according to different modes of toxic action. CA (in aquatic ecotoxicology) and DA (in mammalian toxicology) apply when chemicals trigger similar modes of toxic action. Again, the toxicity pathways give

Authors: Beate Escher, Peta Neale and Frederic Leusch
doi: 10.2166/9781789061987_0119

Table 8.1 Concepts of mixture toxicity.

	Similar Mode of Toxic Action	Dissimilar Mode of Toxic Action
No interaction between chemicals in mixture	Concentration addition (CA) or dose addition (DA)	Independent action (IA)
Interaction between chemicals in mixture	Complex action (may lead to synergy or antagonism)	Dependent action (may lead to synergy or antagonism)

us guidance as to what constitutes similarity – starting with mixtures affecting the same target sites, triggering the same molecular initiating events (MIE) or key events (KE) all the way through to similarity of adverse outcomes.

The concepts of CA/DA and IA provide quantitative and reliable predictions of combined effects from existing information on the activity of individual components of complex mixtures, assuming that the mixture components do not interact. If the mixture components interact, one is faced with complex and dependent action, which may lead to synergy (higher toxicity than expected from CA/DA) or antagonism (lower toxicity than expected from IA).

In this chapter, the major mixture toxicity concepts are introduced and illustrated with examples from carefully crafted mixture toxicity studies. These examples help us appreciate the importance of considering chemicals not one-by-one but rather by looking at their cumulative exposures and combined effect.

A burning issue of international discussion focuses on mixtures at very low concentrations/doses, where individual chemicals are below their observable effect level. Current opinions are split as to whether mixture effects in these cases can be ignored or need to be addressed. Various experimental mixture studies with multiple components using *in vitro* bioassays give us the confidence that even mixture components below their individual detection limits contribute to the mixture effects.

The evidence from studies with bioanalytical tools that allow testing of large number of samples and mixtures at low effect levels seem to point towards linearity of the concentration–response curves (CRCs) at low concentrations/doses (see Figure 7.1). This is mathematically reasonable and appears to fit the experimental evidence, although it should be acknowledged that the detection limit of even an excellent bioassay will hardly ever be below 5% of response (Chapter 12) and it is thus impossible to experimentally investigate linearity below 5% of the response.

The toxic equivalency approach is a special case of CA/DA that has proven to be very useful for the risk assessment of chemicals acting by the same mode of action (MOA). In this chapter, we review the history of this approach and its application for risk assessment and water quality assessment.

The importance of mixtures is undeniable (Drakvik *et al.*, 2020), but mixtures are only slowly making their way into regulatory risk assessment despite very encouraging recent developments (Bopp *et al.*, 2018; Kortenkamp and Faust, 2018; Rotter *et al.*, 2018). It is well recognised that the risk comes from mixtures of chemicals not from individual chemicals but how we assess and regulate mixtures remains a challenge.

8.2 TOXICITY/EFFECTS OF DEFINED MIXTURES

8.2.1 Independent action

Chemicals that act according to different modes of toxic action produce their effects independently. Naturally, one cannot simply add up the effects. Two chemicals that produce 60% effect each, for example, will not result in a total effect of 120% as this is biologically impossible. Rather these chemicals will produce the total sum effect minus the product of the individual effects according to the statistical concept of independent random events (in this case 60% × 60% equals 36%, which results in a combined effect of 120%–36% = 84%) (see Equation 8.1). For a multi-component mixture of n chemicals i, Equation (8.1) expands to Equation (8.2).

$$\text{Effect (mixture)} = \text{effect(A)} + \text{effect(B)} - (\text{effect(A)} \times \text{effect(B)}) \tag{8.1}$$

$$\text{Effect (mixture)} = 1 - \prod_{i=1}^{n} (1 - \text{effect(i)}) \tag{8.2}$$

IA implies that no mixture toxicity or effect will occur if the effects of individual components are below detection limit. It also implies that subsequent exposure events would not lead to effects. IA is based on strict stochasticity of toxicity, which is why IA would in theory also apply to the same agent applied again and again. In reality there is some stochastic component in toxicity, but individual tolerance also needs to be taken into account (Ashauer *et al.*, 2017), so the concept does have some shortcomings for translation into realistic exposure scenarios.

Despite this limited theoretical foundation of IA, the model of IA was successfully applied to predict mixture effects of chemicals with strictly dissimilar MOAs in *Aliivibrio fischeri* (Backhaus *et al.*, 2000) and green algae (Faust *et al.*, 2003).

When applying IA, it is also important to consider what 'no effect' means. The 'no observed adverse effect level' (NOAEL) in mammalian toxicology can be as high as 20% and the 'no observed effect concentration' (NOEC) in ecotoxicology can be as high as 40% depending on the test design (sample size, replication and dose spacing; see Chapter 7). There is therefore a high likelihood that chemicals at the NOAEL or NOEC can still act together to elicit considerable mixture toxicity.

8.2.2 Concentration or dose addition

If chemicals affect the same target site or act through the same MOA, their combined effect is expected to follow the concept of concentration or dose addition. Here, instead of summing up the measured effects, mixture toxicity is determined according to the dose (mammalian toxicology) or concentration (*in vitro* toxicology and aquatic ecotoxicology) causing the effect. For a binary mixture, this can be rationalised as follows: chemical A has an effective concentration causing 50% of maximum effect (EC_{50}) of 12 μg/L and chemical B has an EC_{50} of 20 μg/L. In this case, combination of half of the EC_{50} of A ($C_A = 6$ μg/L) and half of the EC_{50} of B ($C_B = 10$ μg/L) will cause 50% of effect. Any other combination of fractions of the $EC_{50,A}$ and the $EC_{50,B}$, where the fractions add up to 1, will also result in 50% effect (*e.g.*, ¼ of $EC_{50,A}$ and ¾ of $EC_{50,B}$).

A so-called isobologram illustrates the concept of CA for binary mixtures (Figure 8.1). The concentrations are plotted in toxic units TU, which are calculated from Equation (8.3) as the concentration of one mixture component (C_i) divided by its LC_{50} ($LC_{50,i}$). The same equation applies for effect units (EU), which are calculated as the concentration of one mixture component (C_i) divided by its EC_{50} ($EC_{50,i}$).

$$TU_i = \frac{C_i}{LC_{50,i}} \quad \text{or} \quad EU_i = \frac{C_i}{EC_{50,i}} \tag{8.3}$$

A straight line connecting $EU_A = 1$ and $EU_B = 1$ (*i.e.*, $\Sigma EU_i = 1$) corresponds to CA (Figure 8.1). Any combination of EUs that causes 50% effect and $\Sigma EU_i < 1$

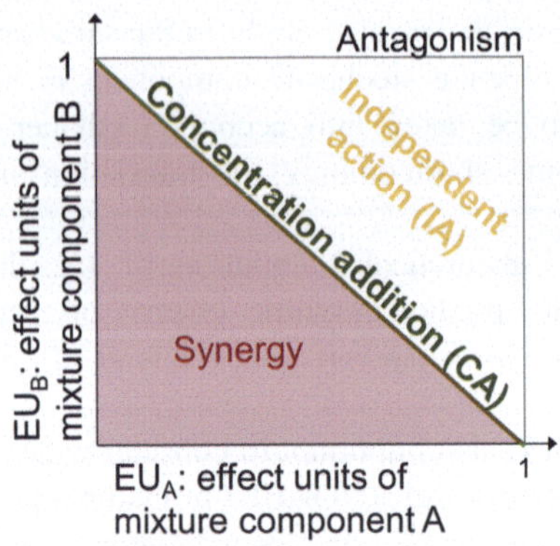

Figure 8.1 Isobologram for a binary mixture.

indicates that a lower concentration is needed to elicit the effect (*i.e.*, synergy); any combination that leads to 50% effect and $\Sigma EU_i > 1$ indicates IA or antagonism.

For multi-component mixtures with n components i, each in a fraction (p_i) of the total concentration, CA translates to Equation (8.4) for the EC_y of the mixture (EC_y(mixture)).

$$EC_y\ (\text{mixture}) = \frac{1}{\sum_{i=1}^{n} (p_i/ECy_i)} \tag{8.4}$$

With the IA and CA/DA models, we have the reference cases for mixture toxicity established. The relevance of CA/DA has been convincingly demonstrated for estrogenic chemicals in a series of experiments ranging from *in vitro* and low complexity bioassays all the way through to *in vivo* chronic endpoints (reviewed in Kortenkamp, 2007). Both natural estrogens and xenoestrogens acted in a CA/DA manner. While this line of evidence is less comprehensive for other modes of toxic action, there is a large body of literature that substantiates the concept of CA/DA (Kortenkamp *et al.*, 2009).

Initially focused on ecotoxicological endpoints, these mixture concepts are now also more frequently applied in human toxicology studies. All conceptual studies have confirmed that mixture toxicity is clearly higher than the toxicity of individual chemicals, whether CA/DA or IA applies.

8.2.3 Synergistic and antagonistic effects

There are exceptions to the reference cases of CA/DA and IA, and these apply when the mixture components interact. Such interactions often take place in the toxicokinetic phase (Figure 8.2). If one component activates a metabolic enzyme, for example, causing faster detoxification of the other mixture component and

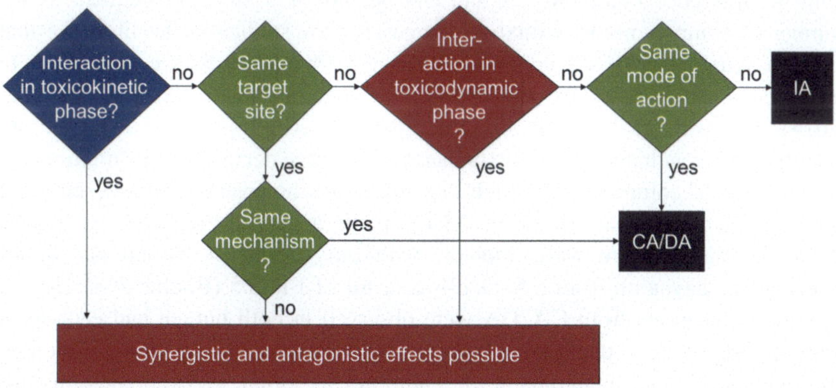

Figure 8.2 Flow chart for the occurrence of synergistic, antagonistic, IA and CA/DA mixture effects.

hence causing the resulting mixture effect to be less than predicted with IA, this is called antagonism. If one compound facilitates uptake of another component or suppresses detoxifying enzymes, interaction in the toxicokinetic phase may lead to synergy, that is, effects are significantly stronger than predicted by the model of CA/DA.

A classic example is piperonyl butoxide (PBO), which is an inhibitor of cytochrome P450, the enzyme that catalyses the phase I oxidation step in metabolism. PBO is often used in pesticide formulations, where it shows synergistic effects with many pesticides, such as the herbicide atrazine. The presence of PBO prevents the detoxification of these pesticides essentially causing them to become more toxic than they would be in a metabolically active organism. There are also cases where the inhibitory action of PBO causes antagonistic effects, for example, with organophosphate insecticides, which require metabolic activation in order to inhibit their target, the enzyme acetylcholinesterase. Finally, some organochlorine pesticides and atrazine induce the activity of cytochrome P450 and thus can act synergistically with organophosphates by enhancing their metabolic activation.

By analogy, in the toxicodynamic phase, interactive effects at the target site may lead to antagonistic or synergistic mixture toxicity (Figure 8.2). CA/DA can generally be expected when chemicals act at the same target site and according to the same mechanism (toxicity pathway). CA/DA has been shown to apply even in cases where only the MOA is similar, despite differences in the molecular pathway.

Interactive effects can also occur at the interface between toxicokinetics and toxicodynamics. An aryl hydrocarbon agonist can, for example, inhibit the activity of estrogenic chemicals by downregulating the expression of the estrogen receptor and inducing the enzymes that metabolise estrogens. In these cases, different target sites may not necessarily directly lead to IA (depicted in Figure 8.2).

While interactive effects are frequently observed for metals, they seem to be rather rare or quantitatively less important for organic chemicals. The deviations from the reference models of IA and CA become less and less frequent as the number of components in a mixture increases. Few studies in the literature that reported synergistic effects could be confirmed when the observed effects were compared with the CA/DA predictions (Kortenkamp *et al.*, 2009). Cedergreen (2014) reviewed the literature on synergistic effects of pesticides, metals and biocides and concluded that most synergistic effects occurred with mixtures of low number of components at high concentrations but were otherwise rare and rather an exception than the norm. Many of the apparently reported synergistic effects in the literature were artefacts with only 6 out of 90 animal studies showing true deviation from CA/DA by a factor of 1 to 3.5 (Boobis *et al.*, 2011).

Where deviations from CA/DA were observed in both human and ecological context, they were typically no larger than a factor of 4 in either direction, towards synergy or antagonism. In addition, IA often gives predictions for mixture effects that are only slightly smaller than CA/DA (also generally falling within less than a factor of 3).

Thus, while it is interesting to examine the subtle differences between the different mixture toxicity scenarios to understand the mechanisms of interaction for diagnostic purposes, CA/DA is a robust and 'realistic worst-case scenario' to default on for the evaluation of environmental mixture toxicity for the purpose of risk assessment. Another advantage of CA/DA is that benchmark toxicity values (EC_{50}, LD_{50}), which are often already available or can be easily measured, are sufficient to calculate mixture predictions. On the other hand, full dose–response relationships need to be established for IA predictions, and complex toxicokinetic/toxicodynamic models need to be developed for predictions of synergy and antagonism. Hence, CA/DA provides a practical and easily applicable model for mixture effects that appears to be robust and sufficiently accurate for most environmental risk assessment applications.

8.2.4 Grouping of chemicals

But where do mixtures start and where do they end? The European Food Safety Agency (EFSA) is favouring an approach that uses '(cumulative) assessment groups', which encompass 'chemical substances that are treated as a group by applying a common risk assessment principle (*e.g.*, dose addition) because these components have some characteristics in common (*i.e.*, the grouping criteria)' (More *et al.*, 2019).

The mixture toxicity concept of CA/DA has been derived by grouping chemicals with clearly identified and identical mechanism of toxicity, but practical experience has demonstrated that even grouping according to the same mode of toxic action will result in CA/DA responses. When looking at the adverse outcome pathways introduced in Chapter 4, any commonality along the toxicity pathway may lead to CA/DA effects (Figure 8.3). The U.S. EPA uses the additional term 'sufficiently similar' for mixture effects elicited on the same target organ or with the same symptoms (Teuschler, 2007).

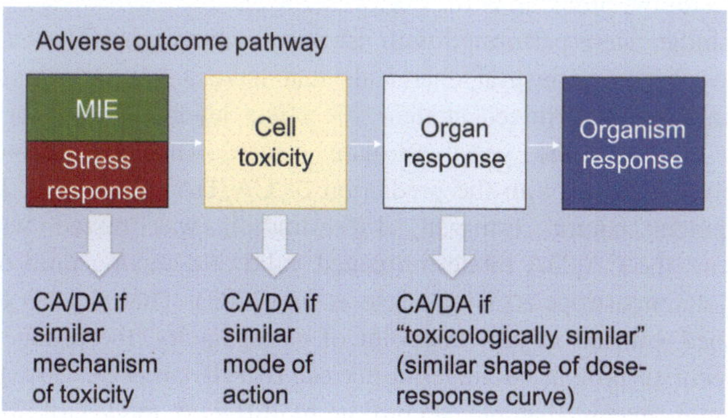

Figure 8.3 Adverse outcome pathways in relation to the mixture toxicity concepts.

Kortenkamp (2020) also cautioned from narrowing the definition of cumulative assessment groups to MIEs and recommended AOP thinking and accounting for converging pathways that still might lead to mixture effects despite not sharing MIEs. Thus, grouping has to go beyond the distinction of similar and dissimilar action and account for joint action along the entire pathway.

For the application of bioanalytical tools, these considerations translate into the expectation that CA/DA will occur in assays indicative of a clearly defined MOA or mechanism of toxicity. Reporter gene assays only express one pathway, so in principle all chemicals that are active in a reporter gene assay should act according to CA/DA. For *in vitro* assays that are indicative of steps further down the toxicity pathway such as genotoxicity assays, the definition of rather broad cumulative assessment groups lead to predictions consistent with CA as demonstrated with various genotoxicants, which differed in their MIE, but showed near CA mixture effects in the micronucleus assay in Chinese hamster ovary-K1 cells (Ermler *et al.*, 2014).

Most importantly, CA/DA is also applicable even if individual chemicals are present at concentrations that do not cause any observable effect, as discussed below.

8.2.5 Something from nothing?

One would intuitively assume that a mixture of compounds, each of which is present at a concentration below that known to cause an effect individually, would not cause any effect. This, however, is not the case when chemicals occur in mixtures.

As early as 1988, Deneer *et al.* (1988) mixed 50 industrial chemicals at concentrations 20–400 times less than the EC_{50} of each compound and observed mixture effects consistent with CA/DA (Table 8.2). Subsequently, researchers systematically investigated and compared the mixture toxicity concepts of CA/DA and IA. When similarly acting chemicals were mixed at their EC01 level (*i.e.*, the 1% effect level, which is not actually measurable in practice) for various ecotoxicological endpoints, the observed effects were typically high in accordance with the predictions for CA (Table 8.2).

Similar studies were performed with estrogenic chemicals (Silva *et al.*, 2002). Eight xenoestrogens (industrial chemicals that have a low estrogenic potency, *e.g.*, bisphenol A) were mixed at their 1% effect level and together caused an appreciable effect in the yeast estrogen screen. The observed effect was quantitatively consistent with the prediction of CA/DA (Table 8.2). Even when the very potent natural estrogen, 17β-estradiol, was mixed with several xenoestrogens, the CA/DA model remained valid at a mixing ratio of 1:50,000 (17β-estradiol:xenoestrogens) (Rajapakse *et al.*, 2002). This *in vitro* observation was confirmed with an *in vivo* endpoint of estrogenicity, the production of the egg yolk precursor protein vitellogenin in male fish (Brian *et al.*, 2005).

Even for chemicals that act according to different modes of toxic action, mixtures of more than 10 compounds at low effect level have shown

Table 8.2 The 'something from nothing' effect for compounds with the same mode of toxic action (MOA).

MOA	Chemicals in Mixture	Bioassay	C_i	Exp. Effect	CA/DA Predict.	Ref
Non-specific toxicity	50 industrial chemicals	Water flea *Daphia magna*	20–400 times lower than EC_{50}	50%	50%	1
Cytotoxicity	10 quinolones	Microtox (biolumine scence inhibition in *Aliivibrio fischeri*)	EC_{01}	54%	72%	2
Respiratory uncouplers	16 phenols	Microtox	EC_{01}	82%	95%	3
Reproduction	18 triazine herbicides	Reproduction of green algae	EC_{01}	47%	44%	4
Binding to ER	8 xeno-estrogens	Yeast estrogen screen	EC_{01}	~25%	~25%	5
Estrogenicity	2 estrogens, 3 xeno-estrogens	Vitellogenin induction (male fathead minnow fish)	NOEC	~58%	~50%	6
Androgen receptor antagonism	13 pesticides and 17 antioxidants, PCP, industrial pollutants	MDA-kb2 assay, suppression of dihydro-testosterone effects	IC_{01}, IC_{10}, IC_{20}	Diverse, CA provided best prediction		7

CA/DA = concentration addition/dose addition; ER = estrogen receptor. 'Exp. effect' = effect measured experimentally, 'CA/DA predict.' = prediction of the CA/DA model, C_i = concentration of component *i*, EC_{50} = effect concentration causing 50% of maximum effect, EC_{01} = effect concentration causing 1% of maximum effect, NOEC = no observed effect concentration, PCP = personal care products, Ref = literature reference. [1]Deneer *et al.* (1988), [2]Backhaus and Grimme (1999), [3]Altenburger *et al.* (2000), [4]Faust *et al.* (2001), [5]Silva *et al.* (2002), [6]Brian *et al.* (2005), [7]Orton *et al.* (2014).

Table 8.3 The 'something from nothing' effect for chemicals with different modes of action (MOAs).

Chemicals in Mixture	Bioassay	C_i	Experimental Effect	Comparison with Predictions	Ref
11 aquatic priority pollutants	Reproduction of green algae	NOEC[b]	64%	$\approx$ IA ($<$CA)	1
16 dissimilarly acting compounds	Reproduction of green algae	6.6–66% of NOEC[b]	18%	$\approx$ IA ($<$CA)	2
Four pesticides with different MOA	E-SCREEN (proliferation of MCF7 cells)	25–100% of NOEC[b]	Significant	$=$ IA	3

C_i = Concentration of component i, NOEC = no observed effect concentration, CA = concentration addition, IA = independent action, Ref = literature reference.

[1]Walter *et al.* (2002), [2]Faust *et al.* (2003), [3]Payne *et al.* (2001).

measurable effects, which were consistent with the prediction obtained from IA (Table 8.3). This is very important as the concept of IA actually implies that if an individual chemical has no effect, the mixture will have no effect. An experimentally determined 'no effect' is, however, not necessarily a true 'zero effect', it is often merely a non-observable effect. If 100 chemicals were mixed at concentration ratios where each would individually elicit a 1% effect, the joint IA effect would be as high as 63%. If the 100 individual chemicals were present at 0.1% effect level, the mixture effect would still reach 9.3%. Thus, even for IA, we can conclude that mixture toxicity has the potential to be of critical importance. These findings imply that there are no 'safe' concentration levels at which chemicals do not contribute to mixture effects (Kortenkamp *et al.*, 2007).

8.3 ASSESSMENT OF CONCENTRATION-ADDITIVE EFFECTS USING THE TOXIC EQUIVALENCY CONCEPT

The concept of toxic equivalency factors (TEF) and toxic equivalent concentrations (TEQ) is an extension of the mixture concept of CA/DA and is only applicable for groups of chemicals with a common MOA. This is a special case of CA/DA and the condition of parallel dose–response curves must be fulfilled. Due to its ease of application and communication it is now widely applied in the risk assessment of chemical mixtures.

The TEF concept was initially developed for the binding of polychlorinated dibenzodioxins (PCDD) to the aryl hydrocarbon receptor (AhR) and associated toxicity endpoints. It was soon expanded to polychlorinated dibenzofurans (PCDF) and coplanar polychlorinated biphenyls (PCB). The reference compound for dioxin-like activity is 2,3,7,8-tetrachlorodibenzodioxin (TCDD), the most potent activator of the AhR discovered to date. The effect of the mixture is then expressed as the concentration or dose of TCDD that would elicit the same effect as the mixture, rather than dealing with the many different toxicity measures for the various chemicals that act according to the same MOA as TCDD.

In Chapter 7, we have learnt that the toxicity of chemicals with the same MOA can be expressed in relation to a reference chemical i as relative effect potency REP_i (Equation 7.16). For the same pair of chemical i and reference chemical, there will be different REP_i values for different toxicity endpoints and test species or cells, as well as for different exposure scenarios (duration, frequency). A variety of REP_i values is therefore available for each chemical i in a range of *in vitro, in vivo* and epidemiological endpoints. For risk assessment a consensus toxic equivalent factor (TEF_i) was derived by a group of experts for a number of chemicals using a large experimental REP_i database and expert knowledge (Van den Berg *et al.*, 2006). The TEF_{TCDD} of TCDD was set by definition as 1, and those of the other dioxins vary from 1 (for penta-CDD) to 0.0003 for octa-CDD. The furans have similar TEFs as the corresponding dioxins (*e.g.*, 2,3,7,8-TCDD vs. 2,3,7,8-TCDF), while the planar PCBs all have TEFs of 0.00003.

According to Equation (8.5), the TEQ of a mixture of chemicals is the sum of the product of the concentration of each component i in the mixture, C_i, and its TEF_i value.

$$TEQ = \sum_{i=1}^{n} C_i \times TEF_i \tag{8.5}$$

As research with dioxins has shown, the TEQ concept has proven to be applicable to whole organism endpoints despite being strictly valid for receptor-mediated toxicity only. The application of the TEQ concept has, therefore, also been recommended for the chemical risk assessment of estrogenic chemicals (Simon *et al.*, 2007), polycyclic aromatic hydrocarbons (PAH) (Nisbet and Lagoy, 1992) and neurotoxic PCBs (Simon *et al.*, 2007). In principle there seems to be no limitation for the application of this useful concept, provided that the chemicals assessed act via concentration-addition and share a common slope of the logarithmic dose–response curve. The toxic units (TU) introduced in Section 8.5.4 rely on the very same assumptions as the TEQ approach with strict CA/DA and similar shape of CRCs.

8.4 MIXTURES IN RISK ASSESSMENT

8.4.1 Concepts

To date, no common worldwide consensus has been reached for chemical risk assessment of mixtures although mixtures are implicitly and/or explicitly addressed in many regulatory documents.

The International Programme on Chemical Safety (IPCS) of the World Health Organization has developed a framework for assessing cumulative risk (IPCS, 2009). This framework accounts for the mixture toxicity concepts discussed in this chapter using the tiered approach depicted in Figure 8.4 (slightly modified

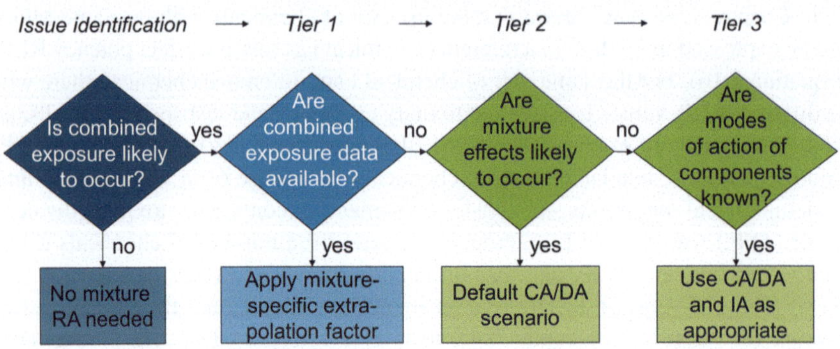

Figure 8.4 Steps for inclusion of mixtures into risk assessment (adapted from IPCS, 2009). CA = concentration addition; DA = dose addition; IA = independent action; RA = risk assessment.

from the report). As a first step (issue identification), it is necessary to assess if combined exposure is likely to occur. The IPCS proposed in tier 1 to adopt a mixture specific extrapolation factor for the derivation of acceptable effect level before moving onto the higher tiers with explicit consideration of mixture effects. In tier 2, if the likelihood of mixture effects has been identified but is yet to be further characterised, CA/DA is used as the default concept. In tier 3, more information is required on the MOAs of the mixture components in order to select the appropriate mixture toxicity models for the risk assessment (Figure 8.4).

If chemicals act by CA/DA, the risk quotients (RQ_i, see Chapter 2 for definition) will add up to a cumulative risk index (RI) for n mixture components i (Equation 8.6). An alternative but synonymous nomenclature is the hazard index (HI) based on hazard quotients (HQ_i). As in the case of RQ, a RI of 1 is the threshold between acceptable ($RI < 1$) and unacceptable risk ($RI > 1$).

$$RI = \sum_{i=1}^{n} RQ_i = \sum_{i=1}^{n} \frac{\text{exposure level}_i}{\text{acceptable effect level}_i} \tag{8.6}$$

Different measures for exposure and acceptable effect levels may apply for ecological and human health risk assessment (*e.g.*, dose or concentration). For Equation (8.6) to be valid, the exposure and acceptable effect levels must have the same units.

If TEFs are available, the RI can be derived from the TEQ of the mixture divided by the acceptable effect level of the reference chemical using Equations (8.7) or (8.8).

$$RI = \sum_{i=1}^{n} \frac{\text{TEQ}}{\text{acceptable effect level}_{\text{reference chemical}}} \tag{8.7}$$

$$RI = \frac{\sum_{i=1}^{n} \text{TEF}_i \times \text{exposure level}_i}{\text{acceptable effect level}_{\text{reference chemical}}} \tag{8.8}$$

8.4.2 Do we have account for mixture effects in risk assessment?

A useful indicator of whether mixture effects play a role in risk assessment is the maximum cumulative ratio (MCR), which is the ratio between the observed cumulative toxicity and the maximum toxicity caused by one chemical (Equation 8.9).

$$MCR = \frac{\text{cumulative toxicity}}{\text{maximum toxicity from one chemical}} \tag{8.9}$$

Mathematically, the MCR can be expressed as the ratio of the RI to the RQ_{max}, that is, the maximum RQ relates to the chemical with highest exposure and/or effect (Equation 8.10).

$$MCR = \frac{RI}{RQ_{max}} \tag{8.10}$$

If the MCR equals to 1, the mixture toxicity is caused solely by one mixture component. If all n mixture components contribute to the cumulative toxicity, the MCR will reach n. In the former case (MCR = 1), cumulative risk assessment would not be required, whereas in the latter case (MCR close to n), cumulative risk assessment would be imperative.

The MCR is also a measure of the fraction of toxicity equivalents that derive from the most toxic component in the mixture, for example, an MCR of 2 indicates that the most toxic chemical is responsible for 50% of the mixture effect and an MCR of 1.1 indicates that the most toxic component causes approximately 90% of the mixture effect.

Reality lies somewhere between these extremes. Price and Han (2011) calculated the MCR values for the pesticide water concentrations found in more than 4000 samples collected by the U.S. Geological Survey (USGS) in the 1990s. Each sample was analysed for up to 83 pesticides, of which up to 30 were actually detected. The obtained MCR values for chronic human health ranged from 1 to 6. A similar result was found for chronic effects in fish (representing ecological risk). For the water samples with $RI < 1$, the MCR values were generally higher than for those with $RI > 1$. This result indicates that in many cases where individual water samples exceeded a RI of 1, the mixture toxicity was driven by only a few components. Conversely, the findings also imply that when low hazard was associated with a water sample and MCR was higher, it was not possible to identify individual culprits. Therefore, the joint effect of many chemicals needs to be considered in order to evaluate the overall hazard.

A similar study performed on the mixture risk of 26 pharmaceuticals in wastewater treatment plant effluents yielded MCRs between 1.2 and 4.2 (Backhaus and Karlsson, 2014). A study for agricultural contamination by pesticides and veterinary pharmaceuticals (Holmes *et al.*, 2018) predicted that in less than 4% of the scenarios the MCR would be >2 for cases with $RI > 1$ but that the MCR would increase with decreasing RI. The MCR also helped to identify the small fraction of mixtures out of over 3000 surface water samples where the single chemical RQ would have largely underestimated the mixture risk (Vallotton and Price, 2016).

The MCR was also applied to evaluate the human health risk for anti-androgenicity of phthalates in a retrospective study that analysed 27 years of exposure data in 24-h urine samples from the German Environmental Specimen Bank (Apel *et al.*, 2020). While RI decreased from 1.8 to 0.2 from 1985 to 2015, the MCR actually increased during this time from 1.5 to 1.9. This means that the overall mixture hazard decreased but the role of mixtures has become more important during that time.

8.4.3 Mixtures in chemicals regulations

The IPCS framework for the risk assessment of combined exposure to multiple chemicals (Meek *et al.*, 2011, 2014) provides the guiding principles for most approaches on how to deal with mixtures in risk assessment.

The IPCS has been instrumental in developing and supporting harmonised approaches for the risk assessment of mixtures of dioxin-like chemicals, where the TEF concept is applied (Van den Berg *et al.*, 2006). The TEFs defined by the WHO working group are internationally accepted and have found their way into many national regulations. Furthermore, TEFs are a cornerstone of the Stockholm Convention, which is a global treaty to protect human health from exposure to persistent organic pollutants (POPs). The WHO has also adopted the IPCS framework for assessing cumulative risk for mixture of chemicals in drinking water (WHO, 2017a).

The globally harmonised system for the classification and labelling of chemicals (GHS) employs a summation method that classifies a mixture according to its relative amounts of already classified chemicals or on the basis of a similar mixture that is already classified (United Nations, 2009). Only components that exceed 1% on a mass basis need to be included in the evaluation, unless there is indication that a component present at a lower concentration is toxicologically relevant.

The U.S. has implemented mixtures in its regulatory frameworks as early as 1986 (U.S. EPA, 1986). The RI approach has been used to evaluate the risk posed by air pollutants and hazardous waste. The cumulative risk assessment of pesticides by the U.S. EPA makes explicit use of the mixture toxicity concepts, employing CA/DA for pesticides that share a common mode of toxic action and IA for those that do not (U.S. EPA, 2002a).

Similarly, in Australia and New Zealand, the CA/DA model has been recommended as the method for assessing whether or not the water quality criteria are exceeded (Australian Government, 2018a).

In the European Union's chemical regulation, REACH (EP&EC, 2006a), some industrial mixtures are explicitly regulated under the name 'substances', which refers to single chemicals and products that are mixtures of unknown and variable composition (*e.g.*, petroleum hydrocarbons, surfactants) and multi-component substances. Such mixtures are treated as if they were individual chemicals in the risk assessment process with toxicity testing on the mixtures. Furthermore, the read-across approach of REACH (European Chemicals Agency, 2017) allows implicit introduction of mixtures in risk assessment.

In the future, one can expect further development in mixture assessment and its application to different environmental and health regulations in Europe, indicated by the 'Chemicals Strategy for Sustainability towards a Toxic-free Environment' (European Commission, 2020), which explicitly calls for the introduction of 'mixture assessment factors' in REACH and for accounting for mixture effects in other chemicals regulations.

A mixture assessment factor (MAF) (Bopp *et al.*, 2018) is an extrapolation factor that would be used on a single chemical's RQ_i to account for mixture effects by multiplying the RQ_i by the MAF to obtain the mixture RI. The MCR analysis (Price and Han, 2011) described above would yield a MAF of 2–17, which implies that the mixture scenario is dependent on the exposure situation and is probably site-specific (Bopp *et al.*, 2018). Van Broekhuizen and Traas (2016) argued that only 5–10 chemicals dominate the mixture effect under most scenarios and therefore proposed a MAF of 10 for environmental risk. Although

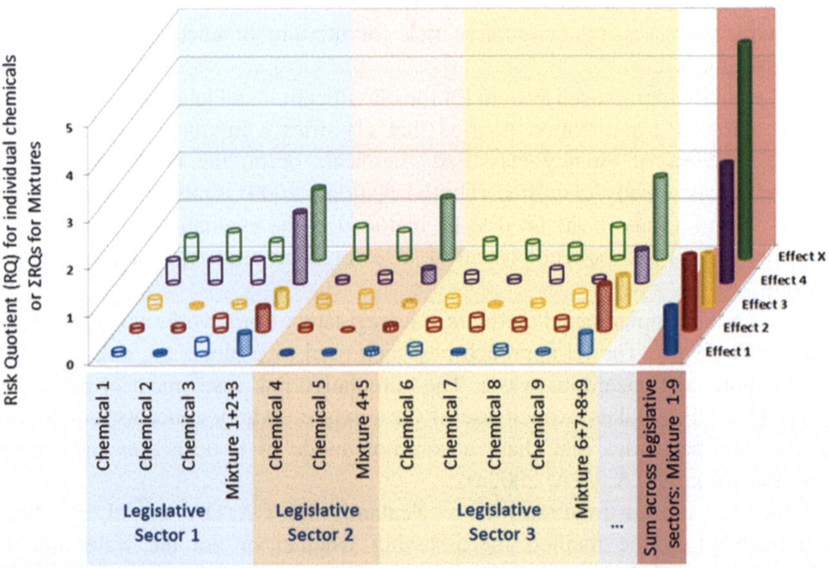

Figure 8.5 Mapping of chemicals and their mixtures to the risks they pose for various toxicological effects. For each nine chemicals the individual risk quotients RQ_i are presented for different types of effect (*e.g.*, hepatotoxicity, neurotoxicity, etc.). Chemicals are grouped according to the legislative sector they are regulated under (*e.g.*, REACH, pesticides, cosmetics, food contaminants, etc.). The risk index RI, that is, the sum of the risk quotients RQ_i is illustrated for mixtures within each sector and in the last column for the cross-sectorial mixture. Reprinted from Bopp *et al.* (2019). Regulatory assessment and risk management of chemical mixtures: challenges and ways forward. *Critical Reviews in Toxicology*, **49**(2): 174–189. BY-NC-ND licence © 2019 European Union. Published by Informa UK Limited, trading as Taylor & Francis Group.

this is a very intuitive approach, in practice it will be more difficult to define MAFs because they will not only depend on the number of chemicals included in their derivation and their relative effect potency but also on the RQ_i of the single chemical as well as on the exposure scenario. If the mixture effect can be accounted for completely by the known chemicals and only few are mixture effect drivers, a MAF of around 10 would be reasonable. However, the MAF might be substantially higher for those exposure scenarios and biological endpoints where the known chemicals only explain a small fraction of the effect.

Another challenge that remains for mixture risk assessment is that of regulatory silos, that is, different legislative sectors producing different guidance for the same chemicals or chemicals of a cumulative assessment group that are expected to act together in mixtures (Kortenkamp and Faust, 2018; Bopp *et al.*, 2019). Regulation of mixture risk will have to include not only multiple types of effects for multiple chemicals but also to bridge across different regulatory spaces as is illustrated in Figure 8.5.

8.5 MIXTURES AND WATER QUALITY

8.5.1 What type of mixture effects occur in water samples?

While it has been established that the mixtures of chemicals at concentration levels that are well below observable effect levels may produce substantial mixture toxicity (Section 8.2.5), there are very few studies in the literature that tackle designed mixtures with many components mixed in the concentrations as they typically occur in water samples. Translation of the results from designed mixture toxicity experiments to real water samples that may contain thousands of chemicals at exceedingly low concentrations, therefore, remains uncertain, although existing evidence points to chemicals acting together at low concentrations in a concentration-additive manner even if they do not show any effect on their own.

Early mixture toxicity studies on guppy fish demonstrated that mixtures of chemicals at very low concentrations could be described by the CA/DA model for the underlying baseline toxicity, independent of the specific MOA these chemicals produce at higher concentrations (Hermens *et al.*, 1985). Warne and Hawker (1995) took this concept a step further arguing that with an increasing number of chemicals at equitoxic concentrations in a mixture, CA/DA was the more likely outcome, and demonstrated that the results of a large number of experimental studies confirmed this hypothesis.

Two types of mixture studies have been undertaken to evaluate how chemicals act together in mixtures. The first is to combine water samples with single reference chemicals for the given bioassays. This was done for a mixture of wastewater and the herbicide reference chemical diuron as well as a baseline toxicant in the combined algae test (Escher *et al.*, 2008a). The isobolograms derived from the TUs for photosynthesis inhibition for the mixture of a wastewater treatment plant effluent and the herbicide diuron (Figure 8.6a) and for

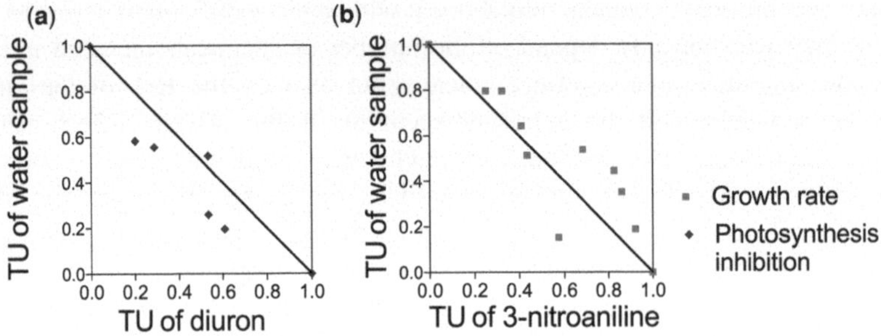

Figure 8.6 Isobologram of a 'binary' mixture of (a) wastewater treatment plant effluent and the herbicide diuron and (b) wastewater treatment plant primary effluent and the baseline toxicant 3-nitroaniline. Data from Escher *et al.* (2008a).

growth rate of a wastewater treatment plant primary effluent and the baseline toxicant 3-nitroaniline (Figure 8.6b) confirmed that the water samples acted in a concentration-additive manner with reference chemicals.

The second approach is to mix chemicals at the concentrations detected in water samples and test for their interaction by comparing the predicted mixture effects with the measured mixture effects. Junghans *et al.* (2006) mixed 25 pesticides in typical exposure scenarios in field runoff and CA predicted the mixture effects best, although IA predictions differed only by a factor of 1.3.

In a study that focused on the oxidative stress response detected with AREc32, pharmaceuticals and pesticides were mixed in groups of 5 and 10 each and together in equipotent concentration ratios and in concentration ratios of their drinking water guideline values, making up a mixture of 5, 10, 15 and 20 components (Escher *et al.*, 2013). All experimental mixtures showed very similar effects as the predictions by CA in the AREc32 assay, confirming that CA is a robust prediction model for chemicals acting according to the same MOA. In a way, the similarity of MOA was forced in this example because the AREc32 reporter gene assay only reacts to one MOA. Nevertheless, it is still encouraging to see that mixture effects of environmentally realistic mixtures of very diverse types of chemicals could be well described by simple mixture models.

Tang *et al.* (2014) evaluated samples along the entire wastewater treatment train from WWTP influent to advanced treated water with three bioassays, including non-specific toxicity towards *A. fischeri*, photosynthesis inhibition in green algae and oxidative stress response in AREc32. They detected four to 50 chemicals in the different samples and tested them alone, as mixtures in six groups (endocrine disrupting chemicals, iodinated contrast media, antibiotics, pharmaceuticals, pesticides and others) and all mixed together. There was excellent agreement between measured mixtures within groups and predictions from single chemicals

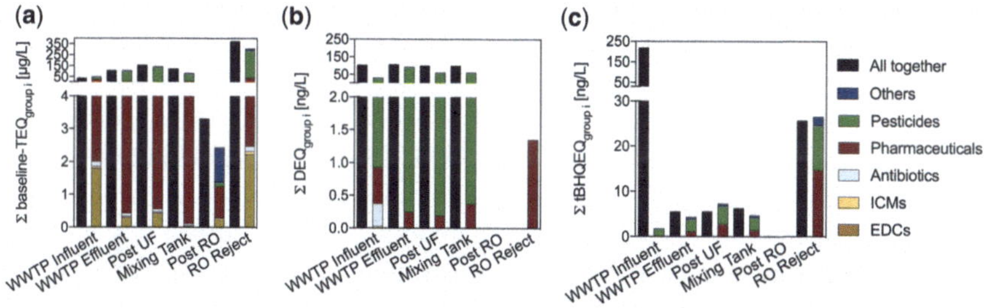

Figure 8.7 Measured BEQ of the designed mixtures in comparison with the cumulative BEQs from component-based mixture prediction of the six chemical groups: (a) non-specific toxicity towards *Aliivibrio fischeri* (baseline toxicity equivalent concentrations, baseline-TEQ), (b) photosynthesis inhibition in green algae (diuron equivalent concentrations, DEQ), (c) oxidative stress response in AREc32 (*t*-butylhydroquinone equivalent concentrations, *t*BHQ EQ). WWTP = Wastewater treatment plant; UF = ultrafiltration; RO = reverse osmosis; EDC = endocrine disrupting chemicals; ICM = iodinated contrast media. Reprinted with permission from Tang and Escher (2014). Which chemicals drive biological effects in wastewater and recycled water? *Water Research*, **60**: 289–299. © 2014 Elsevier.

(not shown) and between experimental mixtures of each of the six groups summed up in comparison with experimental mixtures of all chemicals (Figure 8.7). Although the chemicals in the groups were not all acting according to the same MOA, the steps from individuals to groups to mixtures of all detected chemicals increase our confidence that chemicals combined in mixtures as they occur in real water samples act together to trigger measurable effects and that these effects could be predicted by the summation of bioanalytical equivalent concentrations (BEQ), which is effectively CA.

Note, however, that this strong agreement between experimental and predicted mixture effect is limited to designed mixtures. If the effects of the designed mixtures were compared with the effects of the entire water samples, then only photosynthesis inhibition show a good agreement (Tang and Escher, 2014). For the other two endpoints, the detected chemicals explained less than 3% of the effect in the entire water sample due to the presence of a large number of undetected bioactive chemicals (Escher *et al.*, 2013; Tang and Escher, 2014). This issue will be discussed further in Section 8.5.2 and Chapter 13.

Another example is a study that tested over 200 designed mixtures of two to 14 chemicals mixed in concentrations as they occurred in water samples taken during rain events in small creeks that showed a very high diversity of chemical burden and mixture compositions (Neale *et al.*, 2020a). The activation of AhR in AhR CALUX, the activation of peroxisome proliferator-activated receptor gamma (PPARγ) in PPARγ GeneBLAzer and the oxidative stress response in AREc32 all followed very well the model of CA. CA even applied to the cytotoxicity of the three cell lines (Escher *et al.*, 2020b). This example is presented in more detail as case study 6 in Chapter 14.7.

Altenburger *et al.* (2018) took a different approach where just two defined 12-component mixtures were tested in 19 *in vitro* bioassays. The idea was that each bioassay would respond to one of more of the mixture components and the challenge was to confirm the associated mixture prediction against the background of inactive chemicals. This ambitious study confirmed CA in many bioassays but also pointed out issues with respect to solubility or masking of effects by cytotoxicity in reporter gene assays.

More of these types of studies with complex mixtures should be encouraged to obtain a better picture of what really happens in water samples with potentially thousands of chemicals present at very low concentrations. Currently, evidence is merely anecdotal, but a better conceptual understanding of the mixture effects under these scenarios will greatly enhance future water quality assessment.

8.5.2 How much of the measured effects in water sample can be explained by known and detected chemicals?

The answer to this question is not straightforward. It depends on the one hand on the type of water sample and on the other, on the bioassay that is used to evaluate a given water sample.

For some types of water samples (*e.g.*, industrial effluents, contaminated sites, accidents), a few – often well-known – chemicals explain the majority of the effect. Here, effect-directed analysis (EDA, see Chapter 13, Section 13.3.2) might identify the dominant risk drivers. In other water types (urban wastewater, surface water), which are impacted by a large number of very diverse chemicals, the identified chemicals often only explain a minute fraction of the observed effect. In these cases, bioassays are particularly valuable to estimate the full extent of toxicity or effect in a sample.

How much of the activity can be explained by the detected chemicals also depends on the selected bioassay. In Chapter 13.4, we will introduce a classification of bioassays into two categories based on the type of responsive chemicals and their degree of specificity. From observation alone, we can already say that some assays respond very strongly to a small number of potent chemicals. For whole organism tests, photosynthesis inhibition in algae or immobilisation of *Daphnia magna* by insecticides are good examples. Herbicides are 10,000 to 100,000 times more potent than other chemicals for inhibition of photosynthesis and hence the mixture effect is often completely dominated by herbicides unless a sample does not contain any herbicides (Tang and Escher, 2014; de Baat *et al.*, 2018; Glauch and Escher 2020). Herbicides are not only used in agriculture but also in urban applications, so they are quite ubiquitous and general mixture risk drivers. Insecticides show a similarly high specificity, but their mixture dominance occurs mainly in agricultural streams because *D. magna* reacts also quite sensitively to other toxicants.

For reporter gene assays, hormone receptor transactivation assays are another good example. Natural hormones are up to a million or more times more potent in reporter gene assays than most other chemicals and even xenostrogens are only two to four orders magnitude more potent than non-specifically acting chemicals. Any water sample with the slightest contribution of wastewater, treated or not, or manure from animals will respond mainly to the natural estrogens (Könemann *et al.*, 2018) and mixture effects can typically be well explained by detected estrogenic chemicals.

It is a very different case for reporter gene assays for receptors of lower specificity, such as the pregnane X receptor (PXR) and PPARγ, and for KE-based bioassays, such as those responding to adaptive stress responses, and for apical endpoints, such as cytotoxicity and fish embryo toxicity. Here, the mixture effects of the detected chemicals can only explain a small fraction of the entire water sample's toxicity or effects.

8.5.3 Mixture effects at very low effect levels (<10%)

With the exception of wastewater, most water samples do not cause an effect unless the organic micropollutants are extracted and enriched prior to dosing in a bioassay (Chapter 7.4). CRCs are linear at low effect levels up to 30% of effect and can be described by one single parameter, the slope of the CRC (Chapter 7.3.6). At even lower effect levels, below 10% of effect, the predictions for CA and IA overlap perfectly and can be described by Equation (8.11), which is a joint CA/IA mixture model (Escher *et al.*, 2020b).

$$\begin{aligned} \text{Effect (mixture)} &= \sum_{i=1}^{n} p_i \times \text{slope}_i \times C_{tot} \\ &= \text{slope}_{mixture} \times C_{tot} \end{aligned} \tag{8.11}$$

Several case studies demonstrated the validity of this equation in designed mixture studies for cytotoxicity and reporter gene activation (Escher *et al.*, 2020b). Even more importantly, most component-based predictions of mixture effects in water from analytical data were within the applicability range of this equation (<10% effect). This means that in the future, we do not need to worry whether detected chemicals have the same MOA or not, we can simply apply the joint CA/IA mixture model. This effectively means that TU summation is valid for chemicals with similar and dissimilar MOAs provided that resulting effect levels do not exceed 10% in theory and 30% in practice.

8.5.4 Component-based prediction of mixture toxicity in water

In a component-based risk assessment, the risk index RI (Equation 8.7) of combined exposure to multiple chemicals is calculated from the exposure and effect data of the individual components. A similar approach can also be taken for predicting the mixture toxicity in a water sample by weighting the measured concentrations by their effect potency and summing them up over all mixture components. This is analogous to the TU/EU approach introduced as Equation (8.3) in Section 8.2.2, only without the condition that the concentrations are a fraction of the LC_{50}/EC_{50}; instead, they are the measured concentrations, and a TU/EU exceeding 1 would imply that there is a risk.

The sum of the TUs is defined by Equation (8.12) and analogously for EU by Equation (8.13). We give this component-based TU/EU the index 'chem' because the TU/EU can also be directly measured with a bioassay in a water sample as TU_{bio} (Equation 7.18) and EU_{bio} (Equation 7.20) as outlined in Chapter 7.5.

$$TU_{chem} = \sum_{i=1}^{n} TU_i = \sum_{i=1}^{n} \frac{C_i}{LC_{50,i}} \tag{8.12}$$

$$EU_{chem} = \sum_{i=1}^{n} EU_i = \sum_{i=1}^{n} \frac{C_i}{EC_{50,i}} \tag{8.13}$$

The component-based approach is the most traditional approach when evaluating mixtures in water quality. The reader has probably wondered why this approach was not introduced earlier in this chapter. There is a reason for it. The component-based approach is often 'sold' as characterising true mixture effects but as we have learnt above, it only constitutes the mixture effects of the detected chemicals that have available toxicity/effect data. If we include 20 chemicals in the approach and find that five dominate the TU, will this mean that these five are the true risk drivers or could there be unknown chemicals that contribute to the mixture effects? Comparison of TU_{chem} with TU_{bio} as outlined in Chapter 13 will shed light on the mixture effects elicited by unidentified chemicals and chemicals below their detection limit.

Also, we have learnt that the component-based approach only holds for CA, but it has been used widely irrespective of the MOAs of the components. The most recent insights that realistic mixtures in water mostly obey CA and that CA and IA merge to one single model at low effect levels justify the application of component-based approaches, but one must always be aware that there is no guarantee that the entire mixture toxicity is captured.

One of the earliest mixture studies showed that the TU_{chem} of 12 organics and 11 metals in fish samples were not a good predictor of the integrity of the fish community (Dyer *et al.*, 2000). Nevertheless, the component-based approach had a revival in the last decade, often without verification by comparison with ecological quality.

Water quality monitoring concentrations are now often directly reported as TU_{chem}. While this is evidence of good intentions, the results are often obscured because the choice of the effect data is not always transparent, often relying on predicted effect data or mixed experimental/modelled effect data. If well performed, component-based mixture predictions can help us understand the importance of mixture effects and the contribution of known chemicals, but we always have to keep in mind the bias due to the selection of analytes in the chemical analysis. Malaj *et al.* (2014) demonstrated how the number of acute-risk chemicals was related to the overall risk in a study that included over 200 chemicals measured at over 4000 sites in European rivers.

The TU_{chem} of a diverse set of wastewater treatment plant effluent across Europe derived from algae, daphnid and fish toxicity data for 26 pharmaceuticals were dominated by less than 10 pharmaceuticals overall. In each sample as few as two to four pharmaceuticals explained most of the mixture effect (Backhaus and Karlsson, 2014). Even when evaluating the monitored concentrations of 110 chemicals in an Asian river basin, the top 10 mixture components explained more than 80% of TU_{chem} for acute and 95% of TU_{chem} for chronic effects (Liu *et al.*, 2020). TU_{chem} of 75 detected compounds in the Lake Victoria Basin in Kenya indicated chronic risk with 16 chemicals being risk drivers (Kandie *et al.*, 2020).

Gustavsson *et al.* (2017) considered chemicals below the LOD in TU_{chem} analysis of 141 pesticides in Swedish streams. They either left out chemicals below the LOD, used LOD/2 or estimated likely concentrations below LOD. Using LOD/2 most likely overestimated the TU_{chem} by two orders of magnitude and leaving out chemicals below their LOD probably came closer to a realistic TU_{chem}. Interestingly, while often only one to three pesticides dominated the mixture risk in each sample, these pesticides varied from site to site, so that 83 of 141 pesticides would need to be included to capture 95% of the risk at all sites.

The concept of mixture toxic pressure assessment (Posthuma *et al.*, 2008) relies on multi-substance potentially affected fraction (msPAF) (Klepper *et al.*, 1998; Posthuma *et al.*, 2008) and is an expansion of the component-based approach that also integrates species-sensitivity distributions. The msPAF is the lower fifth percentile of the cumulative density function of the log-normal distribution of TU_{chem}. Only three pharmaceuticals contributed to the acute mixture toxic pressure while 8 (of totally 54) contributed to the chronic mixture toxic pressure in seven Swedish rivers (Lindim *et al.*, 2019). The mixture toxic pressure approach has great potential, especially as quality-controlled SSDs are available for more than 12 000 chemicals (Posthuma *et al.*, 2019). Case studies with this dataset illustrated the utility of this approach and identified new previously untagged priority pollutants.

It is important to note that these concepts work well within the domain of analysed chemicals (*i.e.*, TU_{chem} of one water sample compared to TU_{chem} of another water sample), but that for some water samples and/or bioassays the measure response (TU_{bio}) can be significantly greater than TU_{chem}, as will be discussed in more detail in Chapter 13.

8.6 CONCLUSION

Risk assessment focusing on single chemicals can only be justified if only one chemical of a mixture is toxic, while all other compounds are inert, or their mixture effect is no larger than the toxicity of the dominant component. This condition is only met occasionally for contaminated sites and industrial effluents. In contrast, wastewater and treated water usually contain thousands of compounds including many unknown chemicals and transformation products. While under these conditions, it is virtually impossible to fully understand and evaluate the complex mixture interactions, bioanalytical tools may provide valuable information about the overall mixture effect. It is futile to attempt to resolve all chemicals in any given water sample, but a smart combination of chemical and bioassay analysis combined with mixture toxicity modelling will give us a wealth of information for risk characterisation.

Regulatory chemical risk assessment is ready for implementation of mixtures and various approaches are under consideration in different regions of the world, with the biggest challenge being to decide which chemicals to group together as mixtures.

Despite of all scientific progress described in Section 8.5, there has been no formal uptake for mixtures in water quality legislation. The adoption of effect-based methods would by definition be an acceptance of the importance of mixture effects (Brack *et al.*, 2019). The Water Framework Directive has been exploring the use of effect-based methods with a detailed technical report (Wernersson *et al.*, 2015) but at present this approach to include mixtures in water quality assessment has not reached legal status. A major hindrance might be the lack of accepted effect-based trigger (EBT) values. As outlined in Chapter 13, much scientific progress has been made over the past years on the derivation of EBTs for various water types in a scientific context – this hopefully will lead to increased regulatory uptake in the near future.

Chapter 9

In vitro assays for the risk assessment of chemicals

9.1 INTRODUCTION

The last two decades have seen unprecedented progress in transforming risk assessment from animal testing to an approach that relies on alternative methods comprised of *in vitro* assays and *in vitro* to *in vivo* extrapolation methods. This paradigm shift was initiated by the U.S. National Research Council's strategy to modernise toxicity testing with high-throughput pathway-based methods (NRC, 2007) and the parallel implementation of an integrated testing strategy (ITS) in the European Chemicals Regulation REACH (EP&EC, 2006a, b). Provided that an *in vitro* method is fully validated and there is an (ideally mechanistic) *in vitro* to *in vivo* extrapolation model available, *in vitro* data can now be used in quantitative risk assessment (Blaauboer, 2008).

The field of water quality assessment has profited enormously from the progress made in chemical risk assessment because large numbers of *in vitro* bioassays have been developed and validated and effect data for single chemicals have become publicly available that serve not only for the elucidation of toxicity and risk assessment but also for interpreting water testing results and for linking bioassay responses to chemical analysis.

In this chapter we give a brief overview of the developments in this area, starting with an introduction of the regulatory background for the application of *in vitro* assays and computational methods. In different regulatory environments these novel tools that replace animal testing are referred to as 'alternative test methods' or 'new approach methods'. We give an overview of international developments

Authors: Beate Escher, Peta Neale and Frederic Leusch
doi: 10.2166/9781789061987_0143

with a focus on the 'Integrated Testing Strategy' in the European Union and the 'Toxicity Testing in the 21st Century' strategy in the U.S. (National Academies of Sciences, 2017).

Then we focus on the CompTox strategic initiative of the U.S. EPA that led to the ToxCast program as well as the Tox21 initiative, which has provided (and continues to provide) a huge set of easily accessible *in vitro* activity data for use by researchers and regulators. We outline the main applications of these data for elucidation of toxicity pathways and prioritisation of chemicals for in-depth risk assessment. Further, quantitative *in vitro* to *in vivo* extrapolation (QIVIVE) models serve for human health risk assessment but can also be used for ecological risk assessment.

We conclude with a detailed description of exposure in cell-based bioassays because this will become of high relevance for improving the existing QIVIVE models and will give more confidence in the application of *in vitro* bioassays in risk assessment but also for environmental and biomonitoring applications.

9.2 APPLICATION OF NEW APPROACH METHODS IN REGULATION

9.2.1 Alternatives to animal testing methods

The 3Rs to Replace, Reduce and Refine laboratory animal testing have a long tradition since the 1950s (Hartung, 2010). *In vitro* assays are not a direct substitute for animal testing but are the basis of an alternative test method. An alternative test method for the replacement of an animal test is the combination of an *in vitro* test system and a predictive *in silico* model. The prediction model is an unambiguous algorithm for converting *in vitro* data into predictions of toxicological endpoints in animals or humans.

A response in an *in vitro* assay does not always result in harm to a whole organism but can provide valuable information as to whether an organism-level effect is possible or even likely. A clear advantage of *in vitro* systems is that they can be derived from human cell lines making interspecies extrapolation obsolete (Figure 9.1).

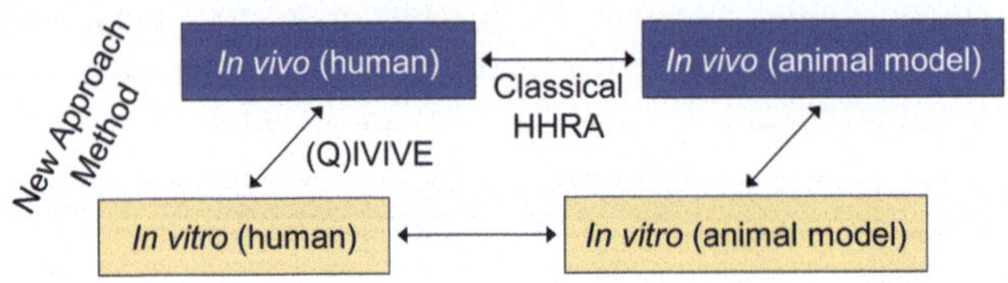

Figure 9.1 The *in vitro–in vivo* parallelogram. HHRA = human health risk assessment; (Q)IVIVE = (quantitative) *in vitro* to *in vivo* extrapolation.

It is imperative that *in vitro* to *in vivo* extrapolation models account for the toxicokinetic processes that describe the pathway from external to internal dose, which are influenced by bioavailability, uptake and excretion, and metabolism of chemicals in the body (Blaauboer, 2010). *In vitro* models are available for every one of these toxicokinetic and toxicodynamic processes. In the future we can expect to see the development of whole 'virtual organisms' that are comprehensive toxicokinetic–toxicodynamic models parameterised with *in vitro* data.

In 1991, the European Centre for the Validation of Alternative Methods (ECVAM) was founded to support the regulatory application of alternative test methods. ECVAM's important role is emphasised by the Directive on protection of animals used for scientific purposes (EP&EC, 2010), which aims to minimise the use of test animals and sets ECVAM up as the EU reference laboratory.

The U.S. counterpart, the Interagency Coordinating Committee on the Validation of Alternative Methods (ICCVAM), is an interagency committee of the U.S. government composed of representatives from 15 federal regulatory and research agencies. ICCVAM conducts technical evaluations of new, revised and alternative test methods with regulatory applicability, and promotes the scientific validation and regulatory acceptance of test methods that more accurately assess the safety and hazards of chemicals and products and that refine, reduce or replace animal use.

The U.S. EPA made a commitment in 2019 to reduce animal testing and funding by one third by 2025 and to eliminate it entirely by 2035. To achieve this goal, they support the development of 'new approach methods' (NAM), which is a general term that refers to any non-animal-based approaches that can be used to provide information in the context of chemical hazard and risk assessment.

In 2009, the United States, Japan, the EU and Canada signed a Memorandum of Cooperation establishing the 'International Cooperation on Alternative Test Methods' (ICATM) to enhance international cooperation and coordination on the validation of non-animal or reduced-animal toxicity testing methods. Since then, South Korea, Brazil and China have joined in. A global harmonisation is believed to help the regulatory acceptance. The Organisation for Economic Cooperation and Development (OECD) published a guidance document on good *in vitro* method practices (OECD, 2018) and is committed to the implementation of the 3R-principles.

9.2.2 Integrated testing strategy in the European Union

Reliance on animal *in vivo* toxicity data to extrapolate safe effect levels has made the risk assessment process very slow and costly and has resulted in incomplete evaluations and data gaps. Testing methods that use mammals or other

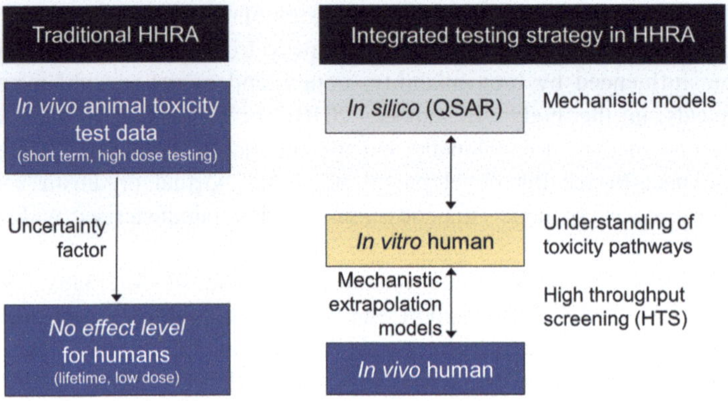

Figure 9.2 Paradigm shift in human health risk assessment (HHRA) to integrate alternative test methods. QSAR = quantitative structure–activity relationship.

vertebrates carry the additional issue of animal ethics. Even then, the relevance of animal data introduces a level of uncertainty and requires interspecies extrapolation to derive human safe levels. These issues have been recognised in the EU chemical legislation REACH, which provides a tiered approach that uses non-animal alternative test methods. In an integrated testing strategy (ITS), *in silico* and *in vitro* approaches are applied for all assessment endpoints. Both of these approaches can be based on animal and/or human cell models (van Leeuwen and Vermeire, 2007). Figure 9.2 exemplifies ITS using the example of effect assessment in human health risk assessment.

Read-across from similar chemicals is the first step in ITS, where *in silico* methods (*i.e.*, computer-based) are used to predict the toxicity of the test chemical based on the toxicity of other chemicals that share structural or physicochemical properties. For example, quantitative structure–activity relationships (QSAR) provide one approach that can be used to predict the toxicological effects of untested chemicals – provided their properties fall within the applicability domain of the chosen QSAR model. The OECD has made several QSAR models available to the scientific community in the form of a free software suite (OECD QSAR Toolbox, van Leeuwen *et al.*, 2009).

In the next step of ITS, *in vitro* bioassays are applied (Figure 9.2). *In vitro* tests can provide a substantial advantage over *in vivo* testing, including the possibility to test toxicity on cells derived from humans (not test animals), lower variability, better experimental control, higher sensitivity, shorter duration and lower financial and ethical cost than whole-animal tests. Despite the great potential of *in vitro* methods, their practical application is presently limited in the EU regulation to screening and priority setting, as well as classification and labelling.

9.2.3 Toxicity testing in the 21st century (Tox21) strategy in the United States

The U.S. National Research Council's strategy to modernise toxicity testing with high-throughput pathway-based methods (NRC, 2007) was realised with two major strategic programmes: ToxCast and Tox21. The U.S. EPA formed the National Center for Computational Toxicology (NCCT) and developed the Toxicity ForeCaster (ToxCast) project for advanced toxicity testing and modelling. The Tox21 Initiative (Collins *et al.*, 2008) is a collaboration between the NCCT of the U.S. EPA, the National Toxicology Program (NTP) of the National Institute of Environmental Health Science (NIEHS), the National Center for Advancing Translational Sciences (NCATS) of the National Institutes of Health (NIH) and the Food and Drug Administration (FDA).

It is the goal of Tox21 to (1) identify mechanisms of chemically induced biological activity, (2) prioritise chemicals for more extensive toxicological evaluation and (3) develop predictive models of *in vivo* biological response (Shukla *et al.*, 2010). To achieve this goal, Tox21 has set up a process to develop high-throughput screening (HTS) assays, to test chemicals in these assays, and to make the data continuously accessible in public databases (Figure 9.3). Nominated assays underwent a review process before they were optimised for robotic quantitative HTS (qHTS) format in 1536-well plates. After robotic validation, almost 10 000 chemicals (the so-called 'Tox21 10 K library') were run in this qHTS format. An R data processing and evaluation pipeline ToxCast Analysis Pipeline (tcpl, more information in Chapter 7.3.4) was used to derive toxicity benchmark concentrations. After application of the data within the consortium, all results including the raw data were made publicly available in databases that can easily be accessed, for example, via the CompTox Chemistry Dashboard (U.S. EPA, 2020).

As of 2020, the Tox21 10 K library has been run in over 50 bioassays, mainly assays on molecular initiating events (MIE), focussing on nuclear receptors, and key events (KE), focussing on stress response pathways, generating over 85 million data points. ToxCast included only 300 chemicals in the first phase, which were screened with 700 assay endpoints, and expanded in the second phase to 1000 chemicals screened in approximately 1000 assay endpoints. These data are now available for developing predictive models and running screening level risk assessment – a few illustrative examples will be presented below. They are also, of course, available for iceberg mixture modelling (Chapter 13), enabling translation between chemical and bioassay results. Ongoing experimental developments of Tox21 are HT transcriptomics (HTTr) and high-content imaging of cultured cells with multiple fluorescent probes for HT phenotyping (Thomas *et al.*, 2019).

At present, *in vitro* data have to be validated with *in vivo* data. It is Tox21's expressed focus to better curate and characterise legacy *in vivo* toxicity studies so they can be efficiently used for validation (Thomas *et al.*, 2018).

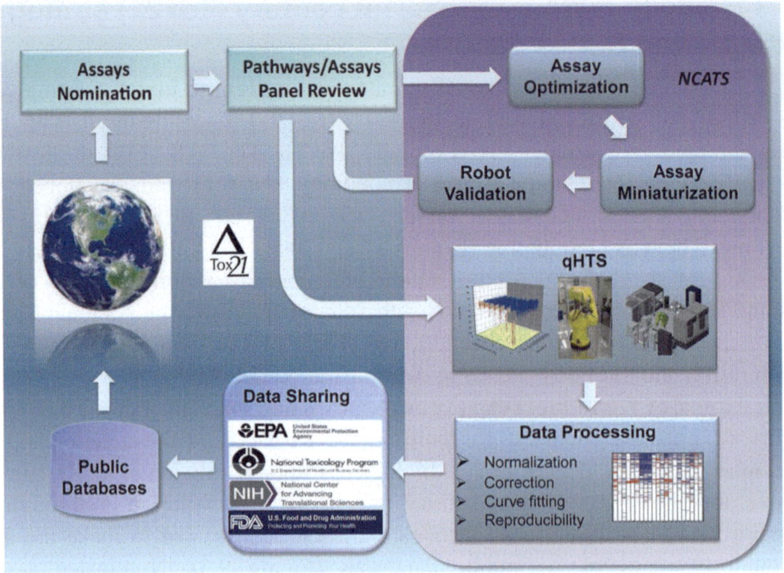

Figure 9.3 The set-up of the Tox21 HTS bioassay profiling platform. NCATS = National Center for Advancing Translational Sciences; qHTS = quantitative high-throughput screening. Figure reprinted with permission from Sakamuru *et al.* (2020). Profiling the Tox21 chemical library for environmental hazards: applications in prioritisation, predictive modelling, and mechanism of toxicity characterisation. In: *Big Data in Predictive Toxicology*, Editors Neagu and Richarz, pp. 242–263. © 2020. The Royal Society of Chemistry.

Although the HTS assays are the core of Tox21 and ToxCast, the computational models developed and refined are necessary to make best use of this wealth of experimental data and apply them for mechanistic toxicology and risk assessment.

9.3 APPLICATION OF *IN VITRO* ASSAYS IN RISK ASSESSMENT

9.3.1 A paradigm shift in human health risk assessment

Initial applications of NAM were geared at replacing individual elements in the risk assessment process but did not question the risk assessment paradigm itself. Judson and co-authors from the U.S. EPA initiated an entirely new risk assessment process, which is based on the direct inclusion of *in vitro* information into quantitative risk assessment (Judson *et al.*, 2011). They termed the proposed framework 'high throughput risk assessment' because it relies on HTS *in vitro* data targeting specific toxicity pathways. This is now recognised as next-generation risk assessment (NGRA) and is detailed in the next sections. Given the importance of

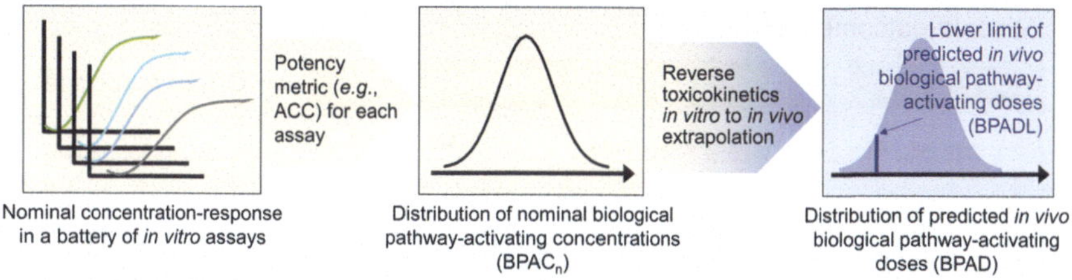

Figure 9.4 Framework of high-throughput chemical risk assessment. Modified and reprinted with permission from Villeneuve *et al.* (2019). High throughput screening and environmental risk assessment – state of the science and emerging applications. *Environmental Toxicology and Chemistry*, **38**(1): 12–26. © 2019. John Wiley and Sons.

the initial formulation of the revised risk assessment paradigm and that the essential elements remain valid, we are starting with looking back in time and present the earliest version before we move on to the state of the art.

In NGRA, a chemical is tested for activity in a large number of HTS *in vitro* assays (Figure 9.4 left). The benchmark concentration derived from concentration–response assessment in these assays is termed 'biological pathway altering concentration' (BPAC). BPACs are typically log-normally distributed if as many pathways as possible are investigated (Figure 9.4, middle). On the exposure side, external doses are linked to concentrations in blood with physiologically based toxicokinetic (PBTK) models.

The distribution of cellular concentrations can then be combined with distributions of BPAC using probabilistic re-sampling techniques such as Monte Carlo to derive a distribution of doses that perturb biological pathways (Figure 9.4, right). A 'lower limit of biological pathways altering dose' (BPADL) is derived from a low percentile of this distribution (*e.g.*, 10% percentile) and treated as the no effect level (NEL) in the effect assessment step of risk assessment.

This generic initial concept has been vastly expanded in the last 10 years (Wetmore, 2015; Sipes *et al.*, 2017; Bell *et al.*, 2018). Many modifications and improvements were introduced, and easily accessible computation tools have been made available to the public. In the following we discuss the individual steps before we outline the NGRA.

9.3.2 Quantitative adverse outcome pathways

HTS of perturbations in cellular pathways using a large test battery of *in vitro* assays has allowed the identification of molecular targets and crucial biological pathways that are linked to adverse effects *in vivo,* and subsequently can be used as molecular biomarkers (Andersen *et al.*, 2010; Martin *et al.*, 2010; Knudsen *et al.*, 2011). The HTS data together with extensive data mining in existing literature have produced an

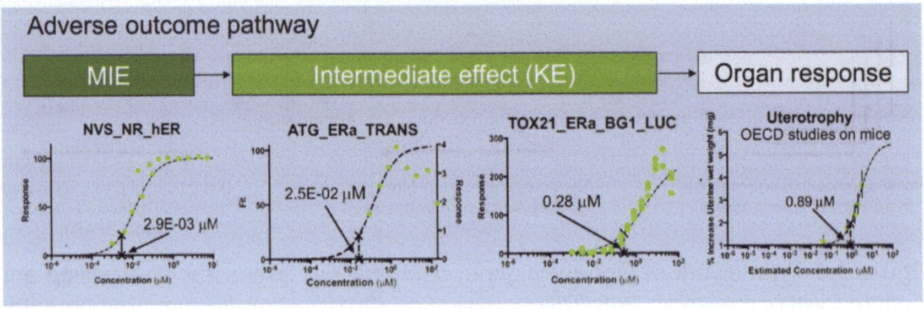

Figure 9.5 Putative adverse outcome pathway (AOP) for uterotrophy elucidated quantitatively with Tox21 bioassays. Modified after Bell *et al.* (2018). *In vitro* to *in vivo* extrapolation for high throughput prioritization and decision making. *Toxicology In Vitro*, **47**: 213–227. © 2018. Elsevier. KE = key event; MIE = molecular initiating event.

unbiased database of chemicals to be prioritised for in-depth toxicological characterisation and stitching together of AOPs (Judson *et al.*, 2009).

Figure 9.5 is an illustration of how the Tox21 HTS assays helped develop a quantitative AOP using the example of uterotrophy, which is characterised *in vivo* by increased uterine weight (Bell *et al.*, 2018). The likely MIE is binding to the estrogen receptor (assessed by Novascreen receptor binding assay NVS-NR-hER) and intermediate effects can be quantified by the activation of the estrogen receptor through reporter gene assay such as the Attagene ERa-trans and the Tox21-ERa_BG1_Luc.

A quantitative association could be made by QIVIVE analysis, indicating that the concentration needed to activate the HTS assays translated as an internal blood concentration agreed very well with the estimated blood concentrations in the *in vivo* assay on mice (Bell *et al.*, 2018). The assumptions made for the QIVIVE in this example is that the nominal active concentrations in the HTS assays are equal to the plasma concentration *in vivo*.

9.3.3 Quantitative *in vitro* to *in vivo* extrapolation

QIVIVE is often directly integrated into the risk assessment but for didactive purposes, we are introducing its underlying concept independently.

The active concentration in a HTS assay, for example, the activity concentration at cut-off (ACC) or the 50% activity concentration AC_{50} (see Chapter 7 for definition of these benchmark concentrations), is adjusted by a conversion factor to obtain an equivalent administered dose (EAD) in units of mg/kg/day for the particular biological endpoint (Figure 9.6). The conversion factor is derived by reverse dosimetry (Bell *et al.*, 2018). An arbitrary dose, for example, 1 mg/kg/day is divided by the associated predicted plasma concentration in humans at this dose level. The predicted plasma concentration can be expressed

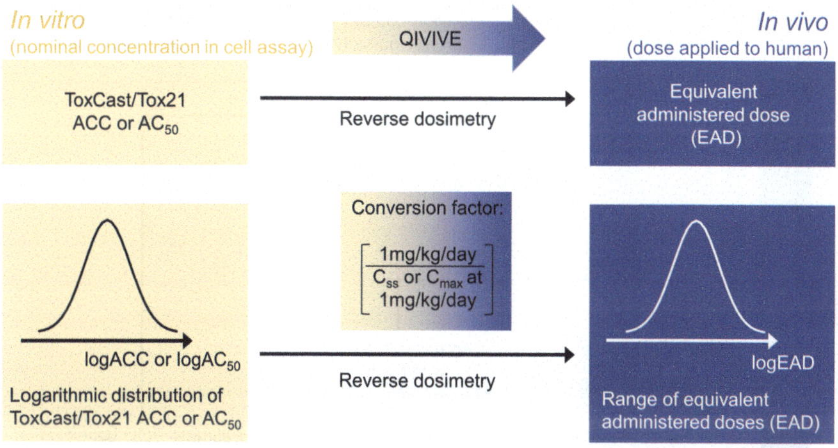

Figure 9.6 Principle of quantitative *in vitro* to *in vivo* extrapolation (QIVIVE). C_{SS} = steady-state plasma concentration; C_{max} = maximum plasma concentrations.

either as steady-state concentration C_{ss} or maximum concentration C_{max} and can be predicted by classical PBTK-models that are well-established models in pharmacology and toxicology. The new high-throughput toxicokinetic models described in more detail in Section 9.3.4.2 may also serve for this purpose.

Using a range of bioassays yields a range of EADs (Figure 9.6). The EAD or the distribution of EADs can be compared to doses *in vivo* or a lower percentile of the EAD distribution used in the risk assessment process as outlined below.

Implementation of bioassay exposure considerations as described in Section 9.4 might further improve the QIVIVE models if the exposure in *in vitro* assays is described by freely dissolved concentrations instead of nominal concentrations (Honda *et al.*, 2019).

9.3.4 Next-generation risk assessment

NGRA combines a systematic HTS-based tiered hazard assessment with high-throughput toxicokinetics on the exposure side. The idea behind NGRA is not to predict animal toxicity by *in silico* and *in vitro* assays but rather relies on the fact that effects on a molecular and cellular level occur at lower concentrations than would be expected from exposure of the population investigated and would hence be protective.

9.3.4.1 Hazard assessment in next generation risk assessment

Thomas *et al.* (2019) have outlined a roadmap to implement HT hazard assessment in NGRA. Existing HTS assays and new assay developments will be implemented in a tiered hazard assessment strategy (Figure 9.7). The first tier of the hazard

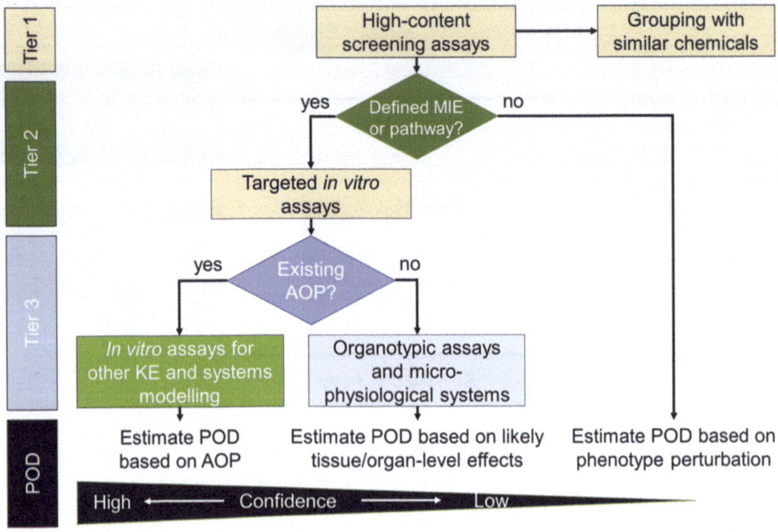

Figure 9.7 Tiered testing framework for hazard characterisation in Tox21 as a component of next-generation risk assessment (NGRA). Simplified from Thomas *et al.* (2019). POD = point of departure; AOP = adverse outcome pathway; KE = key event; MIE = molecular initiating event.

assessment is comprised of the mentioned high content screening assays (HTTr and HT cellular phenotyping), which are presently under development, and may in the meantime be covered by existing multiplexed assays or larger interrogative test batteries. As one outcome of this tier, new chemicals can be grouped with chemicals of a known biological target or pathway.

One of the existing problems is that many chemicals are biologically promiscuous, that is, trigger many pathways. This is especially true at concentrations near cytotoxicity, where a cytotoxicity burst phenomenon can often be observed, which is the non-specific activation of multiple pathways (Judson *et al.*, 2016).

Therefore, only the most sensitive pathways will move forward to a second tier, where more in-depth and targeted *in vitro* assays will be performed to determine if the tested chemical can be associated with an existing AOP. In this case, tier 3 testing can move forward to AOP modelling and probing associated KEs. If no AOP exists, focus will be on organotypic assays and microphysiological endpoints (Thomas *et al.*, 2019). The eventual aim of the process outlined in Figure 9.7 is to produce a point of departure (POD) for risk assessment. The process produces PODs with a decreasing degree of confidence going from left to right and will therefore require increasing extrapolation factors in risk assessment.

9.3.4.2 Exposure assessment in next generation risk assessment: high-throughput toxicokinetics

Toxicokinetic model predict blood and plasma concentrations from the administered dose. Reverse toxicokinetic models are needed for QIVIVE to go from assay concentrations to the administered dose. Numerous physiologically based toxicokinetic models (PBTK) models are available, but few are amenable to high throughput due to the large number of required input parameters. Moxon *et al.* (2020) have recently proposed a tiered approach for implementing PBPK models in higher throughput.

High-throughput toxicokinetic (HTTK) models use experimental *in vitro* or predicted data on the hepatic clearance of the chemical of interest and its plasma protein binding to predict the plasma concentration in humans using a simple three-compartment toxicokinetic model composed of liver, gut and the rest of the body (Sipes *et al.*, 2017). Comparison with *in vivo* data has confirmed the suitability of these HTTK models (Wambaugh *et al.*, 2018). HTTK modelling is thus able to provide exposure assessment in NGRA.

9.3.4.3 Case studies on next-generation risk assessment

There are already some early examples of NGRA emerging. A case study on dermal risk of coumarin in personal care products relied on two assay panels covering nuclear receptors and cell stress, an immunomodulary screening assay, genotoxicity assessment with ToxTracker and HTTr (Baltazar *et al.*, 2020). Using this comprehensive first-tier assessment, POD distributions of nominal concentrations were derived from the various HTS assay panels applied. Interestingly, despite coumarin showing extensive metabolism, the PODs from metabolically competent and incompetent cells did not show substantial differences. Different exposure scenarios assuming application of face cream or body lotion that contained the products using the PBPK model by Moxon *et al.* (2020) yielded C_{max} plasma concentrations that were two to three orders of magnitude lower than the most sensitive HTS endpoints, the enzymatic assay of carbonic anhydrase type 1 and a HT transcriptomics assay with a HepaRG cell line. The authors concluded that NGRA was a valuable addition to existing risk assessment for this compound and confirmed that there was a substantial margin of safety and hence minimal concern considering the investigated exposure scenarios (Baltazar *et al.*, 2020). They also stated that 'there is not yet agreement on how large a margin of safety derived in an NGRA needs to be to assure human safety', emphasising that these first case studies are not an end but an encouraging start of new developments.

9.3.5 Applications of new approach methods for environmental risk assessment

Although HTS assays of Tox21 and ToxCast are primarily targeting human health endpoints, they have also great potential for applications in monitoring and ecological risk assessment (Schroeder *et al.*, 2016). Given that the cellular toxicity pathways are quite conservative and applicable to many biological organisms, it seems conceivable to apply HTS data in ecological risk assessment in a similar way as for human health risk assessment.

Applications for environmental monitoring are described throughout this book. The Attagene multiplex assays were actually applied for water quality monitoring (Escher *et al.*, 2014, Section 13.2.2) before being included as one of the core assays in Tox21. The GeneBLAzer battery came into practical applications in water quality monitoring at about the same time as it was implemented in Tox21. *In vitro* bioassays have a longer tradition in water quality testing, but with the parallel development of HTS assays, traditional assays have been adapted to higher throughput. The available data from single chemical screening has helped justify the choice of bioassays and anchored them better in toxicity pathways and AOPs. The abundance of HTS data that is now publicly available can be used for mixture modelling and iceberg modelling, linking chemical analysis to *in vitro* bioassays as described in more detail in Chapter 13.

There is also a desire to replace animal testing in risk assessment. The HTS assays may well serve this purpose. As outlined in Figure 9.8, we can use HTS data to train predictive models and use the data and the models for prediction of *in vivo* effect concentrations (Villeneuve *et al.*, 2019). As for human health we need to apply QIVIVE models and integrate not just one endpoint but a whole battery of HTS assays to integrate multiple pathway perturbations and link them ideally to established AOPs. Cross-species extrapolation can then draw on established models, but species-specific QIVIVE models might also be able to shed light on the mechanistic differences in species sensitivity of chemical stressors.

Similar to QIVIVE for humans, one can envision an approach that applies the same principles to relate effect concentrations *in vitro* to critical concentrations in fish plasma by assuming that the freely dissolved concentration in the *in vitro* assay equals the unbound concentration in the fish plasma (Figure 9.9). Then reverse bioconcentration modelling can translate the plasma concentration back to external exposure concentration.

Finally, quantitative AOPs (qAOP) can be developed as is illustrated in Figure 9.10 for an AOP in female fathead minnows that is triggered by the MIE inhibition of the cytochrome P450 19A aromatase (Conolly *et al.*, 2017). Three quantitative prediction models were developed: the first describes the KE happening on hypothalamic–pituitary–gonadal axis with aromatase inhibition decreasing the transformation of testosterone to 17β-estradiol (E2), which leads to a reduced synthesis of the egg yolk protein precursor vitellogenin; the second

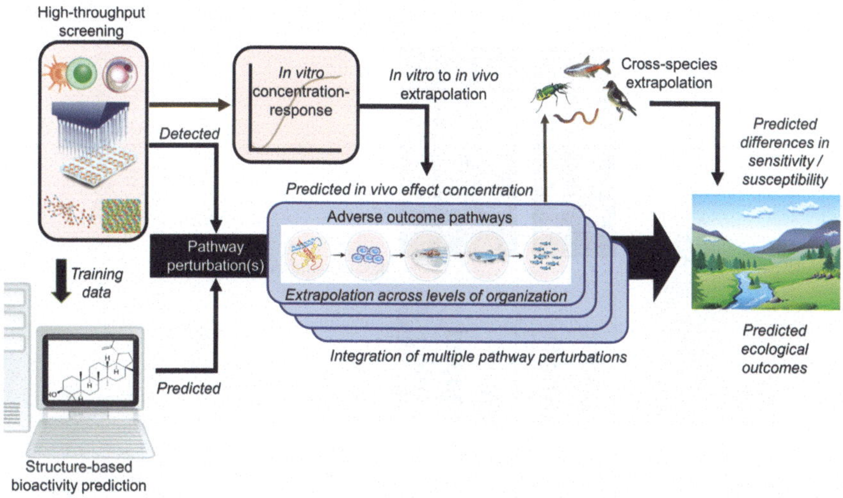

Figure 9.8 Potential use of HTS data and integration for environmental risk assessment. Reprinted with permission from Villeneuve *et al.* (2019). High throughput screening and environmental risk assessment – state of the science and emerging applications. *Environmental Toxicology and Chemistry*, **38**(1): 12–26. © 2019. John Wiley and Sons.

describes the resulting reduction in fecundity expressed as eggs produced per day; which in the third eventually leads to a declining population.

This qAOP model is not chemical-specific and could be translated to other untested aromatase inhibitors (Conolly *et al.*, 2017). The strong quantitative links

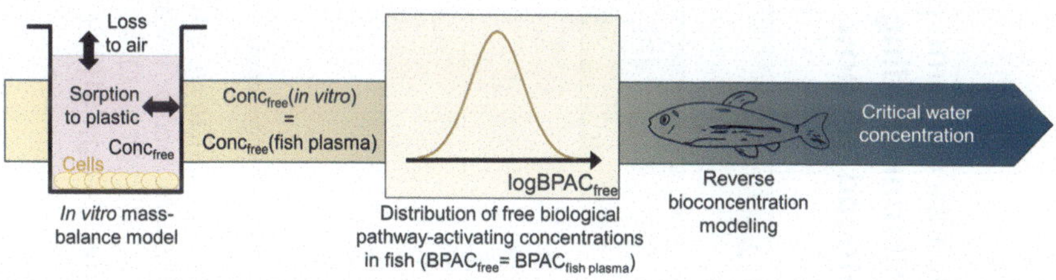

Figure 9.9 Extrapolating *in vitro* concentration–response to equivalent *in vivo* effect in fish analogously to the QIVIVE model for Judson *et al.* (2011) presented in Figure 9.4. Plasma concentrations in fish estimated based on the free fraction of the active chemical concentration in the assay test well. Reverse bioconcentration modelling can then be used to estimate the water concentration that would yield the equivalent internal dose. Modified and reprinted with permission from Villeneuve *et al.* (2019). High throughput screening and environmental risk assessment – state of the science and emerging applications. *Environmental Toxicology and Chemistry*, **38**(1): 12–26. © 2019. John Wiley and Sons.

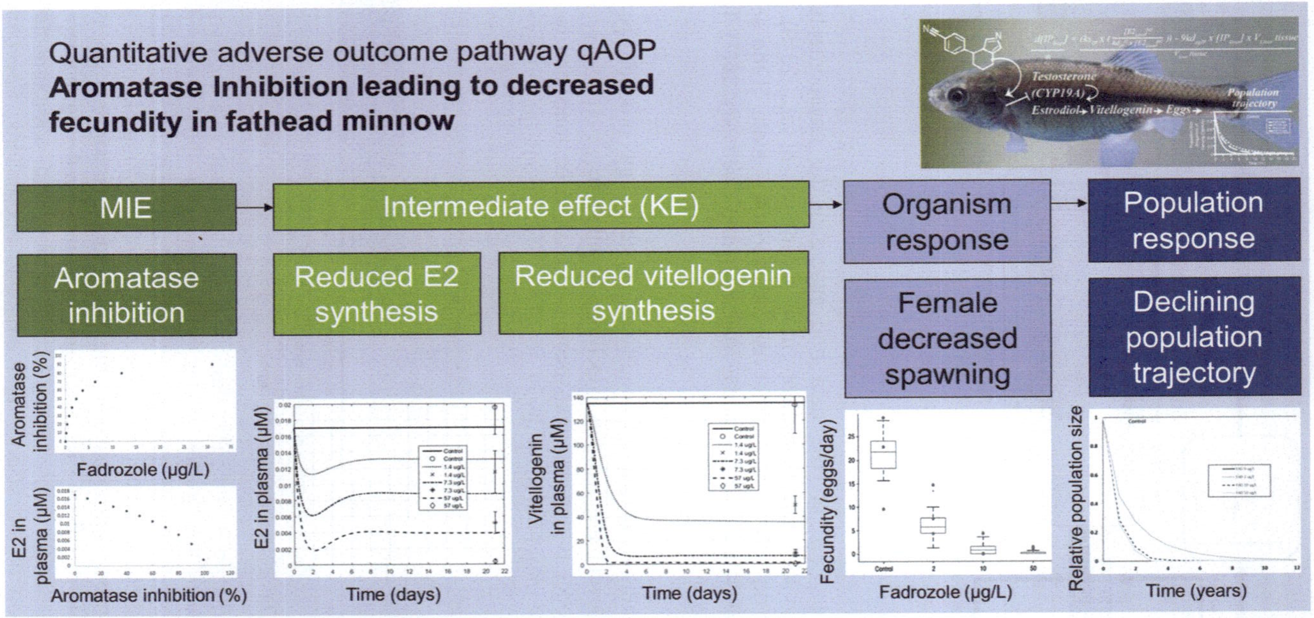

Figure 9.10 Development of a quantitative adverse outcome pathway (qAOP) in fathead minnow on the example of the anti-breast cancer drug fadrozole. MIE = molecular initiating event; E2 = 17β-estradiol; KE = key event. Reprinted after modification with permission from Conolly *et al.* (2017). Quantitative adverse outcome pathways and their application to predictive toxicology. *Environmental Science & Technology*, **51**(8): 4661–4672. © 2017. American Chemical Society.

between KE and adverse outcome give us confidence in using MIE- and KE-based *in vitro* assays in risk assessment and regulation.

9.4 EXPOSURE IN *IN VITRO* BIOASSAYS

When using *in vitro* data for human health risk assessment, the limitations of *in vitro* systems need to be clearly understood and stated to achieve a meaningful risk assessment. One of the most critical issues is exposure.

Cell-based *in vitro* assays are mostly performed in 96- or 384-well plates (Figure 9.11), although the ultra-HTS 1536-well plate format is now also used in robotic platforms such as the Tox21 programme. Most cells adhere to the bottom of the well, with a few cell lines floating freely in suspension. Medium containing serum proteins and other nutrients is added to the well to ensure that sufficient nutrition is available for sustained cellular growth. Only a portion of chemicals added to the well is taken up by the cells. A large fraction is bound to medium components such as serum proteins and lipids. Another fraction of chemicals can be lost due to water–air transfer, and some may sorb to the plastic plate (Figure 9.11).

The chemical fraction that is taken up by cells can be metabolised and/or distributed to target and non-target sites (Figure 9.11). Storage lipids are non-target sites, while macromolecules such as membrane lipids, proteins (enzymes and receptors) and DNA are target sites for the MIE to occur. Both the

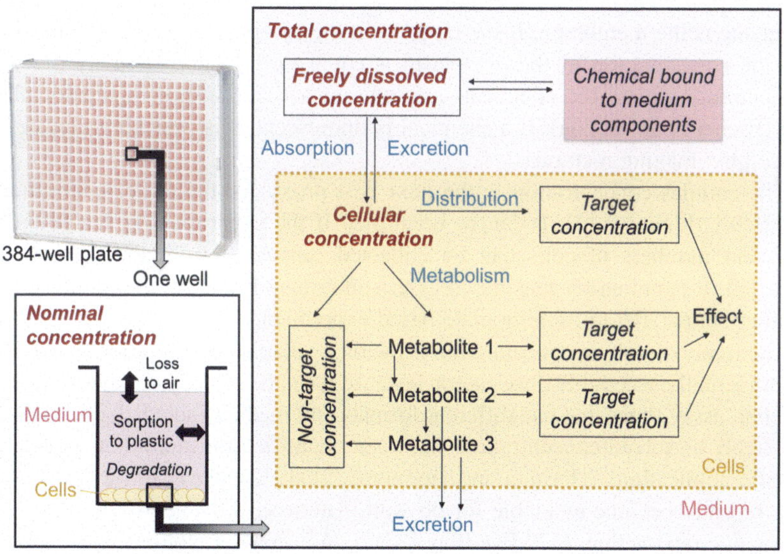

Figure 9.11 Exposure and toxicokinetic processes relevant in a cell-based *in vitro* assay.

initially dosed chemical and its metabolites can contribute to the toxic effect. In some instances, chemicals are activated by metabolism and thereby become more potent than the dosed chemical. More often though, metabolites are more water soluble and hence more easily excreted than the parent chemical. To account for both possibilities, assays with cells of low metabolic capacity are often run once in the presence of an oxidising agent or an isolated liver enzyme mixture (*e.g.*, S9 liver enzyme preparation) and once without this addition. If the observed effects differ, one can conclude that metabolic activation and/or detoxification plays a role. More on metabolism is discussed in Section 9.4.3.

There exist fairly simple mass balance models for stable chemicals that do not degrade in the medium nor are metabolised (Armitage *et al.*, 2014; Fischer *et al.*, 2017). Absorption and excretion process are relatively fast in cells (Fischer *et al.*, 2018a), hence, steady state can be assumed for many chemicals. For very hydrophobic chemicals more than 99% will be sorbed to serum proteins in the medium, whereas very hydrophilic chemicals will remain predominantly in the water phase rather than accumulate in the cell. The dose-metric used to describe exposure in cell assays therefore matters.

9.4.1 Dose-metrics in cell assays

The target concentration is the concentration of a chemical present at the target site, which initiates a response (Table 9.1, Figure 9.11). It is also called the biologically effective dose (BED) or biologically effective concentration. It is almost impossible to measure the target site concentration but there are models to estimate it if the target site is the membrane. If the target is in the cytoplasm, we can often assume that the concentration in the cytoplasm is equal to the external freely dissolved concentration unless the chemicals are ionic or active transport processes play a role (Escher *et al.*, 2020a). If a chemical is fluorescent, it is possible to locate it in the cell by imaging methods.

The cellular concentration is the next best proxy for the target concentration (Table 9.1, Figure 9.11). It can be quantified if the experiment is scaled up and sufficient numbers of cells can be collected, separated from the medium and extracted. It is not easily feasible to assess and quantify the cellular concentration routinely in parallel to running plate-based experiments.

The freely dissolved concentration is the bioavailable concentration and the best experimentally accessible dose-metric because it can be directly compared between different assay types that use different formats and media (Table 9.1, Figure 9.11). The freely dissolved concentrations has been quantified for more than a decade in specific applications (Heringa and Hermens, 2003; Kramer *et al.*, 2012) but has only recently become available for 96-well plate assays thanks to versatile solid-phase microextraction (SPME) fibres with low coating volume (Henneberger *et al.*, 2019; Huchthausen *et al.*, 2020). SPME fibres coated with polymers and C18 that sorb charged chemicals have very recently become available in

Table 9.1 Typical dose-metrics for cell-based bioassays.

Dose-Metric	Definition	Unit	Measurement/Model
Target concentration /BED	Concentration at target site (membrane, cytoplasm, proteins)	$mol/kg_{membrane}$ or $mol/L_{cytoplasm}$	Only modelled, qualitatively with imaging methods
Cellular concentration	Total concentration in the cell	$mol/10^6$ cells	Measured after separation of cells and extraction with solvent
Freely dissolved concentration	Concentration in the surrounding medium that is not bound to proteins	mol/L_{medium}	Measured with solid-phase microextraction (SPME)
Total concentration	Concentration in cells and medium	mol/L_{medium} (volume of cells negligible)	Measured after total extraction with solvent
Nominal concentration	Total amount of chemical divided by the volume of exposure medium	mol/L_{medium}	Calculated from added amount

Adapted from Groothuis *et al.* (2015). BED = biologically effective dose.

multiplexed plate format, allowing in the future more routine and HT quantification of freely dissolved concentrations.

It is common practice to use the nominal concentration in cell-based bioassays, although strictly speaking the total concentration would provide a better measure because it corrects for irreversible loss processes to the air and to the plastic of the plates and degradation in the medium (Table 9.1, Figure 9.11).

Loss to air is a problem in well-plate format that is often underestimated. This process might not only lead to loss of chemicals but semi-volatile chemicals can even cross over and contaminate neighbouring wells (Birch *et al.*, 2019). Only chemicals with a medium–air partition constant $K_{medium/air} > 10^4$ are fully retained during 24 hours of exposure at 37°C (Escher *et al.*, 2019). Chemicals just below this cut-off are the ones likely to contaminate neighbouring wells (Escher *et al.*, 2019). The $K_{medium/air}$ is not only dependent on the Henry constant K_H or the air–water partition constant K_{aw}, which are in turn a function of vapour pressure and solubility, but also dependent on the medium-water partition constant $K_{medium/water}$. Hydrophobic chemicals that bind strongly to proteins and lipids of the medium are better retained in the bioassay well than hydrophilic chemicals with the same K_{aw}. Up to 20% of the Tox21 chemicals might have been partially lost during experiments even with protein-rich medium, for example, when 10% serum is supplemented (Escher *et al.*, 2019), but not all Tox21 assays are run with such protein-rich conditions so losses might have been even higher. Differences in medium composition therefore have an influence on the loss processes in cell assays.

For water quality testing loss to the air is less of a problem as most water pollutants are not volatile. In addition, the sample preparation process involves several blow-down steps so most residual volatile chemicals would be removed prior to bioassay testing in any case.

In contrast, sorption to the plastic plate materials (mostly polystyrene) is a problem that is often overrated under standard cell assays conditions. Organic chemicals do indeed have rather high polystyrene–water partition constants $K_{PS/w}$, but still about one to three orders of magnitude smaller than the corresponding octanol–water partition constant K_{ow} (Fischer *et al.*, 2018b). Given the large mass-to-volume ratio of polystyrene, the loss would be huge if equilibrium were attained – but the diffusion coefficients of organic chemicals in polystyrene are very small. Therefore, the chemicals hardly have the time to partition into the polystyrene during the duration of a typical HTS cell assay, but instead sorb to the surface and penetrate only a few micrometres. Nevertheless, the loss due to binding to the plate material is very much dependent on the test conditions. The fish embryo assay, for instance, is typically conducted in pure water without supplements and the losses to a polystyrene 24-well plate are huge for chemicals of high hydrophobicity (high K_{ow}) and not negligible for chemicals of medium hydrophobicity ($3 < \log K_{ow} < 4$) (Figure 9.12a). It is therefore not

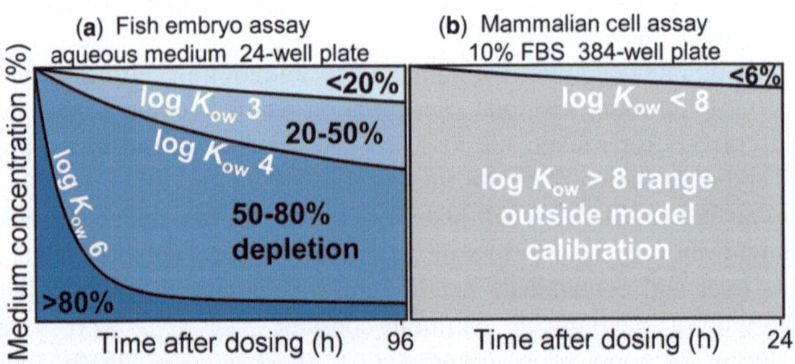

Figure 9.12 Depletion of the medium concentration by sorption to the polystyrene of a well-plate assay. (a) Substantial and chemical-dependent depletion in a fish embryo toxicity assay during 96 h in a 24-well plate. (b) Negligible loss in a mammalian cell assay in medium supplemented with 10% fetal bovine serum (FBS) during 24 h in a 384-well plate. K_{ow} = octanol–water partition constant. Reprinted with permission from Fischer *et al.* (2018b). Application of experimental polystyrene partition constants and diffusion coefficients to predict the sorption of organic chemicals to well plates in *in vitro* bioassays. *Environmental Science & Technology*, **52**: 13511–13522. © 2018. American Chemical Society.

recommended to run the fish embryo assay in polystyrene plates: glass or glass-coated plates reduce the loss to the plate material. In contrast, a cell assay in medium supplemented with 10% fetal bovine serum will hardly be affected by sorption to the polystyrene because the medium proteins and lipids retain the chemical and make is less available for sorption to the well plates (Figure 9.12b). There is no hydrophobicity-dependence of the loss because the partitioning to medium components and to polystyrene are proportional to each other. If serum-free medium is used, the stabilising effect of the medium is much smaller.

Finally, the nominal concentration is the concentration calculated from the amount of chemical added to the assay (Table 9.1, Figure 9.11) and is the one reported in most publications and databases, including Tox21 and ToxCast. It is reasonable to use nominal concentrations for relative comparisons as long as all exposure conditions, number of cells, and medium composition remain constant. Hence, appropriate QA/QC (Chapter 11) is vital when using nominal concentrations as the dose-metric.

9.4.2 Serum-mediated passive dosing

The medium components stabilise the system and reduce possible loss processes. The cellular uptake does not lead to depletion because the freely dissolved concentration can be replenished by desorption from the medium proteins and lipids. This process has also been termed serum-mediated passive dosing (SMPD, Fischer *et al.*, 2019). SMPD can be somewhat deceptive as it also decreases the sensitivity of the bioassay because a lower fraction of total chemicals is available for cellular uptake. The cell concentration at steady state cannot exceed the partition constant between cell and water, and hence the cellular concentration is much smaller than in serum-free medium.

The relative effect potency (REP) derived from nominal concentrations is actually much closer to the REP for cellular concentrations than if derived from freely dissolved concentrations (Fischer *et al.*, 2017). This is intuitively easy to understand because partitioning between cells that are made up of proteins and lipids and medium that is made up of proteins and lipids is fairly uniform across a wide range of chemicals (Jahnke *et al.*, 2016). The freely dissolved fraction is dependent on the hydrophobicity of the chemicals with very low freely dissolved concentrations of hydrophobic chemicals versus much higher freely dissolved concentrations of hydrophilic chemicals at the same nominal concentration (Figure 9.13). For chemicals with log $K_{ow} > 3$, the cellular concentration is lower but directly proportional to the nominal concentration; even for chemicals with log K_{ow} below 3, the decrease of cellular concentration is not excessive.

Serum-mediated passive dosing works well in practice for water quality testing, particularly considering that a water extract contains thousands of chemicals. Thus, solubility of individual mixture components rarely poses a problem, especially

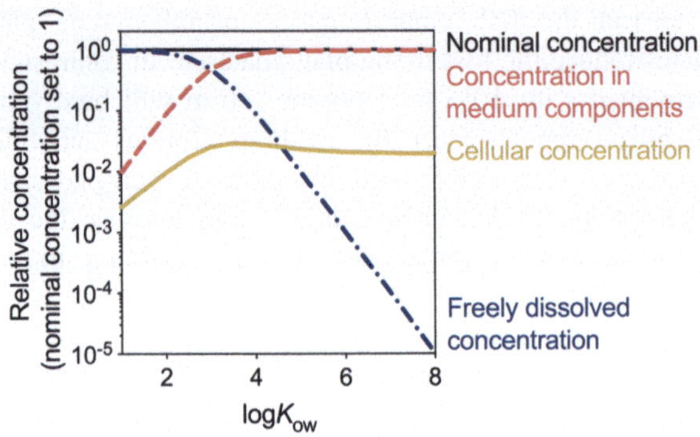

Figure 9.13 Relative concentrations at concentration nominal concentration in medium, cells and freely dissolved. Calculated with a mass balance model for 56 000 cells in 100 μL medium supplemented with 10% FBS (Armitage *et al.*, 2014). Reprinted with permission from Jahnke *et al.* (2016). Strategies for transferring mixtures of organic contaminants from aquatic environments into bioassays. *Environmental Science & Technology*, **50**(11): 5424–5431. © 2016. American Chemical Society. K_{ow} = octanol–water partition constant.

given that a water extract contains few very hydrophobic but many hydrophilic and many charged chemicals.

While media containing proteins and lipids stabilise the bioassay system, the disadvantage is that if the composition of the medium changes and cell numbers are variable, then the nominal concentration will not be directly related to the biologically effective cellular concentration. Therefore, it is very important to characterise the media used to ensure as little variability as possible and to use the same assay components and protocols across all water sample types and single chemicals.

If the serum is replaced by non-animal alternatives, the impact of lowering the sorptive capacity of the medium needs to be accounted for. Likewise, for serum-free set ups of cell assays, cell-free assays and whole organism assays with invertebrates and fish embryos, dosing and exposure assessment needs to be carefully set up to avoid any artefacts due to depletion of the sample. In these circumstances, passive dosing from a polymer might be a viable alternative. In passive dosing, the chemical or sample is partitioned into a polymer. The loaded polymer, once inserted in the bioassay well, slowly releases the chemical to the medium from which it can be taken up by the cells (Smith *et al.*, 2015; Smith and Schäfer, 2016). In case of adherent cells, the set-up might be modified, that the cells do not touch the passive dosing device, for example, by growing the cells on inserts hung into the dosing plate (Kramer *et al.*, 2010). Passive dosing is especially important for assays like the 96-h fish embryo toxicity assay, where even in glass vials the uptake of chemical into the organism might lead to depletion of the aqueous concentration (Seiler *et al.*, 2014; Vergauwen *et al.*, 2015). Passive dosing may also overcome solubility limitations when testing hydrophobic chemicals (Smith *et al.*, 2010).

9.4.3 Metabolism in cell-based bioassays

Toxicokinetics are more complex in tissues, organs and whole organisms than in cells. The liver has the most metabolically active cells. However, the principles that govern an organism's toxicokinetics and the processes of cellular uptake are the same and the liver cell line HepG2 is often used to simulate liver metabolism. The liver can generate circulating metabolites and this process can be simulated *in vitro*. In addition, metabolites can be formed in the medium by adding a liver enzyme mixture (*e.g.*, S9 fraction) to the cell medium of *in vitro* assays to simulate hepatic metabolism (Figure 9.14, extracellular approach). Alternatively, if cells express metabolising enzymes, metabolites can also be generated inside the cell (Figure 9.14, intracellular approach). Not all cells lines are metabolically active. Cancer-cell derived reporter gene assays are often considered to have limited metabolic capacity, but this really depends on the cell type. For instance, the HepaRG cell line derived from a human hepatocellular carcinoma has very active metabolic enzyme systems constitutively expressed (Guillouzo *et al.*, 2007). Exposure to chemicals might also induce metabolic enzymes in reporter gene assays that are not constitutively expressed, which has to be considered in these approaches (Fischer *et al.*, 2020). It is also possible to retrofit cells with metabolising enzymes which has potential to make *in vitro* assays more realistic (Thomas *et al.*, 2019).

9.5 BASELINE TOXICITY AND SPECIFICITY OF RESPONSE

Baseline toxicity is the minimum effect any chemical elicits and can therefore serve as an anchor to relate effects to (Chapter 4). Baseline toxicants were shown to have an average constant membrane concentration of 69 $\text{mmol/L}_{\text{membrane}}$ (95% confidence interval 49–89) across several cell lines (Escher *et al.*, 2019). Mass balance models can be used to translate this critical membrane concentration back to the nominal inhibitory concentration that reduces cell viability and growth by 10% $IC_{10,nom}$ (Equation 9.1).

$$\mathrm{IC}_{10,\mathrm{nom}} = 69\,\mathrm{mM}\cdot\frac{\mathrm{Vf}_{\mathrm{lipid,cell}}}{\mathrm{f}_{\mathrm{membrane,cell}}}\cdot\left(\frac{\mathrm{V}_{\mathrm{cell}}}{\mathrm{V}_{\mathrm{medium}}+\mathrm{V}_{\mathrm{cell}}}+\frac{K_{\mathrm{medium/water}}}{K_{\mathrm{cell/water}}}\right) \quad (9.1)$$

The only determinants going into this equation are the volume fraction of the lipids in the cell $\mathrm{Vf}_{\mathrm{lipid,cell}}$, the fraction of chemicals in the membranes of the cell $\mathrm{f}_{\mathrm{membrane,cell}}$, the volume of cell $\mathrm{V}_{\mathrm{cell}}$ and medium $\mathrm{V}_{\mathrm{medium}}$ and the partition constants between cell and water $K_{\mathrm{cell/water}}$ and medium and water

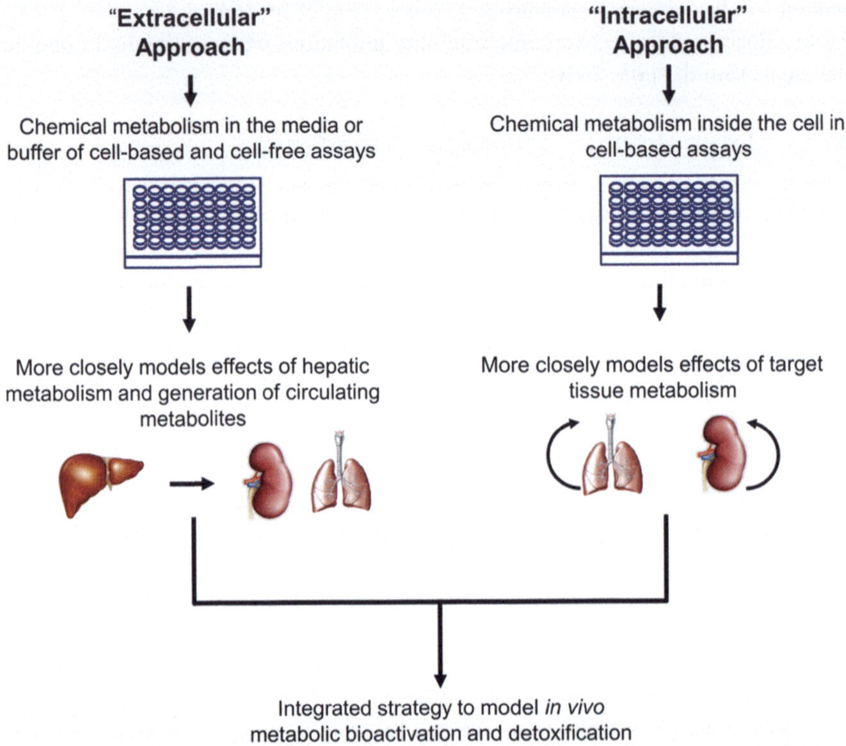

Figure 9.14 Integrated strategy to model *in vivo* bioactivation and detoxification in a diverse range of *in vitro* assays using either the extracellular or the intracellular approach. Reprinted with permission from Thomas *et al.* (2019). The next generation blueprint of computational toxicology at the U.S. Environmental Protection Agency. *Toxicological Sciences*, **169**(2): 317–332. Published by Oxford University Press on behalf of the Society of Toxicology. This work is written by U.S. Government employees and is in the public domain in the U.S.

$K_{medium/water}$. Details of the derivation of this mass balance model and typical cell parameters are given in Escher *et al.* (2019). The $IC_{10,nom}$ increases with increasing hydrophobicity and depends on the serum content (Figure 9.15a). The empirical relationship between log K_{ow} and $\log(1/IC_{10,nom})$ was previously fitted as a linear regression in the range $0.6 < \log K_{ow} < 4.3$ (Escher *et al.*, 2019). Experimental data agreed rather well with the non-linear relationship predicted by the mass balance model, which extends the applicability range to higher hydrophobicity.

Any specifically acting chemical has cytotoxic concentrations below baseline toxicity. The toxic ratio (TR) is a measure of how much more cytotoxic a chemical is compared to its baseline toxicity (Equation 9.2, Figure 9.15b).

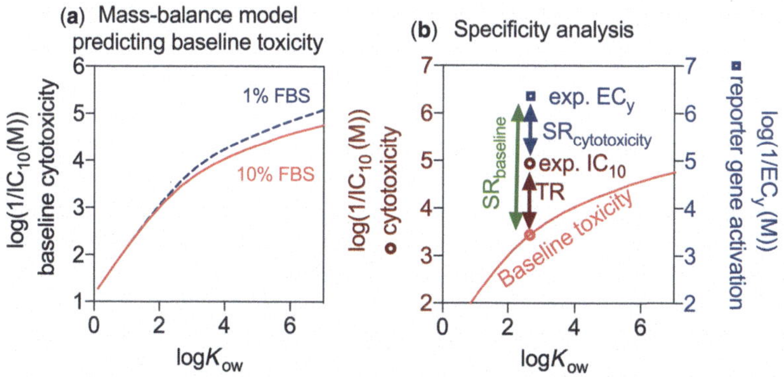

Figure 9.15 (a) Relationship between hydrophobicity expressed as octanol–water partition constant log K_{ow} and nominal inhibitory concentration that reduced cell viability and growth by 10% log $IC_{10,nom}$ predicted with the mass balance model described in Escher *et al.* (2019). (b) Derivation of the specificity ratio $SR_{cytotoxicity}$ (Equation 9.3) and $SR_{baseline}$ (Equation 9.4) from the experimental cytotoxicity IC_{10} and the predicted baseline cytotoxicity IC_{10} and the measured effect concentration for effect y, EC_y. Effect y may refer to 10% or induction ratio IR 1.5.

A TR $\geq$ 10 is associated with specific or reactive toxicity (Maeder *et al.*, 2004).

$$TR = \frac{IC_{10,\text{baseline}}}{IC_{10}} \quad (9.2)$$

Cytotoxic chemicals with a high TR are often reactive chemicals or chemicals where the MIE or KE detected by the reporter gene assays cannot be differentiated from the final adverse outcome at the cellular level because the effect leads directly to the adverse outcome of reduced cell viability. Chemicals with a high TR also often exhibit multiple modes of action. Hence a TR analysis, although hardly ever performed, could be very important for the interpretation of bioassay results. It makes a difference if a chemical is toxic because it is just hydrophobic (baseline toxicity) or if it exhibits specific or reactive toxicity on the apical endpoint of cytotoxicity.

We define the specificity ratio (SR) as the ratio between the experimental cytotoxicity (IC_{10}) and effect concentration (EC_{10} or $EC_{IR1.5}$) (Escher *et al.*, 2020b). As is illustrated in Figure 9.15b, the specificity ratio can either relate to the experimental IC_{10} ($SR_{cytotoxicity}$, Equation 9.3) or to the predicted $IC_{10,\text{baseline}}$ ($SR_{baseline}$, Equation 9.4).

$$SR_{cytotoxicity} = \frac{IC_{10}}{EC_{10}} \quad \text{or} \quad SR_{cytotoxicity} = \frac{IC_{10}}{EC_{IR1.5}} \quad (9.3)$$

$$SR_{baseline} = \frac{IC_{10,\text{baseline}}}{EC_{10}} \quad \text{or} \quad SR_{baseline} = \frac{IC_{10,\text{baseline}}}{EC_{IR1.5}} \quad (9.4)$$

$SR_{baseline}$ is a measure of the specificity of a chemical acting in a given bioassay. The higher $SR_{baseline}$ the more specific the chemical. We will use the $SR_{baseline}$ in Chapter 13 to classify bioassays and to derive effect-based trigger values.

$SR_{cytotoxicity}$ is a measure of the selectivity of a chemical in a given bioassay. If all chemicals tested have a low selectivity, the assay has little value for evaluation of chemicals and water samples. Reporter gene assays for nuclear receptors related to endocrine effects often have very high selectivity for individual chemicals. What kills the cell is not estrogenicity but genotoxicity, so the reporter gene and the cause of cytotoxicity are not directly connected.

Assays for adaptive stress response are generally less selective because the dysfunction and induction of the adaptive stress response is eventually what kills the cell (*e.g.*, due to excess reactive oxygen species). An extreme example of low selectivity is a reporter gene assay indicative of the activation of the p53 pathway, which is an adaptive stress response to DNA damage that triggers repair, cell cycle arrest and apoptosis. Chemicals that activate p53 are often cytotoxic with a high TR. This also has an impact on activation of the reporter gene. The gap between activation of p53 and cytotoxicity is small, even for single reactive chemicals (Stalter *et al.*, 2016a). When testing disinfected water samples, the activation was in most cases masked by cytotoxicity (Yeh *et al.*, 2014) making this assay unsuitable for water quality testing.

If $SR_{baseline}$ is high but $SR_{cytotoxicity}$ is low, the effect is mainly due to cytotoxicity, not the specific endpoint. If both $SR_{baseline}$ and $SR_{cytotoxicity}$ are low, the chemical is a baseline toxicant with respect to cytotoxicity and not specifically acting. In both cases, when the specific effect occurs at concentrations close to cytotoxicity, the cytotoxicity burst phenomenon can lead to expression of specific endpoints, which are then just an experimental artefact (Judson *et al.*, 2016; Fay *et al.*, 2018; Escher *et al.*, 2020c).

The combination of experiments and modelling helps to better understand the responses that can be measured with *in vitro* assays and can scrutinise effect data to be included in mixture models or other interpretation frameworks.

9.6 PRACTICAL CONSIDERATIONS FOR DOSING OF CHEMICALS

The considerations on baseline toxicity and specificity can help us also in practice. When planning an experiment with single chemicals or defined mixture of chemicals, the information in this section can be synthesised to plan a successful experiment. Our recommendation is to plan the highest concentration at maximum solubility in the medium and to consider the implications of baseline toxicity. The medium solubility S_{medium} can be calculated from Equation (9.5) (Fischer *et al.*, 2019).

$$S_{medium} = S_{water} \cdot K_{medium/w} \tag{9.5}$$

The aqueous solubility S_{water} can be retrieved from databases such as the CompTox Chemistry Dashboard (U.S. EPA, 2020) and the medium–water partition constant $K_{medium/w}$ (Equation 9.6) can be approximated from the protein–water partition constant $K_{BSA/w}$ and the liposome–water partition constant $K_{lip/w}$ and the fraction of FBS added to the medium β_{FBS}. Evidently, the composition of FBS is somewhat variable with respect to protein and lipid content but to simplify the Equation (9.6) we have assumed an average composition (Fischer *et al.*, 2019). In turn, log $K_{BSA/w}$ (Equation 9.7) and log $K_{lip/w}$ (Equation 9.8) can be simply predicted from the log K_{ow} for neutral chemicals.

$$K_{medium/w} = 0.046 \cdot \beta_{FBS} \cdot K_{BSA/w} + 0.0015 \cdot \beta_{FBS} \cdot K_{lip/w} + 0.9525 \cdot \beta_{FBS} + (1 - \beta_{FBS}) \tag{9.6}$$

$$\log K_{BSA/w} = 0.71 \cdot \log K_{ow} + 0.42 \tag{9.7}$$

$$\log K_{lip/w} = 1.01 \cdot \log K_{ow} + 0.12 \tag{9.8}$$

With these equations it is possible to build a fairly robust estimate of medium solubility, which is much higher than the aqueous solubility for hydrophobic chemicals (Fischer *et al.*, 2019).

One must also consider the conditions under which the medium can act as a chemical reservoir. The depletion of the freely dissolved concentration should be <5% over the 24 hours duration of the assay. If we consider non-volatile and stable chemicals and the main loss being binding to the polystyrene of the well plate, then we need between 3% and 10% FBS as a minimum to assure non-depletive conditions (Table 9.2).

We illustrate the concept on the example of serum-supplemented medium in Table 9.2. There is a trend of replacing FBS with animal-free products to make *in vitro* assays truly free of animal-derived products. If such alternative media are used, they should be chemically defined to assure consistent outcomes and synthetic proteins could be added to the desired level of retaining capacity.

A final consideration is to test up to baseline toxicity. Remember that the baseline toxicity IC_{10} can be calculated using Equation (9.1) and estimated from the log K_{ow} in Figure 9.15a. As baseline toxicity is the minimum toxicity every chemical has, baseline toxicity occurs at the highest possible concentration. Reporter gene

Table 9.2 Standard set-up of an HTS *in vitro* cell assays that assures that depletion of the freely dissolved concentration is < 5% over the 24-h assay duration (Fischer *et al.*, 2019).

Plate format	96-well	384-well	1536-well
Medium volume (μL)	120	40	6
Cell number	10 000	5000	2000
Required % FBS	≥3	≥5	≥10

activation is only valid if it occurs at a lower concentration than baseline toxicity and any specific cytotoxic effect also occurs at lower concentrations (Escher *et al.*, 2020c). Hence, baseline toxicity should always be predicted for individual chemicals, compared with medium solubility and ideally used as the highest test concentration. More practical recommendations for dosing can be found in Fischer *et al.* (2019).

9.7 CONCLUSIONS

In this section we have explored the exciting advances in HTS assays over the last 10 years – a new chapter that did not exist and could not have been told in the first edition of this book.

The new European chemicals legislation REACH, ToxCast and Tox21 in the U.S., and researchers and agencies worldwide have made it possible to do what was considered impossible two or three decades ago: assessing the risk of chemicals to the environment and to human health based on *in vitro* methods and *in silico* models. This is only possible because of sound scientific work anchored against classical animal-based methods for validation purposes. We have outlined the first applications to give an idea of what lies ahead in the exciting field of NGRA.

We are not at the end of the story but still at the beginning. Despite breath-taking advances there remain a lot of scientific questions to be solved. The first generation of HTS assays, mainly receptor binding assays and reporter gene assays, have limitations and new tools are emerging that improve individual assays but also go in entirely new directions such as the assays using HTTr and HT phenotyping.

A downside of HTS relates to limitations of dosing and exposure assessment. SPME methods to quantify the freely dissolved concentration as proxy of the BED are at present limited to single chemicals and defined mixtures. There is always a trade-off between exactly characterising the dosing in bioassays versus running hundreds of samples in screening mode. Since we will never know the concentrations of all chemicals in a water extract, it is futile to aim and characterising exposure concentrations in cell-based assays for water extracts. Instead, we have to ensure that there is a robust relationship between the nominal and the cellular dose.

The progress in applying HTS for risk assessment also paved the way for applications in monitoring. *In vitro* assays have long been used for water quality monitoring, but the stronger scientific underpinning will help their acceptance in regulation in the future.

Chapter 10

Current bioanalytical tools for water quality assessment

10.1 INTRODUCTION

In vitro bioassays that respond to relevant initiating triggers in a toxicity pathway or are linked to a known mode of toxic action with a defined health outcome have the potential to be useful tools for water quality assessment. Some assays are, however, not suitable for screening of water samples. Assay robustness and specificity in the presence of water matrix components and other chemicals must be characterised and validated prior to their implementation as monitoring tools. This chapter reviews *in vitro* bioassays and well plate-based *in vivo* assays that have already been applied for water quality monitoring in drinking water, surface water, wastewater and recycled water (up to September 2020). Specifically, we focus on assays adapted to high-throughput screening (*e.g.*, run in a 96-well or 384-well plate) because high throughput is essential for large numbers of samples in routine water quality monitoring. Furthermore, we only considered studies where water samples were extracted by solid-phase extraction (SPE), passive sampling or liquid–liquid extraction (LLE), rather than testing unenriched or native water samples, which is covered in Chapter 3 on whole effluent toxicity testing. Native water samples may contain metals, salts and other inorganics, in addition to micropollutants, meaning that the response in a native water sample cannot be attributed to organic micropollutants alone. It should be noted that despite our best efforts this list may not include all deployed assays, but it should nevertheless provide a good perspective on the types of assays currently in use for water quality monitoring. Assays that have so far been limited to chemical

Authors: Beate Escher, Peta Neale and Frederic Leusch
doi: 10.2166/9781789061987_0169

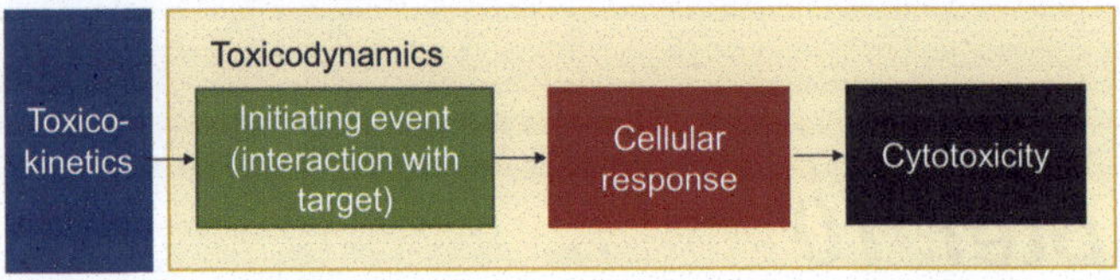

Figure 10.1 Toxicity pathway in a cell (modified from Chapter 4).

risk assessment of individual compounds or mixtures of known composition were excluded.

As explained in Chapter 4, cells can be thought of as models of whole organisms that simulate many crucial processes. Using a cell system ensures, for example, that relevant toxicokinetic processes are accounted for in the bioassay. Cellular responses can then be visualised along the three steps of the cellular toxicity pathway, from the molecular interaction to the cellular effect (Figure 10.1).

Some bioassays are capable of detecting the initiating event, that is, the interaction of the micropollutant with its biological target. The initiating event is considered a measure of potential effect, as later repair and defence mechanisms may be able to prevent an adverse outcome. From a precautionary risk perspective, the potential to do harm is an important assessment endpoint. Often, however, it is not possible to measure the initiating event directly. In such cases, a later step in the cellular toxicity pathway can be quantified as an assay endpoint, either using a natural biomarker or a reporter gene product. Finally, cytotoxicity (cell death or decreased growth) can be quantified as an integrative parameter for all toxicity pathways in a cell (Figure 10.1). In addition to *in vitro* assays, well plate-based *in vivo* assays can capture effects from multiple cellular toxicity pathways resulting in an adverse outcome measured by apical effects such as mortality, growth and development impairment (Wernersson *et al.*, 2015).

Bioassays based on different stages of cellular toxicity pathways including induction of xenobiotic metabolism, receptor-mediated effects, adaptive stress responses and apical effects have been widely applied to evaluate different water extracts (e.g., Escher *et al.*, 2014; Rosenmai *et al.*, 2018; Alygizakis *et al.*, 2019). Many of the bioassays included have also been applied to support health risk assessment of chemicals (Wetmore, 2015; Bell *et al.*, 2018) and this information is useful to estimate how the *in vitro* endpoints are connected to health consequences. However, it must be stressed that the use of *in vitro* bioassays to water samples does not, at this stage, allow prediction of likely health risk.

10.2 PRINCIPLES OF CELL-BASED BIOASSAYS

The cells used in bioanalytical tools can be primary cells or immortalised cell lines. Primary cells are excised from living organisms (*e.g.*, primary hepatocytes are isolated from liver tissue) and are usually more representative of 'natural' cellular function in the organism. Primary cells have a limited lifespan *in vitro* and eventually stop dividing. Not only does this reduce the ethical benefit of using bioanalytical tools, but it also increases the variability of the results as different batches of cells may be sourced from different individuals. Immortalised cell lines, on the other hand, have been either accidentally or deliberately mutated to proliferate indefinitely. This means that a constant supply of identical cells can be produced to ensure minimal variability between different experiments using the same cell lines. Due to mutations, gene expression can change over time even in immortalised cells. It is therefore important to keep the passage number (an indicator of the age of the cell culture) within 40–80 passages from the source cell, depending on the cell type.

A variety of cells and cell lines can be used in cell-based bioassays. Almost any organism can be used, including humans, other mammals such as mice and monkeys, simple eukaryotes such as plants and yeast, or prokaryotic organisms such as bacteria. Higher forms of life are complex multi-cellular organisms with many different cell types, many of which can be cultured *in vitro* and used as models for cell-based bioassays.

Growth rate and cell viability can be determined in all cell-based bioassays as a measure of non-specific cytotoxicity. The measurement is carried out either by direct counting of cells using a specialised cell counter, a flow cytometer, a haemocytometer, microscope imaging methods or through indirect measurement of cellular activity such as metabolic and mitochondrial activity, active transport mechanisms and/or cell membrane permeability.

In some instances, the initiating event causes a specific detectable cellular response in native cells. This is called a biomarker and can be a change of state or the production of a specific chemical or protein. The production of vitellogenin in fish liver cells upon exposure to estrogenic compounds is a typical example of

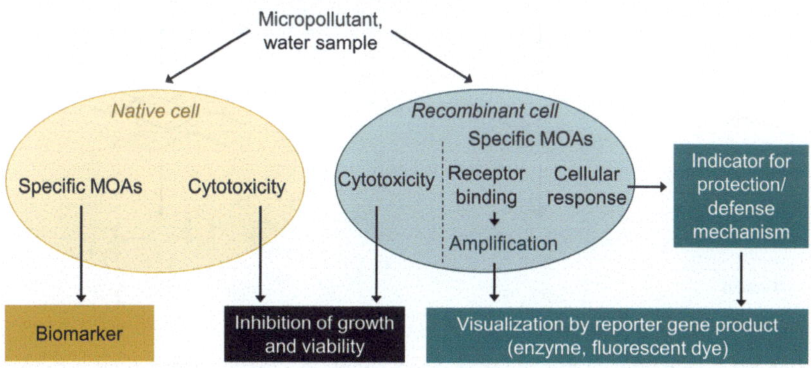

Figure 10.2 Types of cell-based assays and assessment endpoints. MOA = mode of action.

a biomarker. Often the native response is difficult to measure, and cells can be genetically modified into recombinant cells that produce a more visible (and hence measurable) effect in response to the initiating event (Figure 10.2).

The reporter gene assay is one example of genetic engineering used to enhance our visualisation of the cellular response. In a reporter gene assay, a gene encoded with an easily detectable product such as a fluorescent protein or enzyme (*e.g.*, luciferase or β-galactosidase) is paired with a promoter region specific for the mode of action to be investigated. When the initiating event triggers the cellular response, the reporter gene is transcribed to messenger RNA, which is then translated as a fluorescent protein or enzyme that can be measured by fluorescence or enzymatic assays (Figure 10.3).

The production of the reporter product by the cell's genetic machinery is proportional to the induction of the receptor: the stronger the stimulation, the more reporter is produced. In some recombinant cells, multiple copies of both the promoter and the reporter are included to improve sensitivity. A typical example of a reporter gene technique is the AhR CAFLUX assay (Nagy *et al.*, 2002). This assay uses a mouse hepatoma cell (Hepa1c1c7) stably transfected with a plasmid containing enhanced green fluorescent protein (EGFP) as reporter protein downstream of a promoter consisting of four dioxin-response elements (DREs). These recombinant cells produce EGFP upon exposure to dioxin-like compounds, and the amount of EGFP is dependent on the amount of aryl hydrocarbon receptor (AhR) stimulation from the sample. Reporter gene

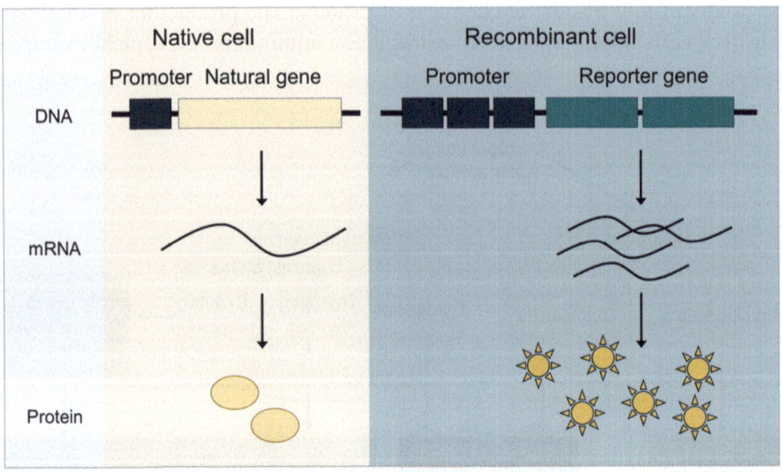

Figure 10.3 Principle of the reporter gene assay. A plasmid containing a reporter gene downstream of the natural gene promoter is inserted into a recombinant cell, and activation of the promoter results in production of a fluorescent reporter protein.

assays indicative of different endpoints are commonly used for water quality assessment; indeed, the majority of assays reviewed in this chapter are reporter gene assays.

10.3 BIOASSAYS INDICATIVE OF XENOBIOTIC METABOLISM

The presence of chemicals can induce biotransformation processes in cells to metabolise, detoxify or in some cases bioactivate chemicals (Omiecinski *et al.*, 2011). Some important xenobiotic metabolism receptors include the AhR, the peroxisome proliferator-activated receptor (PPAR), the pregnane X receptor (PXR) and the constitutive androstane receptor (CAR). Assays indicative of xenobiotic metabolism may not result in cell death, but instead can act as sensitive indicators of the presence of chemicals, with many receptors considered capable of binding a wide range of chemicals. To date, only a few studies have applied assays indicative of CAR to environmental extracts (Escher *et al.*, 2014), with effects detected in wastewater, surface water and drinking water extracts using a yeast-based CAR assay or the Attagene-multiplexed assay (Blackwell *et al.*, 2019). Therefore, this section will focus on assays indicative of the three xenobiotic metabolism receptors commonly applied to water samples: AhR, PPAR and PXR.

10.3.1 Aryl hydrocarbon receptor

The AhR is a ligand-dependent transcription factor that is necessary for virtually all of the toxicity of halogenated aromatic hydrocarbons such as polychlorinated and brominated dibenzo-*p*-dioxins and biphenyls as well as polycyclic aromatic hydrocarbons (PAHs). It activates target genes encoding for the metabolic enzymes CYP1A1, CYP1B1 and NADPH-quinone oxidoreductase (NQO1) but there is also cross-talk with Nrf2, the master regulator of the antioxidant response, and the hypoxia-inducible factor HIF1α. Its activation contributes to carcinogenicity because cytochrome P450 monooxygenase (CYP) can convert many of its ligands to reactive intermediates capable of interacting with and causing DNA damage. Assays indicative of activation of AhR are traditionally applied to detect the presence of dioxin-like chemicals, but recent studies have shown that many environmental chemicals can also activate the AhR pathway (Martin *et al.*, 2010).

Some of the more common reporter gene assays that have been applied to evaluate activation of AhR in water extracts are provided in Table 10.1, along with the reported concentration causing 10% effect (EC_{10}). Generally, AhR CAFLUX (mouse H1.G1.1c3 and rat H4.G1.1c2), AhR CALUX (rat H4L1.1c4) and H4IIE-luc are similarly sensitive to individual chemicals, with EC_{10} values in the low ng/L range for reference compound 2,3,7,8-tetrachloro-*p*-dibenzodioxin

Table 10.1 Common cell-based reporter gene assays applied to evaluate aryl hydrocarbon receptor (AhR) activity in water extracts.

Assay	Cell Line	Detection Method	TCDD EC_{10} (M)	TCDD EC_{10} (ng/L)	EC Reference
AhR CAFLUX	H1.G1.1c3	Fluorescence	8.89×10^{-13}[a]	0.29	König *et al.* (2017)
AhR CAFLUX	H4.G1.1c2	Fluorescence	5.22×10^{-13}[a]	0.19	König *et al.* (2017)
AhR CALUX	H4L1.1c4	Luminescence	5.92×10^{-13}	0.19	Nivala *et al.* (2018)
AhR HepG2	HepG2	Luminescence	6.22×10^{-11}[a]	20	Rosenmai *et al.* (2018)
H4IIE-luc	H4IIE	Luminescence	1.60×10^{-13}	0.05	Lee *et al.* (2015)

[a]EC_{10} value converted from EC_{50} value assuming a slope of the log-logistic concentration response curve of 1.

(TCDD). It should be noted that there are other AhR CALUX cell lines used, for example, H1L6.1c2 (Mehinto *et al.*, 2017) and H4L1.1c2 (Daniels *et al.*, 2018), but these have not been as widely applied as the others to date. The human HepG2-based AhR reporter gene assay is around two orders of magnitude less sensitive than the other reporter gene assays (Rosenmai *et al.*, 2018) and has only been applied in a limited number of studies.

In addition to the assays in Table 10.1, the PAH CALUX assay, which uses H4IIE rat hepatoma cells, has also been applied to wastewater effluent (Alygizakis *et al.*, 2019) and surface water (de Baat *et al.*, 2019a) extracts. The assay reference compound is benzo[*a*]pyrene (EC_{50} 3.0×10^{-9} M (Pieterse *et al.*, 2013)), with all results expressed as benzo[*a*]pyrene equivalent concentrations (B[*a*]P EQ). Other studies have also applied yeast-based activation of AhR assays, such as the yeast dioxin screen (YDS). The reference compound for the YDS is β-naphthoflavone, with an EC_{50} of 3.0×10^{-8} M (Stalter *et al.*, 2011). An assay indicative of activation of AhR, AhR_LUC, was also included in the U.S. EPA ToxCast database. This is based on the human HepG2 cell line (He *et al.*, 2011). TCDD was not measured in ToxCast, but the EC_{10} values of common chemicals run in both AhR_LUC and AhR CALUX were generally within an order of magnitude (Neale *et al.*, 2020a).

In addition to reporter gene assays, the native metabolic enzyme activation can also be quantified in cell lines. The ethoxyresorufin-O-deethylase (EROD) assay measures specific CYP enzyme activity as an indicator of specific CYP isoforms induction, namely CYP1A. This assay was first developed in 1974 (Burke and Mayer, 1974). Early applications of the EROD microplate assay for water quality testing include the study by Huuskonen *et al.* (1998), who assessed the toxicity of lake water receiving paper mill effluents. Louiz *et al.* (2008) demonstrated, using a fish liver cell line (PLHC-1), that PAHs show higher potency after 4 h of exposure, whereas dioxins caused similar responses whether exposed for 4 or 24 h. In this way, the assay becomes somewhat specific to chemical groups by varying the exposure duration. Although the cell-based assay has been further improved and applied to different cell lines (Heinrich *et al.*, 2014), it is still too variable for routine monitoring. Ongoing improvements in the method may in the near future make this assay a more suitable one.

A summary of reported AhR activity in wastewater, surface water, recycled water and drinking water is provided in Table 10.2. A number of studies have evaluated activation of AhR in passive sampler extracts (Jarošová *et al.*, 2012; Hamers *et al.*, 2018; de Baat *et al.*, 2020), but only studies that have applied SPE are included in Table 10.2. Based on the more sensitive reporter gene assays, the TCDD EQ covered the range of 0.1–3.3 ng_{TCDD}/L in wastewater influent, 0.007–1.2 ng_{TCDD}/L in treated wastewater, 0.004–0.36 ng_{TCDD}/L in recycled water and 0.002–0.21 ng_{TCDD}/L in surface water. The reported wastewater treatment plant (WWTP) removal efficacy ranged from 13% to 90% (Jalova *et al.*, 2013; Nivala *et al.*, 2018). TCDD EQ in drinking water ranged from <0.004 to 0.17 ng_{TCDD}/L. Based on EC_{10} values, effects were detected after 0.7–0.8 times enrichment in wastewater influent, between 0.8 and 31 times enrichment in wastewater effluent and 0.4 and 89 times enrichment in surface water.

10.3.2 Peroxisome proliferator-activated receptor γ

PPAR is also a transcription factor that belongs to the superfamily of nuclear receptors and is involved in the regulation of glucose and lipid metabolism, not so much in xenobiotic metabolism (Scarsi *et al.*, 2007). As the name indicates, the main function of PPAR is the delivery of peroxisomes, which are important for fatty acid oxidation and thus relevant for lipid metabolism. There are three isoforms of PPAR – PPARα, PPARβ (also called δ) and PPARγ, which are encoded by different genes, show different tissue expressions and perform slightly different functions. PPARα is expressed predominantly in metabolically active tissues such as liver and kidney cells where its ligands include fatty acids, hypolipidemic drugs and xenobiotics (Seimandi *et al.*, 2005). PPARγ is the key receptor in maintaining glucose and lipid homeostasis and its activation increases the insulin resistance of the cell (Scarsi *et al.*, 2007). To date, most studies have applied assays indicative of binding to PPARγ to environmental water extracts, with only a few studies applying assays indicative of PPARα (Escher *et al.*, 2014; Alygizakis *et al.*, 2019). Therefore, this section will focus on PPARγ, with

Table 10.2 Summary of reported aryl hydrocarbon receptor (AhR) activity as TCDD EQ in different environmental water extracts.

Matrix	Assay	Activation of AhR TCDD EQ (ng_{TCDD}/L)	Reference
Wastewater influent	AhR CAFLUX (H1.G1.1c3)	1.1–1.8	1
	AhR CALUX (H4L1.1c4)	0.25–0.27	2
	AhR reporter gene assay	187–386	3
	H4IIE-luc	$<$0.1–3.3	4–5
Wastewater effluent	AhR CAFLUX (H1.G1.1c3)	0.007–1.2	1, 6–9
	AhR CAFLUX (H4G1.1c2)	0.063–0.10	10
	AhR CALUX (H4L1.1c4)	0.12–0.13	2, 11
	AhR reporter gene assay	97–168	3
	H4IIE-luc	$<$0.05–0.7	4, 8, 12, 13
	PAH CALUX	52–242[b]	14
	YDS	16–158[c]	15
Recycled water	AhR CAFLUX (H1.G1.1c3)	$<$0.007[a]–0.36	6–9
	H4IIE-luc	$<$0.004	5, 8
Surface water	AhR CAFLUX (H1.G1.1c3)	0.01–0.19	1, 8, 16
	AhR CAFLUX (H4G1.1c2)	0.002–0.16	10, 16, 17
	AhR CALUX (H4L1.1c4)	0.002–0.21	11, 18, 19, 20
	AhR reporter gene assay	53	21
	H4IIE-luc	0.009[a]–0.18	8, 13
	YDS	$<$320–602[c]	22
Drinking water	AhR CAFLUX (H1.G1.1c3)	0.024[a]–0.17	1, 8
	AhR reporter gene assay	45–52	21
	H4IIE-luc	$<$0.004[a]	8

NB: Only studies that have applied SPE or LLE are included.
[a]TCDD EQ calculated using EC_{10} values from Table 10.1.
[b]PAH CALUX reference compound is benzo[*a*]pyrene.
[c]YDS reference compound is β-naphthoflavone.
References: [1](Macova *et al.*, 2011); [2](Nivala *et al.*, 2018); [3](Lundqvist *et al.*, 2019b); [4](Jalova *et al.*, 2013); [5](Lee *et al.*, 2015); [6](Macova *et al.*, 2010); [7](Reungoat *et al.*, 2010); [8](Escher *et al.*, 2014); [9](Jia *et al.*, 2015); [10](Neale *et al.*, 2017c); [11](Mueller *et al.*, 2021), [12](Loos *et al.*, 2013), [13](Maier *et al.*, 2016); [14](Alygizakis *et al.*, 2019); [15](Stalter *et al.*, 2011); [16](König *et al.*, 2017); [17](Neale *et al.*, 2015b); [18](Müller *et al.*, 2018); [19](Neale *et al.*, 2018a); [20](Neale *et al.*, 2020a); [21](Lundqvist *et al.*, 2019a); [22](Brettschneider *et al.*, 2019).

Table 10.3 Common cell-based reporter gene assays applied to evaluate peroxisome proliferator-activated receptor (PPARγ) activity in water extracts.

Assay	Cell Line	Detection Method	Rosiglitazone EC_{10} (M)	Rosiglitazone EC_{10} (ng/L)	EC Reference
PPARγ CALUX	U2OS	Luminescence	1.00×10^{-8}	3600	Gijsbers *et al.* (2011)
PPARγ GeneBLAzer	HEK293H	Fluorescence	3.30×10^{-10}	118	Jia *et al.* (2015)

two assays frequently applied, PPARγ CALUX and PPARγ GeneBLAzer (Table 10.3). The EC_{10} value for reference compound antidiabetic pharmaceutical rosiglitazone was over an order of magnitude lower for PPARγ GeneBLAzer. PPARγ activity has been detected in wastewater influent, wastewater effluent and surface water, with activity in recycled water and drinking water below the limit of detection (Table 10.4). Activity in wastewater influent ranged from 500 to 936 $ng_{rosiglitazone}$/L rosiglitazone EQ, with wastewater effluent ranging from 83 to 1680 $ng_{rosiglitazone}$/L rosiglitazone EQ (Table 10.4). The studies that have

Table 10.4 Summary of reported peroxisome proliferator-activated receptor (PPARγ) activity reported as rosiglitazone EQ in different environmental water extracts.

Matrix	Assay	PPARγ Activity Rosiglitazone EQ ($ng_{rosiglitazone}$/L)	Reference
Wastewater influent	PPARγ CALUX	500–800	1
	PPARγ GeneBLAzer	719–936	2
Wastewater effluent	PPARγ CALUX	<32–640	1, 3
	PPARγ GeneBLAzer	<59–1680	2, 4, 5
Recycled water	PPARγ CALUX	<119[a]	4
	PPARγ GeneBLAzer	<59[a]	4
Surface water	PPARγ CALUX	<119[a]	4
	PPARγ GeneBLAzer	0.59–1092	5–9
Drinking water	PPARγ CALUX	<119[a]	4
	PPARγ GeneBLAzer	<1.2[a] to <59[a]	10

NB: Only studies that have applied SPE or LLE are included.
[a]EC_{10} converted to Rosiglitazone EQ using EC_{10} values provided in Table 10.3.
References: [1](Bain *et al.*, 2014); [2](Nivala *et al.*, 2018); [3](Alygizakis *et al.*, 2019); [4](Escher *et al.*, 2014); [5] (Mueller *et al.*, 2021); [6](König *et al.*, 2017); [7](Müller *et al.*, 2018); [8](Neale *et al.*, 2018a); [9](Neale *et al.*, 2020a); [10](Albergamo *et al.*, 2020).

evaluated the removal of PPARγ during wastewater treatment found between 69% and >94% removal (Bain *et al.*, 2014; Nivala *et al.*, 2018). The rosiglitazone EQ in surface water varied from 0.6 to 1092 $ng_{rosiglitazone}$/L, with sites downstream of WWTPs having the highest effect. Based on EC_{10} values, effects were detected after 0.2–0.3 times enrichment in wastewater influent, between 0.09 and >30 times enrichment in wastewater effluent and 0.2 and 97 times enrichment in surface water.

10.3.3 Pregnane X receptor

PXR is a promiscuous nuclear receptor with a large ligand binding pocket that can help protect the cell by triggering detoxification pathways (Grimaldi *et al.*, 2015). PXR controls the transcription of a large array of genes encoding for phase I metabolic enzymes, especially the CYP3A family, which plays an important role in drug metabolism. Two reporter gene assays have been applied to evaluate PXR activity in water extracts, HG5LN hPXR and PXR CALUX. The reference compound for HG5LN hPXR is the pharmaceutical SR12813 (EC_{10}: 1.58×10^{-8} M (Neale *et al.*, 2015b)), while the reference compound for PXR CALUX is the pharmaceutical nicardipine. Effect concentrations for the industrial compound di(2-ethylhexyl)-phthalate (DEHP) for both assays were presented in Escher *et al.* (2018), with a slightly lower value for HG5LN hPXR (Table 10.5). PXR activity was detected in all water samples tested, except for recycled water after reverse osmosis and advanced oxidation (RO/AO) (Escher *et al.*, 2014). The reported PXR activity in wastewater effluent was 3.8–4.7 $\mu g_{SR12813}$/L SR12813 EQ or 20–240 $\mu g_{nicardipine}$/L nicardipine EQ, while the activity in surface water ranged from <0.02 to 2.3 $\mu g_{SR12813}$/L SR12813 EQ (Table 10.6). Based on the EC_{10} values in the HG5LN hPXR assay, effects in wastewater were detected after around two times enrichment in wastewater effluent, between 3 and 30 times enrichment in surface water and 2.5 times enrichment in drinking water.

Table 10.5 Common cell-based reporter gene assays applied to evaluate activation of the pregnane X receptor (PXR) in water extracts.

Assay	Cell Line	Detection Method	DEHP[a] EC_{10} (M)	DEHP EC_{10} (µg/L)	EC Reference
HG5LN hPXR	HG5LN (HeLa)	Luminescence	2.77×10^{-7}	108	Escher *et al.* (2018)
PXR CALUX	U2OS	Luminescence	3.97×10^{-7}	155	Escher *et al.* (2018)

[a]DEHP = Di(2-ethylhexyl)-phthalate.

Table 10.6 Summary of reported pregnane X receptor (PXR) activity in different environmental water extracts.

Matrix	Assay	PXR Activity	Reference
Wastewater effluent	HG5LN hPXR	3.8–4.7 μg/L SR12813 EQ[a]	1
	PXR CALUX	20–240 μg/L nicardipine EQ	2
Recycled water	HG5LN hPXR	<0.66–0.98 μg/L SR12813 EQ[a]	1
Surface water	HG5LN hPXR	<0.02–2.3 μg/L SR12813 EQ	1, 3, 4
Drinking water	PPARγ CALUX	3.2 μg/L SR12813 EQ[a]	1

NB: Only studies that have applied SPE or LLE are included.
[a]EC_{10} converted to SR12813 EQ using EC_{10} values provided in Table 10.5.
References: [1](Escher *et al.*, 2014); [2](Alygizakis *et al.*, 2019); [3](Neale *et al.*, 2015b); [4](Neale *et al.*, 2018a).

10.4 BIOASSAYS INDICATIVE OF HORMONE RECEPTOR-MEDIATED EFFECTS

Hormonal pathways are essential for processes related to growth, sexual development, metabolism and homeostasis. Endocrine-disrupting chemicals, including synthetic hormones, industrial chemicals and pesticides can interfere with hormonal systems by interacting with hormone receptors (le Maire *et al.*, 2010). This includes activating or inhibiting hormone receptors. To date, most of the research has focused on the estrogen receptor (ER), followed by the androgen receptor (AR). However, other relevant nuclear receptors include the glucocorticoid receptor (GR), progesterone receptor (PR), thyroid receptor (TR), mineralocorticoid receptor (MR), retinoic acid receptor (RAR) and retinoid X receptor (RXR). Further information about the sensitivity of assays indicative of ER, AR, PR, GR and TR can be found in the review by Leusch *et al.* (2017). In addition to the more commonly applied hormone receptors, this section will also focus on assays indicative of MR, RAR and RXR. Due to the large number of studies that have applied assays indicative of hormone receptor-mediated effects we decided to focus on the more commonly applied assays.

10.4.1 Estrogen receptor

10.4.1.1 Estrogen receptor agonism

Nuclear estrogen receptors ERα and ERβ are important for the growth and homeostasis of the uterus and mammary glands, as well as bones and the cardiovascular system (le Maire *et al.*, 2010). The majority of assays applied to environmental water extracts focus on ERα (*e.g.*, ERα CALUX and ERα GeneBLAzer), although the T47D-KBluc assay uses the T47D cell line, which expresses both ERα and ERβ (Wilson *et al.*, 2004).

Estrogenic activity is the most commonly studied endpoint in water extracts. As many different assays have been applied in the literature, we focused on assays that have been applied to water samples in four or more studies. The exception was the embryonic zebrafish assay EASZY, which has only been included in two studies, but was included to represent a whole organism assay. A summary of the included activation of ER assays is provided in Table 10.7, with similar sensitivity across the mammalian reporter gene assays. The EC_{10} values of the reference compound 17β-estradiol (E2) vary between 0.13 ng/L for T47D-KBluc and 2.1 ng/L for HeLa-9903. The yeast estrogen screen (YES) and EASZY are less sensitive. Further information about the sensitivity of many of these assays can be found in Leusch *et al.* (2017).

Not only does the sensitivity differ between different bioassays for the reference compound E2 but there are also differences in the relative effect potency (REP) of different compounds. The REP relates the effect of an individual chemical to the effect of the assay reference compound with REP values close to 1 indicating that the individual chemical has a similar potency to the reference chemical and REP <1 indicating the individual chemical is less potent than the reference chemical (Chapter 7). The natural estrogen estrone (E1) has an REP of 0.02 in ERα CALUX but an REP of 0.10 in ERα GeneBLAzer (Escher *et al.*, 2018) (Figure 10.4). Therefore, E1 is more potent relative to E2 in ERα GeneBLAzer than ERα CALUX.

Estrogenic activity is reported as 17β-estradiol EQ (EEQ) in units of ng_{E2}/L in wastewater influent, wastewater effluent, recycled water, surface water and drinking water in Table 10.8. Focusing on the mammalian reporter gene assays, the estrogenic activity expressed as EEQ in wastewater influent ranged from 0.5 to 122 ng_{E2}/L, while the EEQ were mostly between 0.1 and 10 ng_{E2}/L in treated effluent. Estrogenic activity is typically well removed during wastewater treatment, with 80% to >99% removal efficacy reported in the literature for a variety of WWTPs (Jugan *et al.*, 2009; Jalova *et al.*, 2013; Bain *et al.*, 2014; Hamilton *et al.*, 2016; Houtman *et al.*, 2018; Nivala *et al.*, 2018). Estrogenic activity was mostly below the limit of detection in recycled water. Estrogenicity varied greatly in surface water, with EEQ from 0.005 up to 190 ng_{E2}/L, with factors such as proximity to wastewater effluent discharges impacting the observed effects. Finally, low estrogenic activity (EEQ < 0.01 ng_{E2}/L) was often detected in treated drinking water, with one study from China finding EEQ of 5.3 ng_{E2}/L in treated drinking water (Shi *et al.*, 2018). A number of studies have measured estrogenic activity in both source water and treated drinking water, with 39%–99% removal efficacy observed (Escher *et al.*, 2014; Huang *et al.*, 2016; Lv *et al.*, 2016; Xiao *et al.*, 2016; Conley *et al.*, 2017b; Xiao *et al.*, 2017; Shi *et al.*, 2018; Neale *et al.*, 2020b). Based on the EC_{10} values, estrogenic activity was detected after 0.1–6.4 times enrichment in wastewater effluent, 0.09–145 times enrichment in surface water and 20–110 times enrichment in drinking water.

Table 10.7 Common assays applied to evaluate estrogenic activity in water extracts.

Assay	Cell Line/Test System	Detection Method	17β-Estradiol EC_{10} (M)	17β-Estradiol EC_{10} (ng/L)	EC Reference
Yeast reporter gene					
YES	Yeast	Absorbance	3.75×10^{-11}[a]	10.2[a]	Escher *et al.* (2008b)
Mammalian reporter gene					
ERα CALUX	U2OS	Luminescence	7.13×10^{-13}	0.19	Jia *et al.* (2015)
ERα GeneBLAzer	HEK293T	Fluorescence	9.87×10^{-12}	2.7	Nivala *et al.* (2018)
HeLa-9903	HeLa	Luminescence	7.78×10^{-12}[a]	2.1[a]	Valcarcel *et al.* (2018)
MELN	MCF-7	Luminescence	2.42×10^{-12}	0.66	Neale *et al.* (2015b)
MVLN	MCF-7	Luminescence	3.16×10^{-12}[a]	0.86[a]	Shue *et al.* (2009)
T47D-KBluc	T47D	Luminescence	4.63×10^{-13}[a]	0.13[a]	Liu *et al.* (2018)
Cell proliferation					
E-SCREEN	MCF7	Absorbance (cell viability measured using CellTiter (MTS))	8.18×10^{-13}[a]	0.22[a]	Macova *et al.* (2010)
Whole organism					
EASZY	Embryonic zebrafish	Fluorescence	EC_{50} 6.20×10^{-10}	EC_{50} 168	Brion *et al.* (2019)

[a]Presented EC_{10} value converted from EC_{50} value assuming a slope of the log-logistic concentration response curve of 1.

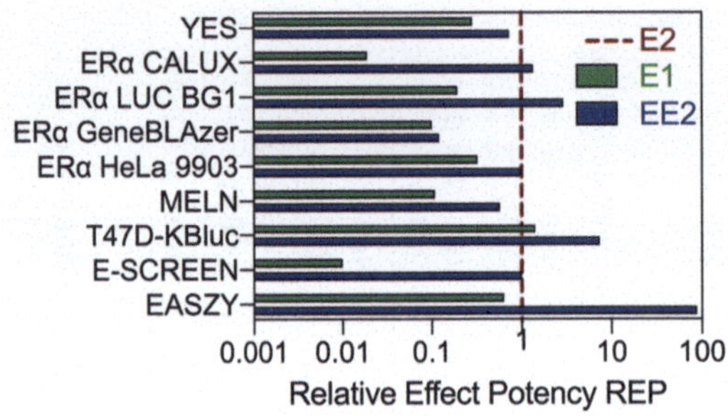

Figure 10.4 Relative effect potency (REP) of estrone (E1) and 17α-ethinylestradiol (EE2) in different assays indicative of estrogenic activity. Data available in Appendix A of Escher *et al.* (2018), for T47D-KBluc in Bermudez *et al.* (2012) and for E-SCREEN in Soto *et al.* (1995).

10.4.1.2 Estrogen receptor antagonism

In contrast to estrogenic activity, anti-estrogenic activity is much less studied in environmental water extracts. Three assays commonly applied to evaluate anti-estrogenic activity include the yeast anti-estrogen screen (YAES) and the mammalian reporter gene ERα CALUX and ERα GeneBLAzer (Table 10.9). Based on the reference compound tamoxifen, the ERα CALUX was the most sensitive assay, with an effect concentration causing a suppression ratio (SPR) of 20% (EC_{SPR20}) value of 0.56 μg/L. Cytotoxicity masked anti-estrogenic activity in wastewater influent, while many treated wastewater effluent samples were below detection (Table 10.10). Using the YAES assay, three studies found tamoxifen EQ (TMX EQ) ranging from 0.7 to 97 $\mu g_{tamoxifen}$/L in treated effluent (Conroy *et al.*, 2007; Fang *et al.*, 2012; Archer *et al.*, 2020). Anti-estrogenic activity was either low or below detection in surface water, while no anti-estrogenic activity was detected in drinking water.

10.4.2 Androgen receptor

10.4.2.1 Androgen receptor agonism

The androgen receptor (AR) is expressed in a range of tissues and has implications for the development and maintenance of a number of systems, including the reproductive, immune, musculoskeletal and cardiovascular systems (Davey and Grossmann, 2016). Although a number of assays have been applied to evaluate androgenic activity, the assays most commonly applied to water extracts include the yeast androgen screen (YAS) and the mammalian reporter gene assays AR CALUX, AR GeneBLAzer and MDA-kb2 (Table 10.11). Based on the reference compound dihydrotestosterone (DHT), the mammalian reporter gene assays were more sensitive compared to YAS. Further information about assay sensitivity can be found in Leusch *et al.* (2017).

Table 10.8 Summary of estrogenic activity reported as 17β-estradiol EQ (EEQ) in different environmental water extracts.

Matrix	Assay	Activation of ER EEQ (ng_{E2}/L)	Reference
Wastewater influent	YES	4.4–86	1–5
	ERα CALUX	0.5–122	3, 6–9
	ERα GeneBLAzer	11–24	10
	MELN	15–94	11, 12
	MVLN	5.4–124	2, 13
	E-SCREEN	<0.02–225	14–16
Wastewater effluent	YES	<0.1–91	1–5, 17–25
	ERα CALUX	<0.006–22.9	3, 6–9, 19, 22, 25–31
	ERα GeneBLAzer	0.03–151	10, 19, 28, 32–36
	hERα-HeLa-9903	0.03–24	19, 28, 37
	MELN	0.04–24	11, 12, 28, 38, 39
	MVLN	0.1–24	2, 13, 40, 41
	T47D-KBluc	15	42
	E-SCREEN	<0.02–34	14–16, 19, 37, 43–46
	EASZY	<6.3–673	39, 47
Recycled water	YES	<0.34[a]	19
	ERα CALUX	<0.008[a]–0.17	19, 27, 29
	ERα GeneBLAzer	<0.14[a]–2.6	19, 34
	hERα-HeLa-9903	<0.08[a]	19
	E-SCREEN	<0.007[a] to <0.06	15, 16, 19, 45, 46
Surface water	YES	<0.07–324	1, 3, 17–19, 21, 24, 48–55
	ERα CALUX	<0.015–129	3, 8, 19, 28, 30, 31, 56–59
	ERα GeneBLAzer	0.005–39	19, 28, 32, 33, 35, 36, 60–68
	hERα-HeLa-9903	<0.02–2.0	19, 28, 69
	MELN	0.003–23	11, 12, 28, 38, 39, 65, 70–73

(*Continued*)

Table 10.8 Summary of estrogenic activity reported as 17β-estradiol EQ (EEQ) in different environmental water extracts (*Continued*).

Matrix	Assay	Activation of ER EEQ (ng_{E2}/L)	Reference
	MVLN	1.2–32	40, 74
	T47D-KBluc	<0.03–190	42, 75, 76, 77
	E-SCREEN	0.005–85	15, 16, 19, 37, 43, 62, 77, 78
	EASZY	<2.0–30	39, 47
Drinking water	YES	0.02–1.4	19, 51, 53, 54, 79
	ERα CALUX	<0.008[a]–5.3	19, 30, 56, 59
	ERα GeneBLAzer	<0.03–0.04	19, 33, 66, 80
	hERα-HeLa-9903	<LOD to 0.35[a]	19, 81
	MELN	<0.3	12
	T47D-KBluc	<0.025–0.11	42, 76, 79
	E-SCREEN	<0.007[a]	16, 19

NB: Only studies that have applied SPE or LLE are included.
[a]EC_{10} converted to 17β-estradiol equivalent concentrations EEQ using EC_{10} values provided in Table 10.7; LOD: limit of detection.
References: [1](Archer *et al.*, 2020); [2](Kusk *et al.*, 2011); [3](Murk *et al.*, 2002); [4](Stalter *et al.*, 2011); [5](Zhang *et al.*, 2011); [6](Bain *et al.*, 2014); [7](Houtman *et al.*, 2018); [8](Roberts *et al.*, 2015); [9](Välitalo *et al.*, 2017); [10](Nivala *et al.*, 2018); [11](Cargouet *et al.*, 2004); [12](Jugan *et al.*, 2009); [13](Jalova *et al.*, 2013); [14](Hamilton *et al.*, 2016); [15](Leusch *et al.*, 2014a); [16](Macova *et al.*, 2011); [17](Aerni *et al.*, 2004); [18](Escher *et al.*, 2008b); [19](Escher *et al.*, 2014); [20](Fang *et al.*, 2012); [21](French *et al.*, 2015); [22](Gehrmann *et al.*, 2018); [23](Huggett *et al.*, 2003); [24](Pawlowski *et al.*, 2003); [25](Zeng *et al.*, 2016); [26](Alygizakis *et al.*, 2019); [27](Jia *et al.*, 2015); [28](Könemann *et al.*, 2018); [29](Leusch *et al.*, 2014b); [30](van der Linden *et al.*, 2008); [31](Houtman *et al.*, 2020); [32](González *et al.*, 2020); [33](Leusch *et al.*, 2018b); [34](Mehinto *et al.*, 2015); [35](Mehinto *et al.*, 2016); [36](Mueller *et al.*, 2021); [37](Henneberg *et al.*, 2014); [38](Miege *et al.*, 2009); [39](Neale *et al.*, 2017c); [40](Furuichi *et al.*, 2004); [41](Jarošová *et al.*, 2014b); [42](Medlock Kakaley *et al.*, 2020); [43](Bicchi *et al.*, 2009); [44](Körner *et al.*, 2001); [45](Macova *et al.*, 2010); [46](Reungoat *et al.*, 2010); [47](Brion *et al.*, 2019); [48](Brettschneider *et al.*, 2019); [49](Chen *et al.*, 2016); [50](Huang *et al.*, 2016); [51](Lv *et al.*, 2016); [52](Vermeirssen *et al.*, 2005); [53](Xiao *et al.*, 2016); [54](Xiao *et al.*, 2017); [55](Zhao *et al.*, 2011); [56](Brand *et al.*, 2013); [57](Jia *et al.*, 2019); [58](Scott *et al.*, 2014); [59](Shi *et al.*, 2018); [60](Daniels *et al.*, 2018); [61](Hashmi *et al.*, 2018); [62](König *et al.*, 2017); [63](Mehinto *et al.*, 2017); [64](Müller *et al.*, 2018); [65](Neale *et al.*, 2018a); [66](Neale *et al.*, 2020b); [67](Neale *et al.*, 2020a); [68](Scott *et al.*, 2018); [69](Prochazkova *et al.*, 2018); [70](Mnif *et al.*, 2012); [71](Neale *et al.*, 2015b); [72](Serra *et al.*, 2020); [73](Toušová *et al.*, 2017); [74](Shue *et al.*, 2009); [75](Conley *et al.*, 2017a); [76](Conley *et al.*, 2017b); [77](Liu *et al.*, 2018); [78](Oh *et al.*, 2006); [79](Van Zijl *et al.*, 2017); [80](Albergamo *et al.*, 2020); [81](Valcarcel *et al.*, 2018).

Table 10.9 Common assays applied to evaluate anti-estrogenic activity in water extracts.

Assay	Cell Line/Test System	Detection Method	Tamoxifen EC_{SPR20}[a] (M)	Tamoxifen EC_{SPR20} (μg/L)	EC Reference
Yeast reporter gene					
YAES	Yeast	Absorbance	6.00×10^{-7}	223	Conroy *et al.* (2007)
Mammalian reporter gene					
ERα CALUX	U2OS	Luminescence	1.50×10^{-9}	0.56	Jia *et al.* (2015)
ERα GeneBLAzer	HEK293T	Fluorescence	5.86×10^{-6}	2177	Neale *et al.* (2020b)

[a]SPR = suppression ratio.

Androgenic activity in wastewater, surface water and drinking water extracts is summarised in Table 10.12. Between 30 and 350 ng/L DHT EQ was detected in wastewater influent, with low or no androgenic activity typically present in wastewater effluent. The reported WWTP removal efficacy ranged from 95% to >99.9% (Jalova *et al.*, 2013; Bain *et al.*, 2014; Houtman *et al.*, 2018), explaining

Table 10.10 Summary of anti-estrogenic activity reported as tamoxifen EQ (TMX EQ) in different environmental water extracts.

Matrix	Assay	Anti-estrogenic Activity TMX EQ ($\mu g_{tamoxifen}$/L)	Reference
Wastewater influent	ERα GeneBLAzer	Cytotoxic	1
Wastewater effluent	YAES	0.7–97	2–4
	ERα CALUX	<0.07[a]–110	5–7
	ERα GeneBLAzer	<2177[a], cytotoxic	1, 8
Recycled water	ERα CALUX	<0.04[a]–4.4	5, 7, 9
Surface water	YAES	<50	10
	ERα CALUX	<0.04[a]–6	5, 11–13
	ERα GeneBLAzer	<1–2.7, cytotoxic	8, 14, 15
Drinking water	ERα CALUX	<0.04[a]	5
	ERα GeneBLAzer	<22[a] to <109[a]	8, 14

NB: Only studies that have applied SPE or LLE are included.
[a]EC_{SPR20} values converted to TMX EQ using EC_{SPR20} values provided in Table 10.9.
References: [1](Nivala *et al.*, 2018); [2](Archer *et al.*, 2020); [3](Conroy *et al.*, 2007); [4](Fang *et al.*, 2012); [5](Escher *et al.*, 2014); [6](Gehrmann *et al.*, 2018); [7](Jia *et al.*, 2015); [8](Leusch *et al.*, 2018b); [9](Leusch *et al.*, 2014b); [10](Zhao *et al.*, 2011); [11](Daniels *et al.*, 2018); [12](Jia *et al.*, 2019); [13](Scott *et al.*, 2014); [14](Neale *et al.*, 2020b); [15](Scott *et al.*, 2018).

Table 10.11 Common assays applied to evaluate androgenic activity in water extracts.

Assay	Cell Line/Test System	Detection Method	DHT EC_{10} (M)	DHT EC_{10} (ng/L)	EC Reference
Yeast reporter gene					
YAS	Yeast	Absorbance	2.86×10^{-10}	83	Sohoni and Sumpter (1998)
Mammalian reporter gene					
AR CALUX	U2OS	Luminescence	1.00×10^{-10}	29	Jia *et al.* (2015)
AR GeneBLAzer	HEK293MSR	Fluorescence	1.40×10^{-10}	41	Leusch *et al.* (2017)
MDA-kB2	MDA-MB-453	Luminescence	3.12×10^{-11}	9.1	Neale *et al.* (2017c)

DHT = dihydrotestosterone.

the low activity in treated effluent. Based on the mammalian reporter gene assays, only low androgenic activity was detected in surface water, with 0.25–12 ng/L DHT EQ reported (Table 10.12). Based on the EC_{10} values of DHT in Table 10.11, this means samples would need to be enriched between 1.8 and 140 times in the assay to detect an effect in surface water. No androgenic activity was detected in recycled water, with only one study detecting androgenic activity in drinking water at 0.13 ng/L DHT EQ (Brand *et al.*, 2013). Based on the DHT EC_{10} for AR CALUX in Table 10.11, this equates to an EC_{10} of relative extraction factor (REF) 223, meaning the sample would need to be enriched over 200 times to elicit 10% of activation of AR.

10.4.2.2 Androgen receptor antagonism

The EC_{SPR20} values of the reference compound flutamide in Table 10.13 are similar in the yeast and mammalian reporter gene anti-androgenic assays. Reported anti-androgenic activity in environmental extracts is provided in Table 10.14. Anti-androgenic activity was highest in wastewater effluent, with flutamide EQ between 0.5 and 360 $\mu g_{flutamide}$/L reported for the mammalian reporter gene assays (Table 10.14). Based on the flutamide EC_{SPR20} values in Table 10.13, this indicates between 0.8 and 600 times enrichment in the assay would be required to detect an effect. Anti-androgenic activity in surface water expressed as flutamide EQ ranged from 0.3 to 257 $\mu g_{flutamide}$/L, while no anti-androgenic activity was detected in drinking water or recycled water.

Table 10.12 Summary of androgenic activity reported as dihydrotestosterone (DHT) EQ in different environmental water extracts.

Matrix	Assay	Androgenic Activity DHT EQ (ng_{DHT}/L)	Reference
Wastewater influent	AR CALUX	<0.62–350	1–6
	AR GeneBLAzer	50–83[a]	7
Wastewater effluent	YAS	<2.8[a]–138	8, 9
	AR CALUX	<0.02–2.7	1–6, 8, 10–14
	AR GeneBLAzer	<2, cytotoxic	7, 8, 15
	MDA-kB2	<0.77–12	8, 16, 17
Recycled water	YAS	<2.8[a]	8
	AR CALUX	<0.97[a] to <30	3, 8, 12, 13
	AR GeneBLAzer	<2.0[a]	8
	MDA-kb2	<0.34[a]	8
Surface water	YAS	<2.8[a]–69	8, 9, 18, 19
	AR CALUX	<0.08–12	3–5, 8, 11, 14, 20, 21
	AR GeneBLAzer	0.29–4.9[a]	8, 15, 22–27
	MDA-kb2	<0.02–4.8	8, 16, 17, 23, 28, 29
Drinking water	YAS	<2.8[a]	8
	AR CALUX	<0.08–0.13	8, 20
	AR GeneBLAzer	<0.1 to <2[a]	8, 15, 26, 30
	MDA-kb2	<0.34[a] to <0.77	8, 16

NB: Only studies that have applied SPE or LLE are included.
[a]EC_{10} converted to DHT EQ using EC_{10} values provided in Table 10.11.
References: [1](Bain *et al.*, 2014); [2](Houtman *et al.*, 2018); [3](Leusch *et al.*, 2014a); [4](Roberts *et al.*, 2015); [5](Šauer *et al.*, 2018); [6](Välitalo *et al.*, 2017); [7](Nivala *et al.*, 2018); [8](Escher *et al.*, 2014); [9](French *et al.*, 2015); [10](Gehrmann *et al.*, 2018); [11](Houtman *et al.*, 2020); [12](Jia *et al.*, 2015); [13](Leusch *et al.*, 2014b); [14](van der Linden *et al.*, 2008); [15](Leusch *et al.*, 2018b); [16](Medlock Kakaley *et al.*, 2020); [17](Neale *et al.*, 2017c); [18](Huang *et al.*, 2016); [19](Zhao *et al.*, 2011); [20](Brand *et al.*, 2013); [21](Scott *et al.*, 2014); [22](Hashmi *et al.*, 2018); [23](König *et al.*, 2017); [24](Müller *et al.*, 2018); [25](Neale *et al.*, 2018a); [26](Neale *et al.*, 2020b); [27](Scott *et al.*, 2018); [28](Conley *et al.*, 2017a); [29](Toušová *et al.*, 2017); [30](Albergamo *et al.*, 2020).

Table 10.13 Common assays applied to evaluate anti-androgenic activity in water extracts.

Assay	Cell Line/Test System	Detection Method	Flutamide EC_{SPR20} (M)	Flutamide EC_{SPR20} (μg/L)	EC Reference
Yeast reporter gene					
YAAS	Yeast	Absorbance	7.50×10^{-7} [a]	207[a]	Stalter *et al.* (2011)
Mammalian reporter gene					
AR CALUX	U2OS	Luminescence	1.10×10^{-6}	304	Jia *et al.* (2015)
AR GeneBLAzer	HEK293MSR	Fluorescence	5.50×10^{-7} [a]	152[a]	Leusch *et al.* (2017)
MDA-kB2	MDA-MB-453	Luminescence	2.07×10^{-7}	57	Neale *et al.* (2017b)

10.4.3 Glucocorticoid receptor

10.4.3.1 Glucocorticoid receptor agonism

The glucocorticoid receptor (GR) is a corticosteroid receptor that controls the actions of glucocorticoids, and a wide range of environmental contaminants can interfere with glucocorticoid activity (Zhang *et al.*, 2019). Mammalian reporter gene assays have been applied to evaluate glucocorticoid activity in environmental water extracts (Table 10.15), with GR CALUX and GR GeneBLAzer most commonly applied. The pharmaceutical dexamethasone serves as the assay reference compound, with the lowest EC_{10} reported for GR GeneBLAzer. The mammalian reporter gene CV-1 GR assay has also been recently applied to detect glucocorticoid activity in different water samples (Conley *et al.*, 2017a; Medlock Kakaley *et al.*, 2020), but no EC value for dexamethasone was available, so this assay was not included in Table 10.15.

Glucocorticoid activity in wastewater influent expressed as dexamethasone EQ ranged from 37 to 121 $ng_{dexamethasone}/L$, with between 1.8 and 628 $ng_{dexamethasone}/L$ reported in wastewater effluent (Table 10.16). The studies that evaluated WWTP treatment efficacy found between −7% and 66% removal (Bain *et al.*, 2014; Roberts *et al.*, 2015; Houtman *et al.*, 2018), indicating much poorer removal for glucocorticoid activity compared to estrogenic activity and androgenic activity. As a result, glucocorticoid activity was frequently detected in surface water, with dexamethasone EQ between 9 and 170 $ng_{dexamethasone}/L$ in an effluent impacted river (Daniels *et al.*, 2018). Based on the GR GeneBLAzer dexamethasone EC_{10} value in Table 10.15, this equates to 0.5–9 times

Table 10.14 Summary of anti-androgenic activity reported as flutamide EQ in different environmental water extracts.

Matrix	Assay	Anti-androgenic Activity Flutamide EQ ($\mu g_{flutamide}/L$)	Reference
Wastewater influent	AR CALUX	5.2–26	1
	AR GeneBLAzer	Cytotoxic	2
Wastewater effluent	YAAS	16–3190	3–5
	AR CALUX	0.48–360	1, 4, 6–10
	AR GeneBLAzer	<15, cytotoxic	2, 11
	MDA-kb2	$<4.4^a$	7
Recycled water	AR CALUX	$<20^a$ to <300	7, 10
	MDA-kb2	$<1.9^a$	7
Surface water	YAAS	20–935	12
	AR CALUX	3.3–257	1, 7, 9, 13, 14
	AR GeneBLAzer	<0.51–90, cytotoxic	11, 15–18
	MDA-kb2	<0.12–9.2^a	7, 15, 19
Drinking water	AR CALUX	$<20^a$	7
	AR GeneBLAzer	<0.7 to $<1.5^a$	11, 17
	MDA-kb2	$<1.9^a$	7

NB: Only studies that have applied SPE or LLE are included.

[a]EC_{SPR20} values converted to Flutamide EQ using EC_{SPR20} values provided in Table 10.13.

References: [1](Šauer *et al.*, 2018); [2](Nivala *et al.*, 2018); [3](Fang *et al.*, 2012); [4](Gehrmann *et al.*, 2018); [5](Stalter *et al.*, 2011); [6](Alygizakis *et al.*, 2019); [7](Escher *et al.*, 2014); [8](Houtman *et al.*, 2018); [9](Houtman *et al.*, 2020); [10](Jia *et al.*, 2015); [11](Leusch *et al.*, 2018b); [12](Zhao *et al.*, 2011); [13](Jia *et al.*, 2019); [14](Scott *et al.*, 2014); [15](König *et al.*, 2017); [16](Müller *et al.*, 2018); [17](Neale *et al.*, 2020b); [18](Scott *et al.*, 2018); [19](Toušová *et al.*, 2017).

Table 10.15 Common assays applied to evaluate glucocorticoid activity in water extracts.

Assay	Cell Line	Detection Method	Dexamethasone EC_{10} (M)	Dexamethasone EC_{10} (ng/L)	EC Reference
GR CALUX	U2OS	Luminescence	8.00×10^{-10}	314	Jia *et al.* (2015)
GR GeneBLAzer	HEK 293T	Fluorescence	2.08×10^{-10}	82	Nivala *et al.* (2018)
GR Switchgear	HT1080	Luminescence	5.00×10^{-10}	196	Jia *et al.* (2015)

enrichment in the assay. Glucocorticoid activity has not been reported in drinking water (Table 10.16).

10.4.3.2 Glucocorticoid receptor antagonism

Only two assays, the GR CALUX and GR GeneBLAzer, have been applied in the literature to evaluate anti-glucocorticoid activity, with the GR GeneBLAzer much more sensitive compared to GR CALUX based on the reference compound mifepristone EC_{SPR20} values (Table 10.17). Anti-glucocorticoid activity was only detected in surface water in two studies (König *et al.*, 2017; Jia *et al.*, 2019), with mifepristone EQ reported between 2.5 and 610 $ng_{mifepristone}$/L. Anti-glucocorticoid activity was not detected in wastewater effluent, recycled water or drinking water, with cytotoxicity masking the effect in wastewater influent (Table 10.18).

10.4.4 Progesterone receptor

10.4.4.1 Progesterone receptor agonism

Two assays, the progesterone receptor (PR) CALUX and PR GeneBLAzer, have been applied to evaluate progestogenic activity in environmental water extracts, with both assays having similar EC_{10} values for the synthetic hormone levonorgestrel (Table 10.19). Progestogenic activity in water extracts has been reported in different equivalent concentrations, including progesterone EQ, levonorgestrel EQ, promegestone EQ and Org2058 EQ. To assist with comparison, the results from the literature were converted to levonorgestrel EQ based on published potency data. Up to 3.2 ng/L levonorgestrel EQ was detected in wastewater influent, while between 0.43 and 7.1 ng/L levonorgestrel EQ was detected in treated effluent (Table 10.20). A number of studies have found increased progestogenic activity after wastewater treatment (Roberts *et al.*, 2015; Houtman *et al.*, 2018), while Bain *et al.* (2014) found between 12% and >93% removal efficacy in three WWTPs in Australia. Up to 9.6 ng/L levonorgestrel EQ was detected in surface water from the Netherlands, although progestogenic activity was often below the assay detection limit or masked by cytotoxicity in surface water. No progestogenic activity has been detected in recycled water or drinking water (Table 10.20).

Table 10.16 Summary of glucocorticoid activity reported as dexamethasone EQ in units of $ng_{dexamethasone}/L$ in different environmental water extracts.

Matrix	Assay	Glucocorticoid Activity Dexamethasone EQ ($ng_{dexamethasone}/L$)	Reference
Wastewater influent	GR CALUX	37–121	1–3
	GR GeneBLAzer	<400, cytotoxic	4, 5
Wastewater effluent	GR CALUX	11–628[a]	1–3, 6–12
	GR GeneBLAzer	<8.2[a]–392	5, 7, 8, 10, 13–16
	GR Switchgear	19–24[a]	7, 8
	CV-1 GR	1.8–21	17
Recycled water	GR CALUX	<10[a] to <310	7–9
	GR GeneBLAzer	<4.1[a]–65, <230	7, 8, 15
	GR Switchgear	<9.8[a]–16	7, 8
Surface water	GR CALUX	0.30–34, <500	3, 7, 10–12, 18–20
	GR GeneBLAzer	<1.1–170, cytotoxic	7, 10, 16, 21–27
	GR Switchgear	<9.8[a]	7
	CV-1 GR	<1.2–43	17, 28
Drinking water	GR CALUX	<2.0 to <120	7, 10, 18
	GR GeneBLAzer	<0.82 to <5.8	7, 10, 27, 29
	GR Switchgear	<9.8[a]	7
	CV-1 GR	<1.2	17

NB: Only studies that have applied SPE or LLE are included.

[a]EC_{10} converted to dexamethasone EQ using EC_{10} values provided in Table 10.15.

References: [1](Bain *et al.*, 2014); [2](Houtman *et al.*, 2018); [3](Roberts *et al.*, 2015); [4](Lee *et al.*, 2015); [5](Nivala *et al.*, 2018); [6](Alygizakis *et al.*, 2019); [7](Escher *et al.*, 2014); [8](Jia *et al.*, 2015); [9](Leusch *et al.*, 2014b); [10](Leusch *et al.*, 2018b); [11](van der Linden *et al.*, 2008); [12](Houtman *et al.*, 2020); [13](Chen *et al.*, 2016); [14](Jia *et al.*, 2016); [15](Mehinto *et al.*, 2015); [16](Mehinto *et al.*, 2016); [17](Medlock Kakaley *et al.*, 2020); [18](Brand *et al.*, 2013); [19](Schriks *et al.*, 2013); [20](Toušová *et al.*, 2017); [21](Daniels *et al.*, 2018); [22](Hashmi *et al.*, 2020); [23](König *et al.*, 2017); [24](Mehinto *et al.*, 2017); [25](Müller *et al.*, 2018); [26](Neale *et al.*, 2018a); [27](Neale *et al.*, 2020b); [28](Conley *et al.*, 2017a); [29](Albergamo *et al.*, 2020).

Table 10.17 Common assays applied to evaluate anti-glucocorticoid activity in water extracts.

Assay	Cell Line	Detection Method	Mifepristone EC_{SPR20} (M)	Mifepristone EC_{SPR20} (ng/L)	EC Reference
GR CALUX	U2OS	Luminescence	2.90×10^{-9}	1246	Jia *et al.* (2015)
GR GeneBLAzer	HEK 293T	Fluorescence	1.00×10^{-10}	43	Jia *et al.* (2015)

10.4.4.2 Progesterone receptor antagonism

Two assays, the PR CALUX and PR GeneBLAzer, have been applied to evaluate anti-progestogenic activity in environmental extracts. Based on the reference compound mifepristone EC_{SPR20} value, the PR CALUX was more sensitive compared to PR GeneBLAzer (Table 10.21). Most water extracts either had no

Table 10.18 Summary of anti-glucocorticoid activity reported as mifepristone EQ in different environmental water extracts.

Matrix	Assay	Anti-glucocorticoid Activity Mifepristone EQ ($ng_{mifepristone}$/L)	Reference
Wastewater influent	GR GeneBLAzer	Cytotoxic	1
Wastewater effluent	GR CALUX	<623[a] to <1200	2, 3
	GR GeneBLAzer	<21[a] to <60, cytotoxic	1–4
Recycled water	GR CALUX	<1200 to <1246[a]	2, 3
	GR GeneBLAzer	<40 to <43[a]	2, 3
Surface water	GR CALUX	<50–610, <1246[a]	2, 5
	GR GeneBLAzer	<0.2–2.5, <43[a], cytotoxic	2, 4, 6–8
Drinking water	GR CALUX	<1246[a]	2
	GR GeneBLAzer	<0.49 to <43[a]	2, 4, 8

NB: Only studies that have applied SPE or LLE are included.

[a]EC_{SPR20} values converted to mifepristone EQ using EC_{SPR20} values provided in Table 10.17.

References: [1](Nivala *et al.*, 2018); [2](Escher *et al.*, 2014); [3](Jia *et al.*, 2015); [4](Leusch *et al.*, 2018b); [5](Jia *et al.*, 2019); [6](Daniels *et al.*, 2018); [7](König *et al.*, 2017); [8](Neale *et al.*, 2020b).

Table 10.19 Common assays applied to evaluate progestogenic activity in water extracts.

Assay	Cell Line	Detection Method	Levonorgestrel EC_{10} (M)	Levonorgestrel EC_{10} (ng/L)	EC Reference
PR CALUX	U2OS	Luminescence	3.44×10^{-11}[a]	10.8	Scott *et al.* (2014)
PR GeneBLAzer	HEK 293T	Fluorescence	1.22×10^{-11}[a]	3.8	Leusch *et al.* (2018b)

[a]Presented EC_{10} value converted from EC_{50} value assuming a slope of the log-logistic concentration response curve of 1.

Table 10.20 Summary of progestogenic activity reported as levonorgestrel EQ in different environmental water extracts.

Matrix	Assay	Progestogenic Activity Levonorgestrel EQ ($ng_{levonorgestrel}/L$)	Reference
Wastewater influent	PR CALUX	<0.32–3.2[b]	1–3
	PR GeneBLAzer	Cytotoxic	4
Wastewater effluent	PR CALUX	<0.01–7.1[b], <90	1–3, 5–9
	PR GeneBLAzer	<0.38[b]–5.7, cytotoxicity	4, 5, 8, 10
Recycled water	PR CALUX	<0.01 to <90	5–7
	PR GeneBLAzer	<0.19[b] to <1.4	5, 10
Surface water	PR CALUX	<0.11[c]–9.6[c]	3, 5, 8, 9, 11, 12
	PR GeneBLAzer	<0.04[a]–1.1, cytotoxic	5, 8, 13–18
Drinking water	PR CALUX	<0.11[c] to <2.0	5, 8, 11
	PR GeneBLAzer	<0.04[a] to <0.19[a]	5, 8, 17, 19

NB: Only studies that have applied SPE or LLE are included.
[a]EC_{10} values converted to levonorgestrel EQ using EC_{10} values provided in Table 10.19.
[b]Results presented as progesterone EQ and converted to levonorgestrel EQ using an REP of 0.16 (Sonneveld *et al.*, 2011).
[c]Results presented as Org2058 EQ and converted to levonorgestrel EQ using an REP of 2.14 (Sonneveld *et al.*, 2011).
References: [1](Bain *et al.*, 2014); [2](Houtman *et al.*, 2018); [3](Roberts *et al.*, 2015); [4](Nivala *et al.*, 2018); [5](Escher *et al.*, 2014); [6](Jia *et al.*, 2015); [7](Leusch *et al.*, 2014b); [8](Leusch *et al.*, 2018b); [9](van der Linden *et al.*, 2008); [10](Mehinto *et al.*, 2015); [11](Brand *et al.*, 2013); [12](Scott *et al.*, 2014); [13](Hashmi *et al.*, 2020); [14](König *et al.*, 2017); [15](Müller *et al.*, 2018); [16](Neale *et al.*, 2018a); [17](Neale *et al.*, 2020b); [18](Scott *et al.*, 2018); [19](Albergamo *et al.*, 2020).

Table 10.21 Common assays applied to evaluate anti-progestogenic activity in water extracts.

Assay	Cell Line	Detection Method	Mifepristone EC_{SPR20} (M)	Mifepristone EC_{SPR20} (ng/L)	EC Reference
PR CALUX	U2OS	Luminescence	2.00×10^{-11}	8.6	Jia *et al.* (2015)
PR GeneBLAzer	HEK 293T	Fluorescence	3.00×10^{-10}	129	Nivala *et al.* (2018)

Table 10.22 Summary of anti-progestogenic activity reported as mifepristone EQ in different environmental water extracts.

Matrix	Assay	Anti-progestogenic Activity Mifepristone EQ ($ng_{mifepristone}$/L)	Reference
Wastewater influent	PR GeneBLAzer	Cytotoxic	1
Wastewater effluent	PR CALUX	<0.72–17	2–4
	PR GeneBLAzer	<2, cytotoxic	1, 5
Recycled water	PR CALUX	<8.6[a]	3, 4
Surface water	PR CALUX	<8–32,000	3, 6
	PR GeneBLAzer	<0.5–4.2	5–10
Drinking water	PR CALUX	<8.6[a]	3
	PR GeneBLAzer	<0.1 to <1.3[a]	5, 9

NB: Only studies that have applied SPE or LLE are included.
[a]EC_{SPR20} values converted to mifepristone EQ using EC_{SPR20} values provided in Table 10.21.
References: [1](Nivala *et al.*, 2018); [2](Alygizakis *et al.*, 2019); [3](Escher *et al.*, 2014); [4](Jia *et al.*, 2015); [5](Leusch *et al.*, 2018b); [6](Scott *et al.*, 2014); [7](König *et al.*, 2017); [8](Müller *et al.*, 2018); [9](Neale *et al.*, 2020b); [10](Scott *et al.*, 2018).

response or were masked by cytotoxicity in PR CALUX or PR GeneBLAzer when run in antagonist mode (Table 10.22). Only one study reported anti-progestogenic activity in wastewater effluent (Alygizakis *et al.*, 2019), with 9 of 12 samples active, while two studies detected anti-progestogenic activity in Australian surface waters (Scott *et al.*, 2014; Scott *et al.*, 2018).

10.4.5 Thyroid receptor

10.4.5.1 Thyroid receptor agonism

A number of assays have been applied to evaluate thyroid receptor (TR) activity in environmental water extracts including yeast reporter gene assays, mammalian reporter gene assays, cell proliferation assays and a whole organism assay (*Xenopus* eleutheroembryonic thyroid assay, XETA) using embryonic *Xenopus laevis* (Table 10.23). Based on reference compound triiodothyronine (T3), the reporter gene assays were the most sensitive. However, effects in mammalian reporter gene assays have only been observed in wastewater influent, with T3 EQ of 25 ng_{T3}/L reported in French wastewater using the PC-DR-LUC assay (Table 10.24). All other reporter gene assays did not detect thyroid activity in wastewater effluent, surface water or recycled water. In contrast, T3 EQ between 1100 and 1340 ng_{T3}/L were detected in wastewater effluent using the XETA, with 960 ng_{T3}/L detected in surface water (Välitalo *et al.*, 2017; Leusch *et al.*, 2018a). This suggests that the XETA, which incorporates toxicokinetic processes and other non-TR-mediated thyroid effects, may be more suitable to evaluate thyroid activity in water extracts than mammalian reporter gene assays.

10.4.5.2 Thyroid receptor antagonism

Three assays have been applied to evaluate anti-thyroid activity in environmental water extracts, the yeast reporter gene yeast two-hybrid assay and the mammalian reporter gene assays GH3.TRE-Luc and TRβ GeneBLAzer. The anti-thyroid activity assay reference compound is the pharmaceutical amiodarone hydrochloride (AH), with the EC_{50} values lower for the mammalian reporter gene assays (Table 10.25). Anti-thyroid activity expressed as amiodarone hydrochloride equivalent concentrations (AH EQ) was detected in wastewater influent (60–422 $\mu g_{AH}/L$), wastewater effluent (13–35 $\mu g_{AH}/L$) and surface water (3.3–16 $\mu g_{AH}/L$) using the yeast two-hybrid assay (Table 10.26). In contrast, only wastewater effluent had a response in TRβ GeneBLAzer, with none of the samples having a response in GH3.TRE-Luc in antagonist mode.

10.4.6 Mineralocorticoid receptor

Similar to the GR, the mineralcorticoid receptor (MR) is a corticosteroid receptor that controls the action of mineralocorticoids (Zhang *et al.*, 2019). Currently, there is only one assay used to assess mineralocorticoid activity in water extracts, the HG5LN-hMR, which can be run in both agonist and antagonist modes. The reference compound in the agonist mode is the hormone aldosterone with an EC_{50} of 9.80×10^{-10} M (Leusch *et al.*, 2018b), while the pharmaceutical spironolactone is the antagonist reference compound with an EC_{50} of 2.89×10^{-9} M (Bellet *et al.*, 2012). To date, mineralocorticoid activity has not been detected in wastewater, surface water or drinking water (Bellet *et al.*, 2012; Creusot *et al.*, 2014; Leusch *et al.*, 2018b). However, anti-mineralocorticoid activity has been detected in wastewater influent, wastewater effluent and surface water (Table 10.27), with spironolactone EQ between up to 0.9 and 2.3 $\mu g_{spironolactone}/L$ reported. Anti-mineralocorticoid activity was below the limit of detection in drinking water.

Table 10.23 Common assays applied to evaluate the thyroid receptor (TR) activity in water extracts.

Assay	Cell Line/Test System	Detection Method	Triiodothyronine (T3) EC_{10} (M)	Triiodothyronine (T3) EC_{10} (ng/L)	EC Reference
Yeast reporter gene					
Yeast two-hybrid	Yeast	Absorbance	2.60×10^{-8}	17,000	Li *et al.* (2008)
Mammalian reporter gene					
TRβ CALUX	U2OS	Luminescence	8.60×10^{-12}	5.6	Jia *et al.* (2015)
TRβ GeneBLAzer	HEK293T	Fluorescence	6.00×10^{-11}	41	Leusch *et al.* (2017)
GH3.TRE-Luc	GH3	Luminescence	6.67×10^{-12}[a]	4.3[a]	Leusch *et al.* (2018a)
PC-DR-LUC	PC12	Luminescence	2.00×10^{-11}[a]	13[a]	Jugan *et al.* (2009)
Cell proliferation					
T-Screen	GH3	Fluorescence (cell viability measured using Alamar Blue (resazurin))	2.80×10^{-10}	182	Jia *et al.* (2015)
Whole organism					
XETA	Embryonic Xenopus	Fluorescence	EC_{50} 4.50×10^{-9}	EC_{50} 3000	Leusch *et al.* (2018a)

[a]EC_{10} value converted from EC_{50} value assuming a slope of the log-logistic concentration response curve of 1.

Table 10.24 Summary of thyroid receptor (TR) activity reported as triiodothyronine (T3) EQ in different environmental water extracts.

Matrix	Assay	Thyroid Activity T3 EQ (ng_{T3}/L)	Reference
Wastewater influent	PC-DR-LUC	<20–25	1
	T-Screen	190–204	2
Wastewater effluent	TRβ CALUX	<0.14[a] to <21[b]	3–5
	TRβ GeneBLAzer	<2.3[b]	5
	GH3.TRE-Luc	<10[b] to <25	5, 6
	PC-DR-LUC	<20	1
	T-Screen	<5.5 to <6.1[a]	3, 4
	XETA	<LOD to 1340	5, 7
Recycled water	TRβ CALUX	<0.14[a] to <5.6	3, 4
	T-Screen	<5.5 to <6.1[a]	3, 4
Surface water	Yeast two-hybrid TR	<14–43	8
	TRβ CALUX	<0.14[a] to <3.5[b]	3, 5
	TRβ GeneBLAzer	<1.2[b]	5
	GH3.TRE-Luc	<5.3[b] to <20	5, 6
	PC-DR-LUC	<20	1
	T-Screen	<6.1[a]	3
	XETA	960[b]	5
Drinking water	TRβ CALUX	<0.14[a] to <3.5[b]	3, 5
	TRβ GeneBLAzer	<1.2[b]	5
	GH3.TRE-Luc	<0.05[b] to <1.3	5, 6
	PC-DR-LUC	<20	1
	T-Screen	<6.1[a]	3
	XETA	<300[b]	5

NB: Only studies that have applied SPE or LLE are included.
[a]EC_{10} values converted to triiodothyronine (T3) EQ using EC_{10} values provided in Table 10.23.
[b]Results presented as thyroxine (T4) EQ and converted to T3 EQ using the REP from Leusch *et al.* (2018a). LOD: limit of detection.
References: [1](Jugan *et al.*, 2009); [2](Kusk *et al.*, 2011); [3](Escher *et al.*, 2014); [4](Jia *et al.*, 2015); [5](Leusch *et al.*, 2018a); [6](Leusch *et al.*, 2018b); [7](Välitalo *et al.*, 2017); [8](Chinathamby *et al.*, 2013).

10.4.7 Retinoic acid receptor and retinoid X receptor

Only a handful of studies have applied assays indicative of the retinoic acid receptor (RAR), where the effect is expressed as all-*trans* retinoic acid EQ (ATRA EQ), and

Table 10.25 Common assays applied to evaluate anti-thyroid activity in water extracts.

Assay	Cell Line/Test System	Detection Method	Amiodarone Hydrochloride EC_{50} (M)	Amiodarone Hydrochloride EC_{50} (μg/L)	EC Reference
Yeast reporter gene					
Yeast two-hybrid	Yeast	Absorbance	3.10×10^{-5}	21,000	Li *et al.* (2008)
Mammalian reporter gene					
TRβ GeneBLAzer	HEK293T	Fluorescence	7.30×10^{-6}	5000	Leusch *et al.* (2018a)
GH3.TRE-Luc	GH3	Luminescence	8.40×10^{-6}	5700	Leusch *et al.* (2018a)

Table 10.26 Summary of anti-thyroid activity reported as amiodarone hydrochloride EQ (AH EQ) in different environmental water extracts.

Matrix	Assay	Anti-thyroid Activity AH EQ (μg_{AH}/L)	Reference
Wastewater influent	Yeast two-hybrid	60–422	1
Wastewater effluent	Yeast two-hybrid	13–35	1
	TRβ GeneBLAzer	350	2
	GH3.TRE-Luc	<1700	2
Surface water	Yeast two-hybrid	3.3–16	1
	TRβ GeneBLAzer	<28	2
	GH3.TRE-Luc	<870	2
Drinking water	TRβ GeneBLAzer	<28	2
	GH3.TRE-Luc	<87	2

NB: Only studies that have applied SPE or LLE are included.
References: [1](Li *et al.*, 2011); [2](Leusch *et al.*, 2018a).

the retinoid X receptor (RXR with 9-*cis*-retinoic acid EQ. For example, wastewater effluent, surface water and drinking water had no response in the RXR CALUX and HELN-RARa-RXR assays (Leusch *et al.*, 2018b). In contrast, some surface water extracts from Serbia had a response in the RAR GeneBLAzer with ATRA EQ <0.02–0.15 ng_{ATRA}/L and in RXR GeneBLAzer with 9-*cis*-retinoic acid EQ of 7 $ng_{9\text{-}cis\text{-retinoic acid}}$/L reported (König *et al.*, 2017). This equates to an effect after 41–170 times enrichment in the assay for RAR GeneBLAzer and 240 times

Table 10.27 Summary of reported anti-mineralocorticoid activity reported as spironolactone EQ in different environmental water extracts.

Matrix	Assay	Anti-mineralocorticoid Activity Spironolactone EQ ($\mu g_{spironolactone}/L$)	Reference
Wastewater influent	HG5LN-hMR	1.3–2.3	1
Wastewater effluent	HG5LN-hMR	<3.1–3.1	2
Surface water	HG5LN-hMR	<0.66–0.91	2
Drinking water	HG5LN-hMR	<0.16	2

NB: Only studies that have applied SPE or LLE are included.
References: [1](Bellet *et al.*, 2012); [2](Leusch *et al.*, 2018b).

enrichment in the assay for RAR GeneBLAzer. RAR activity was also detected in surface water using the yeast two-hybrid RAR assay (ATRA EQ <0.4–8 ng_{ATRA}/L) (Chinathamby *et al.*, 2013) and an *in vitro* reporter gene bioassay using P19/A15 cells (ATRA EQ <10–29 ng_{ATRA}/L) (Javurek *et al.*, 2015). Escher *et al.* (2014) also applied the P19/A15 assay to wastewater, recycled water, surface water and drinking water extracts, with only one wastewater effluent sample inducing 10% effect after 25 times enrichment in the assay.

10.5 BIOASSAYS INDICATIVE OF OTHER RECEPTOR-MEDIATED EFFECTS

In addition to hormone receptor-mediated effects, other relevant specific modes of action include phytotoxicity and neurotoxicity.

10.5.1 Phytotoxicity

Although not directly relevant for human health, several studies have applied algal assays to assess photosystem II (PSII) inhibition in a range of water matrices (Tang and Escher, 2014; Hamers *et al.*, 2018). Most studies have assessed PSII inhibition using the combined algae test (CAT) with PSII inhibition measured after 2 h using imaging pulse-amplitude modulated (PAM) fluorometry using green microalgae *Raphidocelis subcapitata* (formerly named *Selenastrum capricornutum* and *Pseudokirchneriella subcapitata*) (Escher *et al.*, 2008a; Glauch and Escher 2020). The assay reference compound is the herbicide diuron, with reported EC_{50} values ranging from 1.40 to 4.3 $\mu g_{diuron}/L$ (Jia *et al.*, 2015; Allan *et al.*, 2017). In contrast, only two studies used Max-I-PAM with *Chlorella vulgaris*, with EC_{50} values of 16 $\mu g_{diuron}/L$ reported (Macova *et al.*, 2010; Leusch *et al.*, 2014a).

Table 10.28 Summary of studies that have applied photosystem II (PSII) inhibition assays to different environmental water extracts, where PSII inhibition is expressed as diuron EQ.

Matrix	Assay	PSII Inhibition Diuron EQ (μg_{diuron}/L)	Reference
Wastewater influent	CAT (2 h IPAM)	0.07–2.2	1, 2, 3
	Max-I-PAM[a]	0.04–0.23	3
Wastewater effluent	CAT (2 h IPAM)	0.03–1.3	1, 2, 5, 6, 7, 8, 9
	Max-I-PAM[a]	<0.03–0.12	4, 10
Recycled water	CAT (2 h IPAM)	<0.004–1.3	1, 2, 3, 8, 9
	Max-I-PAM[a]	0.02–<0.03	3, 10
Surface water	CAT (2 h IPAM)	0.01–1.3	1, 2, 3, 6, 7, 11
	Max-I-PAM[a]	<0.03–0.06	4
Drinking water	CAT (2 h IPAM)	0.02–0.05	1, 2

NB: Only studies that have applied SPE or LLE are included. CAT = combined algae test.
[a]The Max-I-PAM assay uses *Chlorella vulgaris*, while all other assays used *Raphidocelis subcapitata* (formerly named *Selenastrum capricornutum* and *Pseudokirchneriella subcapitata*).
References: [1](Macova *et al.*, 2011); [2](Tang and Escher, 2014); [3](Glauch and Escher 2020); [4](Leusch *et al.*, 2014a); [5](Escher *et al.*, 2008b); [6](Jia *et al.*, 2015); [7](Neale *et al.*, 2017c); [8](Reungoat *et al.*, 2010); [9](Tang *et al.*, 2014); [10](Macova *et al.*, 2010); [11](Allan *et al.*, 2017).

Consequently, it appears that *R. subcapitata* is more sensitive. PSII inhibition is commonly expressed in diuron EQ (DEQ), with 0.04–2.2 μg_{diuron}/L DEQ detected in wastewater influent, 0.03–1.3 μg_{diuron}/L DEQ detected in wastewater effluent and 0.01–1.3 μg_{diuron}/L DEQ detected in surface water (Table 10.28). Based on the EC_{10} value, this translates to an effect being detected after 0.3–99 times enrichment. Low activity was reported in drinking water (0.02–0.05 μg_{diuron}/L DEQ) and reverse osmosis-treated recycled water (<0.004–0.05 μg_{diuron}/L DEQ) (Table 10.28). The observed effects in water extracts were primarily explained by PSII-inhibiting herbicides, with other chemicals only having a minor contribution to the observed effect (Tang and Escher, 2014; Neale *et al.*, 2017c; Glauch and Escher 2020). Therefore, the PSII inhibition assay could be applied in cases where agricultural activities may potentially impact source water quality.

10.5.2 Neurotoxicity

Neurotoxicity assays applied for ecotoxicology and water quality monitoring have been recently reviewed (Legradi *et al.*, 2018). Inhibition of acetylcholinesterase (AChE) is the most common *in vitro* endpoint for specific neurotoxicity measured in water quality assessment. AChE is the enzyme responsible for recycling the neurotransmitter acetylcholine and its inhibition indicates interference with neurotransmitter metabolism. The AChE inhibition assay was first developed by Ellman *et al.* (1961) and is a useful tool for detection of neurotoxic insecticides such as the organophosphates and carbamate insecticides. The assay was later optimised for use with environmental samples by Hamers *et al.* (2000), who validated the technique with rainwater samples. It has been used to assess neurotoxicity in diverse types of environmental water extracts (Macova *et al.*, 2011; Toušová *et al.*, 2017). Organophosphate insecticide parathion is commonly used as the assay reference compound and the reported parathion EC_{50} values range from 26.3 to 120 μg/L (Escher *et al.*, 2008b; Macova *et al.*, 2010). AChE inhibition was detected in wastewater influent, wastewater effluent, recycled water, surface water and drinking water extracts (Table 10.29), but the assay does suffer from false-positives that limit its application. For example, Neale and Escher (2013) found that dissolved organic carbon (DOC) concentrations as low as 2 mg of carbon/L (mg_C/L) caused quenching in the assay. DOC can be co-extracted during sample processing, so the AChE inhibition assay is not recommended for DOC-rich samples, such as wastewater or surface water.

Additional *in vitro* neurotoxicity assays that are currently available, although most are still to be validated for water quality assessment, include neuronal and glial cell viability assays using SK-N-SH (and derivatives, such as SH-SY5Y cells) and C6 cells. These assays further include precursor cell differentiation

Table 10.29 Summary of studies that have applied the enzymatic acetylcholinesterase (AChE) inhibition assay to different environmental water extracts, where the AChE inhibition is expressed as parathion EQ.

Matrix	AChE Inhibition Parathion EQ ($\mu g_{parathion}/L$)[a]	Reference
Wastewater influent	4.4–6.0	1
Wastewater effluent	<0.04[b]–3.9	1–8
Recycled water	<0.06–1.2, <2.6[a]	1, 3–5, 8
Surface water	0.11–0.27, <2.6[a]	1–3, 6, 7
Drinking water	0.28 to <2.6[a]	1, 3

NB: Only studies that have applied SPE or LLE are included.
[a]Parathion EQ calculated using the parathion EC_{10} in Neale *et al.* (2017c); [b]Chlorpyrifos EQ converted to Parathion EQ based on relative effect potency (REP) of 4.05 (Neale and Escher, 2013).
References: [1](Macova *et al.*, 2011); [2](Escher *et al.*, 2008b); [3](Escher *et al.*, 2014); [4](Leusch *et al.*, 2014b); [5](Macova *et al.*, 2010); [6](Neale and Escher, 2013); [7](Neale *et al.*, 2017c); [8](Reungoat *et al.*, 2010).

and apoptosis assays using neuroblastoma cells, glial maturation (myelination) in U-373MG human astrocytoma cells, neurotransmitter receptor profiles in neuroblastoma cells and interference with neurotransmitter enzymes or postsynaptic receptors (reviewed in Atterwill *et al.*, 1994; Costa, 1998; Tiffany-Castiglioni *et al.*, 2006; Coecke *et al.*, 2007). The neuroblastoma assay was developed for detection of natural neurotoxins such as cyanotoxins and paralytic shellfish toxins (Kogure *et al.*, 1988; Jellett *et al.*, 1992; Manger *et al.*, 1993; Manger *et al.*, 1995). This assay is relevant for testing of recreational waters such as lakes and drinking water withdrawn from surface water. Although the neuroblastoma assay has been mostly tested with freeze-dried samples, it is becoming an increasingly applied tool for testing of water quality (Wood *et al.*, 2006; Cetojevic-Simin *et al.*, 2009; Campora *et al.*, 2010; Kerbrat *et al.*, 2010). For a xenobiotic to be toxic to the central nervous system, it must first cross the blood–brain barrier. This process can be modelled *in vitro* using immortalised brain endothelial cells, for example, SV-HCEC, HBEC-51 or BB19 cells (Prieto *et al.*, 2004), but is yet to be trialled with water samples.

10.5.3 Other assays

In addition to the above studies, new bioassays have been developed specifically to detect modes of action relevant to pharmaceuticals. For example, Bernhard *et al.* (2017) have developed beta-blocker and non-steroidal anti-inflammatory drug assays, with a limit of detection of 2 μg/L metoprolol EQ and 0.5 μg/L diclofenac EQ, respectively. The assays were applied to SPE-enriched wastewater effluent, with 3.2–4.2 μg/L metoprolol EQ and 3.5 μg/L diclofenac EQ detected. Recent studies have applied a TGFα-shedding assay to assess the biological activity of pharmaceuticals that bind to G protein-coupled receptors (Zhang *et al.*, 2018a; Ihara *et al.*, 2020). SPE-enriched wastewater from Japan and the UK were applied, with most samples found to be angiotensin (AT1), dopamine (D2), adrenergic (β1), acetylcholine (M1) and histamine (H1) receptor antagonists (Zhang *et al.*, 2018a). These assays have not been widely applied to date.

10.6 BIOASSAYS INDICATIVE OF REACTIVE TOXICITY

Reactive toxicity occurs when chemicals form covalent bonds with DNA, proteins and membrane lipids. The field of *in vitro* cancer research has been active for several decades and a few bioassay tools applied for water quality assessment have been adapted from the medical field. Genotoxicity assays include those detecting cytogenetic damage (*i.e.*, structural DNA damage, *e.g.*, the Comet assay) or mutagenicity (*i.e.*, introduction of mutations, *e.g.*, the Ames test). Surface water and wastewater have received early attention in terms of bioanalytical screening of genotoxicants. Disinfection by-products (DBPs), which form due to the reaction of disinfectants such as chlorine and chloramine with organic matter in drinking and swimming pool water, are under increasing scrutiny for their contribution to genotoxicity and carcinogenicity. DBPs are reactive chemicals and are responsive in a number of assays indicative of reactive toxicity (Stalter *et al.*, 2016a). To date, there have been >400 studies on genotoxicity of drinking water (Cortes and Marcos, 2018; DeMarini, 2020).

Most genotoxicity and mutagenicity assays can be run either with or without rat liver S9 fraction, which is used to simulate metabolic activation. Epigenetic carcinogens are chemicals that are not DNA-reactive and can cause cancer via non-genotoxic mechanisms such as alteration of gene expression. In chemical risk assessment, cell transformation assays using BALB/c 3T3 mouse fibroblast cells closely model the various stages of *in vivo* carcinogenesis and provide an integrated measure of carcinogenicity via multiple mechanisms including both genotoxic and non-genotoxic pathways (Combes *et al.*, 1999). To our knowledge, these tests have not yet been used in water quality assessment and in the following detailed sections we focus on genotoxicity and mutagenicity and briefly discuss reactivity towards proteins and oxidative stress.

10.6.1 Genotoxicity

The umuC assay, which is also known as umu or SOS/umu, is used to assess DNA damage via the inducible SOS response (Table 10.30). SOS genes are repressed under normal conditions but are released in the presence of DNA damage to help repair any damage (Michel, 2005). A number of *Salmonella typhimurium* strains including TA1535/pSK1002, NM2009, NM3009 and NM5004 and the *Escherichia coli* SOS Chromotest have been applied to water extracts (Escher *et al.*, 2014; Han *et al.*, 2016). However, most studies have used TA1535/pSK1002 either with or without metabolic activation, so this section will focus on the TA1535/pSK1002 strain. The reference compound without metabolic activation is usually 4-nitroquinoline *N*-oxide (4NQO) ($EC_{IR1.5}$: 9.47×10^{-8} M (Macova *et al.*, 2011)), while 2-aminoanthracene (2AA) ($EC_{IR1.5}$: 2.42×10^{-7} M (Tang *et al.*, 2014)) is often used as the reference compound with metabolic activation.

The SOS Chromotest (Quillardet *et al.*, 1982) and SOS *umu/umuC* (Oda *et al.*, 1985) assays were developed in the 1980s and are commonly used tools to screen environmental water for genotoxicity (Table 10.30). Both SOS techniques respond to genotoxicants through colorimetric detection of the SOS response, which is induced by DNA damage. In the early 1990s, the SOS *umu* and Chromotest assays were optimised for high throughput screening of surface waters (Reifferscheid *et al.*, 1991; Langevin *et al.*, 1992). More recently, the Vitotox kit was developed for detection of SOS response by luminescence (van der Lelie *et al.*, 1997; Verschaeve *et al.*, 1999). The Vitotox assay is also in use for water quality screening (Pessala *et al.*, 2004).

The Comet assay (also known as single cell gel electrophoresis (SCGE) assay) is another popular technique for detection of reactive toxicity in polluted waters

Table 10.30 Selection of *in vitro* assays used for assessment of genotoxicity in water samples, which target chlorinated by-products, aromatic amines, PAHs, haloacetic acids and other disinfection by-products (DBP).

Assay for DNA Damage	Cell Type	Endpoint	Reference
SOS response assays: umuC assay (also called *umu* and SOS/*umu*), *umu* microtest and SOS Chromotest	Bacterium *S. typhimurium* TA 1535/pSK1002	Induction of the *umu* operon (SOS response) activates β-galactosidase, which can metabolise the substrate to a coloured product for colorimetric measurement	1–8
Cytotoxicity in SOS defective *E. coli*	Bacterium *E. coli* (several K12 AB and KL strains)	Colony formation	9, 10
Vitotox assay (kit for detection of SOS response)	Bacteria genetically modified *S. typhimurium* (TA 104 *rec*N2–4 strain)	SOS response, which induces luminescence (the TA 104 *pr1* strain, which constantly expresses lux genes is used as positive control)	11, 12
Comet assay (also known as SCGE assay)	A variety of mammalian (incl. human) cells, also fish liver cells (zebrafish *Danio rerio* and rainbow trout *Oncorhynchus mykiss* RTL-W1, RTH-149)	Measures DNA double-strand breaks in single cells (single-strand breaks in some variants). Staining technique, fluorescence. Image analysis results in an output image resembling a comet. The body of the comet represents undamaged cells and the tail, the damaged cells	13–18
Alkaline yeast comet/SCGE	Yeast *Saccharomyces cerevisiae* DLH3	Same as normal comet but appears to be more sensitive than mammalian cell line	19

(*Continued*)

Table 10.30 Selection of *in vitro* assays used for assessment of genotoxicity in water samples, which target chlorinated by-products, aromatic amines, PAHs, haloacetic acids and other disinfection by-products (DBP) (*Continued*).

Assay for DNA Damage	Cell Type	Endpoint	Reference
Micronucleus formation measured by flow cytometry (FCMN or FCMMN assay)	Non-secreting human lymphoblast (WIL2-NS)	Micronucleus formation, measured by flow cytometry	20,21
Propidium iodide (PI) staining and flow cytometry	Mammalian and human cell lines can be used	PI is fluorogenic and binds stoichiometrically to nucleic acid. DNA content can be quantified via fluorescence	22,23
GreenScreen EM (yeast reporter gene assay)	Yeast *S. cerevisiae* transfected with a plasmid incorporating γEGFP3	DNA damage, or rather the resulting DNA repair, which induces the GFP	24,25
SCE induction	Chinese hamster lung (CHL) cells	SCE is measured by a fluorescence staining technique	26–27

References: [1](ISO13829); [2](Oda *et al.*, 1985); [3](Hu *et al.*, 2007); [4](Cao *et al.*, 2009); [5](Reifferscheid *et al.*, 1991); [6](Langevin *et al.*, 1992); [7](Quillardet *et al.*, 1982); [8](White *et al.*, 1996); [9](Aleem and Malik, 2003); [10](Aleem and Malik, 2005); [11](van der Lelie *et al.*, 1997); [12](Verschaeve *et al.*, 1999); [13](Ostling and Johanson, 1984); [14](Rydberg and Johanson, 1978); [15](Singh *et al.*, 1988); [16](Schnurstein and Braunbeck, 2001); [17](Plewa *et al.*, 2002); [18](Wagner and Plewa, 2008); [19](Miloshev *et al.*, 2002); [20](Laingam *et al.*, 2008); [21](NWC, 2011); [22](Nicoletti *et al.*, 1991); [23](Riccardi and Nicoletti, 2006); [24](Cahill *et al.*, 2004); [25](Keenan *et al.*, 2007); [26](Ohe *et al.*, 2009); [27](Perry and Wolff, 1974).

(Table 10.30). This assay relies on the differences in migration behaviour between intact and damaged DNA in an electric field (Rydberg and Johanson, 1978; Ostling and Johanson, 1984). Initially, the Comet assay was only able to detect double-strand breaks but was later optimised for detection of single-strand breaks (Singh *et al.*, 1988). The first applications of the Comet assay for water quality analysis were carried out in 2001 to assess the genotoxicity of rivers in Germany and China (Schnurstein and Braunbeck, 2001; Zhong *et al.*, 2001). Plewa and coworkers have developed both bacterial and mammalian assays to test for DBP genotoxicity. The mammalian assay applies a CHO cell line in the SCGE Comet assay run alongside the CHO microplate cytotoxicity assay (Plewa *et al.*, 2002; Plewa *et al.*, 2004a). This CHO cell adaptation of the assay has been utilised for assessment of recreational waters (Liviac *et al.*, 2010; Plewa *et al.*, 2011).

The GreenScreen is a microplate assay that measures DNA repair (as a consequence of DNA damage) in yeast transfected with green fluorescent protein (GFP) (Cahill *et al.*, 2004). The GreenScreen has been applied for assessment of genotoxicity in industrial effluents (Keenan *et al.*, 2007).

The ability to detect micronucleus formation by flow cytometry (Laingam *et al.*, 2008) has been exploited to detect this response in human lymphocytes (WIL2-NS) exposed to a variety of water matrices, including treated sewage, reclaimed water and drinking water (Leusch *et al.*, 2014b).

There are several additional *in vitro* assays to detect genotoxic carcinogens (Table 10.30), including thymidine kinase and hypoxanthine guanine phosphoribosyltransferase mutation assays, sister chromatid exchange (SCE) assays and chromosomal aberration assays (reviewed in Combes *et al.*, 1999; Kowalski, 2001), and new assays and protocols are regularly developed (Corvi and Madia, 2017) but these have generally seen less to no application in water quality monitoring.

The umuC assay has been applied to wastewater, surface water, recycled water and drinking water (Table 10.31). Wastewater influent and wastewater effluent were the most responsive, with many samples not having a response up to the maximum REF in surface water or highly treated recycled water (*e.g.*, RO or ozone and biological-activated carbon (O_3/BAC)). Increased genotoxicity after disinfection in an Australian drinking water treatment plant (DWTP) was observed by Neale *et al.* (2012). Genotoxicity increased from 4NQO EQ (−S9) of 0.05 $\mu g_{4NQO}/L$ and 2AA EQ (+S9) of 0.18 $\mu g_{2AA}/L$ at the inlet to 4NQO EQ (−S9) 0.43 $\mu g_{4NQO}/L$ and 2AA EQ (+S9) of 1.29 $\mu g_{2AA}/L$ at the outlet. The effect at the outlet was observed after 23–33 times enrichment, with genotoxicity without metabolic activation more sensitive. This suggests that with sufficient enrichment, the umuC assay can be used to assess DBP formation during drinking water treatment. Tap water also induced a response in the umuC assay without S9 after 29–65 times enrichment (Stalter *et al.*, 2016b). Focusing on single DBPs, the addition of S9 did not increase genotoxicity and in some cases

Table 10.31 Summary of studies that have applied the umuC assay indicative of genotoxicity to environmental water extracts, where genotoxicity with metabolic activation (+S9) is expressed as 2-aminoanthracene (2AA) EQ and genotoxicity without metabolic activation (−S9) is expressed as 4-nitroquinoline *N*-oxide (4NQO) EQ.

Matrix	Strain	Genotoxicity +S9 2AA EQ ($\mu g_{2AA}/L$)	Reference
Wastewater influent	TA1535/pSK1002	2–19[a]	1–3
Wastewater effluent	TA1535/pSK1002	<0.2–4.6[a]	2, 3, 4–8
Recycled water	TA1535/pSK1002	<0.2–4.3	2–4, 7, 8
Surface water	TA1535/pSK1002	0.18–0.38[a], <1.6[a]	2, 4, 6, 9, 10
Drinking water	TA1535/pSK1002	<0.34[b]–1.29, <1.6[a]	2, 4, 10
Matrix	**Strain**	**Genotoxicity −S9 4NQO EQ ($\mu g_{4NQO}/L$)**	**Reference**
Wastewater influent	TA1535/pSK1002	0.48–6.9[c]	1–3
Wastewater effluent	TA1535/pSK1002	0.12–5.8[c]	2–4, 6–8, 11
Recycled water	TA1535/pSK1002	<0.05–0.72[c]	2–4, 7, 8, 10
Surface water	TA1535/pSK1002	0.01–0.69	2, 4, 9, 10, 12, 13
Drinking water	TA1535/pSK1002	<0.05–0.61[c]	2, 4, 10, 14

NB: Only studies that have applied SPE or LLE are included.
[a]2AA EQ calculated using 2AA $EC_{IR1.5}$ of 46.7 ng/L (Tang *et al.*, 2014).
[b]Results presented as benzo[*a*]pyrene (B[*a*]P) EQ and converted to 2AA EQ using the REP from Macova *et al.* (2011);
[c]4NQO EQ calculated using 4NQO $EC_{IR1.5}$ of 18 ng/L (Macova *et al.*, 2011).
References: [1](Lee *et al.*, 2015); [2](Macova *et al.*, 2011); [3](Tang *et al.*, 2014); [4](Escher *et al.*, 2014); [5](Escher *et al.*, 2008b); [6](Fang *et al.*, 2012); [7](Jia *et al.*, 2015); [8](Macova *et al.*, 2010); [9](Farre *et al.*, 2013); [10](Neale *et al.*, 2012); [11](Reungoat *et al.*, 2010); [12](Han *et al.*, 2016); [13](Sun *et al.*, 2017); [14](Stalter *et al.*, 2016b).

reduced the potency (Stalter *et al.*, 2016a), suggesting many DBPs are direct genotoxicants.

10.6.2 Mutagenicity

The bacterial Ames assay for detection of mutagens was employed for water quality testing soon after its publication in 1975 (Ames *et al.*, 1975) (Table 10.32). Even then, the assay was often applied directly, without extraction, to a wide range of water types including surface water (Pelon *et al.*, 1977; Vankreijl *et al.*, 1980), ozonated recycled water (Gruener, 1978), coal gasification process water (Epler *et al.*, 1978), drinking water (Simmon and Tardiff, 1976; Nestmann *et al.*, 1979;

Table 10.32 Selection of *in vitro* assays used for assessment of mutagenicity in water samples, which target chlorinated by-products, aromatic amines, PAHs, haloacetic acids and other DBPs.

Assay for Mutagenicity	**Cell Type**	**Endpoint**	**Reference**
Ames test (and modified Ames test)	Bacterium *S. typhimurium* (many strains incl. TA98, TA100 and 98NR)	Number of histidine revertants	1–4
Mutatox assay	Bacterium *Aliivibrio fischeri*	Genotoxic damage such as frame-shift mutations or base-substitution point mutations and more, which induce a dark variant of *A. fischeri* to regain its luminescence	5
Alternative mutagenicity test	Yeast *S. cerevisiae* D7 diploid strain	Formation of 'mutagen-specific' colonies on selective media	6

References: [1](Ames *et al.*, 1975); [2](Maron and Ames, 1983); [3](Barrueco *et al.*, 1991); [4](Kado *et al.*, 1986); [5](Ulitzur *et al.*, 1980); [6](Zimmermann *et al.*, 1975).

Cheh *et al.*, 1980), marine water (Kurelec *et al.*, 1979), pulp and paper mill effluents (Bjorseth *et al.*, 1979; Carlberg *et al.*, 1980) and different wastewaters (Rappaport *et al.*, 1979; Saxena and Schwartz, 1979). The *Salmonella* preincubation assay is a modified Ames mutagenicity assay (Kargalioglu *et al.*, 2002; Plewa *et al.*, 2004b), which is run in conjunction with the *Salmonella typhimurium* microplate cytotoxicity assay and was applied to assess DBPs.

Most of the studies on mutagenicity testing of water quality simply reported a positive or negative response (Berninger *et al.*, 2019; Albergamo *et al.*, 2020), without determining an effect concentration (EC) or bioanalytical equivalent concentration but there is also guidance on how to interpret such data (Roubicek *et al.*, 2020). Table 10.33 includes only studies that have reported an EC value, with studies reporting the effect concentration inducing a revertant ratio of 1.5 ($EC_{RR1.5}$).

The *S. typhimurium* strains applied in Table 10.33 include TA98, which responds to frameshift mutations, and TA100 and TAmix, which both respond to base pair substitutions (Kamber *et al.*, 2009). Escher *et al.* (2014) found increasing mutagenicity (*e.g.*, lower $EC_{RR1.5}$ values) in treated drinking water compared to source water for all strains tested, with little difference between strains or with or without S9. Effects were detected after 3–14 times enrichment in drinking water, 5–25 times enrichment in surface water and 0.6–83 times enrichment in wastewater effluent (Table 10.33). Furthermore, Ames strains TA98, TA100 and YG7108, the latter of which is responsive to nitrosamines, were tested with and

Table 10.33 Summary of studies that have applied the Ames assay indicative of mutagenicity to different environmental water extracts, where mutagenicity is expressed as an effect concentration inducing a revertant ratio RR of 1.5 ($EC_{RR1.5}$) in units of REF.

Matrix	Strain	$EC_{RR1.5}$ (REF) +S9	Reference
Wastewater effluent	TA98	3.5 to >100	1, 2
	TAmix	2.9–66	1, 2
Recycled water	TA98	12.5 to >100	1, 2
	TAmix	13.7 to >100	1, 2
Surface water	TA98	4.5	1
	TAmix	>30	1
Drinking water	TA98	3.2	1
	TAmix	13.8	1
Matrix	**Strain**	**$EC_{RR1.5}$ (REF) −S9**	**Reference**
Wastewater effluent	TA98	6.3 to >100	1, 2
	TA100	0.6–16	1, 2
	TAmix	6.9–83	1, 2
Recycled water	TA98	>30 to >100	1, 2
	TA100	0.5 to >30	1, 2
	TAmix	35–69	1, 2
Surface water	TA98	14	1
	TA100	25	1
	TAmix	>30	1
Drinking water	TA98	4.6	1
	TA100	5.0	1
	TAmix	4.9	1

References: [1](Escher *et al.*, 2014); [2](Jia *et al.*, 2015).

without metabolic activation in source water and treated drinking water extracts from three French DWTPs (Neale *et al.*, 2020b). However, none of the samples had an effect up to the maximum REF of 200.

10.6.3 Non-genotoxic electrophilic mechanisms

Reactive toxicity can also take place via electrophilic attack by electron-deficient chemicals (electrophiles, *e.g.*, the pesticide atrazine), which can bind to nucleophilic (electron donating) groups on endogenous molecules causing structural damage. An example of a biological nucleophile is the amino acid cysteine in peptides and proteins, and the bases of DNA. Although reactivity with DNA causes genotoxicity and has been subject to intensive research, no bioassays indicative of reactive toxicity caused by binding to cysteine have been reported for water quality assessment. This is an important gap in knowledge as damage to proteins may also lead to adverse effects. Glutathione (GSH) is a small cysteine that protects the cell as an antioxidant and by detoxifying electrophiles. Xenobiotic exposure thus has the potential to cause GSH depletion, which can ultimately lead to protein damage. The naturally high concentration of GSH in cells allows its depletion to be quantified using chemical methods or an enzymatic assay. Although depletion of cellular GSH has been assessed in various cell types and water samples (Table 10.34), it is difficult to interpret the significance of the results.

The role of GSH for cell viability can also be assessed with a differential bacterial assay. The wild type *E. coli* strain MJF276 and its mutant strain MJF335 differ in that, MJF355 lacks the enzymes to synthesise GSH. In a growth inhibition assay, both strains show the same sensitivity, unless the tested chemical is a reactive electrophile. In this case, the mutant strain cannot defend the cell and thus the resulting EC_{50} will be lower than for the wild type (Harder *et al.*, 2003; Richter and Escher, 2005). This assay has been applied for fingerprinting the toxicity of DBPs (Stalter *et al.*, 2016a) but proved to be too insensitive for drinking water testing.

Table 10.34 Selection of *in vitro* assays used for assessment of reactive toxicity towards proteins (GSH depletion).

Assay	Cell Type	Endpoint	Reference
GSH assay	Human liver cells (Hep-G2)	GSH depletion. Fluorimetric quantification of GSH concentration in cells	1, 2
GSH reductase enzymatic recycling assay (modified)	Rainbow trout primary hepatocytes	GSH depletion. Colorimetric measurement of GSH concentration in cells	3–5
Differential bacterial growth inhibition assay	Bacterium *E. coli* (MJF276 and MJF335 (mutant))	Growth inhibition. Comparison of the EC_{50}s of two differential strains of which, the mutant lacks GSH for cellular defence	6, 7

References: [1](Hissin and Hilf, 1976); [2](Marabini *et al.*, 2006); [3](Owens and Belcher, 1965); [4](Baker *et al.*, 1990); [5](Farmen *et al.*, 2010); [6](Harder *et al.*, 2003); [7](Richter and Escher, 2005). GSH = glutathione.

10.6.4 Oxidative stress

A selection of available assays for detection of oxidative stress in water is listed in Table 10.35. The presence of reactive oxygen species (ROS) is used as a warning sign of oxidative stress in water samples (Marabini *et al.*, 2006). ROS can be quantified using a colorimetric method, in which oxidation by ROS causes a substrate added to the cells to fluoresce. This endpoint, however, lacks specificity. In a study on disinfected drinking water, where additional indicators of oxidative stress were measured (Table 10.35), it was shown that the products of lipid peroxidation were present and that antioxidant enzymes were active although no ROS could be quantified (Shi *et al.*, 2009b). ROS can also be measured as the production of free radicals (superoxide and hydroxyl radical) in a cell line exposed to a water sample (Xie *et al.*, 2010). Increased levels of the oxidised form of GSH, the GSH disulphide GSSG, in relation to the concentration of GSH is suggestive of oxidative stress but a study on surface water found no clear correlation between the GSH/GSSG ratio and ROS formation (Neale *et al.*, 2017a).

Table 10.35 Selection of *in vitro* assays used or potentially useful for assessment of oxidative stress in water samples.

Assay	Cell Type	Endpoint	Reference
ROS assay (indirect detection of ROS)	Human liver cells (Hep-G2)	Oxidation of a substrate (DCFH-DA) leads to a fluorescent product	1–4
ROS assay (as above)	Rainbow trout primary hepatocytes	Oxidation of a substrate (DCF-DA) leads to a fluorescent product	5
GSH/GSSG-Glo assay	Human liver cells (Hep-G2	GSH/GSSG ratio	4
Antioxidant response	Hep-G2 cells	Enzyme activity of antioxidase, GSH peroxidase (GSH-Px), superoxide dismutase (SOD)	3, 6, 7
Antioxidant response	Human liver cells (Hep-G2, L-02)	Cellular concentration of lipid peroxidation product malonaldehyde (MDA)	3, 8, 9

References: [1](Wang and Joseph, 1999); [2](Marabini *et al.*, 2006); [3](Shi *et al.*, 2009b); [4](Neale *et al.*, 2017a); [5](Farmen *et al.*, 2010); [6](Flohe and Gunzler, 1984); [7](Oberley and Spitz, 1984); [8](Yagi, 1998); [9](Xie *et al.*, 2010). DCF-DA or DCFH-DA = 2′,7′-dichlorofluorescein diacetate.

Water samples that showed ROS formation also often activated the oxidative stress response (Neale *et al.*, 2017a). A complementary endpoint would therefore be induction of the mammalian cellular defence mechanism against oxidative stress, which is activated by multiple pathways (not only ROS) and is discussed in detail in Section 10.7.1.

10.7 BIOASSAYS INDICATIVE OF ADAPTIVE STRESS RESPONSES

Adaptive stress response pathways are activated to help restore cells back to homeostasis after damage from stressors, including organic chemicals (Simmons *et al.*, 2009). This chapter will focus on three adaptive stress response pathways commonly applied to environmental water extracts: oxidative stress response (Nrf2), p53 response for genotoxicity and NF-κB response for inflammation. Assays indicative of adaptive stress responses hypoxia and heat shock response did not have a response in drinking water, surface water, wastewater or recycled water extracts (Escher *et al.*, 2014).

10.7.1 Oxidative stress response

Chemicals that produce ROS and electrophilic chemicals can induce the oxidative stress response, in particular the antioxidant defence pathway Keap1-Nrf2 (Kobayashi *et al.*, 2009). These chemicals release transcription factor Nrf2 from the negative regulator Keap1, which then translocates to the nucleus and activates the antioxidant response element (Zhang, 2006). Five mammalian reporter gene assays have been applied to evaluate the oxidative stress response in environmental water extracts (Table 10.36). *tert*-Butylhydroquinone (*t*BHQ) is often used as the assay reference compound, with similar $EC_{IR1.5}$ or lowest observed effect concentration (LOEC) at an induction factor of 1.5 for AREc32, ARE GeneBLAzer, Nrf2 CALUX and Nrf2 reporter gene assay. The *t*BHQ $EC_{IR1.5}$ value for Nrf2-MDA-MB was over an order of magnitude higher (*i.e.*, less toxic) and the assay did not report any effect with drinking water, surface water, wastewater effluent or recycled water extracts (Escher *et al.*, 2014; Jia *et al.*, 2015). Consequently, Nrf2-MDA-MB does not appear to be suitable for environmental water extracts.

Effects in drinking water have been detected after 3–102 times enrichment (Table 10.37), with drinking water treated with membrane filtration often not inducing the oxidative stress response up to the maximum tested REF of 100–150 (Albergamo *et al.*, 2020; Neale *et al.*, 2020b). In contrast, an increase in the oxidative stress response has been observed after drinking water disinfection (Neale *et al.*, 2012; Escher *et al.*, 2013; Hebert *et al.*, 2018), indicating that oxidative stress response assays can detect formed DBPs. The $EC_{IR1.5}$ in surface water varied from 0.6 to 93 REF, while effects were observed in wastewater

Table 10.36 Common assays applied to evaluate the oxidative stress response in water extracts.

Assay	Cell Line	Detection Method	tBHQ $EC_{IR1.5}$ (M)	tBHQ $EC_{IR1.5}$ (μg/L)	EC Reference
AREc32	MCF-7	Luminescence	1.32×10^{-6}	219	Escher *et al.* (2012)
ARE GeneBLAzer	HepG2	Fluorescence	2.44×10^{-6}	406	Neale *et al.* (2015b)
Nrf2 CALUX	U2OS	Luminescence	1.00×10^{-6} [a]	166	van der Linden *et al.* (2014)
Nrf2 reporter gene assay	HepG2	Luminescence	2.00×10^{-6}	332	Lundqvist *et al.* (2019a)
Nrf2-MDA-MB	MDA-MB-231-745	Luminescence	3.30×10^{-5}	5490	Jia *et al.* (2015)

[a]LOEC at an induction factor of 1.5. *t*BHQ = *t*-butylhydroquinone.

influent and effluent from $EC_{IR1.5}$ 0.3 to 30 REF and $EC_{IR1.5}$ 2 to 47 REF, respectively (Table 10.37). The reported WWTP removal efficacy ranged from 61% to 85% (Volker *et al.*, 2017; Nivala *et al.*, 2018). Effects in recycled water ranged from $EC_{IR1.5}$ 4 to 94 REF.

10.7.2 p53 response

The p53 response responds to DNA damage by initiating repair proteins, changing the cell cycle or by causing apoptosis (Knight *et al.*, 2009). Two assays, the p53 GeneBLAzer and p53 CALUX, have been applied to environmental water extracts (Table 10.38). The two assays use different reference compounds, making it difficult to compare assay sensitivity, with p53 CALUX results reported in both actinomycin D EQ and cyclophosphamide EQ. Most studies report either no effect or cytotoxicity in p53 response assays (Table 10.39). The p53 response in the Danube River was observed after 65 times enrichment in the p53 GeneBLAzer, which gave a mitomycin EQ of 235 $ng_{mitomycin}$/L (Neale *et al.*, 2015b). p53 activity was also detected in wastewater influent and effluent in Finland using the p53 CALUX after the addition of S9 for metabolic activation (Välitalo *et al.*, 2017), with the results expressed as cyclophosphamide EQ.

10.7.3 NF-κB response

The NF-κB-mediated response to inflammation can be induced by a range of compounds including metals, carcinogens and bacterial products (Ahn and

Table 10.37 Summary of studies that have applied assays indicative of the oxidative stress response to different environmental water extracts, where the oxidative stress response is expressed as the concentration causing an induction ratio of 1.5 ($EC_{IR1.5}$) in units of REF.

Matrix	Assay	$EC_{IR1.5}$ (REF)	Reference
Wastewater influent	AREc32	0.28–4.7	1–4
	Nrf2 reporter gene assay	8.1–30	5
Wastewater effluent	AREc32	1.5–22	1–4, 6–8
	ARE GeneBLAzer	8.9–17	9
	Nrf2-CALUX	4.8	10
	Nrf2 reporter gene assay	47 to >50	5
	Nrf2-MDA-MB	>10	7, 10
Recycled water	AREc32	4.2–94	1, 3, 6, 7
	Nrf2-CALUX	4.8 to >30	10
	Nrf2-MDA-MB	>10 to >20	7, 10
Surface water	AREc32	0.6 to >100	1, 6, 8, 11–17
	ARE GeneBLAzer	6.9 to >490	9, 18, 19
	Nrf2-CALUX	6.9	10
	Nrf2 reporter gene assay	22	20
	Nrf2-MDA-MB	>20	10
Drinking water	AREc32	2.5 to >150	1, 6, 14, 16, 21–23
	Nrf2-CALUX	2.9	1
	Nrf2 reporter gene assay	21–25	20
	Nrf2-MDA-MB	>20	10

NB: Only studies that have applied SPE or LLE are included.
References: [1](Escher *et al.*, 2012); [2](Nivala *et al.*, 2018); [3](Tang *et al.*, 2014); [4](Volker *et al.*, 2017); [5](Lundqvist *et al.*, 2019b); [6](Escher *et al.*, 2013); [7](Jia *et al.*, 2015); [8](Mueller *et al.*, 2021); [9](Neale *et al.*, 2017c); [10](Escher *et al.*, 2014); [11](Farre *et al.*, 2013); [12](Hashmi *et al.*, 2018); [13](Müller *et al.*, 2018); [14](Neale *et al.*, 2012); [15](Neale *et al.*, 2018a); [16](Neale *et al.*, 2020b); [17](Neale *et al.*, 2020a); [18](König *et al.*, 2017); [19](Neale *et al.*, 2015b); [20](Lundqvist *et al.*, 2019a); [21](Albergamo *et al.*, 2020); [22](Hebert *et al.*, 2018); [23](Stalter *et al.*, 2016b).

Aggarwal, 2005). These stimuli destabilise the IkB–NF-κB complex, allowing NF-κB to activate target genes in the nucleus (Gilmore, 2006). Three reporter gene assays have been applied to evaluate the NF-κB response in environmental

Table 10.38 Common assays applied to evaluate the p53 response in water extracts.

Assay	Cell Line	Detection Method	Reference Compound	EC_{10}* or $EC^{†}_{IR1.5}$ (M)	EC_{10}* or $EC^{†}_{IR1.5}$ (μg/L)
p53 CALUX (+/−S9)	U2OS	Luminescence	Actinomycin D	2.00×10^{-9}*	2.5*
p53 GeneBLAzer	HCT-116	Fluorescence	Mitomycin C	4.53×10^{-8}†	15†

water extracts (Table 10.40). The tumour necrosis factor alpha (TNFα) $EC_{IR1.5}$ value for NF-κB GeneBLAzer was 4.5 times lower than the NF-κB reporter gene assay developed by Lundqvist *et al.* (2019b) (Table 10.40), while no reference compound data were available for NF-κB CALUX. Extracts of drinking water, surface water and wastewater were active in the NF-κB GeneBLAzer assay, while only wastewater influent was active in the NF-κB reporter gene assay and no samples were active in NF-κB CALUX (Table 10.41). Although effects are often observed at low enrichment factors in NF-κB GeneBLAzer, the assay may not be suitable to evaluate the effects of micropollutants. Endotoxins activate NF-κB and a recent study showed that co-extracted endotoxins likely explained most of the effects in surface water extracts in the NF-κB GeneBLAzer assay (Neale *et al.*, 2018b).

10.8 BIOASSAYS INDICATIVE OF APICAL EFFECTS

In addition to assays indicative of different stages of the cellular toxicity pathway, cytotoxicity and whole organism assays indicative of apical effects are commonly applied to water quality monitoring. Furthermore, some organisms, such as zebrafish, are used as a model species for human health risk assessment (Bambino and Chu, 2017). Although these assays are often used for direct toxicity assessment, we have focused on assays applied to water extracts, including cytotoxicity, algal growth inhibition and fish embryo toxicity (FET). These assays can be run in well plates. The *Daphnia* immobilisation assay has also been applied to passive sampler extracts (Hamers *et al.*, 2018; de Baat *et al.*, 2019a; de Baat *et al.*, 2020), but to our knowledge not to SPE extracts.

10.8.1 Cytotoxicity

Cytotoxicity assays can be used to monitor cytotoxicity on its own but are mostly included as part of a test battery or for concurrent monitoring of acute toxicity as a quality assurance step for specific toxicity assays.

Table 10.39 Summary of studies that have applied assays indicative of the p53 response to different environmental water extracts, where the p53 response is expressed as the concentration causing an induction ratio of 1.5 ($EC_{IR1.5}$) or equivalent concentration (EQ).

Matrix	Assay	$EC_{IR1.5}$ (REF)	Equivalent Concentration	Reference
Wastewater influent	p53 CALUX +S9		61–6,200 μg/L Cyclophosphamide EQ	1
Wastewater effluent	p53 CALUX	>30	<53–540 μg/L Cyclophosphamide EQ	1, 2
	p53 CALUX +S9	>30		2
	p53 GeneBLAzer	>10, cytotoxic		2, 3
Recycled water	p53 CALUX	>30		2
	p53 CALUX +S9	>30		2
	p53 GeneBLAzer	>20 to >30		2
Surface water	p53 CALUX	>30		2
	p53 CALUX +S9	>30		2
	p53 GeneBLAzer	65 to >450, cytotoxic	<34–235 ng/L Mitomycin EQ	2–6
Drinking water	p53 CALUX	>30		2
	p53 CALUX +S9	>30		2
	p53 GeneBLAzer	>150		2, 6, 7

NB: Only studies that have applied SPE or LLE are included.

References: [1](Välitalo *et al.*, 2017); [2](Escher *et al.*, 2014); [3](Neale *et al.*, 2017c); [4](König *et al.*, 2017); [5](Neale *et al.*, 2015b); [6](Neale *et al.*, 2020b); [7](Hebert *et al.*, 2018).

Table 10.40 Common assays applied to evaluate the NF-κB response in water extracts.

Assay	Cell Line	Detection Method	TNFα $EC_{IR1.5}$ (ng/L)	EC Reference
NF-κB GeneBLAzer	THP-1	Fluorescence	20	Neale *et al.* (2015b)
NF-κB reporter gene assay	HepG2	Luminescence	90	Lundqvist *et al.* (2019b)
NF-κB CALUX	—	Luminescence	—	—

Table 10.41 Summary of studies that have applied assays indicative of the NF-κB response to different environmental water extracts, where the NF-κB response was expressed as the concentration causing an induction ratio of 1.5 ($EC_{IR1.5}$) in units of REF.

Matrix	Assay	$EC_{IR1.5}$ (REF)	Reference
Wastewater influent	NF-κB GeneBLAzer	0.05–0.36	1
	NF-κB reporter gene assay	0.30 to >50	2
Wastewater effluent	NF-κB GeneBLAzer	0.09 to >20	1, 3, 4
	NF-κB reporter gene assay	>50	2
	NF-κB CALUX	>30	3
Recycled water	NF-κB GeneBLAzer	>20	3
	NF-κB CALUX	>30	3
Surface water	NF-κB GeneBLAzer	0.1 to >250	3–8
	NF-κB CALUX	>30	3
Drinking water	NF-κB GeneBLAzer	10 to >500	3, 8, 9
	NF-κB CALUX	>30	3

NB: Only studies that have applied SPE or LLE are included.
References: [1](Nivala *et al.*, 2018); [2](Lundqvist *et al.*, 2019b); [3](Escher *et al.*, 2014); [4](Neale *et al.*, 2017c); [5](König *et al.*, 2017); [6](Neale *et al.*, 2015b); [7](Neale *et al.*, 2018b); [8](Neale *et al.*, 2020b); [9](Hebert *et al.*, 2018).
REF= relative extraction factor.

10.8.1.1 Bacterial toxicity

Bacterial bioluminescence inhibition assays, such as the Microtox and BLT-Screen assays, have been applied to drinking water, surface water and wastewater extracts (Tang *et al.*, 2013b; van de Merwe and Leusch, 2015). This type of assay utilises the light emission in naturally bioluminescent bacteria, for example, *Aliivibrio fischeri* (formerly called *Vibrio fischeri*), *Photobacterium phosphoreum* or *P. leiognathi*, as a measure of overall cellular energy status and health. A decreased light output indicates interference with energy metabolism and overall cellular health, reflecting the combined baseline toxicity of all chemicals in the sample. Bacterial bioluminescence inhibition assays are therefore suitable for screening of overall non-specific toxicity. In addition to the low cost and simplicity of these assays, their applications in water quality testing are wide ranging and well represented in the literature providing a large volume of comparative information. The water types tested with the Microtox, for example, span across effluents of coal gasification (Timourian *et al.*, 1982), oil refineries (Chang *et al.*, 1981), pulp mills (Rosa *et al.*, 2010) and sewage treatment plants (Farré *et al.*, 2002) as well as environmental waters (Dizer *et al.*, 2002) and drinking water (Guzzella *et al.*, 2004). Recent modifications have seen this type of assay adapted to 96-well plate format (Macova *et al.*, 2010).

Although bacterial toxicity assays are simple, they are fast (15–30 minutes) and are responsive to a wide range of organic micropollutants. For example, 50% inhibition of bioluminescence was observed after 0.5–17 times enrichment in wastewater influent, 2–27 times enrichment in wastewater effluent, 8–87 times enrichment in surface water and 3–40 times enrichment in drinking water (Table 10.42). Bacterial toxicity assays can detect the formation of DBPs, with increasing effect (*i.e.*, decreasing EC_{50} values) reported throughout

Table 10.42 Summary of studies that have applied bacterial toxicity assays to different environmental water extracts, where the effect is expressed as the concentration causing 50% effect (EC_{50}).

Matrix	Assay	EC_{50} (REF)	Reference
Wastewater influent	Microtox	0.48–17	1–4
Wastewater effluent	Microtox	3.0–27	1–6
	BLT-Screen	1.6	7
Recycled water	Microtox	10–102	1–3, 5, 6
Surface water	Microtox	8.2–87	1, 2, 6, 8, 9
	BLT-Screen	12	7
Drinking water	Microtox	3.2–40	1, 2, 6, 9, 10
	BLT-Screen	5.8	7

NB: Only studies that have applied SPE or LLE are included.
References: [1](Escher *et al.*, 2012); [2](Macova *et al.*, 2011); [3](Tang *et al.*, 2014); [4](Volker *et al.*, 2017); [5](Macova *et al.*, 2010); [6](Tang *et al.*, 2013b); [7](van de Merwe and Leusch, 2015); [8](Farre *et al.*, 2013); [9](Neale *et al.*, 2012); [10](Stalter *et al.*, 2016b).
REF= relative extraction factor.

DWTPs (Escher *et al.*, 2012; Neale *et al.*, 2012). Although bacterial toxicity assays only provide information about non-specific effects and should be complemented with assays indicative of specific effects, their advantage is that they can also be used in locations that only have access to microbiology laboratory facilities.

In addition to bacterial bioluminescence inhibition assays, luminescent bacterial biosensors indicative of DNA damage, oxidative stress and protein damage have been developed in recent years (Woutersen *et al.*, 2011). There is increasing interest in applying these biosensors as online water quality monitoring tools, although further research is required to improve sensitivity.

10.8.1.2 Mammalian cytotoxicity

Mammalian and human cell-based assays are typically applied to measure specific toxicity in reporter gene assays. As discussed, however, it is crucial when running such assays to also monitor baseline toxicity to ensure that the specific response is not masked by cytotoxicity. The cytotoxicity endpoint is a valuable endpoint in its own rights because it targets all chemicals in a water sample and can actually be applied for specific cell types that are representative and relevant for the particular case of concern. Different types of cells may exhibit differential toxicity due to selective and cell-specific function toxicity (Seibert *et al.*, 1996). A project assessing chemical-induced toxicity from drinking water, for example, should include a measure of cytotoxicity to gastro-intestinal and liver cells, which are the cell types likely to be exposed to the highest doses following drinking water intake.

Most cell viability assays rely on functional assays (Vinken and Blaauboer, 2017), such as mitochondrial activity quantified with the 3-[4,5-dimethylthiazol-2-yl]-2,5-diphenyltetrazolium (MTT) assay, where tetrazolium is reduced to formazan by the mitochondrial enzyme succinate dehydrogenase (Table 10.43). A similar assay principle underlies the Alamar Blue assay, where resazurin is reduced to resorufin by mitochondrial enzymes (Table 10.43). Another cell viability assay principle is the membrane integrity as can be assessed by the neutral red uptake (NRU) or the lactate dehydrogenase (LDH) assay. LDH is a stable enzyme that leaks from the cell in relatively high amounts upon cell plasma membrane damage and can be quantified colorimetrically (Table 10.43). Cell growth inhibition can be measured by absorbance, after staining and with bright-field microscopic imaging (Table 10.43).

10.8.1.3 Fish cell lines

Fish cell lines have been used for chemical and effluent assessment since the early 1990s (Table 10.44). Cytotoxicity is usually measured in fish cells via different staining techniques including Alamar Blue and NRU. The response in fish cell lines was well correlated with acute fish toxicity (Tanneberger *et al.*,

Table 10.43 Selection of human and other mammalian *in vitro* bioassays used for assessment of cytotoxicity in water samples.

Assay	Cell Type	Endpoint	Reference
MTT assay	Various human (*e.g.*, Hep-G2, MELN, HG5LN-hPXR (transfected HeLa cells)) and other mammalian cells (*e.g.*, mouse lymphoma cells $EL_{4.3}$)	Colorimetric measurement of live cells	1–6
Alamar Blue assay (also known as resazurin reduction assay)	Human kidney cells (HK2)	Reduction of substrate (Alamar Blue) by live cells yields fluorescent product	7–8
Caco2-NRU	Human epithelial colorectal adenocarcinoma cells (Caco-2)	Cell viability (measured by NRU test)	9–11
NRU	Human breast cancer cells (MCF-7)	Cell viability (measured by NRU test)	12, 2
LDH leakage, for example, CytoTox 96® Non-radioactive cytotoxicity assay	Human liver cells (HepG2)	Cell viability. Colorimetric measurement of LDH leakage from lysed cells	13–15
Mammalian cell microplate cytotoxicity assay	Chinese hamster ovary cells (CHO AS52)	Cell growth inhibition (measured by absorbance at 595 nm)	16–18
SRB assay	Mouse neuroblastoma cells (neuro-2A); human foetal lung cells (MRC-5)	Cell growth (measured by protein staining)	19, 20
Cell viability and growth	Any adherent cell	Confluency assessed via imaging analysis using an Incucyte S3	21

References: [1](Mosmann, 1983); [2](Zegura *et al.*, 2009); [3](Miege *et al.*, 2009); [4](Shi *et al.*, 2009b); [5](Creusot *et al.*, 2010); [6](Delgado *et al.*, 2011); [7](Page *et al.*, 1993); [8](Bunnell *et al.*, 2007); [9](Borenfreund and Puerner, 1985); [10] (Konsoula and Barile, 2005); [11](NWC, 2011); [12](Ma *et al.*, 2005); [13](Nachlas *et al.*, 1960); [14](Promega, 2009); [15](Marabini *et al.*, 2007); [16](Plewa *et al.*, 2000); [17](Plewa *et al.*, 2002); [18](Plewa *et al.*, 2004a); [19](Skehan *et al.*, 1990); [20](Cetojevic-Simin *et al.*, 2009); [21](Escher *et al.*, 2019). LDH = lactate dehydrogenase, MTT = 3-[4,5-dimethylthiazol-2-yl]-2,5-diphenyltetrazolium, NRU = neutral red uptake, SRB = sulphorhodamine B.

Table 10.44 Selection of fish *in vitro* bioassays used for assessment of cytotoxicity targeting all chemicals in water samples.

Assay	Cell Type	Endpoint	Reference
Alamar Blue assay (also known as resazurin reduction assay)	Rainbow trout (*O. mykiss*) liver and gill cells (*e.g.*, RTL-W1, RTgill-W1), brown bullhead (*Ictalurus nebulosus*, BB-3 cell line)	Reduction of substrate (Alamar Blue) by live cells yields fluorescent product	1–5
CFDA-AM	Rainbow trout (*O. mykiss*) liver and gill cells (*e.g.*, RTL-W1, RTgill-W1), brown bullhead (*I. nebulosus*, BB-3 cell line)	Membrane integrity. The esterase substrate (CFDA-AM) is converted to a fluorescent product by esterases in intact plasma membranes	2–6
NRU assay	Rainbow trout (*O. mykiss*) liver cells (RTL-W1)	Cell viability measured by staining (retention of neutral red)	7–10
PI staining and flow cytometry	Rainbow trout (*O. mykiss*), liver cells	PI is fluorogenic and binds stoichiometrically to nucleic acid. DNA content can be quantified via fluorescence	11, 12

References: [1](Page *et al.*, 1993); [2](Schreer *et al.*, 2005); [3](Schirmer *et al.*, 2001); [4](Grung *et al.*, 2007); [5](Farmen *et al.*, 2010); [6](O'Connor *et al.*, 1991); [7](Borenfreund and Puerner, 1985); [8](Klee *et al.*, 2004); [9](Keiter *et al.*, 2006); [10](Wölz *et al.*, 2008); [11](Zucker *et al.*, 1988); [12](Gagné and Blaise, 1998). CFDA-AM = 5-carboxyfluorescein diacetate acetoxymethyl methyl ester, NRU = neutral red uptake, PI = propidium iodide.

2013), in particular if internal concentrations are compared (Stadnicka-Michalak *et al.*, 2014).

10.8.2 Algal growth inhibition

Algal growth inhibition assays using green microalgae *R. subcapitata* have been applied to wastewater effluent, recycled water and surface water extracts, with only one study testing drinking water extracts (Table 10.45). Most studies applied the CAT, with growth inhibition measured after 24 h based on Escher *et al.* (2008a), while one study used 72 h algal growth inhibition based on the OECD guideline 201 (OECD, 2011). The CAT was responsive to wastewater effluent extracts, with 10% growth inhibition observed at an REF from 0.7 to 13, while

Table 10.45 Summary of studies that have applied algal growth inhibition assays to different environmental water extracts.

Matrix	Assay	EC_{10} (REF)	Reference
Wastewater effluent	CAT (24 h growth)	0.71–13	1–4
Recycled water	CAT (24 h growth)	0.70 to >20	2, 3
Surface water	CAT (24 h growth)	1.3 to >90	1, 2, 4
	Algal growth inhibition	17 to >100[a]	5
Drinking water	CAT (24 h growth)	14	2

The effect was expressed as the concentration causing 50% effect (EC_{50}) in units of REF.
NB: Only studies that have applied SPE or LLE are included. CAT = combined algae test.
[a]EC_{50} reported.
References: [1](Escher *et al.*, 2008b); [2](Escher *et al.*, 2014); [3](Jia *et al.*, 2015); [4](Neale *et al.*, 2017c); [5](Toušová *et al.*, 2017).

10% growth inhibition in surface water occurred at an REF from 1.3 to 51 (Table 10.45). No effects were observed in recycled water treated with RO/AO or O_3/BAC up to the maximum tested REF.

10.8.3 Fish embryo toxicity

The FET assay has only been applied for water quality monitoring in a limited number of studies. Escher *et al.* (2014) applied the 48 h FET assay to drinking water, surface water, wastewater effluent and drinking water, but only one wastewater effluent extract had an effect, with 10% mortality observed at an REF of 5. Furthermore, the 48 h FET assay was applied to surface water extracts from the Danube River, with the EC_{50} value ranging from REF 111 to 665 (Neale *et al.*, 2015b). The EC_{50} values were around an order of magnitude lower in surface water from four river basins in Europe in the 96 h FET assay (Toušová *et al.*, 2017). Both Toušová *et al.* (2017) and Neale *et al.* (2015b) applied large volume SPE, with 50–500 L of water concentrated. The 48 h FET assay has also been applied to surface water and wastewater passive sampler extracts, with styrene divinylbenzene 'Speedisk' samplers proving more responsive than silicone rubber samplers (Hamers *et al.*, 2018).

10.9 CONCLUSIONS

A wide range of *in vitro* and well plate-based *in vivo* assays based on different stages of the cellular toxicity pathway including induction of xenobiotic metabolism, receptor-mediated effects, adaptive stress responses and apical effect have been widely applied for water quality assessment. The majority of studies have focussed on estrogenic activity, followed by other endocrine modes of action. Assays indicative of activation of AhR, activation of PXR, estrogenic activity, oxidative stress response and bacterial toxicity have consistently detected effects in wastewater effluent, surface water and drinking water. Mammalian reporter gene assays tend to be more sensitive than yeast reporter gene assays or whole organism assays indicative of the same endpoint. Over the past decade, there have been significant improvements in quantitative expression of bioassay results moving away from the initially simpler binary 'yes/no' results. This has made it easier to compare bioassay results between studies and across different types of waters, providing a much finer understanding and meaningful interpretation of bioanalytical results as a measure of water quality.

Chapter 11

Quality assurance and quality control (QA/QC)

11.1 INTRODUCTION

Many of the criticisms of *in vitro* bioassays about their perceived lack of reproducibility, reliability and standardisation across different laboratories can be addressed by establishing quality assurance and quality control (QA/QC) checks in all experiments undertaken in the laboratory. The 'reproducibility crisis' in biomedical sciences can only be overcome in *in vitro* toxicology by strict and transparent method validation and continuous QA/QC (Hirsch and Schildknecht, 2019).

In a creative research environment, the idea of creating an additional workload by incorporating additional QA/QC steps in experimental protocols may appear cumbersome. The additional degree of confidence in the data generated is, however, undeniably worth the extra effort. In water quality monitoring, this additional degree of confidence is absolutely necessary to ensure the accuracy and reproducibility of the bioassay results, as these can have significant implications for the utility or authority that manages the (waste)water treatment plant or the receiving watershed where the samples came from.

In the early stages of development, bioanalytical tools were often performed in individual vials, where only a few samples can be tested in each assay run. Although low-throughput is not an issue in initial assay development, the conversion of an assay to a high-throughput microplate format is a prerequisite for adapting that assay to water quality monitoring. This chapter is therefore written for 96- and 384-well plate assays, which are currently the most common

Authors: Beate Escher, Peta Neale and Frederic Leusch
doi: 10.2166/9781789061987_0225

microplate format. Although some details may differ, the concepts and principles discussed are of course applicable to any assay format and can easily be adapted.

The following section gives an introduction to method validation. Once a method has been validated and a standard operating protocol (SOP) produced for routine application, QA/QC procedures need to be put in place to monitor and ensure consistent performance over time. QA/QC implementation is addressed in the subsequent sections.

11.2 METHOD VALIDATION

Method validation is a crucial step in adapting a bioassay from a research tool or application in chemical risk assessment into a bioanalytical tool for water quality assessment. Validation of a bioanalytical method demonstrates that the assay is fit-for-purpose and ensures that the analysis is reliable, consistent and that there is confidence in the result. Determining the capabilities of an assay is also useful for understanding how it performs compared to other methods, whether these are other bioassays or chemical analysis methods. Bioanalytical method validation involves determining a variety of assay performance characteristics such as accuracy, precision, robustness, sensitivity and sample stability (CDER/FDA, 2001; EMEA, 2011). All of these parameters can be determined using relatively simple (albeit time-consuming) test protocols, where a series of samples is analysed repeatedly on different assay runs with different operators over a period of several weeks. Specificity and selectivity are also important attributes for method validation that have been covered in depth in Chapter 9.

11.2.1 Accuracy

Accuracy describes how close a bioassay result is to the true value (Figure 11.1). Accuracy is determined by replicate analysis of samples including at least three different concentrations of the analyte of interest. With *in vitro* bioassays, the analyte is usually a model compound used as a reference toxicant (*e.g.*, 17β-estradiol for bioassays that measure estrogenic activity). The mean value from repeated testing in an accurate *in vitro* assay should be within 15%–30% of the actual value.

11.2.2 Precision

Precision describes the closeness of repeated individual measures of the same sample. Figure 11.1 illustrates how accuracy and precision are connected. Although high accuracy and high precision are the target, a test with high accuracy but lower precision might still be acceptable. Precision of an accurate test might be improved by method optimisation. The combination of high precision with low accuracy is problematic because the bioassay results might

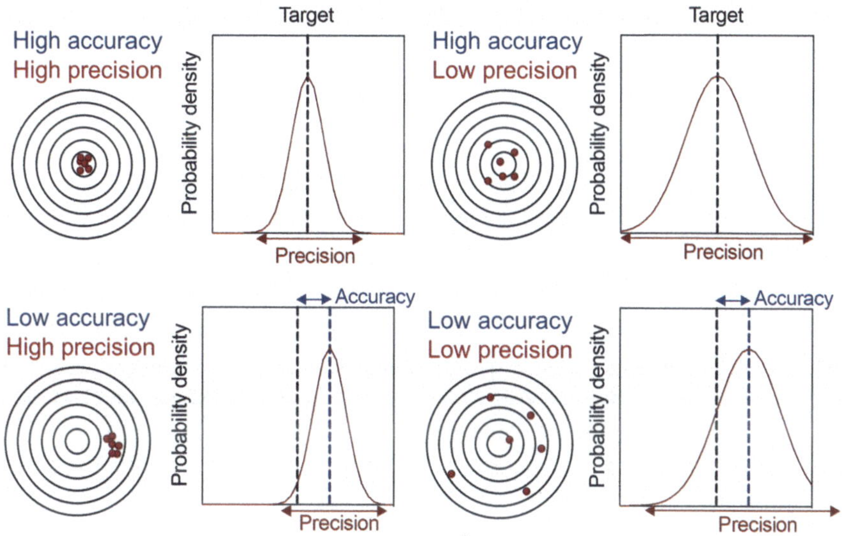

Figure 11.1 Definition of accuracy and precision.

look really good thanks to the high precision, but the values are systematically off the target. Low accuracy combined with low precision are easy to identify and those assays should be avoided.

Precision can be determined at different operating levels, including within-run precision, between-run precision with the same operator (*i.e.*, repeatability) and between-run precision with different operators or even laboratories (*i.e.*, intra- and inter-laboratory reproducibility, respectively). Precision is expressed as the coefficient of variation for repeatability (CVr) (Equation 11.1) and for reproducibility (CVR) (Equation 11.2):

$$\mathrm{CVr} = \sigma_r/\mu_r \tag{11.1}$$

$$\mathrm{CVR} = \sigma_R/\mu_R \tag{11.2}$$

Here, μ_r is the mean and σ_r is the standard deviation of the validation sample results repeated on multiple occasions by the same operator. μ_R is the mean and σ_R is the standard deviation of the validation sample results repeated on multiple occasions by different operators and/or laboratories.

Both of these parameters should be less than 15%–20% for an assay to be deemed repeatable and reproducible.

11.2.3 Robustness

Robustness characterises the sensitivity of a method to operational variations and is a measure of how transferrable the method is to other operators and/or laboratories. Robustness is usually calculated as the ratio (Equation 11.3) of the CVr over the CVR:

$$\text{Robustness index} = \frac{\text{CVr}}{\text{CVR}} \quad (11.3)$$

This robustness index should be between 0.8 and 1.2, and preferably close to 1. A method with low robustness may not yield consistent results between operators and/or laboratories.

11.2.4 Quality

The Z-factor is a measure of assay quality (Zhang *et al.*, 1999). It is an expression of the separation band between positive and negative controls (Equation 11.4), and thus a measure of the effective dynamic range:

$$\text{Z-factor} = 1 - \frac{3 \times (\sigma_p + \sigma_n)}{|\mu_p - \mu_n|} \quad (11.4)$$

Here, μ_p is the mean and σ_p is the standard deviation of the positive control, and μ_n is the mean and σ_n is the standard deviation of negative controls.

A good assay should have a Z-factor between 0.5 and 1. Any Z-factor less than 0.5 indicates a marginal assay, and a Z-factor less than 0 indicates that there is too much overlap between the positive and negative controls for the assay to be useful. The Z-factor is assay-specific and needs to be determined during method development over several plates and independent repeats or if an existing assay is established newly in a laboratory. A further QA/QC measure is to check the Z-factor on every plate to assure that it remains consistently high. It might deteriorate if the reagents expire, if there are issues with pipetting or if other errors occur.

For example, the Z-factors of GeneBLAzer assays normally range between 0.7 and 0.9, and the ERα-GeneBLAzer illustrated in Figure 11.2 has a Z-factor of 0.88.

11.2.5 Matrix interference

An assay might work perfectly fine with single chemicals but that does not mean that it is compatible with complex environmental samples. Most of the assays that are used in water quality assessment have been developed and initially used to screen pure chemicals. When adapting a bioassay from a chemical toxicity testing context to water quality testing one must consider matrix interferences and characterise how much the analysis is affected by the presence of other components in a water sample. Although water is typically extracted prior to dosing in an assay (Chapter 12) dissolved organic carbon (DOC) is partially co-extracted. The co-extracted DOC has little impact on cell-based assays (Neale and Escher, 2014), but can interfere with antagonism measurements (Neale and

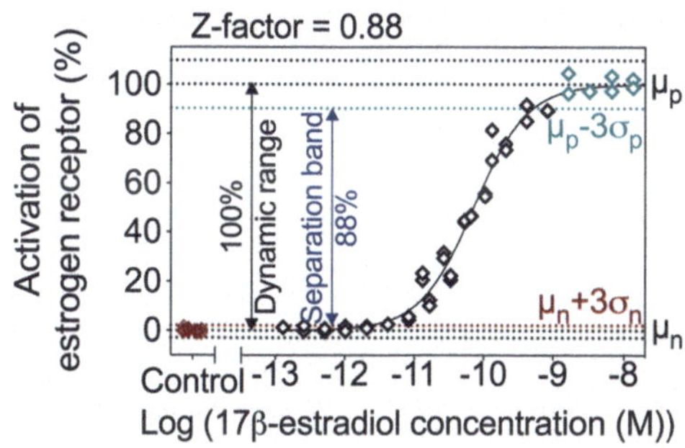

Figure 11.2 Z-factor (Equation 11.4) illustrated on the example of a concentration–response curve (CRC) for 17β-estradiol in the ERα-GeneBLAzer assay.

Leusch, 2015) and has a strongly suppressive effect in cell-free assays, such as the popular acetylcholinesterase inhibition assay (Neale and Escher, 2013).

Solvents (*e.g.*, ethanol, methanol and dimethyl sulphoxide (DMSO)) used during extraction might also interfere with the assay so great care has to be taken to reduce the fraction of solvent or remove the solvent by evaporation prior to testing. For more information on solvent effects see Section 11.3.3.4.

11.2.6 Sensitivity

Sensitivity determines how well an assay responds to varying amounts of target compound(s) and includes determination of the limit of detection (LOD, Equation 11.5), the limit of quantification (LOQ, Equation 11.6) and the calibration curve:

$$\mathrm{LOD} = \mu_n + (3 \times \sigma_n) \tag{11.5}$$

$$\mathrm{LOQ} = \mu_n + (10 \times \sigma_n) \tag{11.6}$$

Here, μ_n is the average response of the negative control and σ_n is the standard deviation of response with the negative control. Typically, the negative controls are unexposed cells. For bioassays, the 'calibration curve' is the concentration–effect curve for the reference and other relevant compounds.

As discussed in Chapter 7 bioassay results are often expressed in effect concentrations EC_y or EC_{IRz} (*e.g.*, EC_{10} or $EC_{IR1.5}$). Therefore, the LOD or LOQ are typically reported not as effects but as the EC of a reference compound, that is, EC_{LOD}. The LOD expressed as a concentration is defined as the concentration of the reference chemical causing three times the standard deviation of the response of the negative control, which corresponds to Equation (11.7) for log-logistic concentration–effect curves and Equation (11.8) for linear concentration–effect curves with % effect and Equation (11.9) with induction ratio (IR) as the effect measure:

$$\log EC_{LOD} = \log EC_{50} - \frac{1}{slope}\log\left(\frac{max - min}{LOD - min} - 1\right) \quad (11.7)$$

$$EC_{LOD} = \frac{LOD}{slope} \quad (11.8)$$

$$EC_{IR,LOD} = \frac{LOD - 1}{slope} \quad (11.9)$$

Figure 11.3 illustrates the derivation of the LOQ and LOD for 17β-estradiol (E2) in the ERα GeneBLAzer assay for activation of the estrogen receptor (ERα). The LOD is 2.2% and the LOQ is 7.3%. The corresponding EC_{LOD} is 0.9 ng_{E2}/L and the EC_{LOQ} amounts to 3.1 ng_{E2}/L. It is interesting to note that the LOQ is close to the 10% effect in this assay, confirming the choice of 10% effect level as statistically significantly different from the negative controls and the EC_{10} as an appropriate effect descriptor.

When it comes to testing water samples, we cannot directly translate the EC_{LOD} and the EC_{LOQ} into the associated bioanalytical equivalent concentrations (BEQ). The BEQ_{LOD} and BEQ_{LOQ} are determined both by the assay's inherent sensitivity and variability, but also by the relative enrichment of a particular sample. An EC_{LOQ} of 3.1 ng_{E2}/L does not mean that estrogenicity can be detected in water samples only if they have 17β-estradiol equivalent concentrations (EEQ) >3.1 ng/L. Water samples can be enriched so the BEQ_{LOQ} for a particular water sample can be much lower. For example, a sample with a relative enrichment factor (REF) of 100 in this assay would have an EEQ_{LOQ} of 0.031 ng_{E2}/L. However, at this high REF, cytotoxicity might interfere and mask the estrogenic effect, which would again decrease the practical detection limit.

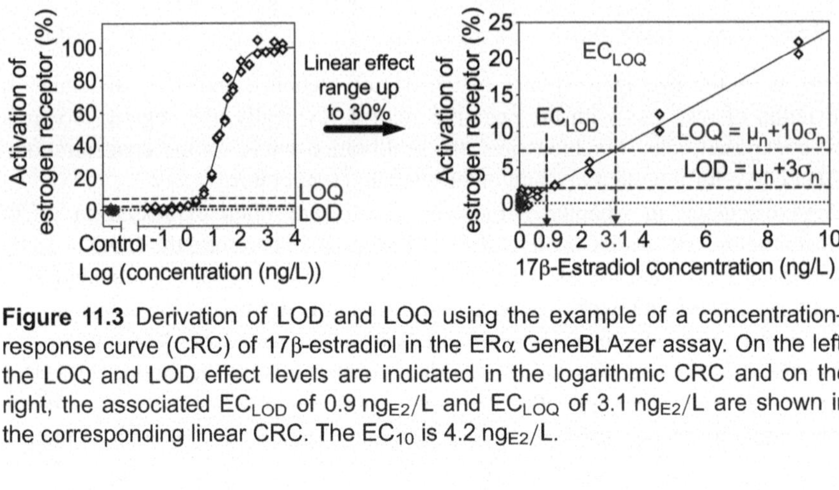

Figure 11.3 Derivation of LOD and LOQ using the example of a concentration–response curve (CRC) of 17β-estradiol in the ERα GeneBLAzer assay. On the left, the LOQ and LOD effect levels are indicated in the logarithmic CRC and on the right, the associated EC_{LOD} of 0.9 ng_{E2}/L and EC_{LOQ} of 3.1 ng_{E2}/L are shown in the corresponding linear CRC. The EC_{10} is 4.2 ng_{E2}/L.

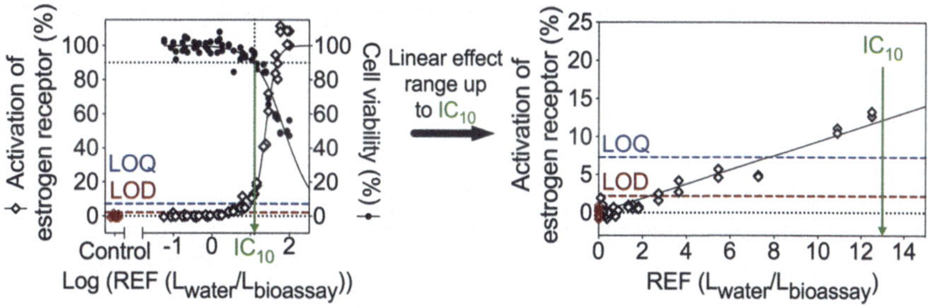

Figure 11.4 Illustration of the derivation of detection limits in water samples on the example of a wastewater treatment plant effluent sample tested in the ERα GeneBLAzer assay (matching the concentration–response curve (CRC) of 17β-estradiol in Figure 11.3). From the cell viability data and IC_{10} of REF 13 can be derived. The LOD and LOQ remain the same at the effect level because they are derived from the negative controls on the same plate as 17β-estradiol in Figure 11.3 but the EC_{LOD} and EC_{LOQ} are related to relative enrichment factors (REF) of the water sample and the detection window is between EC_{LOQ} of REF 7.8 and the IC_{10} of REF 13.

If there is cytotoxicity in a water sample then, as discussed in Chapter 7, all data points above the cytotoxicity threshold must be removed from the analysis. For example, the wastewater treatment plant effluent sample in Figure 11.4 exhibits an IC_{10} of REF 13, thus all response data at REF >13 should be disregarded. Using the filtered data, we can calculate the EC_{LOD} at REF 2.3 and EC_{LOQ} at REF 7.8 using Equations (11.5) and (11.6), respectively, followed by Equation (11.8). The EC_{LOQ} in turn can be translated to an EEQ_{LOQ} of 0.40 ng_{E2}/L using the EC_{LOQ} of E2 of 3.1 ng_{E2}/L divided by the EC_{LOQ} of the sample in REF (in this example, $EEQ_{LOQ} = 3.1\ ng_{E2}/L_{bioassay}/7.8\ L_{water}/L_{bioassay} = 0.40\ ng_{E2}/L_{water}$). The IC_{10} cut-off is at REF 13, so the detectable window is fairly narrow, from REF 7.8 to 13. If a sample does not induce a detectable response even at the highest valid REF tested (*i.e.*, effect <LOQ even at the highest valid REF tested), then the limit of activity in the sample is usually calculated as EEQ_{LOQ} divided by the highest valid REF tested – in the example below assuming no effect up to the IC_{10} cut-off of REF 13, this would be '<0.24 ng_{E2}/L'.

As discussed in Section 11.3.3.5, the extraction blank might pose a problem, and even if it can be subtracted, it will interfere with the detectability of the assay. Especially, if the extraction blank is highly cytotoxic, this will mask the activation of the sample and lower the detectability to a point where no activation can be measured.

11.3 QA/QC IN THE LABORATORY

The concept of QA/QC was initially developed in the manufacturing industry as a set of documents and procedures to ensure consistent product quality. In the laboratory, the application of QA/QC principles ensures that bioassay results are accurate and consistent. This also entails some practical considerations, such as setup, materials and tracking of materials. As a minimum, the laboratory QA/QC steps should consist of analysis replication, inclusion of adequate positive and negative control samples, verification of assay performance with control charts and fixed control criteria, good record-keeping and having appropriate SOPs in place. A series of OECD documents details the principles of Good Laboratory Practice (OECD, 1998), including one specifically developed for *in vitro* studies (OECD, 2004).

11.3.1 Practical considerations

There are some practical considerations to guide how assays can be optimised during development and validation, and how to ensure that QA/QC criteria are consistently met. The following provides a short overview of parameters that can affect the quality of bioassay results (Figure 11.5). More detailed information can be found in relevant chapters of the Assay Guidance Manual (Auld *et al.*, 2020) and in a guidance document on reproducibility of *in vitro* assays (Hirsch and Schildknecht, 2019).

First, a decision on the plate format and the cell model or assay to be used has to be made. Note that assays in 96-well format do not always require the availability of lab automation instrumentation, but these devices become practical for the 384-well plate and indispensable for 1536-well plate formats. In principle, all materials that come into contact with the cells and environmental factors can influence the performance of an assay. Thus, it is worthwhile to plan the optimisation and validation of an assay carefully.

Some important factors are summarised in Figure 11.5. They include the plate material, which can have an influence on the quality of the read out measured with a plate reader or microscope, but also on cell health directly, and in some instances on the bioassay response itself (*e.g.*, some plasticware contains xeno-estrogenic compounds, which can interfere with sensitive reporter gene assays). At the next level, the media used to grow and/or differentiate cells as well to run a specific assay should be selected with care and adapted if necessary. Media components can interfere with measurement of fluorescence (*e.g.*, phenol red) or influence the availability of chemicals to the cells (see also Chapter 9). In addition, optimised seeding densities and incubation times should be determined to ensure the best possible quality of measurements and cell health. Besides these, additional measures such as the use of gas-permeable seals or preincubation of cells at room temperature after seeding before transfer to an incubator (Lundholt *et al.*, 2003) can also be considered to reduce edge effects.

It is also recommended to track the lots of plates, media, supplements and other materials that are used. A lot is a batch of consumables that were manufactured together and have the same lot number. This practice facilitates troubleshooting. Moreover, lot tests should be performed before running large batches of

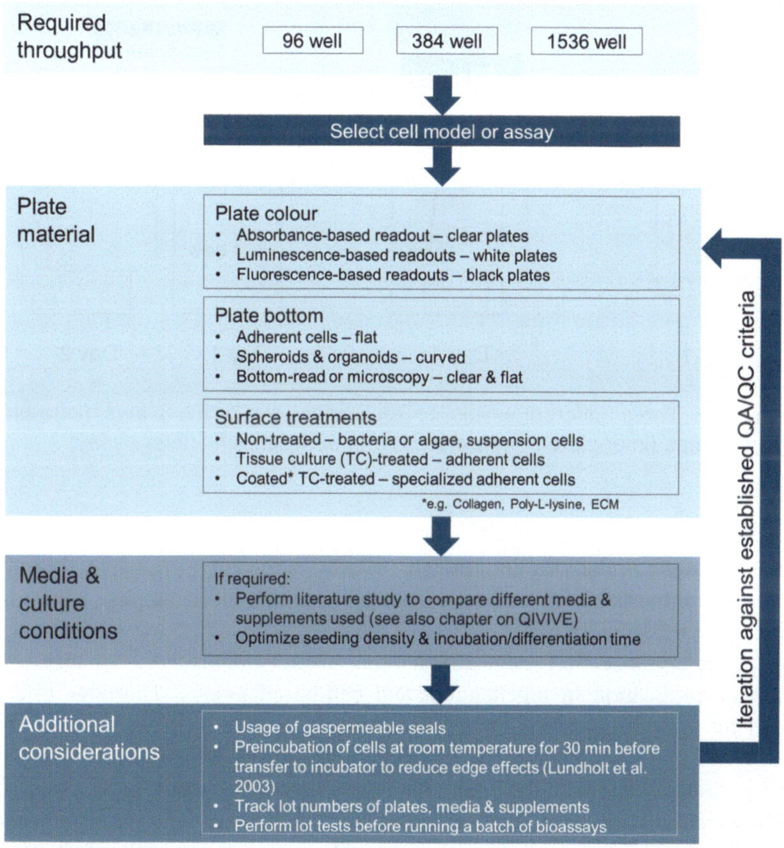

Figure 11.5 Practical considerations for setting up a quality-controlled bioassay. This flowchart provides a summary of factors that influence the QA/QC of bioassays. In practice, this is an iterative process that must balance competing factors such as automation compatibility, cost, availability and performance (adapted from Auld *et al.*, 2020). ECM = extracellular matrix, QIVIVE = quantitative *in vitro* to *in vivo* extrapolation.

bioassays. Based on the results of the lot tests, all materials tested should be reserved in sufficient quantities to run a project or sampling campaign with one lot of the respective materials. Lot tests are also recommended for all materials required during the sample preparation process.

11.3.2 Replication

Replication of experiments is absolutely critical in any type of analysis, including bioassay analysis. Several levels of replication are required for comprehensive QA/QC: within plates, between plates and between runs (Figure 11.6).

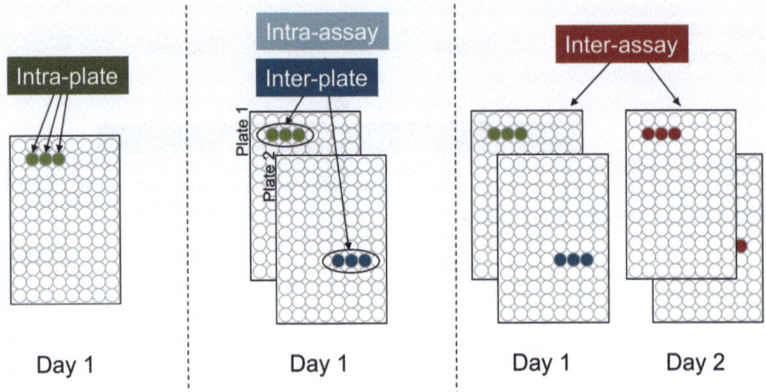

Figure 11.6 Three different levels of replication are within plates (intra-plate), between plates (inter-plate or intra-assay) and between runs (inter-assay).

11.3.2.1 Intra-plate replication

First, the sample is analysed in replicate (usually duplicate or triplicate, depending on the inherent variability of the assay) side-by-side on the same plate ('intra-plate replication' in Figure 11.6).

A phenomenon called 'edge effect' can severely hamper the quality of the intra-plate replication in biochemical and cell-based assays. The edge effect is caused by the design of the plate which results in increased evaporation in the outer wells and also a temperature gradient from the outer to the inner wells. Some options to deal with the edge effect are listed in Section 11.3.1.

The purpose of this first level replication is to determine the variability between different wells. Such variability can be caused by a variety of operator factors (*e.g.*, erratic liquid handling) or by environmental factors (*e.g.*, humidity). The coefficient of variation ($CV = \sigma/\mu$) of these replicate analyses should not exceed a pre-determined CV threshold (this can vary between different assays but is generally set at CV < 10%–15%).

11.3.2.2 Inter-plate replication

Intra-assay replication or inter-plate replication (Figure 11.6) involves analysing the same sample multiple times in the same assay run but at different stages of the run (*e.g.*, once at the beginning of a plate and once at the end or better yet on a different plate altogether). Intra-assay replication verifies that there is no temporal drift during the assay run due to either environmental factors (*e.g.*, an increase in temperature) or instrumental issues (*e.g.*, a spectrophotometer that loses its sensitivity during the experiment). During data analysis, the result for each replicate is compared to ensure that the variability does not exceed a pre-determined level (usually CV < 10%–15%). The average of the intra-assay replicates is then used, again as a single number, as the variability here is still not indicative of the sample variability. Intra-assay replication is only necessary for a small subset of randomly selected samples, usually one sample per assay run.

11.3.2.3 Inter-assay replication

Inter-assay replication (Figure 11.6) involves analysing the same sample in independent assay runs. This verifies that there is no assay drift or bias over time. Samples should always be analysed on at least two independent runs performed on different days. During data analysis, the independent values are compared. If the CV is more than a pre-determined value, which varies between different assays, but is usually no more than 15%–20%, the analysis is performed a third time to obtain a more precise result. This way, the concentration–response curves (CRCs) for all repeats can also be evaluated together. The final result is often reported as the average of the two closest results. The variability of the results at this stage is an indication of variability of the bioanalytical method assuming there is no sample degradation. Although inter-assay variability can be reported as an indication of the confidence in the analytical result, it is not a measure of the true variability of the water sample.

11.3.2.4 True sample replicates

In an appropriately designed monitoring programme, truly independent replicate samples should be taken and analysed. Truly independent replicates are for example water samples taken in triplicate and independently run through the entire extraction process (Chapter 12). The results of the independent samples can be reported as the average $\pm$ standard deviation as in this case the variability does provide a measure of precision. Note, however, that this value includes a variability component from both sample collection and sample processing (*i.e.*, preservation and extraction) and analysis.

11.3.3 Quality control samples

In chemical analysis, it is important to service and maintain the physical detectors of analytical instruments regularly. This 'servicing and maintenance' is even more critical with the biological detectors (*i.e.*, cells) used in bioanalytical tools as these require carefully controlled environmental and culture conditions. Cell culture conditions need to be carefully developed for each test system while keeping in mind good cell culture practices (Bal-Price and Coecke, 2011). Variations in cell culture or practice can cause abnormal behaviour of the test system and produce inaccurate results. Always including and analysing control samples allows detection of such abnormal events and ensures that only data from valid assay runs are analysed and reported.

Quality control samples are a crucial component of QA/QC for bioanalytical tools. A full CRC of the reference compound ('standard curve'), positive and negative control samples, blanks and inter-assay samples should be included in every run. These quality measures are described in more detail below. The results of the quality control samples are compared with control charts (see Section 11.3.4). If any control parameter falls outside the acceptable range, the entire dataset should be rejected and all affected samples re-analysed.

11.3.3.1 Standard curve

A standard curve is a CRC (see Chapter 7) of the reference compound covering the full range of effect from 0% to 100%. For cytotoxicity, calculated as 100% minus the percent cell viability, the minimum effect is by definition 0% (all cells are viable) and the maximum effect is 100% (all cells are dead). With reporter gene assays, the effect is in most cases measured as a colour change (absorbance), relative light units (RLU) or relative fluorescence units (RFU), or by a change of fluorescence. In this case the minimum and maximum might have different absolute values but the % effect of the standard curve must be repeatable. Two typical standard curves for 2,3,7,8-tetrachlorodibenzodioxin (TCDD) in the AhR CALUX assay from two different plates are presented in Figure 11.7. When the RLU were plotted, there were differences in the maximum RLU reached in two plates. However, when the maximum was fitted to the highest response for each plate and the RLU converted to % activation, both curves then overlapped. Note that some RLU data had to be removed due to cytotoxicity interferences (grey bar in Figure 11.7). The same principle applies for assays that are based on RFU measurements.

A standard curve must be included for every assay run and benchmarked against previous curves obtained in the same assay. The standard curve in particular yields four important parameters: the EC_{50}, the slope and the minimum and maximum effects (see Chapter 7).

The standard curve is also often used to derive the BEQ if the reference compound for QA/QC is the same as the reference compound for the BEQ derivation.

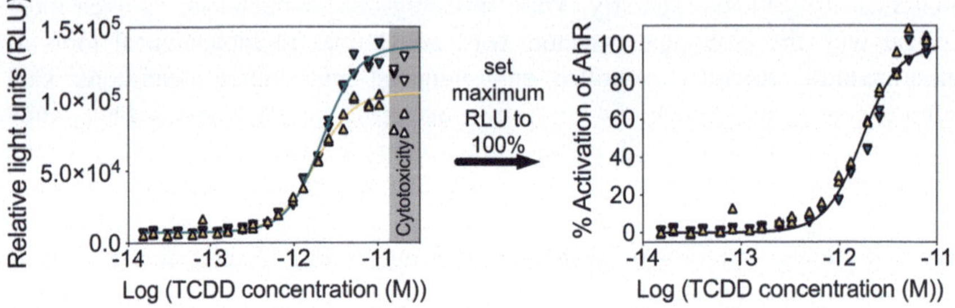

Figure 11.7 Typical standard curves obtained for 2,3,7,8-tetrachlorodibenzodioxin (TCDD) in the AhR CALUX assay.

11.3.3.2 Positive control sample

A positive control sample is a sample that contains a high bioactive test compound other than the reference compound. The positive control should preferably have different physico-chemical properties to the reference compound. For estrogenicity, for example, a good reference compound (standard) would be the natural hormone 17β-estradiol and a good positive control could be the industrial compound 4-nonylphenol.

11.3.3.3 Negative control sample

There are two types of negative control samples: solvent and media negative controls. A solvent negative control is a blank sample that contains an equivalent amount of solvent as the actual water sample extracts, which are usually in ethanol, methanol or DMSO. Even if a solvent extract is blown down prior to dissolving in medium, the negative control should mimic that and should be the equivalent solvent and the same volume blown down and treated exactly the same way as the samples. The negative control provides verification that the solvent itself is not responsible for any of the observed effect.

A medium negative control measures the minimal response in the assay media without any additional influences. Often, the medium negative control is also needed to calculate the response from the raw signals, for example for the autofluorescence correction for fluorescent samples in the β-lactamase assay with a FRET (Förster resonance energy transfer) reagent.

In reality, the response with the solvent and media negative controls should be comparable, otherwise it is an indication that the solvent used to reconstitute the water extract is interfering with the assay.

11.3.3.4 Solvent effects

It is critical to understand the effect of solvents on any bioassay. If solvent interference is observed, it may be necessary to reduce the amount of solvent used in the assay or use a different solvent that does not affect the assay. For testing of water samples, it is recommended to use volatile solvents that can be completely removed by evaporation prior to addition of assay media. It is a common misconception that this process will also remove chemicals provided that blow down is gentle, that is, by a stream of nitrogen gas and not by vacuum. We need to keep in mind that volatile chemicals would have been lost already during the solvent blow-down stage of liquid–liquid or solid-phase extraction (SPE). In addition, *in vitro* assays are run on well plates that are not hermetically sealed and volatile chemicals (or more precisely chemicals with an air–medium partition constant over 10^{-4}) will be partially lost during the 24–48 h incubation at 37°C of a bioassay experiment (Escher *et al.*, 2019). The vial weight of extracts using volatile solvents has to be continuously tracked to assure that no losses are occurring or, if they are, that the solvent is topped up.

Although blow down is recommended for water samples, it might not be applicable for other sample types and pure chemicals and needs to be assessed for each sample type prior to deciding on the sample treatment. For hydrophobic chemicals and extracts from condensed phases such as sediment and tissue, DMSO remains a solvent of choice. Because of its low volatility, it is not necessary to track the vial weight of DMSO extracts as losses from evaporation are negligible. DMSO extracts have a high stability, provided they are kept in the dark because DMSO is subject to oxidation and photodegradation. Some of the DMSO photoproducts are toxic and can produce blank effects. Therefore, DMSO should only be used fresh from ampoules, kept in the dark and not stored in larger solvent bottles on shelves.

The IC_{10} of DMSO depends on the cell type and ranges between 0.5% and 3% of DMSO for common reporter gene assays used in water quality assessment (Table 11.1). No corresponding list is available for other solvents as they partially evaporate during the 24-h incubation at 37°C and no robust IC_{10} can be derived.

DMSO can, however, be problematic in some assays, even at concentration below cytotoxic concentrations. DMSO can reduce the metabolic activity of cells (Ferk and Daris, 2018), stimulate anti-inflammatory activity (Costa *et al.*, 2017), activate the oxidative stress response (Escher *et al.*, 2012) and interfere with the assessment of immunomodulatory effects (Timm *et al.*, 2013). DMSO can also interfere with gene expression at concentrations as low as 0.5% (Sumida *et al.*, 2011), and thus could subtly affect the results of gene expression assays at noncytotoxic concentrations (Leusch *et al.*, 2017). As a matter of fact, even at concentrations where DMSO is typically used as vehicle (0.1%–1.5%) alteration in proteins and DNA, including alteration of DNA topology, have been observed and there was an antioxidant effect as low as 0.5% (Tuncer *et al.*, 2018). At 0.1% DMSO, >2000 genes were differentially expressed in cardiac

Table 11.1 Cytotoxicity of DMSO, quantified by 24 h growth/cell viability via microscopic imaging as $IC_{10} \pm$ standard error of mean, derived by error propagation (Escher *et al.*, 2019).

Cell line	IC_{10} (mM)	IC_{10} (%)
AREc32	92 ± 4	0.7 ± 0.1
AhR CALUX	149 ± 19	1.1 ± 0.1
PPARγ-BLA	303 ± 50	2.2 ± 0.4
AR-BLA	430 ± 159	3.1 ± 1.1
ERα-BLA	172 ± 14	1.2 ± 0.1
PR-BLA	128 ± 8	0.9 ± 0.1
GR-BLA	63 ± 2	0.5 ± 0.1
ARE-BLA	553 ± 112	3.9 ± 0.8

and hepatic microtissue and methylation pattern indicated epigenetic changes (Verheijen *et al.*, 2019).

11.3.3.5 Field and laboratory blanks

Blank samples are important to confirm that sampling and extraction are not responsible for any of the observed activity. Although this is usually investigated during method validation, it is important to include blanks nevertheless in case an unknown parameter changes that may affect the assay results (*e.g.*, a change in the manufacturing process that introduces a chemical in a consumable that can leach into the sample and cause accidental contamination during sampling and extraction).

There are two types of blanks: field and laboratory blanks. A field blank is an ultrapure water sample that is taken to the field and exposed to similar conditions as the actual samples (*e.g.*, temperature variation during travel) and is extracted, concentrated and analysed at the same time and in the same way as the actual samples. A laboratory blank is similar, except that it has not been taken into the field.

For water samples, blanks are expected to be fairly clean and not cause any effects or cytotoxicity >10% up to an REF of 100. To achieve this, great care has to be taken in the extraction process (Chapter 12), high purity solvents have to be used and SPE material needs to be cleaned and conditioned prior to use. In practice, only very clean water samples such as drinking water are enriched to such high REFs, and for more contaminated water it will be sufficient that there are no effects or cytotoxicity >10% up to the highest tested REF of the samples.

Blanks are often more problematic for extracts from condensed phases such as accelerated solvent extracts from soil and sediment or biota. In this case, it might be necessary to apply a blank subtraction for data evaluation. If necessary, this should be performed at the level of effect units (EU = 1/EC, Equation 11.10) or toxic units (TU = 1/IC, Equation 11.11) and a subtraction should only be performed if the blank has an EU_{blank} or TU_{blank} that is less than 50% of the EU_{sample} or TU_{sample} (Jahnke *et al.*, 2018):

$$EU_{blank-corrected} = EU_{sample} - EU_{blank} \tag{11.10}$$

$$TU_{blank-corrected} = TU_{sample} - TU_{blank} \tag{11.11}$$

11.3.3.6 Matrix spike recovery

Sample preparation is discussed in detail in Chapter 12 but a relevant routine QA/QC measure for sample preparation is matrix spike recovery (Denison *et al.*, 2020). A positive control that is also representative for the physico-chemical properties of the targeted chemicals is spiked in a defined concentration to the water sample and undergoes all extraction steps. The matrix spike recovery is calculated using Equation (11.12):

$$\text{Matrix spike recovery} = \frac{EU_{\text{water+spike}} - EU_{\text{water}}}{EU_{\text{spike}}} \tag{11.12}$$

11.3.3.7 Inter-assay sample

An inter-assay sample is a previously analysed positive sample from a similar water matrix as the one under investigation (*e.g.*, wastewater and drinking water). Analysis and quantification of the sample and comparison with previously obtained results confirms that the quantification step is reliable and repeatable and shows that the sample would be assigned the same BEQ (Chapter 7) irrespective of when or by whom it was tested.

11.3.4 Control charts and fixed control criteria

11.3.4.1 Control charts

Control charts (also called Shewart charts) are a useful QA/QC tool to benchmark the results of a bioassay run against those obtained from previous runs. Control charts also help determine if there is a gradual shift in assay performance.

In a control chart, the value of a control parameter (*e.g.*, EC_{50} and slope) is plotted against time (or run number) (Figure 11.8a). The mean value of all previous runs as well as warning and control (or action) limits are all indicated on the graph. The control limits are set as three standard deviations away from the mean. Any value that falls outside these control limits indicates abnormal bioassay behaviour in which case the entire bioassay dataset should be discarded and the affected samples re-tested. The warning limits are set as two standard deviations away from the mean. Any value that falls between the warning and control limits indicates a possible concern about the quality of the data that should be investigated. The whole dataset and other control parameters should then be checked to determine its validity.

Figure 11.8b shows an example of the log EC_{50} of 17β-estradiol in ERα GeneBLAzer over three years. In 2018/2019, the data varied with larger error bars per run and also larger variation between runs, with one plate even falling below the lower warning limit (but still within control range). After optimisation of the pipetting method in 2019, the EC_{50} had lower error bars and also varied much less between the runs performed on different dates. A similar development was seen for $EC_{IR1.5}$ of t-butylhydroquinone (tBHQ) in AREc32 (Figure 11.8c). tBHQ hydrolyses quickly, so the experiment is time-critical and methanol stocks cannot be kept for a long time. Overall, the quality was good but rare individual outliers can be easily identified using these plots.

If a consistent shift of the EC_{50} or $EC_{IR1.5}$ can be observed between plates and over more than five rounds of experiments performed on different dates, this may

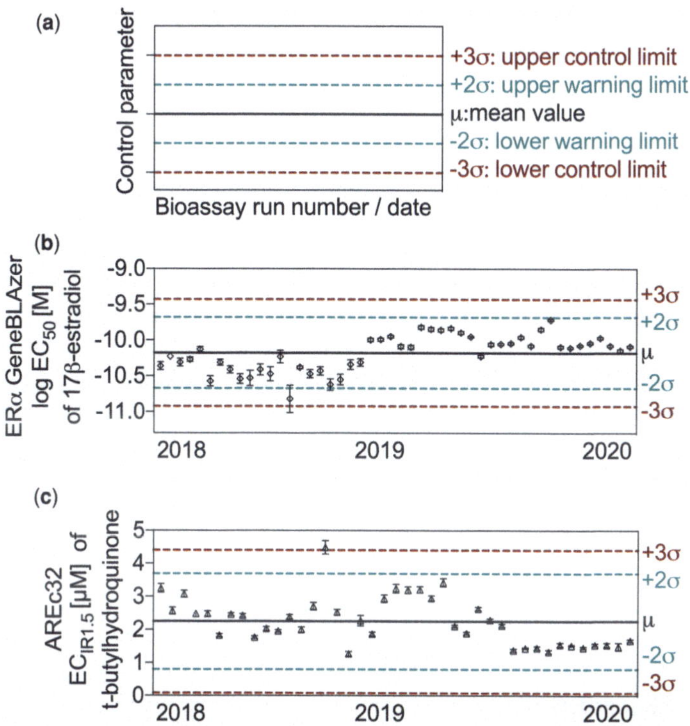

Figure 11.8 (a) Makeup of a control chart showing the mean value, upper and lower warning limits (mean $\pm$ 2σ) and upper and lower control limits (mean $\pm$ 3σ); (b) example control chart for the log EC_{50} of 17β-estradiol in ERα GeneBLAzer and (c) example control chart for the $EC_{IR1.5}$ of t-butylhydroquinone in AREc32.

indicate a gradual shift in assay performance. This shift may be due to a variety of factors but the most common are cell passage number, instrument age (*e.g.*, light bulb on a plate reader) and degrading reagents. The reason for the shift should be investigated as it may be an early warning sign of a potentially larger problem.

11.3.4.2 Fixed control criteria

For some control parameters (typically minimum, maximum and induction ratio), it is more appropriate to determine a minimum or maximum value, called a 'fixed control criteria', rather than a range. The assay run is valid only if the control parameter meets the fixed control criteria. Some parameters are universal: for example, a minimum Z-factor of 0.5. Others depend on the assay: a minimum induction of 6, for example, may be a valid control criterion for luciferase activity measured in a reporter gene assay and an induction of 2 may be the minimum in a cell proliferation assay.

The exact mix of control charts and fixed control criteria will depend on the assay type. What is important is to determine acceptable values and ranges prior to analysis and to use benchmarks to validate (or reject) a bioassay run.

11.3.5 Standardisation and documentation

Accountability is an important aspect of all laboratory work including of course water quality testing. Standardisation of methodology will contribute to QA/QC by limiting the likelihood of error and includes implementation of SOPs. The SOP must include a detailed description of the experimental protocols, its field of application, equipment operation and standardised data analysis methods (*e.g.*, a preformatted Excel template with the appropriate formulae that simply requires a cut-and-paste of the raw data or an automated script like the ToxCast Analysis Pipeline (tcpl) coded in R (github.com/USEPA/CompTox-ToxCast-tcpl)).

Good record-keeping based on the FAIR principles – Findability, Accessibility, Interoperability, Reusability (Wilkinson *et al.*, 2016) – will ensure that all necessary information is available. Such records should include documents such as chain of custody and field observation forms for sampling, information on sample origin and manipulation in centralised sample tracking databases, well-maintained laboratory books with details of experimental manipulations (*e.g.*, cell passage number and operator), and archives of raw and analysed data in a safe location.

11.3.6 Guidelines

There are a number of test guidelines available for *in vitro* assays and water quality monitoring. The series of guidelines for water quality testing from the International Organization for Standardization (ISO) includes mutagenicity testing with the Ames test (ISO 11350, 2012), bioluminescence inhibition of *Aliivibrio fischeri* (ISO11348-3, 1998), algal toxicity (ISO8692, 2004) and the fish embryo toxicity test (ISO15088, 2007). All these guidelines apply direct water and wastewater testing, and mere sample filtration is recommended for sample pre-treatment. Here, the benchmark concentration is the lowest ineffective dilution, which is the highest dilution of the original water sample that is below the LOQ (Equation 11.5).

The newest addition to this series are reporter gene assays for estrogenicity using two types of the yeast estrogen screen (YES, ISO/DIS19040-1, 2017; ISO/DIS19040-2, 2017) and a human cell-based reporter gene assay (ISO/DIS19040-3, 2017). In these guidelines, a choice is given between testing original water samples and testing SPE extracts in DMSO.

There are also OECD technical guidelines for testing of chemicals. The 'Performance-Based Test Guideline (PBTG)' 455 (OECD, 2015) describes the methodology of 'stably transfected transactivation *in vitro* assays to detect estrogen receptor agonists and antagonists (ER TA assays)'. These OECD technical guidelines are intended for chemical testing, but their detailed quality control procedures and assay descriptions may serve as inspiration for water quality testing applications.

11.3.7 High-throughput screening

Automation of cellular and other plate-based assays in high-throughput screening (HTS) setups is particularly important for drug development. With the Tox21 and ToxCast initiatives of the U.S. National Institute of Health and the U.S. Environmental Protection Agency, HTS has also found its way into chemical risk assessment (Chapter 9). 384- and even 1536-well plates and low-volume pipetting platforms are common in HTS. At least partial automation is now also emerging in water quality assessment. Automation poses specific challenges for QA/QC and while a detailed coverage is beyond the scope of this chapter, we refer the reader to a book chapter by Powell *et al.* (2016) for more details.

11.4 CONCLUSIONS

Implementation of QA/QC protocols increases the amount of time and consumables needed to conduct bioassay testing, but the benefits of confidence in the accuracy, reliability and consistency of the data are well worth the effort.

A good bioassay for water quality testing is one that is accurate and precise, both within a run, but also between runs (repeatable) and among different operators and laboratories (reproducible), robust, sensitive and is not affected by matrix interference. For reporter gene assays, the window between the specific effect and effects on cell viability should be sufficiently large to ensure that specific effects are not masked by cytotoxicity.

In the Supplementary Information to this book on www.ufz.de/bioanalytical-tools, we provide additional resources for QA/QC including example spreadsheets for the approaches discussed in this chapter.

Chapter 12

Sampling, sample preparation and dosing

12.1 INTRODUCTION

To ensure that bioassay results are meaningful, it is important to select an appropriate sampling strategy and use suitable sample pre-treatment and processing methods. This is because organic micropollutants are often present at low concentrations in the range of nanogram per litre, particularly in drinking water and clean source waters (Glassmeyer *et al.*, 2017; Troger *et al.*, 2018), so water samples may need to be enriched up to 100 times in the bioassay before an effect can be detected.

Although issues of sample preparation are not specific to bioanalytical tools, good sample preparation is essential for bioanalysis in the same way as it is for chemical analysis. Even the best bioassay cannot rectify problems created by poor sample collection and preparation. High effects and cytotoxicity in the blanks may jeopardise entire monitoring projects. Liquid–liquid extraction (LLE) and solid-phase extraction (SPE) are currently the most common methods applied for preparation of water samples for subsequent bioanalytical measurement. Condensed phases such as sediments or biota samples are also typically extracted with LLE but require additional clean-up of the extracts. These non-aqueous sample types will be discussed in Chapter 15, while we focus in this chapter on water samples.

It is necessary to validate the chosen extraction method using both chemical and bioassay analysis before any serious monitoring is undertaken. Such validation is necessary in order to establish that the extraction procedure provides good and

Authors: Beate Escher, Peta Neale and Frederic Leusch
doi: 10.2166/9781789061987_0245

Figure 12.1 From environment to laboratory and into the well-plate: overview of sampling, sample preparation and dosing.

stable recoveries for a range of environmentally relevant analytes known to induce a biological response in the assay. This prior validation is particularly critical because, unlike chemical analysis where labelled standards can be added to calculate the recovery efficiency of each sample, surrogate standards are generally not added to water samples for bioassay analysis, which is unable to differentiate between a spiked compound and a compound present in the original sample. Other quality assurance/quality control (QA/QC) documentation (such as certificate of analysis of the solvent and/or sorbent, chain of custody and field observation forms) and procedures (such as a field blank and positive control) become very important to compensate for the lack of a spiked standard in every sample. See also Chapter 11 for QA/QC of bioassays.

This chapter provides an overview of sampling strategies and sample pre-treatment and extraction options (Figure 12.1) and includes a decision-making flow chart to help users select appropriate sampling/extraction methods for aqueous samples. Finally, independent of the sample type and sample treatment, dosing into the bioassay is also an important practical consideration (Figure 12.1).

Various case studies that describe several different sampling scenarios across the entire water cycle in combination with diverse bioassay test batteries are described in more detail in Chapter 14.

12.2 WATER SAMPLING STRATEGIES

The sampling strategy depends on the purpose, objectives and sample context (Figure 12.2). If the purpose of a sampling campaign is to assess the product quality of a water treatment plant, with the objective of comparing the effect in the final water to an effect-based trigger value (Chapter 13), then only the product water needs to be collected. In contrast, both source water and product water are required if the purpose is to evaluate treatment process efficacy. Samples can also be collected after intermediate steps throughout the treatment train, such as after advanced oxidation or disinfection if the purpose of the sampling campaign is to identify and understand critical processes.

Composite samples are recommended for wastewater to capture the diurnal variation observed for some micropollutants (Nelson *et al.*, 2011; Petrie *et al.*, 2017), with many studies collecting 24 h composite influent and effluent samples (Körner *et al.*, 2001; Bicchi *et al.*, 2009; Macova *et al.*, 2010; Jalova *et al.*, 2013; Bain *et al.*, 2014; Roberts *et al.*, 2015) and even some seven consecutive 24h

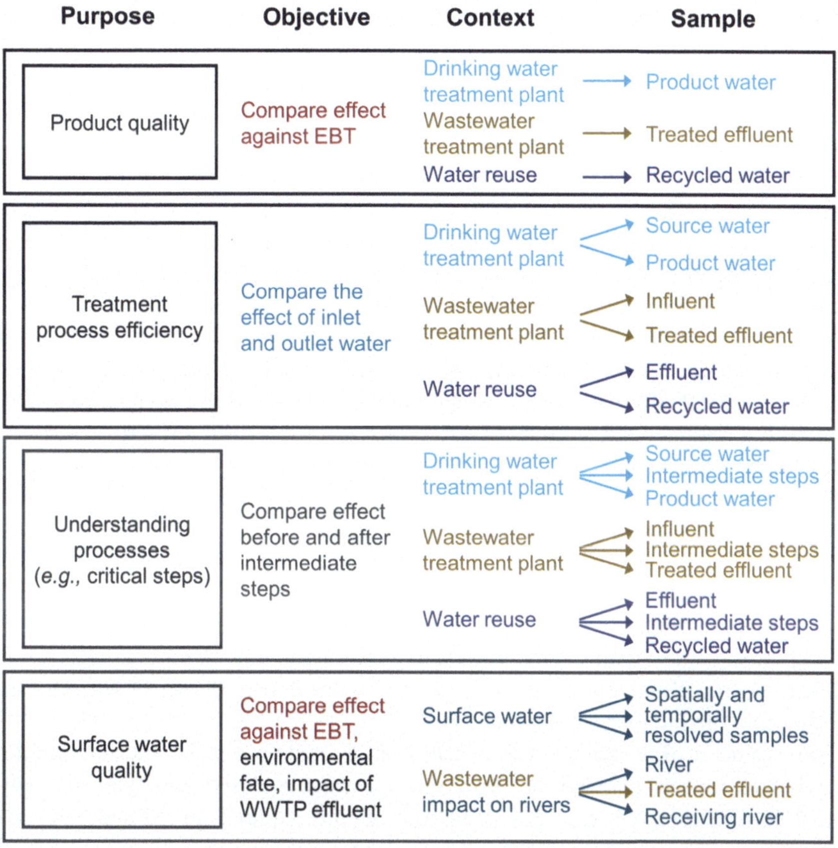

Figure 12.2 Examples of different sampling campaign purposes and objectives for drinking water, wastewater, water reuse and surface water with the required samples for each purpose indicated. WWTP = wastewater treatment plant; EBT = effect-based trigger (Chapter 13).

samples to evaluate potential weekly variations of wastewater quality (Neale *et al.*, 2020d). Grab sampling is suitable for collecting drinking water or recycled water samples if little difference in quality over time is demonstrated.

If temporal dynamics are to be expected, autosamplers allow subsequent sampling of larger number of individual samples to describe diurnal changes in water quality (Mueller *et al.*, 2020), which will be important if photodegradation plays a role or if very dynamic storm events need to be captured (Mueller *et al.*, 2021). Event-driven sampling can be accomplished by setting the autosampler off with a trigger, for example, if the water levels rise by a certain value. In this way, more than 120 samples were collected triggered by rain events in 44 small streams in one summer and they demonstrated a diversity of pollution patterns (Neale *et al.*, 2020a).

Water samples should be collected in solvent washed amber glass bottles, with the bottles stored on ice and in the dark until returning the samples to the laboratory for further processing.

The required volume of water to be collected will depend on the expected level of chemical contamination and thus the need for concentration, with less sample required for wastewater influent and a larger volume needed for drinking water. Using the example of SPE, previous experience from the literature suggests that a common 200 mg/6 cc SPE cartridge can enrich 0.5 L of wastewater influent, 1 L of wastewater effluent or surface water and 2 L of drinking water, recycled water or clean surface water. Double the volume can be applied to larger SPE sorbent masses (*e.g.*, 500 mg cartridges). Assuming the volumes above applied to a 200 mg SPE cartridge and a final extract volume of 0.5 mL, this equates to an enrichment factor (EF) of 1000 for wastewater influent, 2000 for wastewater effluent or surface water and 4000 for drinking water, recycled water or clean surface water. Assuming a bioassay dosing factor (DF) of 0.01 (1% solvent) to 0.001 (0.1% solvent), this results in a maximum relative extraction factor (REF) of 1–10 for wastewater influent, 2–20 for wastewater effluent or surface water and 4–40 for drinking water, recycled water or clean surface water. It is worth noting that some studies have enriched up to REF 500 (König *et al.*, 2017; Hebert *et al.*, 2018) but in these cases the solvent was blown down prior to dosing into the bioassays (Section 12.7).

The number of assays that can be run from a single extract will depend on the number of repeats planned and how much extract is dosed, but typically only small extract volumes (*e.g.*, a few microlitres) are required for 96 and 384-well plate assays so a number of different assays can often be run.

If higher REFs were required, samples can be extracted on multiple SPE cartridges and combined into one final extract (Escher *et al.*, 2014; Neale *et al.*, 2017c). An alternative approach is to use large volume SPE (LVSPE), which has been applied to enrich between 6 (influent) to 500 L (surface water) (Neale *et al.*, 2015b; Välitalo *et al.*, 2017). LVSPE has been applied in a smaller number of studies than SPE and primarily to surface water (König *et al.*, 2017; Toušová *et al.*, 2017). LVSPE allows onsite sampling, but also requires more equipment than other sample extraction methods.

Simon *et al.* (2019) explored the impact of sample volume on effect recovery, with 0.5 and 2 L of wastewater effluent and 1 and 4 L of surface water spiked with a mixture of four estrogenic compounds and 11 pesticides and enriched with SPE using 300 mg LiChrolut EN/RP-18 sorbent. The extracts were analysed in assays indicative of estrogenic activity (ERα CALUX), photosystem II inhibition and algal growth (combined algae test) and bacterial toxicity using *Aliivibrio fischeri*. The average activity in the large volume extracts was between 79% and 104% of the activity in the lower volume extracts, showing that sample volume did not have a significant impact on recovery. Therefore, the sample volumes should not have a large impact on recovery, though it should be noted that some chemicals, such as highly polar chemicals, may not be well recovered by conventional SPE. This is discussed further in Section 12.5.1.

12.3 SAMPLE PRE-TREATMENT OPTIONS

Once an appropriate water sample has been collected, there are a several decisions to be made regarding sample pre-treatment, including sample storage time and sample filtration.

12.3.1 Water sample preservation and storage

If not enriching onsite, the pH of a water sample is commonly adjusted to an acidic pH prior to sample enrichment. Microbial activity can potentially cause micropollutant biodegradation or biotransformation and adjusting to an acidic pH can reduce microbial activity in the sample. Reducing the pH can also improve the extraction of weak acids, with improved recovery of a pharmaceutical cocktail in a bioluminescence inhibition assay at pH 3 compared to pH 7 observed for three different SPE sorbents (Escher *et al.*, 2005). In contrast, Šauer *et al.* (2018) found no difference in androgenic or anti-androgenic activity in wastewater influent in duplicate samples at pH 3 and pH 7.4, respectively, using C18 SPE discs. If the pH of the sampled water had been adjusted to pH 3 or lower, a few millilitres of ultrapure water should be passed over the SPE cartridge after passing the sample to avoid that the elution extract becomes acidic, which could disturb the bioassay.

In addition to pH adjustment, a small number of studies have used 10 mg/L copper(II) sulphate pentahydrate (Conley *et al.*, 2017a) or 1 g/L sodium azide (Mehinto *et al.*, 2015) for sample preservation.

In the case of chlorinated samples, the chlorine residual should be quenched immediately after sampling. Although chlorine will not be extracted by SPE, quenching is important to prevent the formation of additional disinfection by-products (DBPs) and to prevent the chlorine from potentially reacting with the SPE sorbent. Many studies used sodium thiosulphate to quench the chlorine residual (Macova *et al.*, 2011; Escher *et al.*, 2014; Neale *et al.*, 2020b), and others used ascorbic acid (Conley *et al.*, 2017b). Typically, 3.5 mg/L of sodium thiosulphate quenches 1 mg/L free Cl_2 and 5 mg/L ascorbic acid quenches 1 mg/L free Cl_2 (Farre *et al.*, 2013). Hebert *et al.* (2018) found no difference in effects of glass-bottled Evian water controls with and without 20 mg/L sodium thiosulphate in assays indicative of the oxidative stress response, p53 response and NF-κB response. Similarly, sodium thiosulphate controls did not have any effects in assays indicative of hormone receptor-mediated effects (Neale *et al.*, 2020b). As we are not aware of any studies that have assessed any potential effects from quenching with ascorbic acid, we recommend using sodium thiosulphate for quenching as it has been shown not to induce a response in a number of bioassays.

After collection, water samples are commonly stored for no longer than 48 h at 4°C before extraction (Aerni *et al.*, 2004; Cargouet *et al.*, 2004; Fang *et al.*, 2012; Daniels *et al.*, 2018). Alternatively, some studies froze water samples to store for a longer period prior to extraction (Könemann *et al.*, 2018). Jarošová *et al.* (2014b) investigated the impact of sample storage time prior to extraction on estrogenic activity in wastewater effluent, with matching samples stored at 4°C and extracted 48 h and 45 days after collection. Of the seven samples, the estrogenic activity at least doubled in two of the samples, but overall, the difference in effect was small (*e.g.*, 0.7 ng/L after 48 h and 1.7 ng/L after 45 days).

12.3.2 Water sample filtration

Water samples are commonly filtered prior to SPE, with glass fibre filters most frequently used. Glass fibre filters have previously shown to sorb only negligible amounts of estrogens compared to cellulose acetate and nylon filters, with the latter found to adsorb a significant fraction of estrogens from solution (Walker and Watson, 2010). The filter pore size (*i.e.*, the particle size retained with 98% efficacy) used in the literature varies widely from 0.1 to 11 μm, with the majority of studies using filters with a pore size of 0.7–2 μm (Figure 12.3).

The European Union Water Framework Directive (WFD) recommends sample extraction methods that capture the whole water sample, which includes both the dissolved and suspended particulate phases (European Commission, 2009). The suspended particulate matter (SPM) captured by sample filtration is often discarded, but several studies have shown considerable biological activity

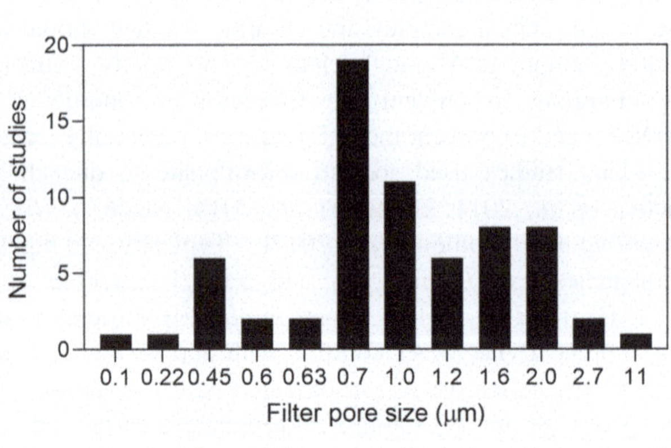

Figure 12.3 Overview of different filter pore sizes in micrometres (μm) used for sample filtration prior to solid-phase extraction (studies published from 2001 to January 2020).

associated with particulate matter (Legler *et al.*, 2003; Hamers *et al.*, 2015; Schulze *et al.*, 2015; Mueller *et al.*, 2021). For example, Dagnino *et al.* (2010) evaluated the estrogenic and aryl hydrocarbon receptor (AhR) activity in both the dissolved and particulate phases from three wastewater treatment plants, with both phases contributing to the estrogenic and AhR load discharged from the plants. A higher fraction of AhR activity was found on SPM (Dagnino *et al.*, 2010). Similarly, Wölz *et al.* (2008) found that SPM from the Neckar River induced AhR-mediated activity in the ethoxyresorufin-O-deethylase (EROD) assay while the corresponding extracted water samples had no effect. These studies suggest that it is important to consider SPM to gain a better understanding of the bioactivity, but the decision to collect SPM will depend on the objectives of the sampling campaign and the studied endpoints, with SPM likely to be important for non-specific toxicity, activation of AhR and binding to the peroxisome proliferator-activated receptor. During a heavy storm event in a small river, the effect load coming from SPM was as high as or higher than in the aqueous phase despite the SPM being present in mass (SPM) to water volume ratios lower than 0.001 during the rain event (Mueller *et al.*, 2021).

If intending to capture the entire water sample, the options are to use LLE, not filter before SPE, or to filter prior to SPE and extract the captured SPM separately using solvents (Ademollo *et al.*, 2012). Some studies only filtered samples that were expected to block the SPE cartridge, such as wastewater influent or river water, due to the high particle content (Gehrmann *et al.*, 2018; Xiao *et al.*, 2016). Similarly, many studies evaluating drinking water did not apply a filtration step (Van Zijl *et al.*, 2017; Hebert *et al.*, 2018; Valcarcel *et al.*, 2018). Könemann *et al.* (2018) found no significant difference in estrogenicity for filtered and unfiltered surface water samples.

To provide guidance on whether to filter or not, U.S. EPA Method 1694 'Pharmaceuticals and Personal Care Products in Water, Soil, Sediment, and Biosolids by HPLC/MS/MS' recommends that aqueous samples containing visible particles should be filtered prior to SPE (U.S. EPA, 2007). Water samples with a turbidity of 5 nephelometric turbidity units (NTU) will visually appear slightly milky or cloudy, while crystal clear water usually has turbidity less than 1 NTU, with the turbidity only detected by instrumental analysis (NHMRC, 2011). Consequently, water samples with a turbidity of 5 NTU or greater should be filtered prior to SPE. In general, drinking water is typically <1 NTU, recycled water can range from <1 to 2 NTU depending on treatment processes and secondary effluent is generally <2 NTU but can increase when sludge is poorly settled. The turbidity of river water can vary greatly, while the turbidity of lakes tends to be more stable. As an example, the turbidity of treated drinking water from Paris ranged from 0.02 to 0.04 NTU, while the water feeding these plants ranged from 2 to 14.3 NTU (Neale *et al.*, 2020b). For comparison, the turbidity of Canadian surface waters ranged from 0.5 to 50 NTU (Cantwell and Hofmann, 2011).

Based on low sorption capacity, glass fibre filters are recommended for filtration of samples with a turbidity of 5 NTU or greater. Although a wide range of filter pore sizes are used within the literature, glass fibre filters between 0.7 and 1.5 μm were recommended for filtration prior to SPE for chemical analysis (ISO11369, 1997; U.S. EPA, 2007; Furlong *et al.*, 2008).

12.4 EXTRACTION OF WATER SAMPLES

12.4.1 Extraction versus testing the entire water sample

In vitro assays typically target complex mixtures of organic micropollutants but not inorganics and metals, which can be comprehensively analysed using chemical methods. Therefore, extraction methods also serve to separate the organic micropollutants from the matrix, inorganics and metals in a water sample (Figure 12.4). This differs from whole effluent toxicity (WET), which considers the whole water sample and stress from organic and inorganic chemicals as well as ionic strength, DOC and pH (see Chapter 3).

Few studies have run unenriched or native water samples in yeast reporter gene and mammalian reporter gene assays (Niss *et al.*, 2018; Abbas *et al.*, 2019; Brettschneider *et al.*, 2019). This is equivalent to WET testing and would incorporate the effect from different components in water including salts, metals and other inorganics, as well as organic micropollutants. Consequently, the effect of organic micropollutants could not be differentiated from other components in

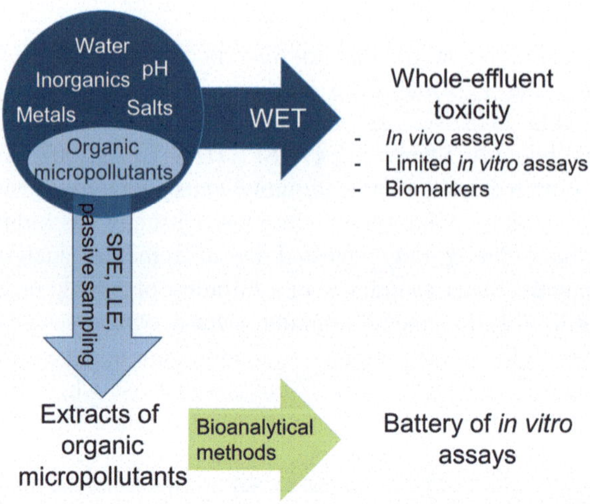

Figure 12.4 Enrichment of samples by solid-phase extraction (SPE), liquid-liquid extraction (LLE) or passive sampling selects organic micropollutants and removes matrix components, such as metals and inorganics.

water. This approach was used for wastewater samples but is unlikely to be able to detect an effect in cleaner water samples, such as highly treated wastewater, drinking water and surface water, which usually need to be concentrated to detect effects. As a minimum pre-treatment for testing native water samples, water must be filtered as described above to remove any particles. For mammalian cell-based bioassays, it is also important to sterile filter the sample using a filter with a 0.22 µm pore size (Niss *et al.*, 2018).

12.4.2 Solid-phase extraction

SPE dates back to the 1970s and relies on passing a water sample through a column packed with a sorbent that will retain the analytes as the water passes through. SPE is the most commonly used method to enrich micropollutants from water prior to bioanalysis. SPE cartridges or discs contain sorbents that retain organic micropollutants, while other components present in water, such as metals, salts and other inorganics, pass through the cartridge or disc, thus simplifying the matrix (Poole, 2003). After drying the cartridges to remove any residual water, the sorbed analytes can then be eluted with solvents, creating a concentrated extract that can be run in the bioassays. SPE has a number of advantages including good recovery of a wide range of contaminants and ability to be automated, although cartridges can clog with samples with a high particulate content as discussed in Section 12.3.2.

Reversed-phase and ion-exchange columns are normally used for aqueous applications. A variety of different sorbents have been fashioned for water samples such as silica-based (*e.g.*, C2, C8 and C18), hydrophilic lipophilic balanced (HLB) co-polymers and graphitised carbon black (GCB). These sorbents are specific for different classes of compounds, allowing selective extraction of analytes from water for chemical analysis. For bioanalytical measurement, sample preparation should be as non-selective as possible because otherwise only a portion of the chemical contaminants in the water would be retained and the mixture composition changed for detection in the bioassay. Co-polymer cartridges such as HLB are therefore widely used in bioanalytical monitoring due to their wide retention spectrum. SPE is commonly used in bioanalysis because it can be easily automated, does not produce emulsions, uses less solvent and usually produces better recoveries than LLE.

The main drawbacks of SPE are that the cartridges can become clogged with samples rich in organic matter (Ademollo *et al.*, 2012). It also has to be considered that the sorbent has a saturable capacity, which corresponds to 1%–5% of the sorbent mass, that is, a 200 mg sorbent bed can retain 2–10 mg of analyte and can become saturated with highly polluted samples causing a breakthrough of un-retained micropollutants.

Given the wide use of SPE, Section 12.5 will provide more information about SPE sorbents and extraction procedures.

12.4.3 Passive sampling

Passive samplers collect micropollutants from the water environment over a longer period of time and allows chemical and bioassay analyses of very low concentrations of chemicals. To date, passive sampling has primarily been applied to surface water (van der Oost *et al.*, 2017a; Toušová *et al.*, 2019). However, uncertainties regarding the volume of water sampled need to be considered and parameters such as temperature and flow velocity can affect the uptake of chemicals into the sampler (Novák *et al.*, 2018). Furthermore, the composition of the chemical mixture taken up into the sampler may differ from the chemical mixture in the water as different chemicals will have different uptake rates into the passive sampler. A number of different types of passive samplers have been applied in the literature, including silicone rubber, Empore discs and polar organic chemical integrative samplers (POCIS), in order to target chemicals with a wide range of hydrophobicity. For example, increased biological activity was found in Empore disc extracts compared to silicone rubber (Novák *et al.*, 2018), and Hamers *et al.* (2018) observed greater effect in Speedisk passive samples, which contain styrene divinylbenzene sorbent, compared to silicone rubber. Based on the chemicals extracted by different passive samplers, de Baat *et al.* (2019a) applied non-polar silicone rubber extracts to assays indicative of activation of AhR, oxidative stress response and pregnane X receptor (PXR), while polar POCIS extracts were tested in assays indicative of hormone receptor-mediated effects.

12.4.4 Liquid–liquid extraction

LLE has been used since the early nineteenth century and relies on the partitioning of the analytes from the water into a solvent that is immiscible with water, for example, dichloromethane, hexane, ethyl acetate or methyl *t*-butyl ether (MTBE). After vigorous shaking, the solvent and the water are allowed to separate into two clearly distinct layers. The solvent layer is collected, and the procedure is repeated twice with the water sample and fresh solvent to achieve the best possible recovery of the analyte from the water. The three solvent batches are combined and concentrated, usually by evaporation of the solvent. The advantage of LLE is that it is relatively simple to perform and it extracts everything that is in water, also SPM. On the downside, however, LLE requires large quantities of solvents, can be time consuming and emulsions can occur between the two liquid phases, impacting the reliability of the extraction process.

Only a few studies have applied LLE for sample enrichment prior to dosing into *in vitro* bioassays (van der Linden *et al.*, 2008; Brand *et al.*, 2013). In these studies, 300 mL of solvent were required per litre of water, which is around 7.5–10 times more than required for SPE (6 cc/500 mg cartridge). Therefore, great care has to be taken and high purity solvents have to be used to assure that there are no blank effects in the bioassays.

Due to the high solvent use, time consuming nature and lack of case studies, LLE is not recommended for sample enrichment of water samples, unless the focus is on total extraction of water plus SPM to capture also particle-bound water pollutants. Note that this can also be achieved with SPE by extracting the SPM retained on the filter. The separate analysis also has the advantage that we can differentiate between the aqueous and sorbed fraction of chemicals in a whole water sample.

12.4.5 Capturing volatile chemicals

Any solvent extraction, passive sampling and conventional SPE sample processing involves a blow down step, meaning that volatile chemicals, such as some DBPs (*e.g.*, trihalomethanes), will not be retained in the final extract. Furthermore, mammalian cell-based assays are incubated at 37°C for often 16–24 h, also potentially resulting in the loss of volatile chemicals, although some bioassays can be adapted to be run without a headspace to prevent the loss of volatile chemicals (Stalter *et al.*, 2013). Stalter *et al.* (2016c) developed a purge and cold-trap method to capture and concentrate volatile DBPs from drinking water. However, the method is tedious and requires extraction onsite or within a very short period of time. Therefore, it is not recommended for routine monitoring, but instead for research purposes. Importantly, volatile DBPs appeared to only have a minor contribution to the overall effects in drinking water (Stalter *et al.*, 2016c), which suggests that we can capture the majority of DBP-associated toxicity with simpler common SPE methods.

12.5 SOLID-PHASE EXTRACTION

As SPE is the most commonly applied extraction method, more detailed information on SPE, including different SPE sorbent options and common SPE procedure, is provided below.

12.5.1 Solid-phase extraction sorbents

A wide range of sorbents have been used for SPE, with Oasis HLB (Waters) the most commonly used. Other common SPE sorbents include Chromabond HR-X (Macherey-Nagel), StrataX (Phenomenex) and octadecyl silica C18. Most of the commonly used sorbents contain a copolymer mix, such as poly (divinylbenzene-*co*-*N*-vinylpyrrolidone), with a hydrophilic monomer to capture polar chemicals and a lipophilic monomer to capture hydrophobic chemicals. However, these sorbents tend to recover a lower fraction of charged chemicals compared to neutral chemicals (Neale *et al.*, 2018; Osorio *et al.*, 2018). Consequently, some studies have applied combinations of multiple sorbents, such as reverse-phase sorbents with ion-exchange materials, in order to capture a wider range of micropollutants, including very polar chemicals and charged chemicals (Aerni *et al.*, 2004; Toušová *et al.*, 2017; Osorio *et al.*, 2018). Other studies have applied multilayer SPE with Oasis HLB and coconut charcoal to improve the recovery of highly polar compounds (Escher *et al.*, 2014; Leusch *et al.*, 2014b). However, a multilayer SPE cartridge with Oasis HLB, Strata-X-CW, Strata-X-AW and Isolute ENV+ (1:1:1.5) and Supelclean EnviCarb was found to cause blank effects at a comparatively low REF of 20 (Neale *et al.*, 2018). Bioassays are not able to differentiate between effects from a sample and effects due to impurities from sample processing, so it is important to select an extraction method with minimal blank effects. In any case, it is important to always include processing controls with ultrapure water or glass-bottled water when enriching water samples to confirm that the observed effects are due to micropollutants in the sample and not related to the SPE sorbent or solvents.

In addition, a number of studies have compared the influence of different SPE sorbents on bioactivity. For example, Rosenmai *et al.* (2018) applied both Oasis HLB (poly(divinylbenzene-*co*-*N*-vinylpyrrolidone) and BondElut ENV (modified styrene divinylbenzene) sorbents to extract wastewater and drinking water samples, with no consistent difference in effect observed. Abbas *et al.* (2019) compared three SPE sorbents, Oasis HLB, Telos C18/ENV and Supelco ENVI-Carb+, at pH 2.5 and 7 and found that Telos C18/ENV at pH 7 was the most effective for wastewater effluent and groundwater, although considerable cytotoxicity was observed, which can mask the effect.

12.5.2 Solid-phase extraction procedure

Prior to extracting the water sample by SPE, it is necessary to condition the SPE cartridge or disc to wet and activate the sorbent bed. Water-miscible methanol, followed by ultrapure water is commonly used for conditioning (Bain *et al.*, 2014; Alygizakis *et al.*, 2019; Lundqvist *et al.*, 2019a). However, if other less polar solvents are used for eluting the cartridge, such as dichloromethane or ethyl acetate, then these solvents must also be used for conditioning. After conditioning is finished, the cartridge must not run dry and the water sample should be immediately percolated through the SPE cartridge.

Once water has been percolated and the chemicals in water sample are sorbed on the SPE sorbent, the cartridge must be completely dried in a vacuum or nitrogen stream. The dried cartridge can be sealed with parafilm and aluminium foil and stored at −20°C until elution (Tang *et al.*, 2014). Other studies have stored dried SPE cartridges for up to 2 weeks at 4°C (Scott *et al.*, 2014). Dried cartridges can also be sent to bioassay laboratories for elution, which is simpler, safer and cheaper than sending litres of unenriched water or solvent extracts.

To elute a wider range of polar and non-polar chemicals, multiple solvents are often used for elution, such as methanol and 1:1 hexane:acetone (Scott *et al.*, 2014; Jia *et al.*, 2015) or methanol and ethyl acetate (Houtman *et al.*, 2018; Müller *et al.*, 2018). Other solvents used in the literature in different combinations include acetonitrile, dichloromethane and MTBE. It should be noted that any impurity in the solvents can potentially cause blank effects in the bioassays. Consequently, it is important to use high purity (*e.g.*, HPLC grade) solvents for conditioning and elution and to limit the volume of solvent used. Based on 500 mg/6 cc Oasis HLB SPE cartridge, 10 mL of each solvent is often used for conditioning and 10 mL of each solvent is used for elution (Scott *et al.*, 2014; Müller *et al.*, 2018). Smaller solvent volumes can be used with smaller SPE sorbent beds.

Several studies have compared the effect of conditioning and elution solvents on bioactivity. For example, Leusch *et al.* (2014a) found no significant difference in the bioanalytical results when using 1:1 hexane:acetone and methanol compared to methanol alone for conditioning and elution. Furthermore, Prochazkova *et al.* (2018) compared the effect of two different solvent conditioning and elution combinations on estrogenic activity in surface water extracts. The first method targeted estrogenic compounds by conditioning and eluting with methanol, while the second method targeted less polar compounds and used ethyl acetate, methanol and 20% 2-propanol for conditioning and ethyl acetate for elution. The different solvents often resulted in different 17β-estradiol equivalent concentration values for the matching samples, but no systematic difference in estrogenicity was observed.

After elution, the solvent is blown to dryness under nitrogen gas and resuspended in a final solvent, such as methanol, dimethyl sulphoxide (DMSO) or ethanol. Murk *et al.* (2002) compared the effect of storage conditions on the same extract dissolved in ethanol and DMSO in the ERα CALUX assay. Initially, there was no difference in effect, but ethanol was found to evaporate quickly when stored at room temperature or 4°C and even evaporated at −20°C within 6 weeks. In contrast, the DMSO stock did not significantly change in activity over the 6-week period when stored at 4°C and −20°C. In Chapter 7, we discussed the cytotoxicity and effects of solvents. DMSO is most problematic. Hence, we still recommend using methanol as the solvent but only if the weight of the vial is continuously monitored during the chain of custody and the solvent is topped up if losses are recorded.

12.5.3 Effect recovery by solid-phase extraction

The recovery of individual chemicals by SPE has been well studied (Osorio *et al.*, 2018; Schulze *et al.*, 2017), but less is known about effect recovery for bioassays. Furthermore, unlike chemical analysis where an internal standard can be added to correct for chemical recovery by SPE, internal standards should not be used for bioanalysis as they may induce an effect in the bioassay that cannot be distinguished from the other micropollutants in the sample.

There are a number of approaches that have been applied in the literature to assess recovery, with most involving spiking a cocktail of chemicals into the water matrix prior to SPE enrichment (Figure 12.5).

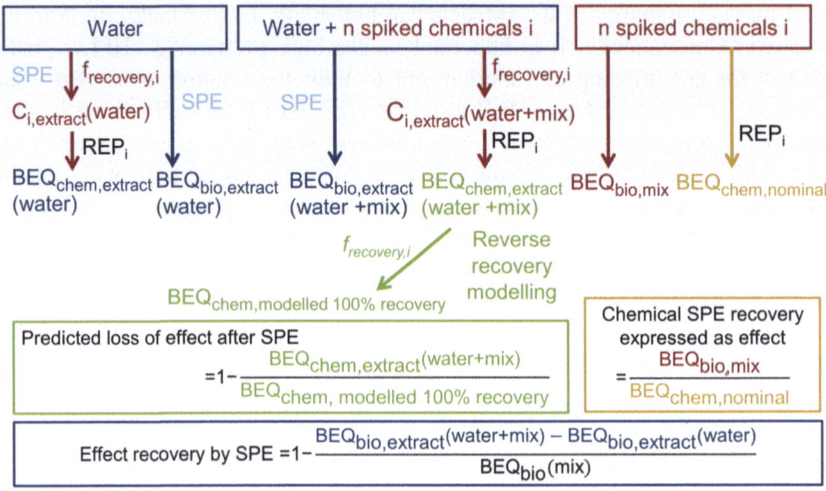

Figure 12.5 Various approaches to evaluate solid-phase extraction (SPE) recovery of chemicals and effects. BEQ = bioanalytical equivalent concentration (definition and equations in Chapter 7), SPE = solid-phase extraction, REP_i = relative effect potency of chemical *i*. Figure adapted from Neale *et al.* (2018). Solid-phase extraction as sample preparation of water samples for cell-based and other *in vitro* bioassays. *Environmental Science: Processes & Impacts*, **20**: 493–504. Reproduced by permission of The Royal Society of Chemistry © 2018.

As it can be difficult to measure the effect of the water alone, many studies compared the effect in the extract, often expressed as a bioanalytical equivalent concentration from bioanalysis ($BEQ_{bio,extract}$) to the predicted effect based on the concentration of chemicals detected in the extract ($BEQ_{chem,extract}$) (Kolkman *et al.*, 2013; Leusch *et al.*, 2010) or the nominal concentration of spiked chemicals ($BEQ_{chem,nominal}$) (Thorpe *et al.*, 2006; Kunz *et al.*, 2017). The ratios $BEQ_{chem,extract}/BEQ_{bio,extract}$ and $BEQ_{chem,nominal}/BEQ_{bio,extract}$ can be used as proxy for effect recovery (Figure 12.5). Furthermore, Abbas *et al.* (2019) attempted to assess SPE recovery by comparing the effect of the native water and SPE extracts in unspiked wastewater and groundwater. Since other components in the native water sample, such as salts, metals and other inorganics, could also have an effect in the bioassay in addition to organic micropollutants, this comparison is difficult to interpret.

To more quantitatively evaluate effect recovery by SPE, it is necessary to consider the effect of the spiked mixture alone, the effect of the extracted sample and the effect of the unspiked water alone (Figure 12.5). Neale *et al.* (2018) evaluated the effect recovery of a mixture of 579 micropollutants spiked into pristine surface water using a suite of bioassays indicative of xenobiotic metabolism, hormone receptor-mediated effects and adaptive stress responses.

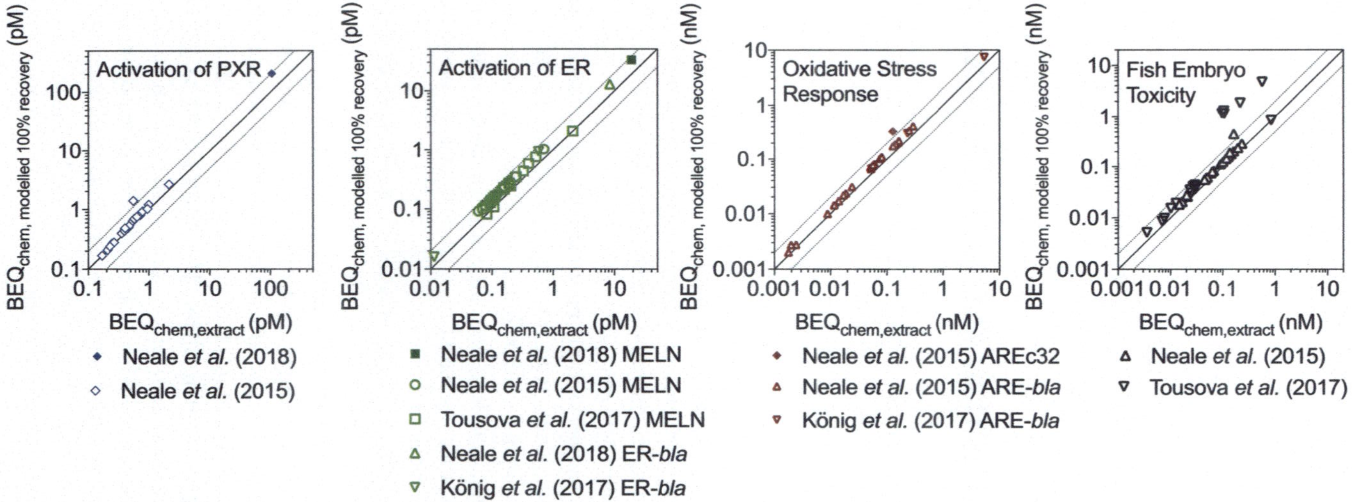

Figure 12.6 Comparison of the bioanalytical equivalent concentrations $BEQ_{chem,extract}$ and $BEQ_{chem,modelled\ 100\%\ recovery}$ for activation of PXR, activation of ER (both ER GeneBLAzer (ER-*bla*) and MELN) and oxidative stress response (both AREc32 and ARE-*bla*) in various studies. The dotted lines indicate a factor of 2 difference between $BEQ_{chem,extract}$ and $BEQ_{chem,modelled\ 100\%\ recovery}$. Figure adapted from Neale *et al.* (2018). Solid-phase extraction as sample preparation of water samples for cell-based and other *in vitro* bioassays. *Environmental Science: Processes & Impacts*, **20**: 493–504. Reproduced by permission of The Royal Society of Chemistry © 2018.

LVSPE with HR-X sorbent was used. Effect recovery was calculated using the effect of the spiked water ($BEQ_{bio,extract}$(water + mix)) minus the effect of the unspiked water ($BEQ_{bio,extract}$(water)) divided by the effect of the mixture stock solution (BEQ_{bio}(mix)). Effect recovery was within a factor of 2 of the optimal 100% recovery for most assays, which suggests that LVSPE is suitable for capturing the majority of active chemicals (Neale *et al.*, 2018). Also, a comparison of $BEQ_{chem,extract}$ and $BEQ_{chem,modelled\ 100\%\ recovery}$ from various studies showed that loss during SPE can be neglected when testing water samples (Figure 12.6).

12.6 SAMPLE COLLECTION AND SAMPLE PROCESSING FLOW CHART

There are a number of decisions to be made regarding sample collection, pre-treatment and enrichment. A decision-making flow chart (Figure 12.7) guides users through some of those key decisions. Once the final sample pre-treatment and processing methods have been selected, it is important to use the same approach for all samples that are to be compared. It is not possible to truly compare changes over time or differences between sites if different sample pre-treatment and processing methods are used as this can affect the chemical mixture in the final extract. Furthermore, where possible, the same bioassay and chemical analysis pre-treatment and sample processing methods should be used to allow greater comparability between the results.

The information used to support sample pre-treatment and processing decisions is often based on user experience, with few studies investigating the impact of different sample processing options on the biological effect. The majority of these studies focus on estrogenic activity, with little known about other endpoints. Furthermore, some of the advice is based on chemical analysis protocols, rather than being specific for bioassays.

One of the least standardised but very important pre-treatment steps is sample filtration (Figure 12.7). A wide range of filter pore sizes are used in the literature, which will affect the amount of SPM retained on the SPE cartridge. However, to our knowledge, the decision regarding which filter size to select or whether to filter or not has not been based on scientific studies but rather on user experience. Therefore, we suggest a uniform approach of filtering samples with a turbidity of greater than 5 NTU using glass fibre filters. The role of SPM as a 'carrier' of toxicity in water is important and hence, a concise definition of what is considered as particulate matter and what is dissolved, including chemicals bound to dissolved organic matter, is vital if one wants to compare different studies and also evaluate the contribution of dissolved versus bound chemicals in water.

Concerning SPE, the volumes that are given in Figure 12.7 are merely indicative as was discussed in Section 12.2.

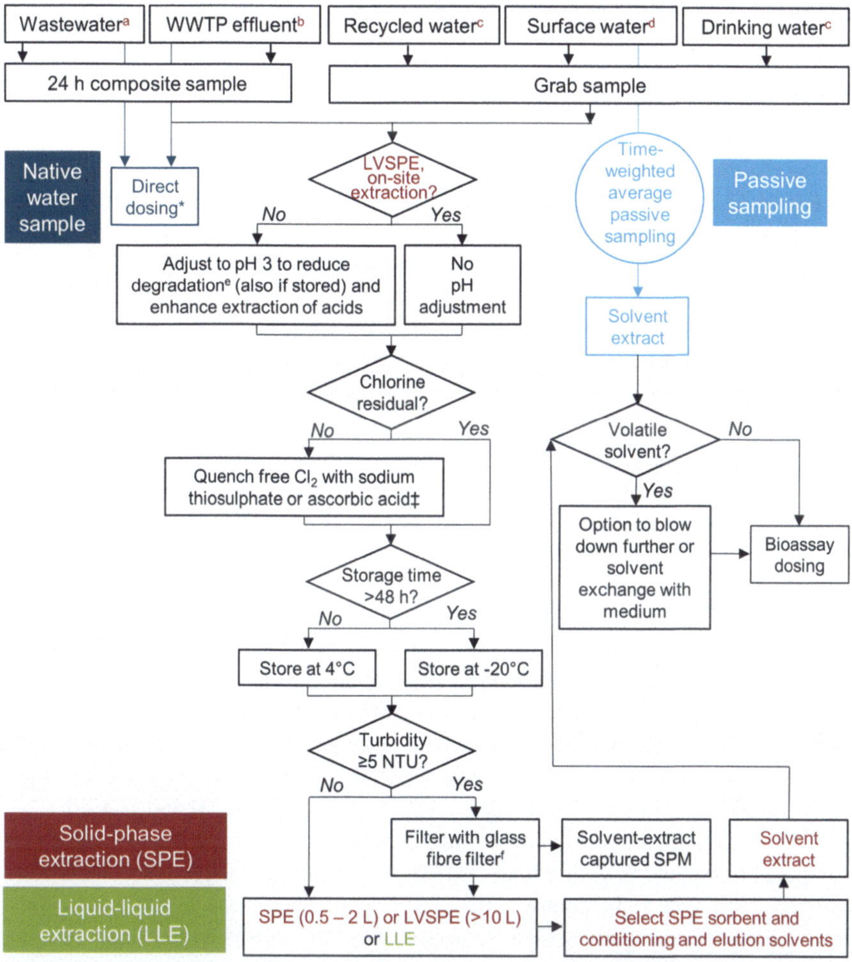

Figure 12.7 Sample pre-treatment and processing decision-making flow chart. The following volumes are based on solid-phase extraction (SPE) materials Oasis HLB, StrataX, Chromabond HR-X cartridges: [a]0.5 L per 200 mg or 1 L per 500 mg SPE cartridge; [b]1 L per 200 mg or 2 L per 500 mg SPE cartridge; [c]2 L per 200 mg or 4 L per 500 mg SPE cartridge; [d]1–2 L per 200 mg or 2–4 L per 500 mg SPE cartridge; [e]sodium azide can also be added for preservation; [f]glass fibre filter with pore size 0.7–1.5 µm recommended. *Not recommended. ‡3.5 mg/L sodium thiosulphate or 5 mg/L ascorbic acid to quench 1 mg/L free Cl_2.

12.7 DOSING INTO BIOASSAYS

The sample extracts can be dosed directly into the bioassay or solvent exchanged to a less toxic solvent by blowing down the elution solvent and resuspending in a final solvent, such as DMSO or methanol. Methanol and DMSO are the most commonly used solvents for bioassay dosing. DMSO can dissolve a wider range of compounds compared to methanol, but it is non-volatile, meaning that the extract cannot be further enriched by blowing down. In addition, as noted in Chapter 7, DMSO is more toxic compared to methanol. In contrast, methanol is volatile, meaning it can be blown down to further enrich the sample.

Figure 12.8 gives an example of the combination of extraction and dosing into well plate-based assays. Typically, a dosing plate is first prepared and then a fraction of the content of this plate is transferred to an assay plate that already contains cells that have been incubated for up to 24 h to achieve adherence to the plate in an appropriate volume of medium, for example, 4000–20,000 cells in 120 μL medium for 96-well plates or 1000–5000 cells in 30 μL medium for 384-well plates. The transfer from the dosing plate to the assay plate can be accomplished manually using a multichannel pipette, which is the typical approach for 96-well plates, or by copying the entire 96-wells of the dosing plate over to a 384-well plate using robotic devices, ideally using a 96-tip pipetting head. This allows quadruplicates on the 384-well plate. As the inter-plate variability (Chapter 11) is often rather low, it is common practice to transfer each dosing plate over only in duplicates by using two 96-well dosing plates per 384-well assay plate.

Using the volumes given in the example in Figure 12.8, the EF for the extraction is 4000 and the dilution for the highest concentration tested in the bioassay is 160, resulting in a dosing factor of 0.00625. The REF value for the highest concentration tested then becomes: $REF = 4000 \times 0.00625 = 25$. The second most concentrated sample would have an REF of $25/2 = 12.5$, while the third most concentrated sample would have an REF of $12.5/2 = 6.25$, and so forth.

In this example, the highest concentration of solvent is 1%. DMSO is more toxic compared to methanol, with a final DMSO concentration of 0.1% recommended in cell-based bioassays. In contrast, up to 1% of methanol can be added to some mammalian reporter gene assays (Leusch *et al.*, 2017). This equates to a dilution in the assay of 1000 for DMSO extracts compared to 100 for methanolic extracts, meaning extracts in DMSO need to be enriched 10 times more than methanolic extracts to achieve the same REF.

The REF can be further increased by exchanging the methanolic extract (or any other extract in a volatile solvent) with cell culture media. This is achieved by adding a volume of methanol to a glass vial (*i.e.*, 2 mL HPLC vial), blowing down to dryness and resuspending in cell culture media, which can be directly transferred to the cells. This allows one to increase the REF in the assay without inducing any solvent effects. This approach has been applied recently to drinking water extracts to help detect effects in relatively clean samples (Hebert *et al.*, 2018; Neale *et al.*, 2020b). It is important to ensure that the sample is well dissolved in the bioassay medium. This is rarely a problem for water extracts but may occur if SPM extracts are tested. It is also important to include solvent controls in the assay

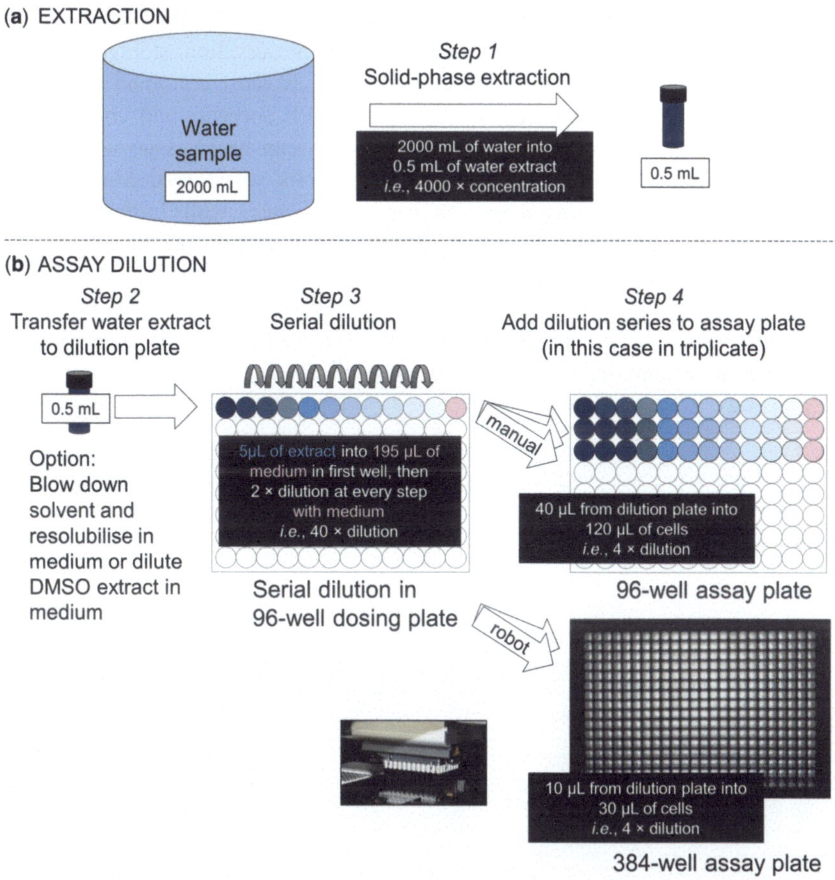

Figure 12.8 Typical sample treatment and dosing into bioassays. (a) Sample extraction and concentration, starting with 2000 mL of water sample concentrated down to 0.5 mL of water extract (*i.e.*, an enrichment factor EF of 4000). (b) The water extract is transferred to a plate and serially diluted. The serial dilution is then transferred to the assay plate in replicates with the option to transfer to a 96-well plate in this case in triplicate or use a 96-tip head to copy over two to four times to a 384-well plate.

to ensure that the solvent itself does not induce a response in the assay and even if the solvent is blown down, the solvent controls have to mimic that step entirely.

12.8 CONCLUSIONS

The whole workflow from sampling to dosing is important to achieve meaningful bioassay responses. The sampling strategy will be dictated by the purpose and objective of the sampling campaign, as well as the sample context. After sample collection, important considerations include sample preservation, storage time and whether to filter or not prior to enrichment. SPE is most commonly used for sample enrichment, but it is important to select SPE sorbents and conditioning and elution procedures with no blank effects. This can be an issue when using enrichment methods adapted from chemical analysis workflows. Many of the currently used methods are based on user experience, with further research required to evaluate the influence of different sample pre-treatment and processing steps on the biological effect.

Further dosing poses a challenge if DMSO is used as a solvent. Although DMSO extracts are easier to handle, extracts in volatile solvents pose the advantage that the solvent can be evaporated prior to dosing into cell assays.

Chapter 13

Design of test batteries and interpretation of bioassay results

13.1 INTRODUCTION

Cell-based bioassays have been applied for water quality assessment for decades. Initially, batteries of assays were put together rather arbitrarily, limited by the availability of particular assays or prior experience. More recently, rational designs of test batteries have been attempted and groups have collaborated to assemble broad test batteries for evaluative purposes (Brack *et al.*, 2019). After years of application, practical monitoring and screening test batteries have emerged that are widely applied. In this chapter, we outline design principles of test batteries for water quality assessment for different purposes.

Due to the complex mixture of chemicals present in environmental water extracts, a single bioassay cannot capture all the effects that may be triggered. Therefore, a test battery of bioassays covering different modes of action is required to detect the effect of as many active chemicals in a water sample as possible. Some studies have proposed bioassay test batteries as a proxy for acute and chronic effects at an ecosystem level (Schweigert *et al.*, 2002; Di Paolo *et al.*, 2016) or based on potential human health effects (Leusch *et al.*, 2014b), while other test batteries have focused on covering relevant modes of action (Escher *et al.*, 2008b; Neale *et al.*, 2017b) or focused on specific modes of action such as genotoxicity (Pellacani *et al.*, 2006) or endocrine disruption (Swart *et al.*, 2011).

Significant attention has been given in the last decade to endocrine disrupting effects, and particularly bioassays for estrogenicity. There was good reason for this focus: estrogens are very potent and the concentrations detected in surface

Authors: Beate Escher, Peta Neale and Frederic Leusch
doi: 10.2166/9781789061987_0265

waters are in the range where developmental and reproductive effects on fish have been observed. It is important, however, to include a variety of bioassays in any test battery to ensure that the spectrum of evaluated endpoints is sufficiently broad to provide a thorough assessment of overall water quality.

As discussed in previous chapters, bioassays can be selected to target either particular groups of chemicals (*e.g.*, photosynthesis inhibition is often very well correlated with herbicide concentration) or toxicity endpoints relevant to the protection goal of interest (*e.g.*, bacterial cytotoxicity assays would be useful to predict the impact of toxicants on microorganisms in wastewater treatment plants). In many instances, both approaches can cross over and lead to similar bioassay selections. A project evaluating the suitability of stormwater reuse for irrigation, for example, would most likely include an assay targeting photosynthesis inhibition, either because it can detect herbicides potentially present in the samples (chemical group motivated) and/or because photosynthesis inhibition is a relevant toxic endpoint for the crops to be irrigated (protection goal motivated). A project considering the suitability of reclaimed water for potable use would certainly include bioassays for genotoxicity, because this is a relevant protection goal for human health, but also because some compounds of concern potentially present in reclaimed water (*e.g.*, disinfection by-products) are genotoxic.

Bioassays that have been applied for water quality monitoring are reviewed in Chapter 10. Here, we introduce how test batteries are set up. It is important that each assay in the battery is validated and that issues that can affect the quality of the data are well understood. Chapter 11 provides some advice on quality assurance and quality control protocols that can help ensure the production of reliable and meaningful data. Chapter 12 gives guidance of sample preparation, which is relevant for the interpretation of bioassay results. It is possible to test the entire water sample (see also Chapter 3 on whole effluent testing) but it is most common practice to extract the organic micropollutants using solid-phase extraction (SPE), liquid–liquid extraction (LLE) or passive samplers. The extraction of complex mixtures of organic micropollutants, leaving behind inorganic components and most of natural organic matter, becomes relevant when interpreting the results of bioassays.

In this chapter, we also show how chemical analysis can be linked to the bioassay results by iceberg modelling, a mixture modelling approach that relies on the concept of concentration-additive effects (Chapter 8). Iceberg modelling leads to the classification of bioassays into two categories. Category 1 assays are highly specific bioassays that are mainly triggered by a limited number of known and generally potent chemicals. For these category 1 bioassays, almost all of the detected effects can typically be explained by known chemicals. A good example of a category 1 assay is a reporter gene assay for the estrogen receptor (ER). In most water samples, >90% of the estrogenic activity will be caused by natural hormones (such as estradiol, estrone and estriol), synthetic hormones (such as ethinylestradiol) and a few industrial xenoestrogens (such as bisphenol A or nonylphenol). Effect-directed analysis (EDA) has been applied successfully to identify unknown contributors to the mixture effects in water samples for category 1 bioassays.

In category 2 bioassays, only a small fraction of the measured effect can typically be explained by known and detected chemicals, which has important implications for the interpretation of the bioassay results. Typical category 2 bioassays are those indicative of adaptive stress responses or apical endpoints.

Many *in vitro* bioassays, particularly mammalian reporter gene assays, are highly sensitive by design and can detect effects in relatively clean waters, such as drinking water and recycled water, especially after sufficient enrichment (Jia *et al.*, 2015; Conley *et al.*, 2017b; Neale *et al.*, 2020b). However, just because an effect is detected in a bioassay does not necessarily mean that the chemical water quality is unacceptable. To help bioassay users differentiate between an acceptable and unacceptable bioassay response, effect-based trigger values (EBTs) have been proposed for both category 1 and category 2 bioassays. An EBT is comparable to a water quality guideline value (GV) but provides a specific threshold for each type and class of bioassay. What is acceptable or not depends on the water type and its usage and should be related to safe concentrations of regulated chemicals and be protective for the target organisms, for example, aquatic life in surface water, humans for drinking water.

13.2 TEST BATTERIES

13.2.1 Test battery design

Ideally, a bioassay test battery for water quality assessment should focus on effects commonly detected in water and include assays that cover a range of relevant modes of action as well as apical effects in whole organisms (Figure 13.1) (Neale *et al.*, 2017b). This helps to ensure that expected effects are covered, while safeguarding against missing any unexpected modes of action.

Test batteries indicative of different stages of cellular toxicity pathways, such as induction of xenobiotic metabolism, receptor-mediated effects, adaptive stress responses and cytotoxicity, have been applied to wastewater, surface water, stormwater, recycled water and drinking water (Escher *et al.*, 2014; Leusch *et al.*, 2014b; Nivala *et al.*, 2018; de Baat *et al.*, 2020). Although an effect at the cellular level may not necessarily result in higher level adverse effects (Ankley *et al.*, 2010), assays indicative of different stages of cellular toxicity pathways can act as sensitive indicators of chemicals. In this respect, they can be used as proxies for chronic effects. An example is endocrine disruption, which would require testing of the whole life cycle and adverse effect on reproduction, both of which are far too time-consuming and expensive for routine testing of water samples. We know from field studies that estrogenic compounds in wastewater can lead to feminisation of male fish, and that a low nanogram per litre level of

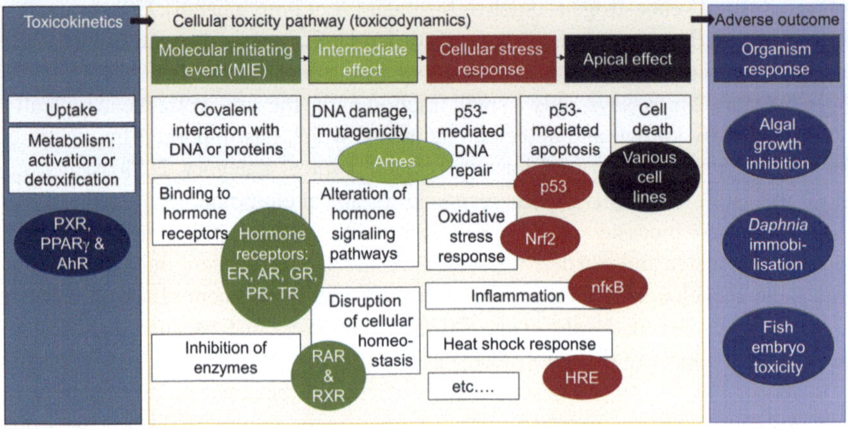

Figure 13.1 Bioassays relevant for water quality monitoring (circles) anchored in crucial steps of the adverse outcome pathway (boxes). ER = estrogen receptor, AR = androgen receptor, GR = glucocorticoid receptor, PR = progesterone receptor, TR = thyroid receptor, ARA = retinoic acid receptor, RXR retinoid X receptor, p 53 =, Nrf2 = NF-E2 related factor 2, nfκB = nuclear factor kappa B, HRE = hypoxia response element. Modified and reprinted with permission from Neale *et al.* (2017b). Development of a bioanalytical test battery for water quality monitoring: Fingerprinting identified micropollutants and their contribution to effects in surface water. *Water Research*, **123**, 734–750. © 2017 Elsevier.

17α-ethinylestradiol can collapse entire fish populations, hence, including an estrogenicity test in a bioassay battery will be protective for this aspect of reproductive chronic effects.

In addition, whole organism assays indicative of apical effects, such as algal growth inhibition, invertebrate and fish embryo toxicity, are a useful complement to cell-based assays as they can integrate effects from multiple toxicity pathways and assure that all bioavailable chemicals are considered. For drinking water, studies with mammals are prohibitive in a routine monitoring context and have only rarely been performed (Narotsky *et al.*, 2013) but the fish embryo toxicity test with *Danio rerio* may serve as a proxy for acute toxicity and reporter gene assays for chronic effect. Hence, drinking and surface waters can be monitored with very similar test batteries, which can be applied across the entire water cycle from wastewater to recycled water, from groundwater to drinking water and beyond (Escher *et al.*, 2014).

In all test batteries, the specific effect measured with a reporter gene assay or a biomarker should be accompanied by cytotoxicity assessment (Chapter 7). This is because cytotoxicity may cause false-negative results by masking the effect or may elicit false-positive results if there is apparent activation due to the 'cytotoxicity burst' phenomenon (Judson *et al.*, 2016; Escher *et al.*, 2020a).

13.2.2 Multiplex bioassays serving as test batteries

In addition to assays indicative of a single endpoint, some studies have applied multiplex high-throughput screening assays such as the Attagene *trans*-FACTORIAL™ and *cis*-FACTORIAL™ assays to drinking water, surface water and wastewater (Escher *et al.*, 2014; Blackwell *et al.*, 2019; Medlock Kakaley *et al.*, 2020). These assays include many targeted endpoints covering different stages of cellular toxicity pathways. For example, Escher *et al.* (2014) found the pregnane X receptor (PXR), ERα, peroxisome proliferator-activated receptor (PPARγ), aryl hydrocarbon receptor (AhR) and the antioxidant response element ARE were the most responsive in wastewater, surface water and recycled water extracts. Blackwell *et al.* (2019) evaluated the surface water of 38 rivers in the USA using this multiplexed battery and also confirmed that the ERα and GR were most prominently activated, together with AhR and PXR. Figure 13.2 shows an exemplary radar plot of blank-normalised area-under-the-curve (AUC) values for one water sample of this study. Medlock Kakaley *et al.* (2020) found that PXR and PXR signalling pathway (PXRE) were the only nuclear receptors activated in the intake to a drinking water treatment plant (DWTP), with no effect in treated water. These results support the findings from individual reporter gene assays, with assays indicative of activation of PXR, AhR and ERα found to be most responsive to a wide range of water types. They also confirm that the iterative strategy over the last few years of implementing various reporter gene assays has narrowed down the pertinent endpoints and has not overlooked important endpoints relevant for water quality. However, compared to targeted reporter gene assays, Blackwell *et al.* (2019) reported that the multiplex assays were less sensitive when applied to the same surface water samples compared to reporter gene assays. Thus, they might be less suitable for surveillance monitoring, where reporter gene assays will remain the preferred choice. Since these multiplexed assays cover a wider range of endpoints, they are especially suitable for screening purposes to assure that new endpoints are not overlooked when a new type of water is investigated.

13.2.3 Routine test batteries for monitoring applications

A practical test battery of at least three or four bioassays representative of effects commonly detected in water samples and aligned with relevant steps of cellular toxicity pathways is recommended for routine water quality monitoring. Although it is possible that other relevant effects may be missed with only three to four bioassays, such a routine test battery can still provide a good measure of the overall water quality and these routine test batteries are useful for benchmarking across different stages of the water cycle if those bioassays were chosen because they are most responsive to the target type of water.

Assay selection will depend both on the context (*e.g.*, water type and treatment type) and the purpose of the sampling campaign (*e.g.*, to assess product quality or

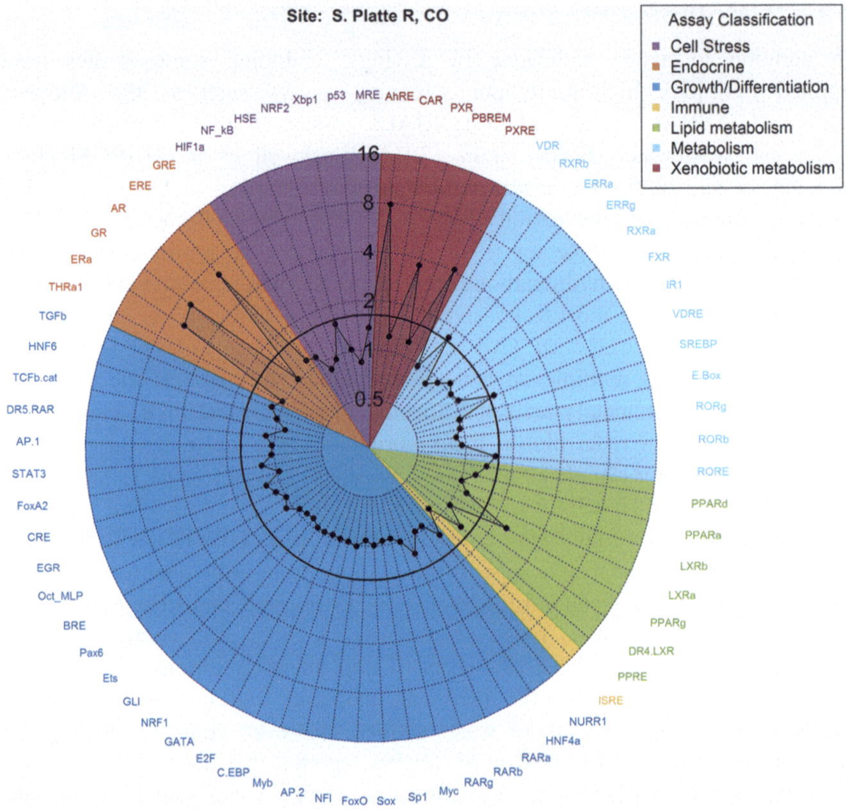

Figure 13.2 Example radar plot showing commonly activated endpoints of the Attagene *cis*-FACTORIAL™ and *trans*-FACTORIAL™ assay in surface water extracts on the example of one site in South Platte River in Colorado. The effect endpoint is the blank-normalised AUC between an REF of 0.04 and 10. The dark circle marks the effect threshold of 1.5. Figure reprinted with permission from Blackwell *et al.* (2019). Potential toxicity of complex mixtures in surface waters from a nationwide survey of United States streams: Identifying *in vitro* bioactivities and causative chemicals. *Environmental Science & Technology*, **53**(2), 973–983. © 2019 American Chemical Society.

treatment process efficacy). In the case of wastewater and water reuse for non-potable use, a minimum test battery should include assays indicative of activation of the AhR, activation of ERα and oxidative stress response. These three endpoints are responsive to extracts of a range of water types, as demonstrated by both individual and multiplexed assays, and represent different stages of cellular toxicity pathways, that is, xenobiotic metabolism, receptor-mediated effects and adaptive stress responses. Furthermore, proposed EBTs are available for these endpoints as outlined below. This recommendation aligns with recommendations for testing surface water quality (Brack *et al.*, 2019), and such harmonisation is important given that rivers are receiving effluent discharges and are often at the same time the source water for DWTPs.

In the context of drinking water treatment or water reuse for potable use, a test battery should include an assay indicative of genotoxicity or mutagenicity in addition to activation of AhR, activation of ERα and oxidative stress response. It is worth noting that oxidative stress response assays can detect increased effects after drinking water disinfection. However, they cannot replace mutagenicity or genotoxicity testing (with traditional bacterial assays such as the Ames test or umuC assay) but are often also triggered by genotoxic chemicals, not only those with direct reactive toxicity.

Although test batteries of three or four assays are recommended as the minimum in most situations, more comprehensive test batteries could include any assay previously found to have a response in water extracts plus whole organism assays indicative of apical effects (*e.g.*, an algal growth inhibition assay or fish embryo toxicity assay). The selection of additional assays may be related to specific water quality concerns. For example, a phytotoxicity assay could be included if raw drinking water is collected from a catchment with significant agricultural activity.

If advanced cell culture laboratories required to run mammalian reporter gene assays are not accessible, simple bacterial toxicity assays, such as the Microtox or BLT-Screen, could be applied as a simple screening tool. Both assays are similarly sensitive and have been applied to wastewater, surface water and drinking water. It should be noted that these assays only provide information about non-specific effects and should be complemented with assays indicative of specific effects when possible, but they can be powerful as indicators of relative chemical water quality (*e.g.*, to measure changes over time or to compare different waters).

13.3 LINKING BIOASSAY RESULTS WITH CHEMICAL ANALYSIS: ICEBERG MODELLING

13.3.1 Iceberg modelling

In Chapter 1 we introduced the spotlight analogy with different analytical and bioanalytical tools shedding light on different parts of the universe of chemicals (Figure 13.3). Cytotoxicity gives a measure of all chemicals that are bioavailable and can be taken up into cells. Thus, each chemical in a water sample or extract contributes to the mixture effects. Specific effects quantified with reporter gene assays shed a bright light only on a small subset of the chemicals that can be captured by cytotoxicity. There is overlap of the chemicals captured by bioassays and the chemicals that can be quantified by target analysis and suspect screening (Figure 13.3).

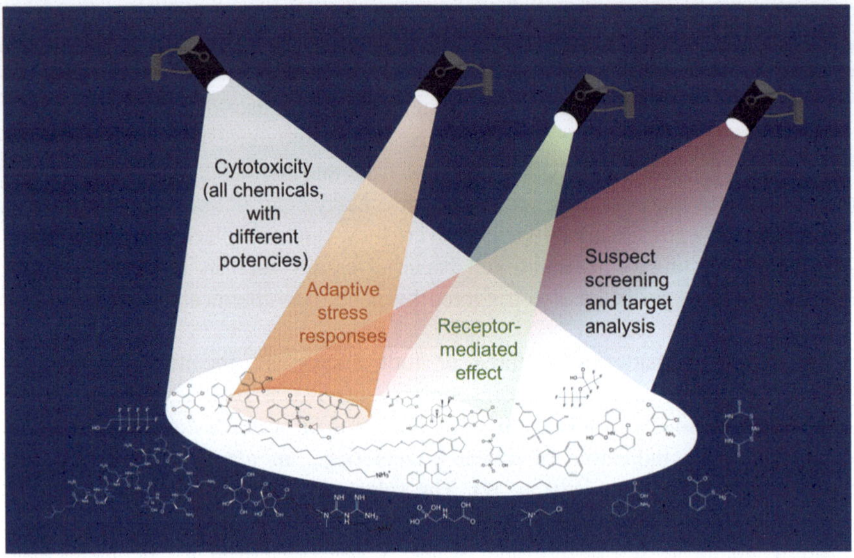

Figure 13.3 Chemical analysis and bioanalytical tools shed light on different parts of the universe of organic micropollutants. This is a simplified version of Figure 1.7 in Chapter 1, which was adapted from Escher *et al.* (2020b).

To quantify this overlap and to estimate how much of the observed cytotoxicity and specific effects can be explained by the detected and quantified chemicals, so-called 'iceberg (mixture) modelling' can be used. In Chapter 7, we have learnt how effect concentrations EC_y can be derived from concentration–response curves describing effect y, which is typically 10% of maximum effect or an induction ratio (IR) of 1.5. For better comparison between assays and samples we have also introduced the so-called 'bioanalytical equivalent concentrations' (BEQ_{bio}) that translate the effect detected in a sample into concentration units of a reference chemical for this bioassay by dividing the EC_y of the reference compound by the EC_y of the water sample (Figure 13.4). The BEQ_{bio} gives the overall effect that a sample can trigger expressed as concentration of a reference compound that would cause exactly this effect.

In Chapter 8 we have learnt about mixture toxicity concepts and that at low effect levels, as they occur in water samples, we can predict the combined effect of mixtures with many components for both similarly acting and dissimilarly acting compounds with a very simple model where the products of the relative effect potencies (REP_i) of each known mixture component i and its concentration (C_i) can be summed up to calculate the predicted mixture BEQ_{chem}. In the iceberg analogy the BEQ_{chem} constitutes the visible tip of the iceberg and the BEQ_{bio} the entire iceberg. The part of the iceberg that is submerged represents the effects from unknown chemicals or those that are present below their chemical limit of

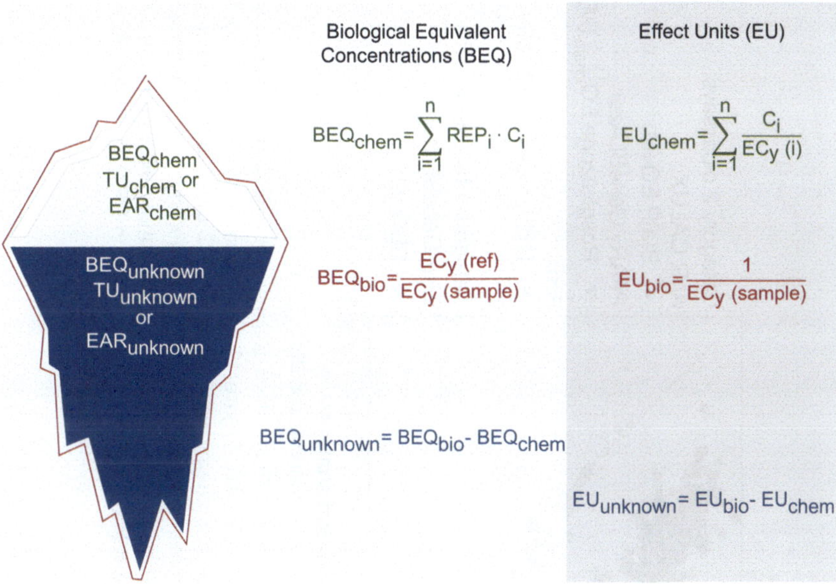

Figure 13.4 Iceberg modelling connecting mixture effects predicted from concentration of n chemicals i (subscript chem) with measured effects (subscript bio). The unknown effect (subscript unknown) can be calculated by subtraction of the measured effect from that predicted from known chemicals. Figure adapted from Villeneuve *et al.* (2019) and Neale *et al.* (2020b). BEQ = bioanalytical equivalent concentrations; EU = effect unit; REP = relative effect potency; C_i = concentration of chemical i; EC_y = effect concentration triggering effect y.

detection. The $BEQ_{unknown}$ can be deduced from subtracting BEQ_{chem} from BEQ_{bio} (Figure 13.4). Similar calculations hold for effect units (EU; Figure 13.4). Analogously to EU, we can define toxic units (TU) for cytotoxicity from inhibitory concentration (IC_y) for cytotoxicity or lethal concentrations for whole organism toxicity (not shown in the figure).

If $BEQ_{bio} = BEQ_{chem}$, then the measured effect can be fully explained by known chemicals. If $BEQ_{bio} > BEQ_{chem}$, then the detected chemicals cannot fully explain the measured effect, either because their concentrations are below the chemical detection limit or because there are additional chemicals that were not measured but contribute to the mixture effect (Figure 13.5a).

Typical results of iceberg modelling are shown in Figure 13.5b. For example, a study on surface water impacted by wastewater treatment plants (Könemann *et al.*, 2018) reported three potent estrogens (17β-estradiol, 17α-ethinylestradiol and estrone) often explained almost all of the measured estrogenic activity EEQ_{bio}, especially at higher EEQ levels. At lower EEQ levels, EEQ_{chem} were slightly lower than EEQ_{bio}. This is likely due to one or more of the potent estrogens falling

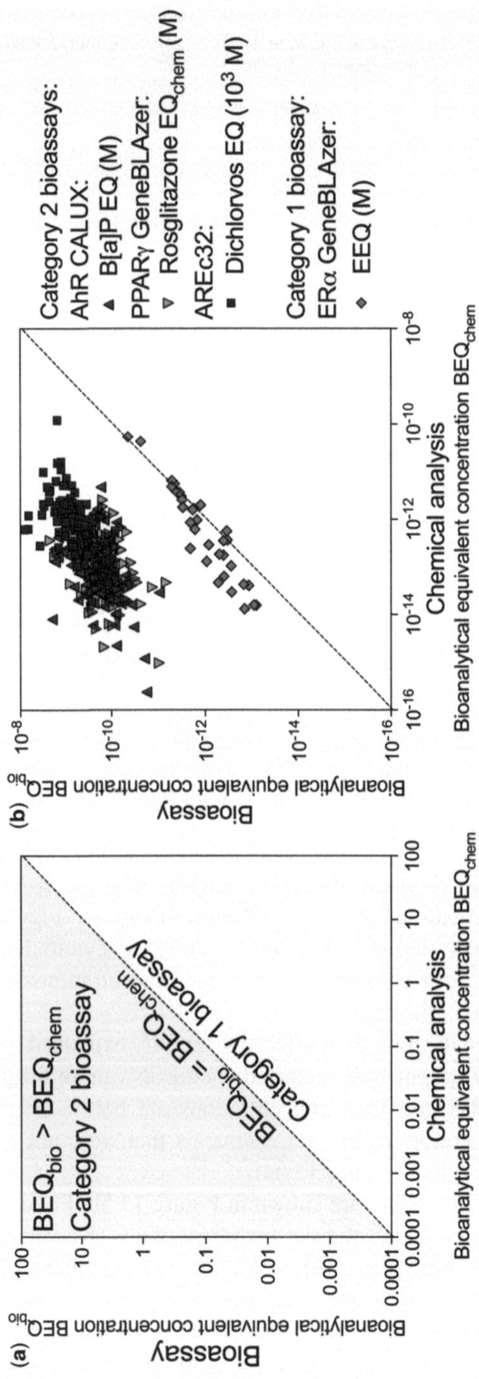

Figure 13.5 (a) Comparison between bioanalytical equivalent concentrations BEQ_{bio} from bioassay measurements and the BEQ_{chem} predicted from the detected chemicals' concentrations and their REP (see Figure 13.4 for equations). For category 1 bioassays, $BEQ_{bio} \cong BEQ_{chem}$ and the effect can be explained by the detected chemicals. For category 2 bioassays, $BEQ_{bio} > BEQ_{chem}$ and most of the effect stems from unknown chemicals. (b) Examples for estrogenicity (EEQ, green diamonds) detected with ERα GeneBLAzer in surface water impacted by WWTP (Könemann *et al.*, 2018) and surface water during rain events (Neale *et al.*, 2020a): benzo[a]pyrene EQ (B[a]P EQ) in AhR CALUX (blue upward triangles), rosiglitazone EQ in PPARγ GeneBLAzer (blue downward triangles) and dichlorvos EQ in AREc32 (red squares).

below the detection limit of the chemical analysis (usually around 0.1–1 ng/L). Similar good agreement between BEQ_{bio} and BEQ_{chem} has also been observed for other hormone receptor activation (Hashmi *et al.*, 2018, 2020) and photosynthesis inhibition in green algae (Tang and Escher, 2014; Glauch and Escher 2020). Bioassays where $BEQ_{bio} \cong BEQ_{chem}$ have been termed 'category 1 bioassays' (Escher *et al.*, 2018).

In contrast, bioassays where $BEQ_{bio} > BEQ_{chem}$ have been termed 'category 2 bioassays' (Escher *et al.*, 2018). For example, less than 1% of BEQ_{bio} can typically be explained by BEQ_{chem} when measuring activation of the AhR in the AhR CALUX, activation of PPARγ in the PPARγ GeneBLAzer or oxidative stress response with the AREc32 assay (Figure 13.5b), even when many chemicals are analysed. To illustrate, Neale *et al.* (2020a) detected 290 chemicals in surface water samples, but almost a fifth of those chemicals (55–58) had no experimental effect data, and thus it was not possible to calculate the BEQ_{chem} contribution for those compounds. Of the remaining >200 chemicals, 40 were active in the AhR CALUX, 20 in the PPARγ GeneBLAzer and 52 in the AREc32, but these still only explained <5% of the BEQ_{bio}. The low fraction of effect explained is not solely due to a lack of data but lays in the type of chemicals and their potency in these bioassays, as will be detailed in Section 13.4.

13.3.2 Effect-directed analysis

So-called 'effect-directed analysis' (EDA) can be used in category 1 bioassays to identify the bioactive components of a complex mixture in a water sample (Houtman *et al.*, 2011; Brack *et al.*, 2016). In EDA, a water sample or extract that is bioactive can be fractionated and the individual fractions re-tested for bioactivity (Figure 13.6). The bioactive fractions then undergo chemical analysis to identify the causative agents. These identified chemicals are also tested in the bioassays, ideally in mixtures in concentration in which they were found to confirm the extract/fraction effect. Several iterations of the process can be performed to narrow in the causative toxicants. The whole process is similar to the toxicity identification evaluation (TIE) protocol in WET testing (Chapter 3), but in this case applied to a cell-based bioassay context.

For category 2 bioassays, EDA is not easily applicable because the bioactivity is typically spread over many fractions and will disappear after fractionation as has been demonstrated on the example of the AREc32 assay for oxidative stress response (Hashmi *et al.*, 2018).

Hashmi *et al.* (2018) identified the causative estrogens in surface water impacted by untreated wastewater as mainly 17β-estradiol, estrone and 17α-ethinylestradiol with some very minor contribution by genistein, daidzein and testosterone. They also showed a larger spectrum of natural and man-made hormones to be responsible for the androgenic effects. Mixtures of the detected chemicals confirmed that mixture effects of the components were concentration-additive in most fractions.

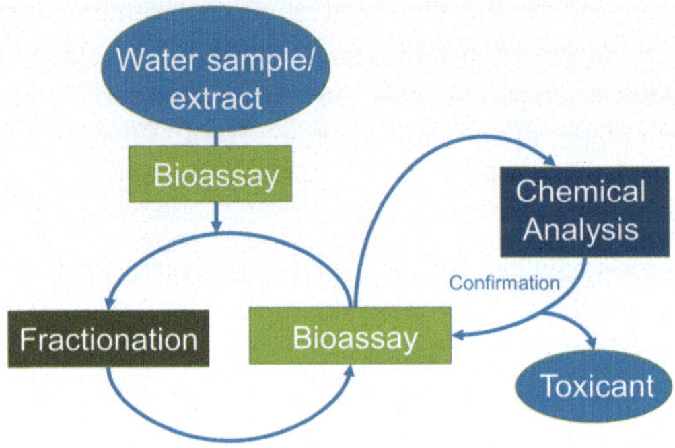

Figure 13.6 Principle of EDA. Modified after and reprinted with permission from Brack *et al.* (2016). Effect-directed analysis supporting monitoring of aquatic environments – An in-depth overview. *Science of the Total Environment*, **544**, 1073–1118. © 2016 Elsevier.

Fractionation can also 'bring out' the activity. For example, activation of hormone receptors may be masked by cytotoxicity in unfractionated samples, or agonistic and antagonistic effects may cancel each other out. The same unfractionated water samples from Hashmi *et al.* (2018) did not show any progestogenic and glucocorticoid activity but bioactivity was detected with reporter gene assays for the progesterone receptor (PR) and glucocorticoid receptor (GR) after fractionation (Hashmi *et al.*, 2020). It proved to be much more challenging to identify the causative chemicals for progestogenic and glucocorticoid activity requiring non-target screening analysis before confirmation of individual components' activity with reporter gene assays (Hashmi *et al.*, 2020).

Classical EDA is quite work-intensive and has been more commonly applied to identify causative chemicals at contaminated sites or in research contexts than for water quality monitoring. This might change in the future with the development of high-throughput methods for EDA. Zwart *et al.* (2018, 2020) developed an EDA platform that combines liquid chromatography (LC) with mass spectrometry (MS) and parallel bioassay detection. The direct comparison of the MS chromatograms with the bioassay responses revealed prominent MS peaks with no bioactivity but also identified smaller peaks with high bioactivity. This method was not only applied to estrogenic, androgenic and glucocorticoid activity (Zwart *et al.*, 2018, 2020) but also to the Ames assay for mutagenicity (Zwart *et al.*, 2020). Here, 1,2,3-benzotriazole, a highly abundant anticorrosion agent, and 2-formyl-1H-pyrrole could be identified as a mutagenic component of wastewater.

In a further optimisation step of the platform for even higher resolution, the fractions were directly split after the LC separation, with an aliquot going to mass spectrometry analysis and the other collected in 384-well plates that could be used directly for the bioassay (Houtman *et al.*, 2020). No concentration–response curve could be run but almost 300 fractions per sample provided something like a 'bioassay chromatogram' (Figure 13.7). The typical culprits

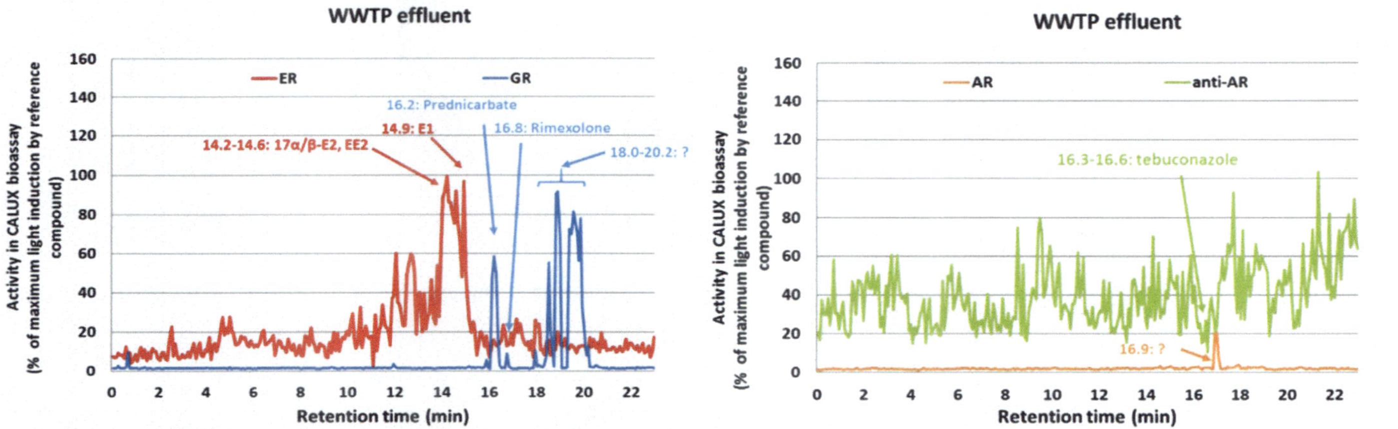

Figure 13.7 Bioassay chromatogram for ER and GR CALUX (left) as we run AR CALUX in agonistic and antagonistic modes (right) with the bioactive components identified by mass spectrometry (MS). AR = androgen receptor, ER = estrogen receptor, GR = glucocorticoid receptor, PR = progesterone receptor. Reprinted with permission from Houtman *et al.* (2020). High resolution effect-directed analysis of steroid hormone (ant)agonists in surface and wastewater quality monitoring. *Environmental Toxicology and Pharmacology*, **80**, 103460. © 2020 Elsevier.

(17β-estradiol, estrone and 17α-ethinylestradiol) could be identified with the estrogen receptor (ER) CALUX assay but also less well-known compounds were revealed as contributing strongly to bioactivity in GR CALUX and AR CALUX, which was also run in antagonistic mode. These early results are quite encouraging for future applications of EDA as a routine method in monitoring.

13.4 CATEGORY 1 AND CATEGORY 2 BIOASSAYS

According to iceberg modelling, category 1 bioassays target highly specific, mainly receptor-mediated effects such as binding to hormone receptors or inhibition of photosynthesis. Category 2 bioassays are those assays covering effect endpoints that are triggered by many more and more diverse chemicals that still exhibit specific effects but with a lower degree of specificity. Category 2 bioassays also include oxidative stress response and reporter gene assays with more promiscuous nuclear receptors, such as AhR and PPARγ.

The specificity ratio $SR_{baseline}$, introduced in Chapter 9 (Equation 9.4), provides another more quantitative way to classify bioassays into these two categories. $SR_{baseline}$ is a quantitative measure of how much more potent the activation of a specific response is in comparison with the baseline cytotoxicity of the same

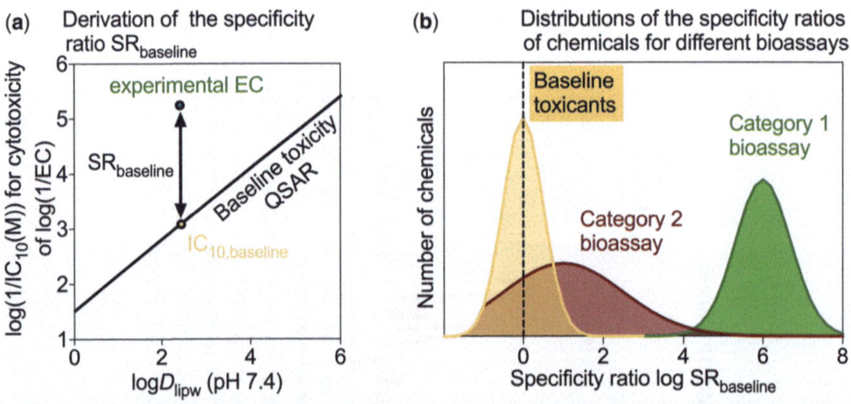

Figure 13.8 (a) Conceptual figure describing the derivation of the specificity ratio $SR_{baseline}$ from experimental effect concentrations (EC) and the inhibitory concentration causing 10% reduction in cell viability ($IC_{10,baseline}$) that was predicted from the quantitative structure activity relationship (QSAR) for baseline toxicity. (b) Conceptual figure of distributions of specificity ratios log $SR_{baseline}$ for highly specific bioassays (category 1) and those that respond to many different chemicals (category 2) compared to distributions of log $SR_{baseline}$ of baseline toxicants. Figure reprinted with permission from Escher and Neale (2021). Effect-based trigger values for mixtures of chemicals in surface water detected with *in vitro* bioassays. *Environmental Toxicology and Chemistry*, **40,** 487–499. © 2021. The Authors. *Environmental Toxicology and Chemistry* published by Wiley Periodicals LLC on behalf of SETAC.

compound (Figure 13.8a). Baseline toxicants have an $SR_{baseline}$ of 1 and the data are logarithmically distributed around log $SR_{baseline}$ of 0 (Escher and Neale, 2021). Chemicals activating category 1 bioassays have high log $SR_{baseline}$ values, often in the range of 4–8, while category 2 chemicals have log $SR_{baseline}$ with medium values and often very broad distributions (Figure 13.8b).

Chemicals active in category 2 assays follow a broad log-normal distribution (over 4–5 log units wide) centred close to a value of log $SR_{baseline} = 0$, as demonstrated in Figure 13.9a–c for the activation of the oxidative stress response, PPARγ and AhR receptors. Category 1 bioassays, in contrast, have a low number of very highly potent chemicals with medians of the log-normal distribution at $SR_{baseline}$ of 4×10^6 (log 6.6) for estrogenicity (Figure 13.9d) and 2×10^4 (log 4.3) for inhibition of the photosystem II in green algae (Figure 13.9e). What

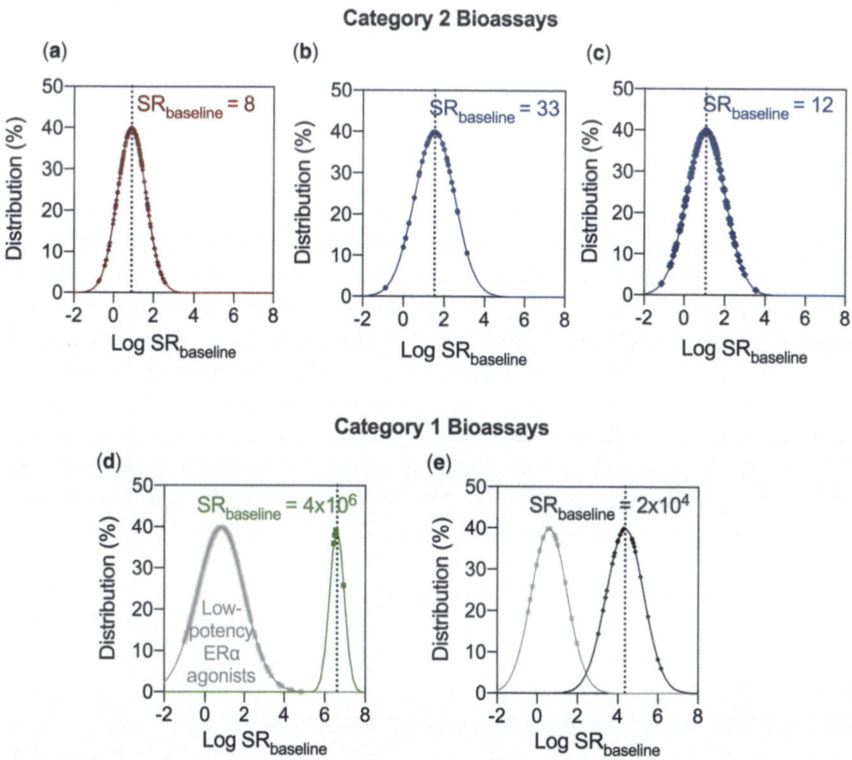

Figure 13.9 Distributions of specificity ratio $SR_{baseline}$ for category 1 and 2 bioassays using experimental data from Escher and Neale (2021) for (a–d), and from Glauch and Escher (2020) for (e). (a) Activation of oxidative stress response ARE, (b) activation of the PPARγ receptor, (c) AhR, (d) estrogen receptor ERα and (e) inhibition of photosynthesis in green algae after 24 h.

makes the picture more complex is that category 1 bioassays additionally respond to a large number of chemicals with very low $SR_{baseline}$ that show a distinctly separate distribution from the high potency chemicals. Those low potency chemicals generally do not play a significant role in iceberg modelling unless they are present at significantly higher concentration (1000 to 1 million times higher) compared to the high potency chemicals.

It must be noted, though, that the classification is dependent on the context. For AhR, there are highly potent ligands, such as polychlorinated dibenzo-p-dioxins and furans and if these were included, AhR would also be classified as a category 1 bioassay. However, those chemicals are very hydrophobic and do not dissolve in water. The AhR assay is therefore classified as a category 2 bioassay for water samples.

The classification criteria for category 1 and 2 bioassays are also relevant for the derivation of EBT values discussed in Section 13.5.

13.5 EFFECT-BASED TRIGGER VALUES

As noted earlier, *in vitro* bioassays (particularly mammalian reporter gene assays) are highly sensitive and can detect effects in relatively clean waters especially after sufficient enrichment. It is critical therefore to provide a tool to water utilities and regulators to determine a value below which a bioassay response is unlikely to produce adverse effects. This threshold is called an 'effect-based trigger value' (EBT), and various approaches have been devised to calculate EBTs for category 1 and category 2 assays. There are principal differences for the two categories because we know the mixture effect drivers in category 1 bioassays, while there are many unknown chemicals contributing to the effects of category 2 bioassays. The two are thus presented next in two sections, one for each category of bioassays. It must be noted that this category 1 and 2 classification is a recent concept, and therefore the reviewed papers did not specifically provide this classification. The approaches were divided into category 1 and category 2 for the purpose of this chapter.

EBTs for surface water and wastewater need to be protective of ecosystem health and are therefore typically derived from safe concentrations for aquatic organisms. After translation to an *in vitro* bioassay, the cell line applied does not necessarily need to be of the origin of an aquatic species because the *in vitro* effects are used as bioanalytical measures and not as ecological effect measures. The same considerations hold for drinking water quality.

13.5.1 Approaches to derive effect-based trigger values for category 1 bioassays

Several different approaches have been applied to derive EBTs for category 1 bioassays, covering both drinking water and surface water (Figure 13.10). Most

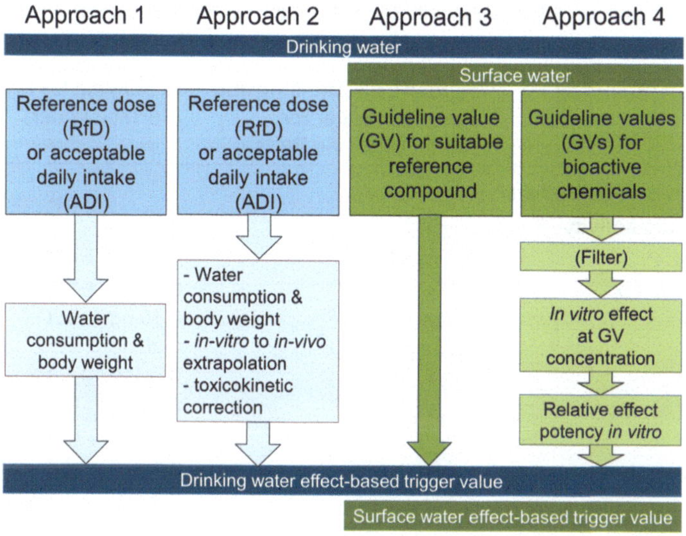

Figure 13.10 Derivation of EBT values of category 1 bioassays for drinking water and surface water by different approaches: (1) from an ADI of a RfD; (2) from an ADI/RfD incorporating toxicokinetic parameters; (3) from an established GV and (4) from an established GV incorporating relative potencies.

rely on chemical GVs as the point of departure. A summary of the currently available EBTs for category 1 bioassays is provided in Table 13.1. Bioassay-specific EBTs, rather than generic EBTs for an endpoint, are provided in Table 13.1 because differences in assay sensitivity and chemical potency can result in different EBTs for assays indicative of the same endpoint.

The simplest approach for drinking water EBTs is to translate the reference dose (RfD) or acceptable daily intake (ADI) of a chemical that also serves as a potent reference compound to water concentrations in a similar fashion as World Health Organisation (WHO) drinking water guidelines are derived (WHO, 2017b), using the average body weight (usually taken as 60–70 kg), the amount of water typically consumed in a day (usually taken as 2 L/day) and a specific allocation for drinking water (10%–20%) to account for other sources of exposure. These are combined to calculate a concentration that is equivalent to the GV for drinking water. The EBT is then simply the bioanalytical equivalent of this concentration (approach 1 in Figure 13.10). Genthe *et al.* (2009) derived an EBT for estrogenic activity from the ADI with this method.

In addition, the concentrations of reference compounds considered safe *in vivo* can be converted to concentrations in *in vitro* assays using differences in the toxicokinetics of different compounds to adjust this direct read-across EBT

Table 13.1 Summary of proposed effect-based trigger values for category 1 bioassays covering human health and ecological health that are currently available in the literature.

Endpoint/Bioassay	Human EBT (Drinking and Recycled Water for Indirect Potable Reuse)	Ecological EBT (Surface Water)
Receptor-mediated effects		
Estrogenic activity		
–[a]	0.2 ng/L EEQ[5]	0.4 ng/L EEQ[1]
ERα CALUX	0.2 ng/L EEQ[2] 3.8 ng/L EEQ[6] 0.25 ng/L EEQ[7]	0.5 ng/L EEQ[3] 0.10 ng/L EEQ[4] 0.28 ng/L EEQ[7] 0.2–0.4 ng/L EEQ[b] [8]
ERα GeneBLAzer	1.8 ng/L EEQ[2]	0.34 ng/L EEQ[4] 0.24 ng/L EEQ[7]
E-SCREEN	0.9 ng/L EEQ[2]	0.1–0.3 ng/L EEQ[b] [8]
YES	12 ng/L EEQ [2]	0.2–0.4 ng/L EEQ[b] [8]
HeLa-9903	0.6 ng/L EEQ [2]	1.0 ng/L EEQ[4] 0.18 ng/L EEQ[7]
MELN		0.37 ng/L EEQ[4] 0.56 ng/L EEQ[7] 0.2–0.3 ng/L EEQ[b] [8]
MVLN		0.1–0.3 ng/L EEQ[b] [8]
ERα-Luc-BG1		0.62 ng/L EEQ[4]
A-YES		0.56 ng/L EEQ[4]
3d YES		0.88 ng/L EEQ[4]
ISO-LYES (Sumpter)		0.97 ng/L EEQ[4]
ISO-LYES (McDonnell)		1.1 ng/L EEQ[4]
pYES		0.5 ng/L EEQ[7]
EASZY (Cyp19a1b-GFP)		2.2 ng/L EEQ[4]
REACTIV (unspiked)		0.80 ng/L EEQ[4]
Androgenic activity		
AR CALUX	11 ng/L DHT EQ[6] 4.5 ng/L DHT EQ[7]	
AR GeneBLAzer	14 ng/L testosterone EQ[2]	
AR CALUX	11 ng/L DHT EQ[6] 4.5 ng/L DHT EQ[7]	
Anti-androgenic activity		
Anti-AR CALUX	4.8 μg/L flutamide EQ[7]	25 μg/L flutamide EQ[3] 14 μg/L flutamide EQ[c] [4]
Anti-AR GeneBLAzer		3.3 μg/L flutamide EQ[c] [4]

(Continued)

Table 13.1 Summary of proposed effect-based trigger values for category 1 bioassays covering human health and ecological health that are currently available in the literature (*Continued*).

Endpoint/Bioassay	Human EBT (Drinking and Recycled Water for Indirect Potable Reuse)	Ecological EBT (Surface Water)
Anti-MDA-kb2		3.5 μg/L flutamide EQ[c] [(4)]
Anti-AR RADAR (spiked)		3.6 μg/L flutamide EQ[c] [(4)]
Glucocorticoid activity		
GR CALUX	150 ng/L dexamethasone EQ[(2)] 21 ng/L dexamethasone EQ[(6)]	100 ng/L dexamethasone EQ[(3)]
Progestogenic activity		
PR CALUX	724 ng/L levonorgestrel EQ[d] [(6)]	
Anti-progestogenic activity		
Anti-PR CALUX		1967 ng/L endosulphan EQ[c] [(4)]
Thyroid activity		
TTR RLBA		0.06 μg/L thyroxine EQ[(4)]
TTR FITC-T4		0.49 μg/L thyroxine EQ[(4)]
XETA (unspiked)		0.62 ng/L triiodothyronine EQ[(4)]
Anti-thyroid activity		
Anti-TR-LUC-GH3		0.60 μg/L bisphenol A EQ[c] [(4)]
Photosynthesis inhibition		
Combined algae test (2 h-PSII)	0.6 μg/L diuron EQ[(2)]	0.07 μg/L diuron EQ[(4)]
Acetylcholinesterase inhibition		
AChE assay	26 μg/L parathion EQ[(2)]	

DHT = dihydrotestosterone; EEQ = 17β-estradiol equivalent concentration;
PSII = photosystem II.
[a]No specific assay indicated.
[b]EBT calculated specifically for wastewater effluent.
[c]A mixture factor (MF) of 100 was applied (strictly speaking this would then apply as category 2 bioassay).
[d]Converted from Org2058 equivalent concentration to levonorgestrel equivalent concentration using REP in Brand *et al.* (2013).
References for EBT: [1](Kunz *et al.*, 2015); [2](Escher *et al.*, 2015); [3](van der Oost *et al.*, 2017b); [4](Escher *et al.*, 2018); [5](Genthe *et al.*, 2009); [6](Brand *et al.*, 2013); [7](Brion *et al.*, 2019); [8](Jarošová *et al.*, 2014a).

(approach 2 in Figure 13.10). Brand *et al.* (2013) derived EBTs using the ADI of a potent reference compound and considered oral bioavailability and the fraction unbound to plasma as indicators of adsorption and distribution to estimate the safe internal concentration. Toxicokinetic data, including bioavailability and fraction unbound to proteins, were also considered for other potent chemicals that act by the same mode of action. This approach was applied to a battery of CALUX assays and used REP specific to these assays.

Although *in vitro* effects can be extrapolated to *in vivo* for risk assessment by so-called quantitative *in vitro* to *in vivo* extrapolations (Wetmore, 2015; Yoon *et al.*, 2015), the derivation of EBTs takes the inverse route, using safe concentration *in vivo* and extrapolating them to the *in vitro* situation. This is typically done for reference compounds, assuming that it is representative for all chemicals with this effect in the bioassay. Both approaches 1 and 2 are applicable for drinking water only.

Many other studies have translated existing chemical GVs, both drinking water GVs and environmental quality standards (EQS), directly into *in vitro* BEQ or *in vitro* bioassay effect thresholds (approaches 3 and 4 in Figure 13.10). These approaches are suitable for, both, drinking water or surface water GVs.

In the simplest way, the GV is directly translated into the BEQ for a given bioassay related to the bioassay's reference compound (approach 3 in Figure 13.10). For example, Kunz *et al.* (2015) proposed to use the 17β-estradiol (E2) annual average environmental quality standard (AA-EQS) of 0.4 ng/L as the EBT to assess whether the risk for adverse reproductive effects was tolerable using *in vitro* assays indicative of estrogenic activity. The E2 AA-EQS was selected rather than the estrone (E1) or 17α-ethinylestradiol (EE2) AA-EQS values because E2 is commonly used to express the results of bioassay studies and because *in vitro* and *in vivo* REP values for E1 and EE2 are expressed relative to E2. Using a similar approach, Leusch *et al.* (2014a) proposed a threshold of 0.1 ng/L EEQ for the E-Screen assay and 0.2 μg/L diuron EQ for photosynthesis inhibition in the green algae *Chlorella vulgaris* based on Australian and New Zealand Guidelines for Fresh and Marine Water Quality (Australian Government, 2018a).

Approach 3 implies that this reference compound is representative for all chemicals causing the specific effect of the bioassay. This is not necessarily the case and therefore a number of studies have determined the *in vitro* effect at the GV concentration using the different potencies of various bioactive chemicals (approach 4 in Figure 13.10).

Jarošová *et al.* (2014a) derived '17β-estradiol equivalent concentration (EEQ) Safe regarding Steroid Estrogens' (EEQ-SSE) values based on the assumption that four potent estrogens, E1, E2, estriol (E3) and EE2, explained most estrogenic activity in wastewater. EEQ-SSE was defined as the EEQ at which no adverse effects should be observed in municipal effluent based on the *in vivo* predicted no effect concentration (PNEC). A literature review was conducted to identify general concentrations of the four estrogens in municipal effluent, along with their REP in common *in vitro* estrogenic activity assays. The long term EEQ-SSE ranged from 0.1 to 0.4 ng/L EEQ. Although based on wastewater effluent, the EEQ-SSE can be applied to surface water receiving effluent after appropriate dilution factors are considered.

Applying approach 4, Escher *et al.* (2015) used the Australian Drinking Water Guidelines (ADWG) (NHMRC, 2011) and the Australian Guidelines for Water Recycling (AGWR) (NRMMC/EPHC/NHMRC, 2008) to derive EBTs for assays indicative of receptor-mediated effects. EC values for individual chemicals with GVs were collected from the literature for each bioassay, then filtered with EC values over an order of magnitude smaller or larger than the GV excluded. This was to prevent extremely more potent or less potent chemicals from skewing and dominating the distribution. The REP of each filtered chemical was calculated, which normalised the potency of the chemical to the assay reference chemicals, then the GV was converted to a BEQ by multiplying the REP and the guideline concentration. The EBT was derived from the 5th percentile of a cumulative distribution of the BEQ values. The 5th percentile was selected to be protective for the majority of chemicals, while still accounting for mixture effects. This process was repeated for each bioassay, with 11 of the 18 assays (mainly category 1 bioassays but also including some category 2 bioassays) having sufficient data to derive a preliminary EBT. The preliminary EBTs were able to differentiate between recycled water for indirect potable reuse, which was below EBTs, and secondary treated effluent, which was above.

Following on from this approach, Escher *et al.* (2018) used current and proposed AA-EQS values to derive EBTs for 48 *in vitro* and *in vivo* assays, with sufficient data available to derive 32 preliminary EBT. Similar to Escher *et al.* (2015), single chemical data were collected from the literature to translate the EQS into a BEQ value. However, rather than using one algorithm for all assays, different approaches were used for different classes of bioassays. For category 1 assays (receptor-mediated effects), the EBT was derived based on the average BEQ of all chemicals, with a filtering step applied if low potency chemicals were included to prevent these chemicals reducing the EBT to unrealistically low levels. In the case of estrogenic activity, where the mixture composition of potent estrogens often had a similar pattern in environmental samples, an exposure-corrected approach was applied. The BEQ was multiplied by the fraction of potent estrogens commonly found in wastewater and surface water of approximately 11% E2, 9% EE2 and 80% E1 (Kase *et al.*, 2018).

The EBTs in Table 13.1 are sorted according to endpoint and bioassay, with different columns for drinking and recycled water (*i.e.*, human health-relevant EBTs) and surface waters (*i.e.*, ecological health-relevant EBTs).

Separate EBTs for drinking water and environmental waters are provided because the difference in the health target (humans consuming drinking water

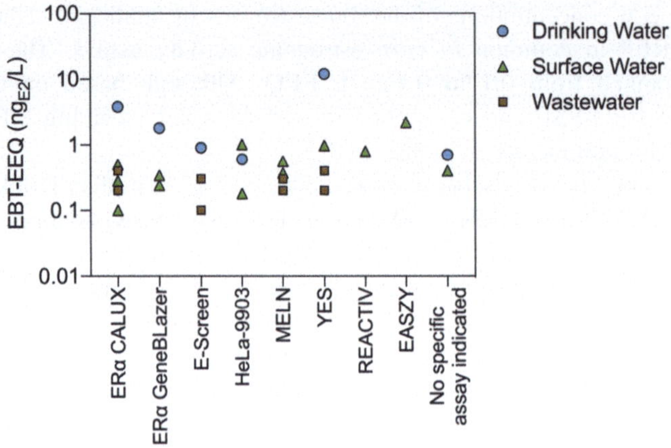

Figure 13.11 Overview of published EBT values for estrogenic activity in units of 17β-estradiol equivalent concentration (EEQ, ng_{E2}/L) for drinking water (circles), surface water (triangles) and wastewater effluent (squares). See Table 13.1 for numerical values and more details.

versus wildlife living in and consuming surface water) causes differences in safe concentrations even if the very same assays can be applied to evaluate both drinking and surface water quality.

As an example, available EBTs for estrogenic activity are shown in Figure 13.11. Surface water EBTs are often lower than drinking water EBTs. This is not unexpected because GVs for surface water tend to be more protective for specifically vulnerable species than the drinking water GVs, which 'only' need to be protective of human health (see Chapter 3). For example, estrogenic chemicals cause adverse effects at very low concentrations in aquatic organisms, whereas terrestrial animals exposed primarily through dietary intake are less adversely affected (Escher *et al.*, 2018). However, differences between EBTs indicative of the same assay and the same water type still exist due to differences in the derivation methods applied.

13.5.2 Approaches to derive effect-based trigger (EBT) values for category 2 bioassays

Directly reading across from chemical GV is only sensible for category 1 bioassays, but for category 2 bioassays, where many chemicals can contribute to the observed effect, the EBT derivation needs to account for mixture effects.

Initial attempts were made to derive the EBTs for category 2 bioassays with the same approach 4 (Figure 13.12) as for category 1 bioassays (Figure 13.10). However, the majority of the chemicals had to be filtered out due to low potency

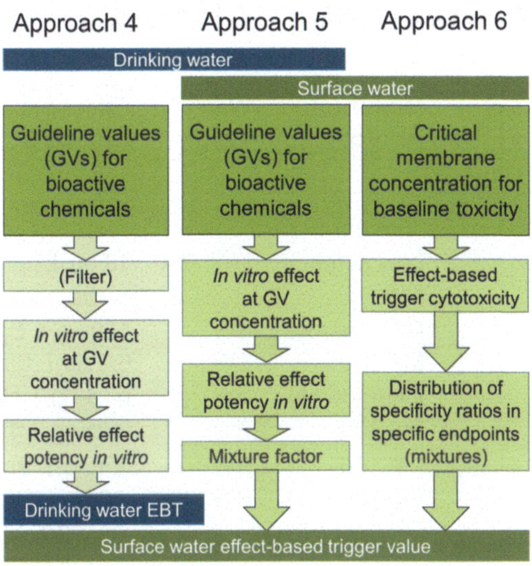

Figure 13.12 Derivation of effect-based trigger values (EBT) of category 2 bioassays for drinking water and surface water by different approaches: (4) from an established GV after filtering out low potency chemicals and incorporating relative potencies (this is the same approach 4 as for category 1 bioassays in Figure 13.10), (5) as approach 4 but adding a mixture factor (MF) and (6) starting from a generic effect-based trigger (EBT) for cytotoxicity divided by median of the specificity ratio of the bioassay.

and the filtering step was not clearly defined. The EBTs derived with this method are marked with 'low potency filter' in Table 13.2.

In the next step a mixture factor (MF) was introduced that depended on the fraction of responsive chemicals in the category 2 bioassay (approach 5 in Figure 13.12). This MF ranged from 10 to 10,000 as shown in Table 13.2 and is discussed in detail further below.

Finally, an alternative approach was proposed that did not read across from GV but took a different point of departure. An EBT for cytotoxicity was derived from negligible disturbance of cell membranes and then adjusted by a measure that accounts for the degree of specificity of the chemicals being active in the given specific endpoints (approach 6 in Figure 13.12).

In line with approach 5, Tang *et al.* (2013b) derived EBTs for bacterial toxicity based on bioluminescence inhibition in *Aliivibrio fischeri* (Microtox assay) by reading across from the ADWG and AGWR. The EBT was derived based on the predicted mixture effect of all chemicals in the guidelines using the model of concentration addition divided by the sum molar concentration of all chemicals in the guidelines. This EBT included an extrapolation factor to account for the number of chemicals included in the derivation, model uncertainties and the

Table 13.2 Summary of proposed effect-based trigger values for category 2 bioassays that are currently available in the literature.

Endpoint/Bioassay	Mixture Approach	Water Type	Effect-based Trigger EBT-BEQ
Xenobiotic metabolism			
AhR activity			
AhR-*cis*-FACTORIAL™	Low potency filter	DW, RW	18 μg/L carbaryl EQ(1)
PAH CALUX	Read-across	SW	150 ng/L B[a]P EQ(2)
	MF 100	SW	6.2 ng/L B[a]P EQ(3)
	intermediate	SW	62.1 ng/L B[a]P EQ(4)
DR CALUX	MF 100	SW	0.05 ng/L TCDD EQ(2)
H4L1.1c4 AhR assay	MF 100	SW	6.4 ng/L B[a]P EQ(3)
	MF 100	SW	4.3 ng/L B[a]P EQ(6)
	SR-method	SW	250 ng/L B[a]P EQ(5)
PPARγ activity			
PPARγ CALUX	Read-across	SW	10 ng/L rosiglitazone EQ(2)
PPARγ-GeneBLAzer	MF 100	SW	36 ng/L rosiglitazone EQ(3)
	MF 100		19 ng/L rosiglitazone EQ(6)
	SR-method		1.2 μg/L rosiglitazone EQ(5)
PXR activity			
PXR-*cis*-FACTORIAL™	Low potency filter	DW, RW	59 μg/L metolachlor EQ(1)
PXR CALUX	Read-across	SW	3.0 μg/L nicardipine EQ(2)
	MF 100		272 μg/L DEHP EQ(3)
	intermediate		corresponding to 54 μg/L nicardipine EQ 5.4 μg/L nicardipine EQ(4)
HG5LN-hPXR		SW	16 μg/L DEHP EQ(3)
Adaptive stress response			
AREc32	Low potency filter	DW, RW	284 μg/L dichlorvos EQ[a](7)
	MF 1000	SW	156 μg/L dichlorvos EQ(3)
	MF 1000	SW	140 μg/L dichlorvos EQ(6)
	SR-method	SW	1400 μg/L dichlorvos EQ(5)
Nrf2 CALUX	Read-across	SW	10 μg/L curcumin EQ(2)
	MF 1000		26 μg/L dichlorvos EQ(3)
ARE GeneBLAzer	MF 1000	SW	392 μg/L dichlorvos EQ(3)

(*Continued*)

Table 13.2 Summary of proposed effect-based trigger values for category 2 bioassays that are currently available in the literature (*Continued*).

Endpoint/Bioassay	Mixture Approach	Water Type	Effect-based Trigger EBT-BEQ
Apical effects in well plate-based in vivo assays			
Bacterial toxicity			
Microtox	MF 10,000	DW	4100–4392 μg/L baseline TEQ[b(8)]
		SW	1264 μg/L baseline TEQ[(3)]
Algal toxicity			
72 h algal growth inhibition	MF 1	SW	0.12 μg/L diuron EQ[(3)]
24 h synchronous algae reproduction	MF 1	SW	0.11 μg/L diuron EQ[(3)]
Combined algae assay (24 h growth)	MF 1	SW	0.13 μg/L diuron EQ[(3)]
Invertebrates			
48 h *Daphnia* immobilisation test	MF 10	SW	15 ng/L chlorpyrifos EQ[(3)]
Fish embryo			
Fish embryo toxicity (48 h)	MF 100	SW	276 μg/L bisphenol A EQ[(3)]
Fish embryo toxicity (96–120 h)	MF 100	SW	183 μg/L bisphenol A EQ[(3)]

Legend for column 'Mixture approach': Low potency filter (approach 4), MF = mixture factor (approach 5), intermediate = value chosen between two other highly divergent EBT, SR-method (approach 6 in Figure 13.12), Read-across = (partial) read-across from *in vivo* data. Water type: DW = drinking water, RW = recycled water, SW = surface water.

[a]Converted to dichlorvos equivalent concentration using dichlorvos EC value in Escher *et al.* (2018).

[b]Converted to baseline toxic equivalent concentration (baseline TEQ) using the virtual baseline toxicant EC value in Escher *et al.* (2018). AhR, aryl hydrocarbon receptor; B[a]P = benzo[a] pyrene; DEHP = di(2-ethylhexyl)-phthalate; PPARγ = peroxisome proliferator-activated receptor gamma; PXR = pregnane X receptor; TCDD = 2,3,7,8-tetrachloro-p-dibenzodioxin. References for EBT: [1](Escher *et al.*, 2015); [2](van der Oost *et al.*, 2017b); [3](Escher *et al.*, 2018); [4](de Baat *et al.*, 2020); [5](Escher and Neale, 2021); [6](Neale *et al.*, 2020a); [7](Escher *et al.*, 2013); [8]Tang *et al.*, 2013b).

acceptable fraction of chemicals present at their GV. This yielded an EBT-EC_{50} of 3 in units of relative enrichment factor (REF) for drinking water and REF 2.8 for recycled water for indirect potable reuse. This means that a drinking water extract would exceed the EBT if it induced 50% bacterial toxicity after less than three times enrichment. Converted to TU, the EBT would be 0.33 TU for drinking water and 0.36 TU for recycled water. The EC_{50} values measured in actual recycled water samples were much higher (*i.e.*, less toxic) than the EBT-EC_{50} at REF 2.8.

A similar approach was applied by Escher *et al.* (2013) for the oxidative stress response assay. A tentative EBT of an effect concentration causing an IR of 1.5 ($EC_{IR1.5}$) at REF 6 was proposed for both drinking water and recycled water. Based on the $EC_{IR1.5}$ value of potent pesticide dichlorvos, this translates to a dichlorvos equivalent concentration (dichlorvos EQ) of 284 μg/L (Table 13.2).

All EBTs for category 2 bioassays are listed in Table 13.2. They are expressed as BEQ, although the reference compound used for a particular endpoint can vary between studies. For example, the EBT for oxidative stress response assays is in units of dichlorvos EQ in Escher *et al.* (2018) but in curcumin equivalent concentrations (curcumin EQ) in van der Oost *et al.* (2017b). If one knows the effect concentrations of the reference compounds in both bioassays (and thus the relative potency of each compound in the assay), these EBTs can be easily converted.

In approach 5, an MF was included to account for the many chemicals present and acting together (remember that for category 2 assays, $BEQ_{chem} \ll BEQ_{bio}$). Initially, an MF of 100 was set for assays indicative of xenobiotic metabolism and 1000 was used for adaptive stress responses (Escher *et al.*, 2018). The MF was multiplied by the average BEQ. The MF values were based on experience with the fraction of effect explained in a particular assay by known chemicals using iceberg modelling. For example, less than 0.1% of the effect was generally explained in the oxidative stress response assay (Escher *et al.*, 2013; Neale *et al.*, 2017b), hence an MF of 1000 was applied.

For apical endpoints and whole organism bioassays, an MF also needs to be included to account for different degrees of susceptibility. Algae react very specifically to herbicides that inhibit the photosystem II, and hence any algal toxicity assay is close to a category 1 bioassay and the MF can be reduced to 1 for most applications. An MF of 10 or smaller was used for *Daphnia magna* because they are especially susceptible to insecticides, but the selectivity is not as high as for herbicides in algae. For the fish embryo toxicity bioassay an MF of 100 was proposed (Escher *et al.*, 2018).

Recognising the subjectivity associated with an MF correction based mainly on expert knowledge, Escher and Neale (2021) recently proposed a method to derive the EBT for category 2 assays from experimental data specific for each bioassay (approach 6 in Figure 13.12), starting with the constant critical membrane concentration for baseline toxicity (minimum toxicity caused by narcosis, see Chapter 4) common to all cells (Escher *et al.*, 2019). By definition, mixtures of baseline toxicants will have very similar cytotoxicity IC_{10} irrespective of the cell line used. If 1% cytotoxicity, an effect that is in practical terms indistinguishable from the negative controls, is accepted as a safe toxicity level, then the EBT-IC_{10} would calculate to an REF of 10 in all cell lines.

For specific endpoints, the generic EBT-IC_{10} of REF 10 was then divided by the median of the distributions of the specificity ratios log $SR_{baseline}$ of the given bioassay (Escher and Neale, 2021). The resulting effect concentration can then be translated to an EBT-BEQ for mixtures in category 2 bioassays by applying the bioassays-specific EC_{10} of the associated reference compound (Figure 13.13).

The EBT-BEQ derived using this approach performed much better than previous ones derived with the MF in differentiating between wastewater, which was generally above the EBT, and surface water, which was generally below the EBT (Figure 13.14).

13.5.3 Approaches to derive effect-based trigger (EBT) values from read-across of *in vivo* data

In addition to the approaches using GV and *in vitro* data alone to derive EBTs, *in vivo* data can also help to strengthen the EBT derivation. Brion *et al.* (2019) derived EBTs for five *in vitro* estrogenic activity assays using experimental *in vitro* and *in vivo* results for 16 surface water and 17 wastewater extracts. The response in the four mammalian reporter gene assays and one yeast reporter gene assay were compared with the whole organism EASZY assay, with true negatives (*i.e.*, samples with *in vitro* activity below the EBT and no response *in vivo*) and true positives (*i.e.*, samples with *in vitro* activity above the EBT and an *in vivo* response) used to determine specificity and sensitivity, respectively. Logistic regression models were applied to determine the maximum sensitivity and specificity cut-off for each assay, and this was used to estimate the EBT. Although this is based on a limited number of samples, the EBTs are within a similar range to other estrogenic *in vitro* assays (Table 13.1). However, this approach is only valid for endpoints with both *in vitro* and *in vivo* assays available.

In an approach that combined *in vivo* effect data with experimental data, van der Oost *et al.* (2017b) applied three different methods to determine BEQ, which were used to derive EBTs for a battery of CALUX assays indicative of xenobiotic metabolism (PAH CALUX, DR CALUX, PPARγ CALUX and PXR CALUX), hormone receptor-mediated effects (ERα CALUX, Anti-AR CALUX and GR CALUX) and adaptive stress responses (Nrf2 CALUX). A list of chemicals was selected based on their available toxicity data and reported concentrations in water, with chemicals with low REP values removed by filtering to prevent them from biasing the EBT (similar to approach 4). *In vivo* toxicity data, including no observed effect concentrations (NOEC), lowest observed effect concentrations (LOEC) and predicted no effect concentration (PNEC) and effect concentration causing 50% effect (EC_{50}) were converted to BEQ using the REP, with an acute-to-chronic ratio of 10 applied to acute data. Safe BEQ, which indicated no risk to the ecosystem, were derived using the lowest BEQ and dividing by an assessment factor based on the endpoint. The 5th percentile BEQ (HC5 BEQ), which indicated low risk, was derived using a BEQ distribution, similar to a

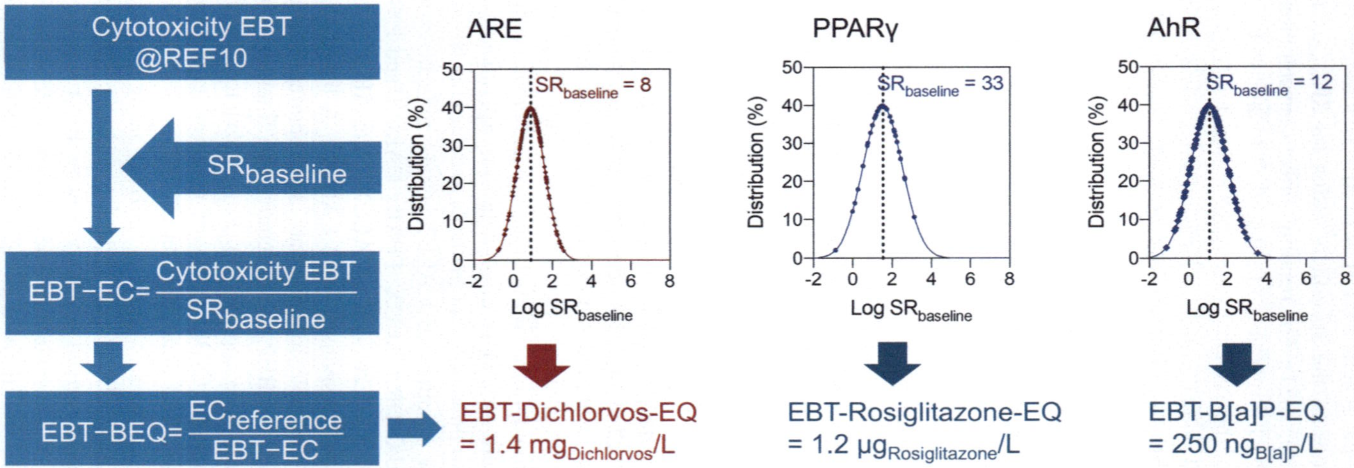

Figure 13.13 Derivation of EBT-BEQ with the 'SR-method' from the cytotoxicity EBT scaled by the specificity ratio $SR_{baseline}$. Figure modified and reprinted with permission from Escher and Neale (2021). Effect-based trigger values for mixtures of chemicals in surface water detected with *in vitro* bioassays. *Environmental Toxicology and Chemistry*, **40,** 487–499. © 2021. The Authors. *Environmental Toxicology and Chemistry* published by Wiley Periodicals LLC on behalf of SETAC.

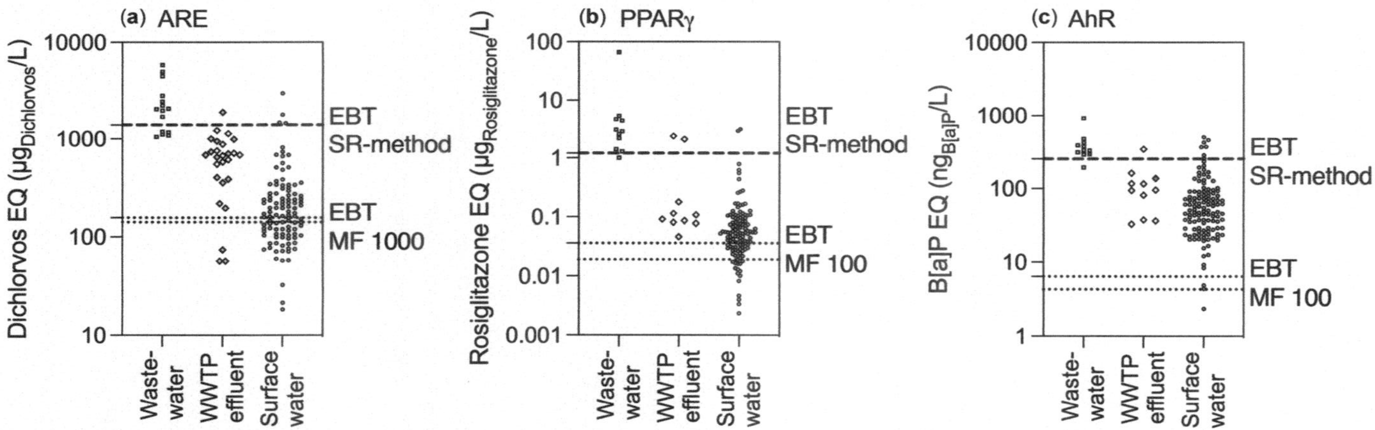

Figure 13.14 New EBT-BEQ using the SR-method (red dashed lines) for category 2 bioassays perform much better than the previously proposed EBT-BEQ derived with an MF of 100 and 1000 (black dotted lines (Escher *et al.*, 2018)) in differentiating between wastewater (above EBT) and surface water (below EBT). Numerical values of the EBT for (a) AREc32, (b) PPARγ GeneBLAzer and (C) AhR CALUX are given in Table 13.2. Figure modified and reprinted with permission from Escher and Neale (2021). Effect-based trigger values for mixtures of chemicals in surface water detected with *in vitro* bioassays. *Environmental Toxicology and Chemistry*, **40,** 487–499. © 2021. The Authors. *Environmental Toxicology and Chemistry* published by Wiley Periodicals LLC on behalf of SETAC.

species sensitivity distribution approach. Finally, bioanalysis of surface water with good ecological status was used to determine background BEQ values. An EBT for each assay was derived based on the evaluation of the safe BEQ, HC5 BEQ and background BEQ values, with different multiplication factors applied based on expert judgement. In the case of non-specific effects (*e.g.*, non-specific toxicity to bacteria and zooplankton), an average acute-to-chronic ratio of 10 and a safety factor of 2 for assumed 50% recovery by SPE or passive sampling was used to determine an EC at an REF of 20 or a TU of 0.05. REF takes into consideration both sample enrichment and dilution in the bioassay, so samples that have a non-specific effect after 20 times or more enrichment would be considered acceptable. A similar approach was taken for genotoxicity (Ames, umuC and p53 CALUX), with EBTs derived for non-specific endpoints reduced by an assessment factor of 10. This gave an EC REF of 200 or a TU of 0.005.

13.6 WHAT TO DO IF A WATER EXTRACT EXCEEDS THE EFFECT-BASED TRIGGER VALUE?

To facilitate the implementation of EBTs for regulatory purposes, there is a need for a framework to determine what steps to take if the measured effect in a water sample exceeds the proposed EBT. Leusch and Snyder (2015) have proposed such a framework with three tiers of screening, targeted analysis and exploration recommended. If the measured effect exceeds the EBT during the screening tier, re-testing is recommended to confirm the results and to determine if this is an on-going issue. A second tier of targeted analysis of known potent chemicals with GVs is then recommended if the re-tested sample still exceeds the EBT. If a detected chemical's concentration exceeds the GV then operators would need to follow the well-established chemical GV exceedance response procedure. If the measured effect is over 10 times higher than the EBT, a number of options are suggested in a third tier including (1) full chemical analysis of all chemicals in the relevant guidelines, (2) EDA to identify the causative chemicals or (3) bench-scale experiments to identify effective treatment methods to reduce effect detected by bioassay even if the identity of the causative chemicals is unknown. It should be noted that options 1 and 2 would not be suitable for category 2 assays where many low potency chemicals can contribute to the effect, even if non-targeted screening approaches can be used to investigate the potential presence of unknown chemicals and transformation products (Brunner *et al.*, 2019).

van der Oost *et al.* (2017b) have also suggested a two-tier system consisting of hazard identification and risk analysis as part of the Smart Integrated Monitoring approach. If the effect of a water sample exceeds the EBTs in the hazard identification tier, chemical analysis and EDA are among the steps recommended in the risk analysis tier.

It is possible to integrate both category 1 and 2 bioassays in an assessment framework outlined in Figure 13.15. This framework follows essentially Leusch

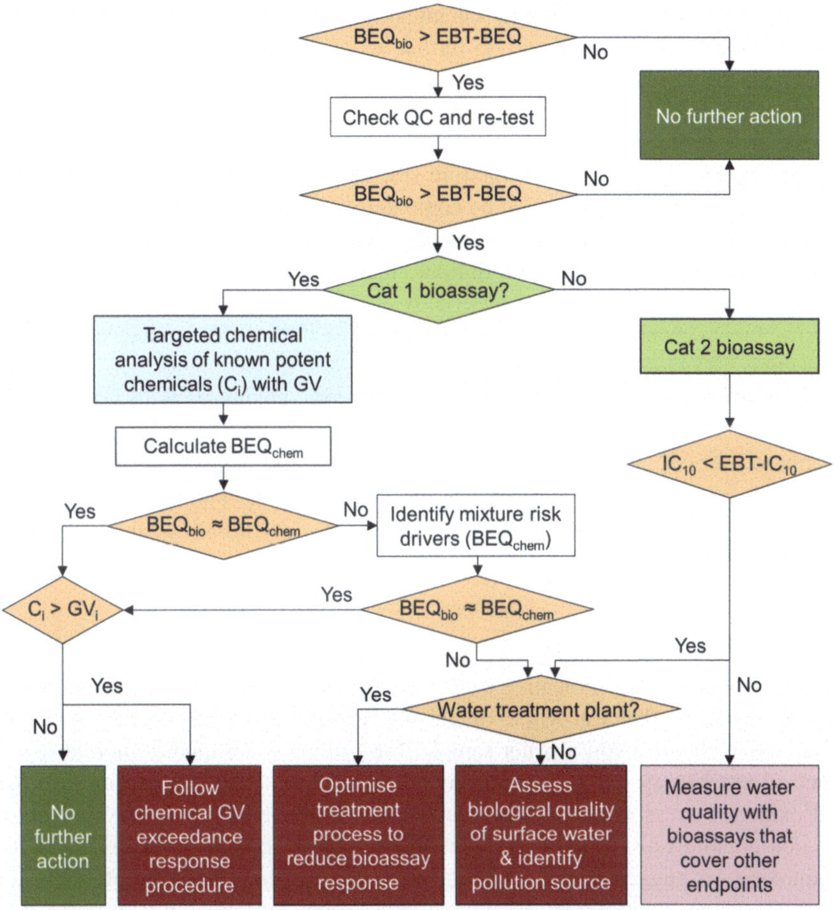

Figure 13.15 Framework for application of EBTs in water quality assessment. Modified from Leusch and Snyder (2015).

and Snyder (2015) for category 1 bioassays and attempts to also integrate category 2 bioassays in a common decision framework.

When an EBT-BEQ is exceeded, it is important to confirm that this result is correct by a quality control check. If the exceedance of the EBT-BEQ is confirmed, then it is important to consider the bioassay category. In the case of category 1 bioassays, targeted chemical analysis of known potent chemicals (C_i) with GVs should be performed to calculate the BEQ_{chem}. If the BEQ_{bio} is close to the BEQ_{chem}, that is, most causative chemicals are known, then determination by chemical analysis of the concentration C_i of individual target chemicals and comparison with the relevant single chemical GV_i will be sufficient to judge water quality. If $C_i < GV_i$ for all chemicals i, then the water is compliant with current chemical guidelines, and no further action is warranted. If $C_i > GV_i$, then the conventional response to chemical GV exceedance is to be followed. Going back a step, if BEQ_{bio} is not close to BEQ_{chem}, then this indicates that an unexpected chemical may be causing the bioassay response. Efforts should be focused on identifying the new mixture risk drivers – if this effort is successful, then their concentrations can also be compared to GVs, if available.

If the applied assay is a category 2 assay, then it becomes important to determine if IC_{10} is lower (*i.e.*, sample is more toxic) than the cytotoxicity EBT (EBT-IC_{10}). Exceedance of the EBT but not of EBT-IC_{10} for non-specific toxicity means that specific bioassays should be checked to determine if they are also exceeding their respective EBTs.

In those cases of category 1 bioassays where the chemicals with GVs cannot explain the BEQ_{bio} and mixture risk drivers cannot be identified, then either the treatment processes need to be optimised to reduce the bioassay response if that is possible, or the biological quality of the surface water should be assessed, and the source of the pollutant should be identified. The same should happen of $IC_{10} <$ EBT-IC_{10} for category 2 bioassays.

13.7 CONCLUSIONS

A battery of bioassays indicative of different stages of cellular toxicity pathways as well as apical effects in whole organisms is recommended to detect the effects of all active chemicals in a water sample. For routine water quality monitoring, a practical test battery of assays indicative of activation of AhR, activation of ERα and oxidative stress response is recommended for wastewater and recycled water for non-potable reuse, while an additional assay indicative of mutagenicity or genotoxicity is suggested for drinking water or recycled water for potable water reuse. The new multiplexed assays are good tools for screening purposes and for selection of bioassays, but they are generally of too low sensitivity to be used for routine monitoring purposes. In Chapter 14 several case studies are outlined that illustrate the applications of the concepts introduced in the present chapter.

There has been substantial harmonisation on how to interpret *in vitro* bioassay results over the last few years. Linking bioassay data to analytically quantified concentrations has become an important tool not only to understand and identify mixture risk drivers but also to get a feeling for the relevance and meaning of different bioassay endpoints. Iceberg modelling has clearly identified that there are two categories of bioassays – category 1 assays where few chemicals dominate the mixture effect ($BEQ_{chem} \approx BEQ_{bio}$) and category 2 assays where many low potency chemicals contribute to the effect ($BEQ_{chem} < BEQ_{bio}$).

EBTs are essential to understand the significance of bioassay results and for the wider acceptance of effect-based monitoring because they can distinguish between acceptable and unacceptable chemical water quality. A number of different approaches have been applied to develop EBTs, including simple translation from ADIs and chemical GVs, incorporation of chemical potency and mixtures, comparison of *in vitro* and *in vivo* responses to determine maximum sensitivity and specificity cut-offs (Brion *et al.*, 2019) and using multiple lines of evidence (van der Oost *et al.*, 2017b). Some of these approaches are only valid for category 1 assays, while other approaches have been developed specifically for category 2 assays.

The majority of EBTs were derived for receptor-mediated effects, with estrogenic activity the most commonly assessed endpoint. There are fewer EBTs available for induction of xenobiotic metabolism, adaptive stress responses and apical effects, with the majority of these EBTs derived for surface water rather than drinking water. EBTs based on surface water GVs are typically lower than drinking water GVs, so surface water EBTs are likely to be protective of drinking water.

Despite a plethora of different approaches with differing requirements for expert opinion being applied, it is remarkable that most of the current EBTs generally end up within a log unit of one another. EBTs are increasingly applied in the literature to benchmark water quality and to evaluate treatment efficacy, giving input for practical frameworks proposed for steps to take should the effect in a sample exceed the EBT. As of 2020, EBTs have only been implemented by the State Water Resources Control Board of the State of California (2019), which proposed monitoring trigger values of 3.5 ng/L EEQ in ERα assays, and 0.5 ng/L TCCD-EQ for AhR assays for recycled water. There is to date no implementation in any other legislation but with their maturation and increasing adoption in research contexts, this will hopefully change in the future.

Chapter 14

Case studies

14.1 INTRODUCTION

The number of publications reporting on the application of bioanalytical tools for water quality assessment is ever increasing. Bioassays have been applied to evaluate the treatment efficacy of wastewater treatment plants (WWTPs), advanced water reclamation plants and drinking water treatment plants (DWTPs). Table 14.1 provides an overview of studies that have applied at least three different assays, including at least one cell-based assay, to assess water quality in different water types. Most of these studies were on wastewater and surface water followed by drinking water (Figure 14.1) with some studies addressing different water types and also entire treatment trains. Hormone receptor-mediated effects were the most popular endpoints, often covering a whole suite of different hormone receptors or different aspects of estrogenicity (Figure 14.1). Genotoxicity assays and other assays covering reactive toxicity were most popular with drinking water, due to the known modes of action of disinfection by-products (DBPs) (Figure 14.1). Apical endpoints from whole organisms testing and cytotoxicity often complemented the test batteries, although cytotoxicity was not always measured despite it being recommended to always assess cytotoxicity in parallel to reporter gene activation measurement.

From the wealth of available publications in Table 14.1, several case studies were selected for more detailed review to highlight particular applications of *in vitro* bioassays in water quality assessment. These case studies cover different water types from wastewater to drinking water and provide examples of how bioassays can be used to describe water quality, assess treatment efficacy and for critical

Authors: Beate Escher, Peta Neale and Frederic Leusch
doi: 10.2166/9781789061987_0299

Table 14.1 Overview of studies published (until September 2020) that have applied test batteries of *in vitro* and *in vivo* assays indicative of at least three different endpoints and at least one cell-based assay to a range of water types.

Water	XM	HRM	RM	RT	ASR	AE	Ref.
WW, SW				3		6	Sanchez *et al.* (1988)
WW		2				1	Gagné and Blaise (1998)
WW				1		2	Castillo *et al.* (2001)
WW		3					Garcia-Reyero *et al.* (2001)
WW, SW			1	2		1	Dizer *et al.* (2002)
WW, SW		3					Murk *et al.* (2002)
WW				1		6	Manusadzianas *et al.* (2003)
WW, SW		5					Pawlowski *et al.* (2003)
WW		2		1		2	Aguayo *et al.* (2004)
SW, DW				2		3	Buschini *et al.* (2004)
SW, DW				4		2	Guzzella *et al.* (2004)
WW				1		2	Klee *et al.* (2004)
WW	1	1		1		4	Pessala *et al.* (2004)
WW				3		1	Rutishauser *et al.* (2004)
WW		1				2	Schiliró *et al.* (2004)
WW						3	Emmanuel *et al.* (2005)
DW				4			Lah *et al.* (2005)
WW	1	1				1	Ma *et al.* (2005)
SW		3					Matsuoka *et al.* (2005)
SW	1	2					Pillon *et al.* (2005)
SW				4		2	Zani *et al.* (2005)
WW, SW		4					Bandelj *et al.* (2006)
WW				3	4		Fatima and Ahmad (2006)
DW				3		2	Guzzella *et al.* (2006)
SW		2				2	Keiter *et al.* (2006)
WW		3	1				Leusch *et al.* (2006)
DW					2	1	Marabini *et al.* (2006)
SW				3		2	Pellacani *et al.* (2006)
WW		2				1	Allinson *et al.* (2007)
WW	1	1		1			Gustavsson *et al.* (2007)
WW		1		2		1	Isidori *et al.* (2007)
DW				3		3	Marabini *et al.* (2007)

(*Continued*)

Table 14.1 Overview of studies published (until September 2020) that have applied test batteries of *in vitro* and *in vivo* assays indicative of at least three different endpoints and at least one cell-based assay to a range of water types (*Continued*).

Water	XM	HRM	RM	RT	ASR	AE	Ref.
WW, SW			1			2	Escher *et al.* (2008a)
WW, SW		1	2	1		2	Escher *et al.* (2008b)
WW				4			Krishnamurthi *et al.* (2008)
WW, SW, DW		3					van der Linden *et al.* (2008)
WW, RW				1		3	Cao *et al.* (2009)
WW, RW		3		1		2	Escher *et al.* (2009)
WW				2		4	Gartiser *et al.* (2009)
SW		3				2	Inoue *et al.* (2009)
DW				3		3	Maffei *et al.* (2009)
WW		2	1				Mahjoub *et al.* (2009)
WW		1				6	Mendonca *et al.* (2009)
WW		3					Shi *et al.* (2009a)
WW, RW				2	4	1	Shi *et al.* (2009b)
WW, RW		2				1	Wu *et al.* (2009)
WW, SW, DW				2		1	Zegura *et al.* (2009)
DW				3			Ceretti *et al.* (2010)
WW	2	4				1	Creusot *et al.* (2010)
RW					2	1	Farmen *et al.* (2010)
WW				1		5	Gartiser *et al.* (2010)
WW		4					Leusch *et al.* (2010)
WW		4					Li *et al.* (2010)
WW, RW	1	1	2	1		1	Macova *et al.* (2010)
WW, RW		2				1	Mnif *et al.* (2010)
WW				2		1	Rodrigues *et al.* (2010)
WW		1		1		4	Stalter *et al.* (2010a)
DW				1	2		Xie *et al.* (2010)
WW	1	1	2	1		1	Macova *et al.* (2010)
WW	1	1	2	1		1	Reungoat *et al.* (2010)
RW		1	1			1	Escher *et al.* (2011)
WW		3					Kusk *et al.* (2011)
WW, RW, SW, DW	1	1	2	1		1	Macova *et al.* (2011)

(*Continued*)

Table 14.1 Overview of studies published (until September 2020) that have applied test batteries of *in vitro* and *in vivo* assays indicative of at least three different endpoints and at least one cell-based assay to a range of water types (*Continued*).

Water	XM	HRM	RM	RT	ASR	AE	Ref.
WW	1	4					Stalter *et al.* (2011)
SW		4					Zhao *et al.* (2011)
WW		8					Bellet *et al.* (2012)
WW		4		1			Fang *et al.* (2012)
SW	1	2					Jarošová *et al.* (2012)
SW	2	2					Mnif *et al.* (2012)
DW				1	1	1	Neale *et al.* (2012)
DW		4					Brand *et al.* (2013)
SW	1	3					Chinathamby *et al.* (2013)
SW				1	1	1	Farre *et al.* (2013)
WW, SW	1	2					Jalova *et al.* (2013)
Stormwater	1	1	1	1	1	2	Tang *et al.* (2013a)
WW	1	6					Bain *et al.* (2014)
SW	1	6					Creusot *et al.* (2014)
WW, RW, SW, DW, stormwater	6	10	2	3	3	3	Escher *et al.* (2014)
WW, RW, SW		2	1	1		1	Leusch *et al.* (2014a)
WW, RW		7	1	1			Leusch *et al.* (2014b)
SW		6					Scott *et al.* (2014)
WW, RW			1	1	1	1	Tang *et al.* (2014)
WW, SW	1	2					Zounkova *et al.* (2014)
WW, RW, GW	3	9	1	2	1	2	Jia *et al.* (2015)
RW, GW	1	2		1			Lee *et al.* (2015)
WW, RW		3					Mehinto *et al.* (2015)
SW	2	1			3	1	Neale *et al.* (2015b)
WW, SW	1	5					Roberts *et al.* (2015)
DW				1	1	1	Stalter *et al.* (2016b)
SW		3					Conley *et al.* (2017a)
SW	3	9			3		König *et al.* (2017)
SW	1	2					Mehinto *et al.* (2017)
WW, SW	1	2	2	1	3	1	Neale *et al.* (2017c)

(*Continued*)

Table 14.1 Overview of studies published (until September 2020) that have applied test batteries of *in vitro* and *in vivo* assays indicative of at least three different endpoints and at least one cell-based assay to a range of water types (*Continued*).

Water	XM	HRM	RM	RT	ASR	AE	Ref.
SW		5	1			2	Toušová *et al.* (2017)
WW		3		1	1	1	Välitalo *et al.* (2017)
SW	3	3			2	3	van der Oost *et al.* (2017)
SW	1	4				1	Daniels *et al.* (2018)
SW		4		1			Gao *et al.* (2018)
WW		4					Gehrmann *et al.* (2018)
WW, SW	1	4	1			3	Hamers *et al.* (2018)
SW		2			1		Hashmi *et al.* (2018)
DW					3		Hebert *et al.* (2018)
WW		5					Houtman *et al.* (2018)
WW, SW, DW		13					Leusch *et al.* (2018b)
SW	2	6			1		Müller *et al.* (2018)
SW	3	4			1		Neale *et al.* (2018a)
WW	2	8			2		Nivala *et al.* (2018)
SW	2	3			3		Novák *et al.* (2018)
SW, DW	2	3			1		Rosenmai *et al.* (2018)
SW		6					Scott *et al.* (2018)
SW, DW		2		2			Shi *et al.* (2018)
DW		3					Valcarcel *et al.* (2018)
WW	4	4			1		Alygizakis *et al.* (2019)
SW	1	2					Brettschneider *et al.* (2019)
SW	3	3			2	3	de Baat *et al.* (2019a)
SW		6					Jia *et al.* (2019)
WW	1	4			2		Lundqvist *et al.* (2019b)
SW	1	3					Toušová *et al.* (2019)
DW	2	4		1	1		Albergamo *et al.* (2020)
SW	4	3	1		2	2	de Baat *et al.* (2020)
WW, SW		4					Houtman *et al.* (2020)
WW, SW, DW		3					Medlock Kakaley *et al.* (2020)
WW, SW	2	1			1		Mueller *et al.* (2020)
SW	2	1			1		Neale *et al.* (2020a)

(*Continued*)

Table 14.1 Overview of studies published (until September 2020) that have applied test batteries of *in vitro* and *in vivo* assays indicative of at least three different endpoints and at least one cell-based assay to a range of water types (*Continued*).

Water	XM	HRM	RM	RT	ASR	AE	Ref.
SW, DW		8		1	3		Neale *et al.* (2020b)
WW, SW		9					Neale *et al.* (2020c)

Xenobiotic metabolism: aryl hydrocarbon receptor (AhR); peroxisome proliferator-activated receptor (PPAR), pregnane X receptor (PXR). Hormone receptor-mediated effects: estrogen receptor (ER) (agonist and antagonist), androgen receptor (AR) (agonist and antagonist), glucocorticoid receptor (GR) (agonist and antagonist), progesterone receptor (PR) (agonist and antagonist), thyroid receptor (TR) (agonist and antagonist), mineralocorticoid receptor (MR) (agonist and antagonist), retinoic acid receptor (RAR), retinoid X receptor (RXR).
Other receptor-mediated effects: phytotoxicity, neurotoxicity.
Reactive toxicity: genotoxicity, mutagenicity.
Adaptive stress responses: oxidative stress response, p53 response, NF-κB response.
Apical effects: bacterial toxicity, algal growth inhibition, fish embryo toxicity.
Water: Water type; WW: wastewater; RW: recycled water (for direct or indirect drinking water augmentation); SW: surface water, DW: drinking water; GW: groundwater.
Bioassay type: XM: Xenobiotic metabolism; HRM: Hormone receptor-mediated effects; RM; Other receptor-mediated effects; RT: Reactive toxicity; ASR: Adaptive stress response; AE: Apical effects; Ref.: Literature reference.

control point verification monitoring. Following on from Chapter 13, all studies used test batteries that covered different stages of the cellular toxicity pathway, with assays relevant for both human and ecosystem health. Furthermore, the observed effects were compared with surface water and drinking water effect-based trigger values (EBT) described in Chapter 13 (Section 13.5) in some case studies.

The first case study used bioassays indicative of hormone receptor-mediated effects, reactive toxicity and adaptive stress responses to evaluate treatment efficacy and DBP formation in DWTPs (Neale *et al.*, 2020b) (Section 14.2). The second case study benchmarked different types of product water for potable and non-potable reuse (Leusch *et al.*, 2014b) (Section 14.3). This study methodically applied assessment endpoints selected from a human health relevance perspective. The third case study applied a bioassay test battery to evaluate treatment efficacy of different constructed wetlands as well as a conventional WWTP (Nivala *et al.*, 2018) (Section 14.4). The fourth case study used bioassays to evaluate the contribution of wastewater effluent to the chemical burden in small streams and applied iceberg modelling to determine which detected chemicals are contributing to the observed effect (Neale *et al.*, 2017c) (Section 14.5). The fifth case study applied a multiplex test battery to surface water across the USA and predicted the mixture effects of the detected chemicals by using the

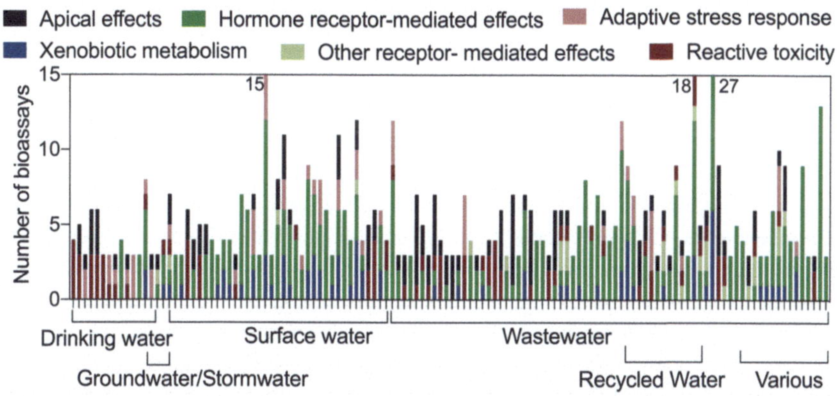

Figure 14.1 Overview of studies published that have applied test batteries of *in vitro* and *in vivo* assays until September 2020 (for details see Table 14.1).

exposure activity ratio (Section 14.6). Finally, our last case study on surface water during rain events compared not only the detected bioassay responses with those predicted by iceberg modelling from the detected chemicals, but also mixed the detected chemicals in the detected concentration ratios to test how chemicals act together in realistic mixtures (Section 14.7).

For additional case studies, the reader is referred to the European technical report on aquatic effect-based tools under the Water Framework Directive, which includes further 13 case studies where *in vitro* and *in vivo* bioassays have been applied in water quality monitoring (Wernersson *et al.*, 2015).

14.2 CASE STUDY 1: TREATMENT OF DRINKING WATER

Source waters that feed DWTPs may contain organic micropollutants, and DBPs may form during treatment processes, such as chlorination. The first case study aimed to evaluate the efficacy of treatment processes in three DWTPs in the Paris area, France, using a test battery focusing on hormone receptor-mediated effects to evaluate micropollutant removal, and reactive toxicity and adaptive stress responses to assess DBP formation (Neale *et al.*, 2020b). Two of the plants, DWTP 1 and DWTP 2, applied pre-ozonation (DWTP 2 only), clarification, sand filtration, ozonation, granular-activated carbon filtration, ultraviolet (UV) and chlorination, with samples collected from the source water, after UV treatment and after chlorination. The third plant, DWTP 3, treated 30% of the water using a similar biological treatment process to the other DWTPs, with 70% of the water treated using nanofiltration. Samples at DWTP 3 were collected from the source water, after biological treatment, after nanofiltration and after chlorination. Samples were collected from all DWTPs over four seasons in 2018.

The water samples were extracted using solid-phase extraction (SPE) and measured in assays indicative of activation and inhibition of the estrogen receptor (ERα), androgen receptor (AR), glucocorticoid receptor (GR) and progesterone receptor (PR), as well as assays indicative of mutagenicity, oxidative stress response, p53 response and NF-κB response. As low effects were expected in the product water, the methanolic extracts were exchanged with cell culture media using the approach described in Chapter 12 (Section 12.6). Concentration–response curves (CRCs) for activation of ERα are shown in Figure 14.2 for source and product water from DWTP 1, demonstrating the large potency differences and the need to apply low-level linear CRCs because the effect threshold of 10% was only reached after enriching 150-fold. Such a high enrichment means that the sample is very clean and has no issues with respect to EBT values, but it is still important to derive an EC_{10} in order to calculate the treatment efficacy of the DWTP.

Of the studied endpoints, the extracts were only active in assays indicative of activation of ER, oxidative stress response and NF-κB response (Figure 14.3). No other hormonal activity was detected in any of the samples, while cytotoxicity often masked the p53 response in source water. Furthermore, none of the source or product water samples were mutagenic in the Ames assay using strains *Salmonella typhimurium* TA98, TA100 and YG7108 (both with and without metabolic activation).

Although estrogenic activity was commonly detected in source water and ranged from 0.17 to 3.98 ng/L 17β-estradiol equivalent concentrations (EEQ), the treatment processes were able to remove the estrogenic activity, with the effect in the product water not detectable at the highest relative enrichment factor (REF) of 100–150 tested (with one exception where the EC_{10} was just above REF 100).

To determine whether the detected low estrogenic effects in the source water were of any concern, the estrogenic effects were compared with the EBT-EEQ of 1.8 ng_{E2}/L for drinking water (Escher *et al.*, 2015), which translated into an EC_{10} of REF 2.4 (Figure 14.3). Since there was no estrogenic effect up to the highest

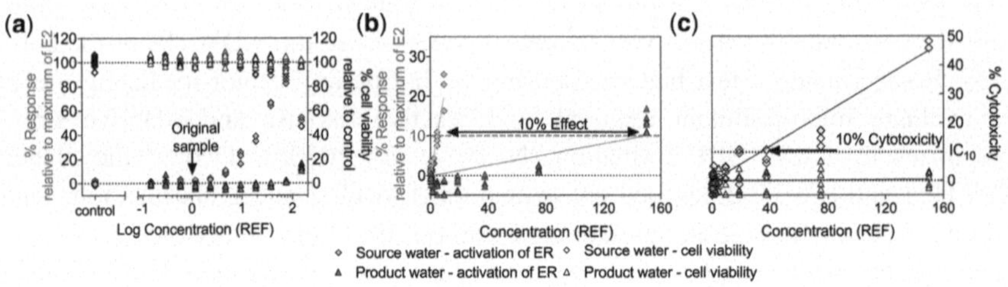

Figure 14.2 Example full concentration–effect curves for activation of the estrogen receptor ERα (filled symbols) and cell viability (empty symbols) for source water (diamond symbols) and product water (triangle symbols) in ERα GeneBLAzer (agonist mode) (a), along with linear concentration–effect curves for activation of ERα (b) and cytotoxicity (c). Data from Neale *et al.* (2020b).

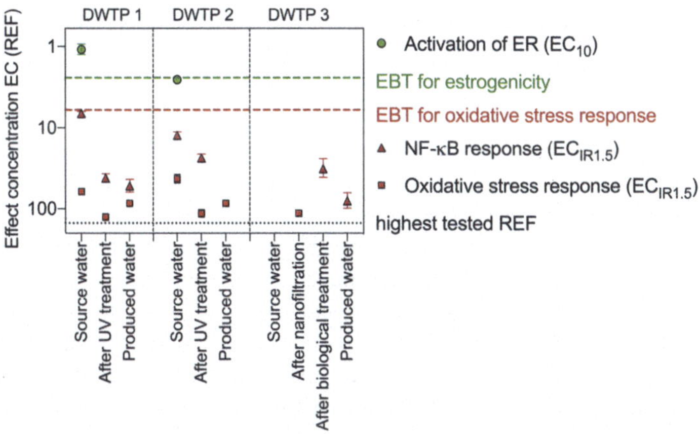

Figure 14.3 Effect concentrations for activation of ER (EC_{10}, orange circles), oxidative stress response ($EC_{IR1.5}$, red squares) and NF-κB response ($EC_{IR1.5}$, red triangles) in units of relative enrichment factor (REF) for all DWTPs in May 2018. Data from Neale *et al.* (2020b).

tested enrichment of REF 100 or 150, the estrogenic effects in the product water must be at least 80 times lower than the effect at the EBT-EEQ.

The oxidative stress response was often highest in the source water, decreased after UV treatment and then increased slightly in the product water after chlorination (Figure 14.3). However, the extracts still needed to be significantly enriched, between 78 and 136 times in the assay, before an effect could be detected in both treated and product water. The effect before and after chlorination was assessed to determine the contribution of formed DBPs to the observed oxidative stress response based on the approach described in Hebert *et al.* (2018). Between 25% and 32% of the oxidative stress response could be attributed to formed DBPs, showing that most of the effect was due to chemicals already present in the source water. The oxidative stress response in the product water was 13–17 times lower than the EBT expressed as $EC_{IR1.5}$ of REF 6 (Escher *et al.*, 2013). Both EBTs were derived using the Australian Drinking Water Guidelines, so are not specific for France. However, the large difference between the observed effect and the proposed EBTs shows the high quality of the final treated water.

14.3 CASE STUDY 2: QUALITY OF RECYCLED WATER

The second case study is a project that applied a battery of bioanalytical tools to provide a measure of human health risk associated with potable use of recycled water (Leusch *et al.*, 2014b). The bioassay selection process was started with a thorough review of the relevant toxicological endpoints associated with negative human health outcomes from drinking water, conducted in parallel with a review of available *in vitro* methods to monitor these toxicological endpoints (see Chapter 10). Based on this assessment, a battery of protection goal-oriented bioassays was developed (Table 14.2).

The battery was first benchmarked against a list of 39 priority chemicals selected based on their likelihood of occurrence in recycled water, the availability of chemical and bioassay methods to detect them and the scientific and social perception of their hazard. The priority chemical list covered a range of chemicals, including natural and synthetic hormones, industrial chemicals, pharmaceuticals and personal care products, veterinary drugs, pesticides and DBPs. This benchmarking allowed chemical/group fingerprinting. Estrogen hormones, for example, were highly estrogenic and anti-androgenic, while also slightly cytotoxic and genotoxic. Pesticides were found to be slightly estrogenic with two of the tested chemicals (chlorpyrifos and diazinon) also neurotoxic. DBPs were slightly genotoxic, cytotoxic and immunosuppressive (Table 14.3). Benchmarking bioassay effects with a wide range of chemicals is valuable when applying an integrated testing strategy scheme, where bioanalytical tools are used to screen samples and direct subsequent chemical analysis.

When benchmarking was complete, grab water samples were collected from several water reclamation plants with a variety of treatment trains and end uses of the reclaimed water as well as tap (drinking) water, bottled water and rainwater collected from a tank in a suburban area. The water samples were extracted and concentrated using SPE prior to chemical analysis (targeting the 39 priority chemicals) and testing in the bioassay battery.

Biological activity was found to decrease with increasing level of treatment from treated wastewater and class A recycled water intended for irrigation (secondary wastewater that has been subjected to UV and/or chlorination, which remove pathogens but do not greatly remove chemical contaminants) to reverse osmosis (RO) treated water, drinking water, bottled water and rainwater (Table 14.4). Treated wastewater and class A water in particular exhibited significant levels of estrogenic and progesterone-like activity, most likely due to natural and synthetic hormones. Overall, the results were in good agreement with chemical analysis and the effect fingerprint matrix. Interestingly, slight estrogenicity and anti-estrogenicity were detected in RO-treated water. No chemicals were identified in the RO-treated samples that could explain this activity. The authors hypothesised that plasticisers in RO membranes may be implicated as these are well known to possess low estrogenic endocrine activity.

The results of the chemical analysis show that advanced water treatment could effectively remove all chemicals tested. The bioassay results confirmed that no toxic 'unknown' chemicals were present in either RO-treated water or drinking water. This case study shows that bioanalytical tools, along with appropriate risk assessment, management and communication, have the capacity to facilitate a significant improvement to the current chemical-by-chemical risk assessment approach and help to communicate risks to the community.

Table 14.2 Bioassay test battery applied for benchmarking the human health risks associated with potable use of recycled water (Leusch *et al.*, 2014b).

Mode of Action	Assay	Toxicological Endpoint	Result Expression
Xenobiotic metabolism			
Hepatotoxicity	HepCYP1A2 (enzymatic activity determined by luciferase precursor assay)	Induction of the multi-function oxidase CYP1A2 in liver cells	Benzo[*a*]pyrene EQ (B[*a*]P EQ)
Hormone receptor-mediated effects			
Estrogenicity and anti-estrogenicity	ERα CALUX (ERα luciferase reporter gene assay)	ERα mediated transcriptional estrogenic effect	17β-Estradiol EQ (EEQ) for agonist; Tamoxifen EQ (TMXEQ) for antagonist
Androgenic activity and anti-androgenic activity	AR CALUX (AR luciferase reporter gene assay)	AR-mediated transcriptional androgenic effect	DHT EQ (DHT EQ) for agonist; Flutamide EQ (Flu EQ) for antagonist
Glucocorticoid activity	GR CALUX (GR luciferase reporter gene assay)	GR-mediated transcriptional glucocorticoid effect	Dexamethasone EQ (Dexa EQ)
Progestogenic activity	PR CALUX (PR luciferase reporter gene assay)	PR-mediated transcriptional progesterone-like effect	Org2058 EQ (Org2058 EQ)
Thyroid activity	TRβ CALUX (TRβ luciferase reporter gene assay)	TRβ-mediated transcriptional thyroid-like effect	Thyroid hormone EQ (T3 EQ)
Other receptor-mediated effects			
Neurotoxicity	AChE assay	Inhibition of acetylcholinesterase (AChE)	Chlorpyrifos EQ (Chlorpy EQ)
Reactive toxicity			
Mutagenicity	Ames TA98 and TA100 test, with and without S9 metabolic activation	Mutagenic potential	Relative genotoxic unit (rGTU)

(*Continued*)

Table 14.2 Bioassay test battery applied for benchmarking the human health risks associated with potable use of recycled water (Leusch *et al.*, 2014b) (*Continued*).

Mode of Action	Assay	Toxicological Endpoint	Result Expression
Genotoxicity	WIL2NS FCMN (flow cytometry micronucleus test)	DNA damage leading to micronucleus formation	Relative genotoxic unit (rGTU)
Adaptive stress response			
Immunotoxicity: stimulation and suppression	THP1-CPA (IL1β production by THP1 cells, measured by ELISA)	Modulation of cytokine IL1β production by monocytes (stimulation or inhibition)	PMA EQ (PMA EQ) for stimulation; Dexamethasone EQ (Dexa EQ) for suppression
Non-specific toxicity			
Cytotoxicity	Caco2 NRU (neutral red uptake assay with Caco-2 cells)	Basal cytotoxicity and decreased viability of gastrointestinal cells	Relative toxic unit (rTU)
Cytotoxicity	WIL2NS TOX (flow cytometry with WIL2-NS cells)	Basal cytotoxicity and decreased viability of white blood cells	Relative toxic unit (rTU)
Cytotoxicity	HepaTOX (resazurin reduction assay with C3A cells)	Basal cytotoxicity and decreased viability of liver cells	Relative toxic unit (rTU)

AR = androgen receptor; CALUX = Chemical Activated Luciferase gene eXpression; CPA = cytokine production assay; DHT = dihydrotestosterone; ER = estrogen receptor; EQ = equivalent concentration; GR = glucocorticoid receptor; Org2058 = 16α-ethyl-21-hydroxyl-19-norpregn-4-ene-3,20-dione; PMA = phorbol-12-myristate-13-acetate; PR = progesterone receptor.

Table 14.3 Effect fingerprinting for different chemical groups.

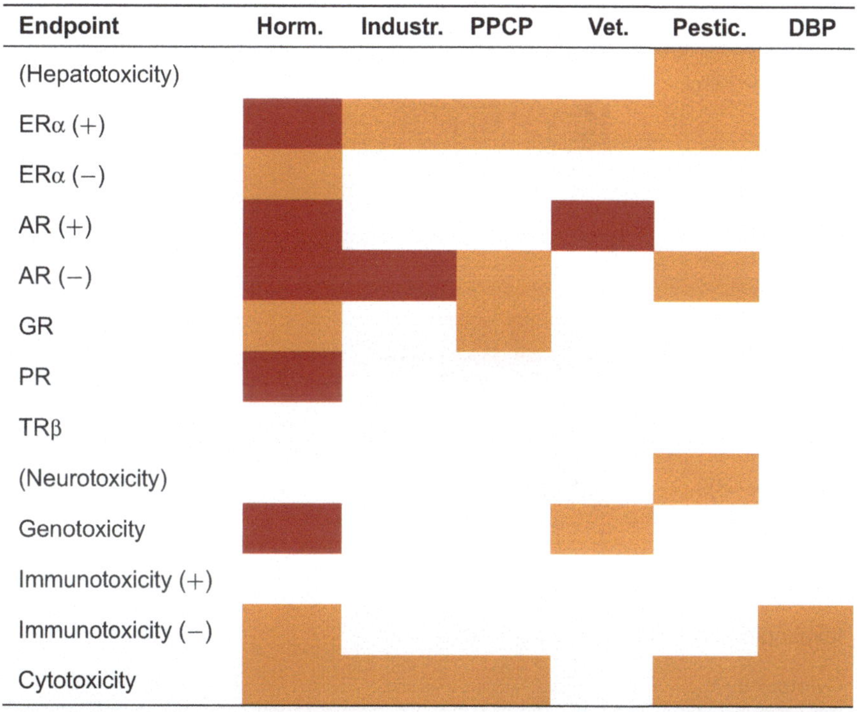

Hepatotoxicity and neurotoxicity are in brackets to highlight that the associated assays are incomplete and/or indirect indicators of toxicity, that is, the 'hepatotoxicity' assay measures liver enzyme induction (not necessarily toxicity), and the 'neurotoxicity' assay measures acetylcholinesterase (a relatively limited measure of total neurotoxicity). (+) indicates agonistic effect. (−) indicates antagonistic effect.

Abbreviations: ER = estrogen receptor; AR = androgen receptor; GR = glucocorticoid receptor; PR = progesterone receptor; TR = thyroid receptor; Horm. = hormones (natural and synthetic); Industr. = industrial chemicals; PPCP = pharmaceutical and personal care products; Vet. = veterinary drugs; Pestic. = pesticides; DBP = disinfection by-products.

The colour coding reflects the potency of the chemicals tested: white = no significant effect, orange = low-to-moderate effect, red = strong effect (Leusch *et al.*, 2014b).

14.4 CASE STUDY 3: WASTEWATER TREATMENT

In this next case study, a battery of bioassays indicative of xenobiotic metabolism, hormone receptor-mediated effects and adaptive stress responses was applied to evaluate effect removal by a conventional WWTP and different constructed wetlands (Nivala *et al.*, 2018). Constructed wetlands are an option for wastewater treatment when centralised treatment is not possible, but few studies had evaluated effect removal by constructed wetlands. Therefore, a test battery covering different stages of cellular toxicity pathways was assembled

Table 14.4 Summary of the toxicity results for the different water samples.

Endpoint	WW	Class A	RO	DW	BW	RW
(Hepatotoxicity)	orange	orange				
ERα (+)	red	red	orange			
ERα (−)			orange			
AR (+)						
AR (−)						
GR	orange	orange				
PR	red	red				
TRβ						
(Neurotoxicity)	red	orange				
Genotoxicity	orange	orange				
Immunotoxicity (+)						
Immunotoxicity (−)	orange	orange				
Cytotoxicity	orange					

Hepatotoxicity and neurotoxicity are in brackets to highlight that the associated assays are incomplete and/or indirect indicators of toxicity, that is, the 'hepatotoxicity' assay measures liver enzyme induction (not necessarily toxicity), and the 'neurotoxicity' assay measures acetylcholinesterase (a relatively limited measure of total neurotoxicity). (+) indicates agonistic effect. (−) indicates antagonistic effect.
Abbreviations: ER = estrogen receptor; AR = androgen receptor; GR = glucocorticoid receptor; PR = progesterone receptor; TR = thyroid receptor; WW = treated wastewater; Class A = Class A recycled water intended for irrigation only;
RO = RO-treated water, potentially for potable use; DW = drinking water; BW = bottled water; RW = rainwater.
The colour coding reflects the biological activity detected in the samples: white = no significant activity, orange = low-to-moderate activity, red = strong activity.

(Table 14.5) and applied to influent and effluent samples from pilot-scale conventional and intensified constructed wetlands, as well as a conventional WWTP. In addition to the bioassays, a suite of indicator chemicals, including pharmaceuticals and food additives, were measured to evaluate chemical removal efficacy. The indicator chemicals were selected based on their biodegradability.

The majority of samples were active in assays indicative of activation of aryl hydrocarbon receptor (AhR), binding to peroxisome proliferator-activated

Table 14.5 Overview of bioassays applied in Nivala *et al.* (2018).

Mode of Action	Assay	Positive Reference Compound
Xenobiotic metabolism		
Activation of aryl hydrocarbon receptor (AhR)	AhR CALUX	2,3,7,8-Tetrachlorodibenzo-*p*-dioxin (TCDD)
Binding to peroxisome proliferator-activated receptor gamma (PPARγ)	PPARγ GeneBLAzer	Rosiglitazone
Hormone receptor-mediated effects		
Activation of estrogen receptor (ER)	ERα GeneBLAzer	17β-Estradiol
Inhibition of estrogen receptor (ER)	ERα GeneBLAzer	Tamoxifen
Activation of androgen receptor (AR)	AR GeneBLAzer	Metribolone (R1881)
Inhibition of androgen receptor (AR)	AR GeneBLAzer	Cyproterone acetate
Activation of glucocorticoid receptor (GR)	GR GeneBLAzer	Dexamethasone
Inhibition of glucocorticoid receptor (GR)	GR GeneBLAzer	Mifepristone (RU486)
Activation of progesterone receptor (PR)	PR GeneBLAzer	Promegestone
Inhibition of progesterone receptor (PR)	PR GeneBLAzer	Mifepristone (RU486)
Adaptive stress response		
Oxidative stress response	AREc32	*tert*-Butylhydroquinone (*t*BHQ)
NF-κB response	NF-κB-GeneBLAzer	Tumour necrosis factor alpha (TNFα)

Cytotoxicity was measured in parallel for all assays.

receptor gamma (PPARγ), activation of ER, oxidative stress response and NF-κB response, with the NF-κB response the most responsive endpoint in the study. In contrast, the effect was often masked by cytotoxicity or there was no effect up to the maximum tested REF for the other endpoints. Consequently, effect removal efficacy could only be calculated for five endpoints.

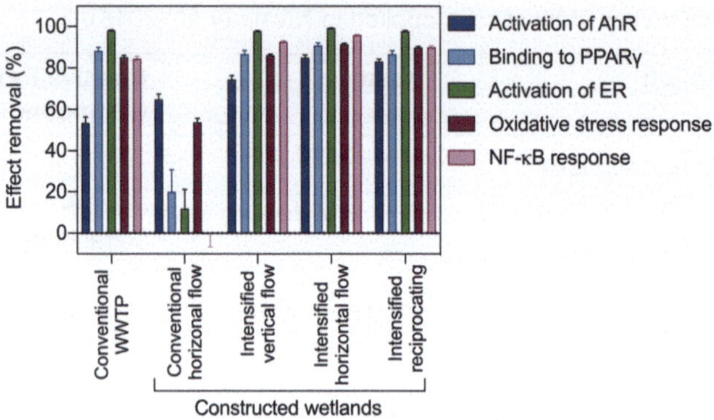

Figure 14.4 Effect removal efficacy (%) after treatment by a conventional wastewater treatment plant and different constructed wetlands. Data from Nivala *et al.* (2018).

Effect removal by intensified wetlands was comparable to or better than the conventional WWTP (Figure 14.4), ranging from 74% to 85% removal of AhR activity to 98% to 99% removal of estrogenic activity. Between 46% and 69% of AhR activity was removed after activated carbon filtration and ozonation in a water reclamation plant (Reungoat *et al.*, 2010), while 80% to >99% removal of estrogenic activity by WWTPs has been reported (*e.g.*, Jalova *et al.*, 2013; Houtman *et al.*, 2018). This shows that intensified wetlands can remove biological activity to a greater extent compared to some conventional WWTPs. In contrast to the intensified wetlands, the conventional horizontal flow wetland without aeration had much poorer removal of biological activity, particularly for PPARγ activity, estrogenic activity and NF-κB activity.

14.5 CASE STUDY 4: SURFACE WATER IMPACTED BY WASTEWATER TREATMENT PLANT EFFLUENT

This next case study applied a bioassay test battery indicative of cellular toxicity pathways as well as apical effects in whole organisms to evaluate the contribution of wastewater effluent to the chemical burden in small streams (Neale *et al.*, 2017c). Water samples were collected from three sites in Switzerland (Birmensdorf, Muri and Reinach) with wastewater effluent and surface water upstream and downstream of the WWTP collected under low flow conditions and enriched using SPE. The extracts were run in a test battery of assays indicative of 13 endpoints (Figure 14.5), with chemical analysis for 405 chemicals conducted in parallel. A mass balance approach was applied to determine the fraction of wastewater effluent downstream of the WWTP, while iceberg modelling (see Section 13.3 in Chapter 13) was used to link chemical analysis and bioassay results.

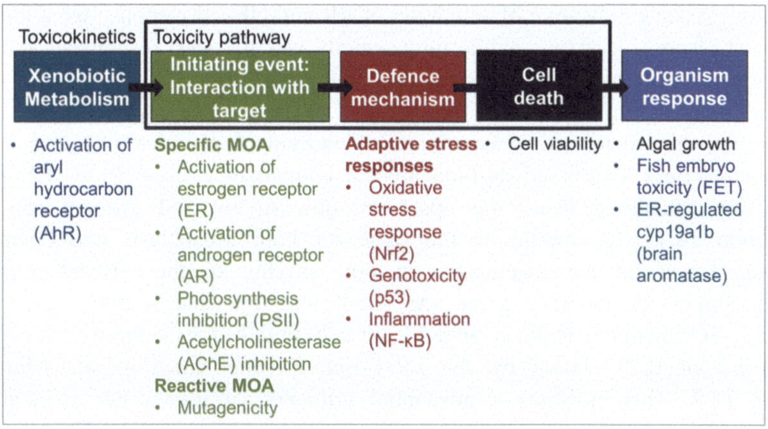

Figure 14.5 Bioassay test battery applied in Neale *et al.* (2017c).

Of the 405 analysed chemicals, 191 were detected at least once. The lowest number of chemicals and lowest sum of chemical concentrations were detected at the upstream sites, with pesticides contributing 43%–90% of the total chemical concentration. In contrast, wastewater effluent had the highest total chemical concentrations, with corrosion inhibitors (*e.g.*, benzotriazoles) and pharmaceuticals the main chemical classes detected. A similar profile was observed downstream of the WWTPs.

For most assays, the response was highest in the effluent samples, followed by the downstream sites, with the lowest response at the upstream sites. This is shown in Figure 14.6, where a small effect concentration (EC) indicates a greater

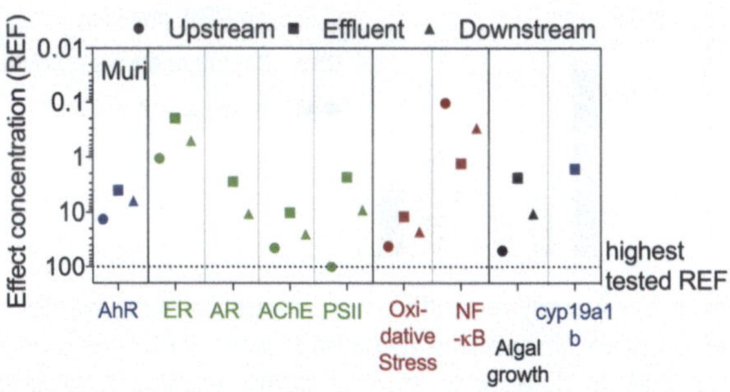

Figure 14.6 Effect concentrations (EC) in units of relative enrichment factor (REF) for the Muri samples. Figure adapted with permission from Neale *et al.* (2017c). Integrating chemical analysis and bioanalysis to evaluate the contribution of wastewater effluent on the micropollutant burden in small streams. *Science of the Total Environment*, **576**, 785–795. © 2017 Elsevier.

response. Assays indicative of activation of ER and NF-κB response were among the most responsive assays, while mutagenicity was the least responsive and the p53 response was masked by cytotoxicity in all samples.

The mean fraction of effluent downstream of the WWTP (f_{eff}) was calculated using both the detected concentration of individual chemicals (C_i) and the biological effect, expressed as bioanalytical equivalent concentration (BEQ_{bio}). This was calculated using the upstream, downstream and effluent samples (Equation 14.1). f_{eff} should be the same for both bioanalysis and chemical analysis based on the assumption of pure mixing of the effluent and no micropollutant degradation given the small temporal and spatial scale. The mean f_{eff} for chemical analysis ranged from 0.20 to 0.30 for the three sites, with a mean f_{eff} of 0.23 shown by the solid line for the sampling site Muri in Figure 14.7. This indicates a substantial influence of wastewater effluent on the downstream site, with a mean dilution of only 4.3-fold. The f_{eff} for bioanalysis ranged from 0.13 for activation of ER to 0.49 for activation of AhR for Muri, with the f_{eff} for activation of AR and 2 h photosystem II (PSII) inhibition similar to the f_{eff} for chemical analysis at 0.22 and 0.24, respectively (Figure 14.7). The high uncertainty for activation of AhR was due to the small differences between $BEQ_{bio,effluent}$ and $BEQ_{bio,upstream}$. Overall, both chemical analysis and bioanalysis gave similar f_{eff} values, showing that the applied mass balance approach could be applied to evaluate the influence of wastewater

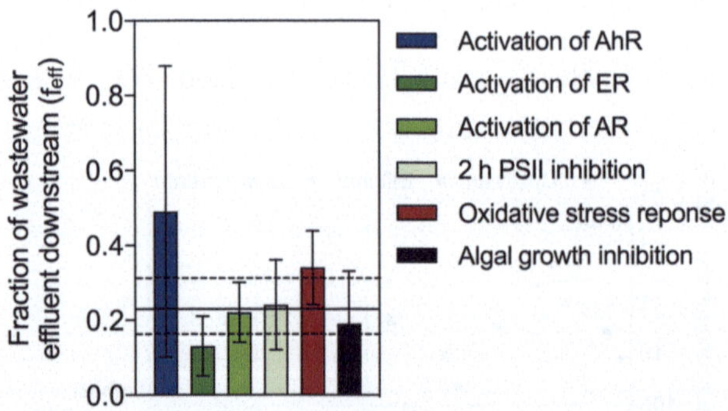

Figure 14.7 Fraction of wastewater effluent downstream of the wastewater treatment plant (f_{eff}) based on both bioanalysis (coloured bars) and chemical analysis (solid horizontal line) for Muri. The error bars indicate standard deviation and were calculated using error propagation. Figure adapted with permission from Neale *et al.* (2017c). Integrating chemical analysis and bioanalysis to evaluate the contribution of wastewater effluent on the micropollutant burden in small streams. *Science of the Total Environment*, **576**, 785–795. © 2017 Elsevier.

downstream:

$$f_{eff} = \frac{C_{i,downstream} - C_{i,upstream}}{C_{i,effluent} - C_{i,upstream}} \quad \text{or} \quad f_{eff} = \frac{BEQ_{bio,downstream} - BEQ_{bio,upstream}}{BEQ_{bio,effluent} - BEQ_{bio,upstream}} \tag{14.1}$$

The contribution of detected chemicals to the observed effect was assessed using iceberg modelling. Between 45% and 108% of 2 h PSII inhibition could be explained by the detected herbicides (Figure 14.8), with the majority of the effect explained by PSII inhibitors diuron and terbuthylazine. A similarly high fraction of the effect was also previously explained by detected chemicals for photosynthesis inhibition (Bengtson Nash *et al.*, 2006; Escher *et al.*, 2011; Tang and Escher, 2014). Up to 30% of the effect could be explained by four chemicals in the activation of AhR assay, with most of the effect explained by fungicide propiconazole. Of the 191 detected chemicals, 135 chemicals were tested in the ToxCast database for the oxidative stress response assay, with 26 of the chemicals reported to be active, that is, an EC could be derived. Despite a large number of detected chemicals being active, only 0.06%–1.9% of the oxidative stress response could be explained by the detected chemicals (Figure 14.8).

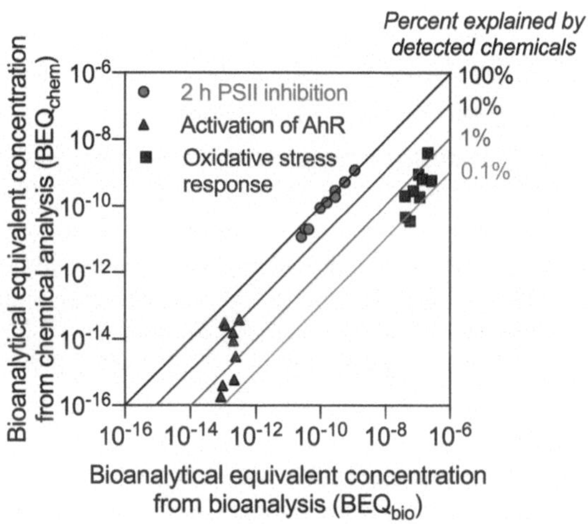

Figure 14.8 Comparison of bioanalytical equivalent concentrations from bioanalysis (BEQ_{bio}) with bioanalytical equivalent concentrations from chemical analysis (BEQ_{chem}) for 2 h PSII inhibition in algae, activation of AhR and oxidative stress response. Figure adapted with permission from Neale *et al.* (2017c). Integrating chemical analysis and bioanalysis to evaluate the contribution of wastewater effluent on the micropollutant burden in small streams. *Science of the Total Environment*, **576**, 785–795. © 2017 Elsevier.

Typically, less than 1% of the oxidative stress response can be explained by detected chemicals (Tang *et al.*, 2014; Yeh *et al.*, 2014; Neale *et al.*, 2015b, 2020a).

The large difference in the fraction of effect explained for the three assays is related to the assay category. The 2 h PSII inhibition assay is an example of a category 1 assay, where a small number of highly potent chemicals are active, PSII-inhibiting herbicides in the current example. Other examples of category 1 assays include assays indicative of receptor-mediated effects, such as activation of ER (Murk *et al.*, 2002; Rutishauser *et al.*, 2004; Könemann *et al.*, 2018) and activation of GR (Schriks *et al.*, 2010b; Jia *et al.*, 2016).

In contrast, the oxidative stress response assay is an example of a category 2 assay, which detects more integrative effects and many low potency chemicals can contribute to the effect. A large number of chemicals can have an effect in the oxidative stress response assay, so measuring more chemicals to increase the percentage of effect explained is not practically feasible. Other examples of category 2 assays include some assays indicative of xenobiotic metabolism, such as activation of pregnane X receptor (PXR) (Creusot *et al.*, 2014), and assays indicative of apical effects, such as fish embryo toxicity (Neale *et al.*, 2015b) and bacterial toxicity (Tang *et al.*, 2013b). This case study helps to highlight the importance of applying bioassays alongside chemical analysis to gain a better understanding of the chemical burden in surface water and wastewater.

14.6 CASE STUDY 5: BENCHMARKING SURFACE WATER QUALITY ACROSS THE USA

Blackwell *et al.* (2019) evaluated the bioactivity of 38 rivers in the USA with the multiplex Attagene FACTORIAL™ battery that included 96 different endpoints (see Chapter 13.2.2 for more information on this assay and its typical response in surface water). PXR- and AhR-related endpoints were most commonly activated in these samples, followed by ERα- and PPARγ-related endpoints (Figure 14.9) and the observed effects varied widely across sampling sites with sites impacted by wastewater high in ERα- and GR-activity while AhR and PXR were instead associated with urban activity but also with natural organics from swamps (Blackwell *et al.*, 2019).

The same samples were also measured with reporter gene assays for ERα, AR and GR (Conley *et al.*, 2017a). More samples were active in the reporter gene assays than in the FACTORIAL™ endpoints, pointing to a higher sensitivity of the reporter gene assays. The active samples showed a good correlation between the estrogenic response in the T47D-KBluc reporter gene assay and ERE_*cis* and ERa_*trans* of the FACTORIAL™ assay (Figure 14.10).

In addition, more than 700 micropollutants were quantified in these samples (Bradley *et al.*, 2017). The mixture effects of the detected chemicals in a water sample C_i were evaluated with the exposure activity ratio $EAR_{mixture}$, which is

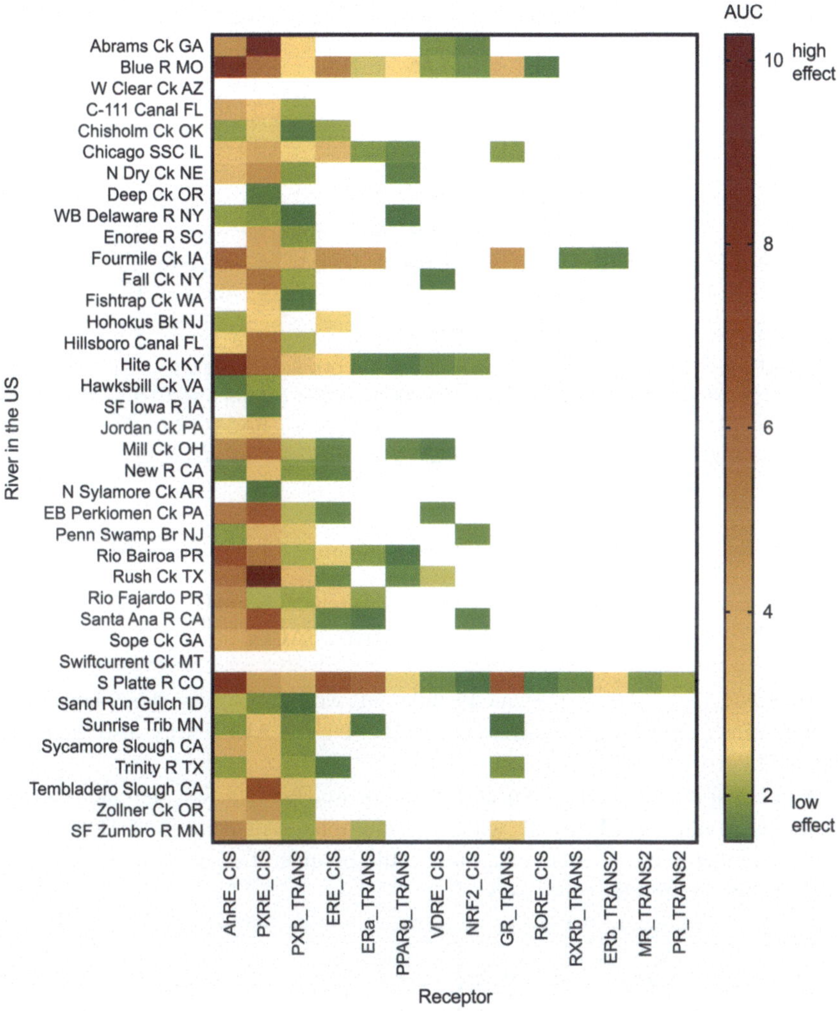

Figure 14.9 Heat map showing effects expressed as AUC for commonly activated endpoints of the Attagene *cis*-FACTORIAL™ and *trans*-FACTORIAL™ assay in surface water extracts. Data from Blackwell *et al.* (2019). AUC = area under the curve.

the ratio of the exposure concentrations and the ECs, in this case the activity concentrations at cut-off (ACC_i) from the ToxCast database (Equation 14.2). The EAR_i of all detected bioactive chemicals were then summed up to obtain the $EAR_{mixture}$. In analogy to the mixture risk quotients, that is, the risk index (Chapter 8), if the $EAR_{mixture}$ exceeds 1, then the detected chemicals are present at concentrations where together they could elicit a bioassay effect. The EAR_i of

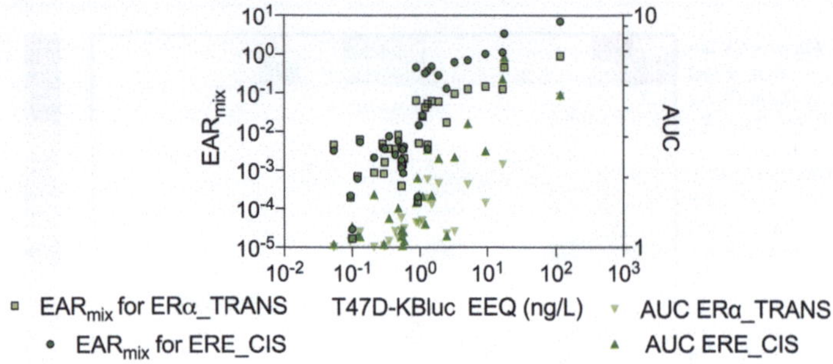

Figure 14.10 Comparison of the EEQ measured with a reporter gene assay (T47D-KBluc) (Conley *et al.*, 2017a) with the predicted exposure activity ratio EAR_{mix} for the Attagene FACTORIAL™ endpoints ERα-*trans* and ERE_*cis* and the measured AUC of the samples in the Attagene FACTORIAL™ endpoints ERα-*trans* and ERE_*cis* (right *y*-axis). Data from Blackwell *et al.* (2019).

the individual compound can be evaluated to identify mixture risk drivers:

$$EAR_{mixture} = \sum_{i=1}^{n} \frac{exposure_i}{activity_i} = \sum_{i=1}^{n} \frac{\text{chemical concentration } C_i}{ToxCastACC_i} \quad (14.2)$$

The $EAR_{mixture}$ for the estrogenicity endpoints were above 1 for those sites where a high effect (expressed as area under the curve, AUC) was also detected (Figure 14.10).

A tiered approach would be useful in future studies: in a first step the relevant endpoints could be identified with the multiplex Attagene FACTORIAL™ assays and then these relevant endpoints could be quantified using more sensitive single-endpoint reporter gene assays.

14.7 CASE STUDY 6: BENCHMARKING SURFACE WATER QUALITY IN SMALL STREAMS DURING RAIN EVENTS

The final case study describes water quality monitoring triggered by rain events of 44 small rivers all across Germany (Neale *et al.*, 2020a). Despite the sites being selected mainly based on anticipated agricultural impact, chemical markers of untreated wastewater and estrogenic effects were high during some rain events, likely indicating sewer overflow. Road run-off also impacted many of the rain events with high concentrations of chemicals typical of tyre wear. The diversity of the events was striking with some sites having two to four events that often showed little similarity with respect to chemical composition, overall concentration, effect magnitude and effect pattern (Neale *et al.*, 2020a).

In this case study, in addition to a battery of bioassays (ERa GeneBLAzer, AhR CALUX, PPARγ GeneBLAzer and AREc32) and chemical analysis of close to

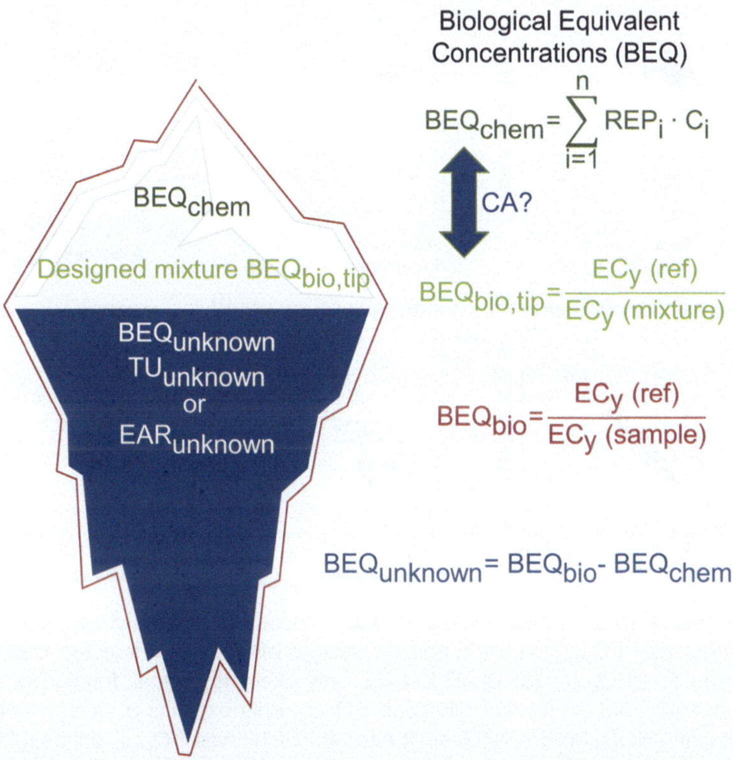

Figure 14.11 Iceberg modelling in this case study was complemented by designed mixture experiments of mixtures of the detected bioactive chemicals (figure adapted from Figure 13.4 in Chapter 13). CA = concentration addition.

400 chemicals, designed mixture experiments were performed with the chemicals that were detected in the samples. The bioanalytical equivalent concentration measured in the designed mixtures $BEQ_{bio,tip}$ was quantified by the defined mixture experiments (Figure 14.11) in addition to just comparing the bioanalytical equivalent concentration from the bioassays BEQ_{bio} with the bioanalytical equivalent concentration from chemical analysis BEQ_{chem} (Chapter 13).

Despite the fact that almost 400 chemicals were included in the analytical method, the predicted mixture effects of the detected chemicals (BEQ_{chem}) only explained between 0.01% and 1% for the category 2 bioassays AhR CALUX, PPARγ GeneBLAzer and AREc32, with only a few samples having a higher fraction of effect explained (Figure 14.12a-c).

If the 2 to 14 chemicals that were, both, detected and active in the tested bioassays were mixed in the concentrations as they occurred in the samples and tested in the bioassays, there was an excellent agreement between mixture effects $BEQ_{chem,tip}$

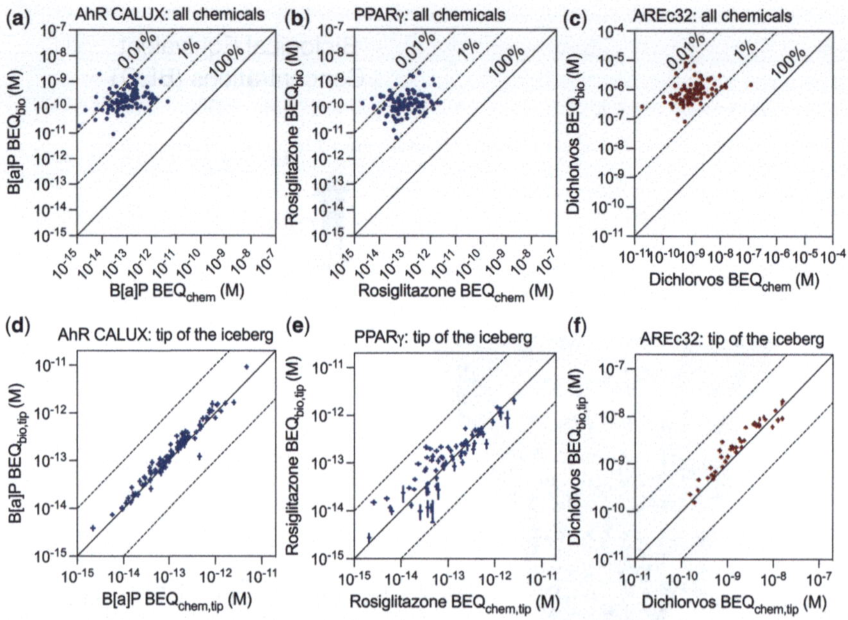

Figure 14.12 (a-c) Comparison of the measured bioanalytical equivalent concentrations BEQ_{bio} and the predicted mixture effect of the detected chemicals expressed as BEQ_{chem}: (a) B[*a*]P EQ_{chem} only explains a small fraction of B[*a*]P EQ_{bio} in AhR CALUX. (b) Rosiglitazone EQ_{chem} only explains a small fraction of Rosiglitazone EQ_{bio} in PPARγ GeneBLAzer. (c) Dichlorvos EQ_{chem} only explains a small fraction of Dichlorvos EQ_{bio} in AREc32. (d-f) Comparison of the predicted mixture effects $BEQ_{chem,tip}$ and the mixture effect of the designed mixtures $BEQ_{bio,tip}$ for (d) AhR CALUX, (e) PPARγ GeneBLAzer and (f) AREc32. Reprinted with permission from Neale *et al.* (2020a). Assessing the mixture effects in in vitro bioassays of chemicals occurring in small agricultural streams during rain events. *Environmental Science & Technology*, **54**(13): 8280–8290. © 2020 American Chemical Society.

predicted with the model of concentration addition of the mixture components and their measured mixture effect $BEQ_{bio,tip}$ (Figure 14.12d-f). These mixture experiments confirmed that the BEQ concept is valid and can be applied to environmental samples independent of the concentration ratio in the mixture and the effect potencies of the individual components.

In this study, we also took a closer look at the known fraction of effects, the 'tip of the iceberg'. If we consider 15 chemicals that have experimental effect data for single chemicals and were detected frequently (but not in all rain event samples), their composition and overall concentrations varied substantially. The AhR CALUX was highly dominated by the urban herbicide diuron at a few sites

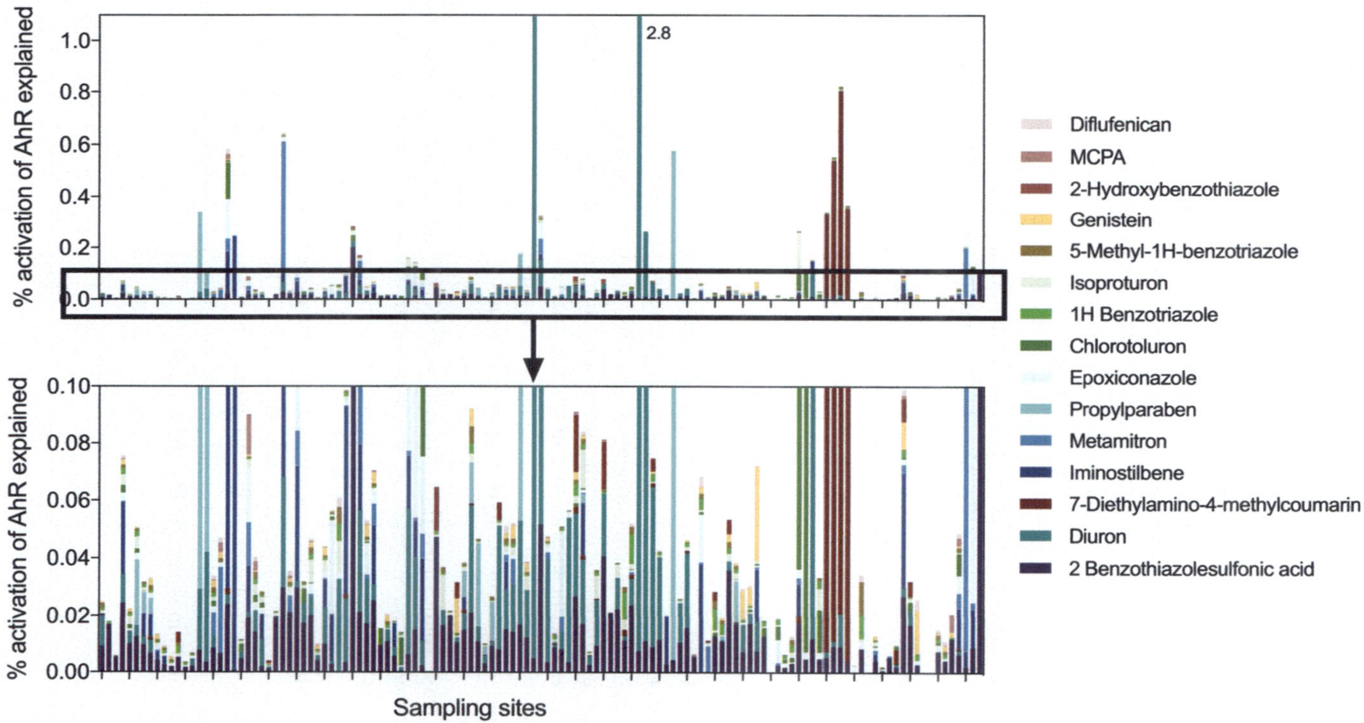

Figure 14.13 Contribution of 15 chemicals to the BEQ_{chem} in the AhR CALUX assay; % activation of AhR explained = BEQ_{chem}/BEQ_{bio}. Data from and detailed description of the sampling sites in Neale *et al.* (2020a).

(Figure 14.13), which could explain as much as 1.1% and 2.8% of the mixture effect of the known chemicals in two of the samples. At another site, 7-diethylamino-4-methylcoumarin, which was consistently present at high concentration during four rain events, alone explained 0.3%–0.8% of the biological effect of the sample. Interestingly, 2-benzothiazolesulphonic acid, a rather inconspicuous transformation product of mercaptobenzothiazole and chemical used on the production of rubber, had a consistently relevant contribution to both activation of the AhR and oxidative stress response. There was a much more variable contribution of many chemicals of diverse origin at most sites, and most other chemicals were fairly specific for one or the other assay (Neale *et al.*, 2020a). This means that although in each sample there were often few chemicals dominating the mixture effects of the known chemicals, the dominant chemicals were in many cases site-specific. Minor contributing chemicals, in contrast, were often found across many samples.

This case study demonstrates the need to complement chemical analysis with bioanalytical tools, especially during random events (such as rain events), where unexpected chemicals might be released and where the bioassays provide a reliable sum parameter for chemical pollution.

14.8 CONCLUSIONS

Bioanalytical tools have been widely applied in the literature, with studies focusing on wastewater, surface, drinking and recycled water. The large number of case studies in Table 14.1 is encouraging, especially the large number of additional applications and diversity of applications since the first edition of this book 10 years ago. Every application demonstrates the progress but also the limitations of the bioanalytical approach to water quality assessment and how computational analyses and concurrent analytical quantification of micropollutants strengthen the studies.

Several case studies covering different water types and different applications were discussed in more detail in this chapter to show how bioassays can be applied to monitor water quality, assess treatment efficacy and evaluate the effectiveness of critical control points. Some of the case studies demonstrated how water quality can be benchmarked against existing EBTs, or how iceberg modelling can help understand drivers of toxicity. The case studies also highlight the advantage of applying bioassays alongside chemical analysis. These case studies can only demonstrate a small part of the potential bioanalytical tools have for the application in water quality monitoring.

Chapter 15

Application of bioanalytical tools beyond water: Sediment and biota

15.1 INTRODUCTION

This book focuses on the application of bioanalytical tools to water samples, but bioanalytical tools are being applied to an ever-increasing range of environmental samples. Essentially, any mixture of micropollutants can be characterised by bioanalytical tools provided it can be extracted unaltered from the sample. As the matrices tested become more varied, the separation of organic micropollutants from the matrix becomes more and more important to ensure that matrix effects do not affect the bioassay results.

Organic chemicals commonly detected in water are also, of course, detected in biota, our food and our bodies (CDC, 2019). We have to look at the environment as an interconnected system where chemicals move between the different compartments of water, air and earth, become enriched along the food chain, and eventually end up in people (Escher *et al.*, 2020).

Condensed matrices (solid samples) such as soil, sediment, plants, biota and tissue samples pose a particular challenge with respect to sampling and sample preparation. While most waters can be tested directly (Chapter 3) or only undergo fairly simple enrichment/extraction without any clean-up (Chapter 12), this is not possible for condensed matrices.

Traditionally, chemicals in such matrices have been extracted using (accelerated) solvent extraction on previously freeze-dried samples. Coextracted lipids and matrix components require tedious clean-ups, including gel permeation chromatography and acid digests, to remove the natural matrix and to isolate the micropollutants of interest. Blood can be relatively easily extracted with a simple protein

Authors: Beate Escher, Peta Neale and Frederic Leusch
doi: 10.2166/9781789061987_0325

precipitation with acetonitrile aided by salting out followed by minimal clean-up, the so-called QuEChERS (quick, easy, cheap, effective, rugged and safe) extraction method (Baduel *et al.*, 2015; Plassmann *et al.*, 2015). QuEChERS has also been applied to food items and fish tissue (Ramalhosa *et al.*, 2009; Forsberg *et al.*, 2011; Norli *et al.*, 2011; Jakimska *et al.*, 2013; Cloutier *et al.*, 2017), which required more extensive clean-up.

Harsh clean-up results in the removal of a substantial fraction of chemicals and potentially alters the composition of the samples. The challenge for condensed matrices is to use sample preparation and clean-up methods that are not discriminating and not destructive. Various methods have been optimised for chemical analysis but most of them have low recoveries or are selective for certain chemical groups and therefore are not suitable for sample preparation for *in vitro* bioassays. Effect-directed analysis (EDA, Chapter 13) that is typically used to fractionate samples to identify mixture effect drivers also has the advantage of partially removing matrices and endogenous chemicals that can mask the effect of pollutants.

In situ and *ex situ* equilibrium passive sampling techniques can be used as a one-step sampling and sample preparation method for sediments and biota. Silicone (polydimethylsiloxane, PDMS) in particular offers many advantages because the partition coefficients are fairly independent of the hydrophobicity of the chemicals for most non-aqueous matrices (Jahnke *et al.*, 2016) and kinetics are reasonably fast in sediments (Li *et al.*, 2013), lipid-rich tissues (Jin *et al.*, 2013) and blood (Jin *et al.*, 2015a). In addition, due to limited matrix interference, extracts from silicone passive samplers can often be used directly in chemical analysis and bioassays without additional or minimal clean-up. The polymer phase may thus serve as a reference phase that is equilibrated with complex media to extract a defined fraction of a sample without changing its composition.

In this chapter, we give a brief summary of applications of bioanalytical tools to non-aqueous matrices, including air and dust, condensed environmental matrices (suspended particulate matter (SPM), sediment and soil), biota (fish tissue, marine mammal tissue) and in human biomonitoring. We focus on studies that have attempted to measure complex mixtures of organic chemicals in an unbiased fashion. In the past, *in vitro* bioassays were applied to quantify dioxin-like and other persistent chemicals in biospecimen after solvent extraction and acid digestion but the true mixture effects in a sample are not accessible with these methods, and we therefore did not include those types of studies here.

15.2 SUSPENDED PARTICULATE MATTER, SEDIMENT AND SOIL

15.2.1 Suspended particulate matter

SPM is probably the most natural follow-up phase after water to investigate with bioanalytical tools because SPM is in intimate contact with the water phase, continuously exchanging chemicals. The importance of sorbed chemicals can be strikingly demonstrated by an example from a study where water and associated particulate matter were sampled during a storm event (Mueller *et al.*, 2021). With SPM concentrations less than 1 g/L, there was about 1000 times less mass of SPM than water, but the associated bioanalytical effect fluxes (BEF) were approximately equal in water and SPM. This suggests that SPM carries a significant chemical load associated with significant bioactivity (Figure 15.1). This is, however, an extreme situation. During dry weather, the SPM concentration is much lower, and the associated effect load of SPM is hardly measurable. But this example shows the potential relevance of SPM. Despite of this, remarkably few studies have addressed SPM and, as we noted in Chapter 12, not enough attention has been given to filtration when developing water extraction methods.

15.2.2 Sediments

The hazard potential of sediments (Burton *et al.*, 2000) can be assessed after total extraction of sediments and submitting them to batteries of *in vitro* bioassays (Boehler *et al.*, 2017). Given the complexity of the matrix and mixture of contaminants in sediment, EDA has been often applied to identify bioactive components in sediments (Brack *et al.*, 1999, 2005; Li *et al.*, 2018, 2019). Interestingly, the *in vitro* effects in a polluted river site were dominated by the polar fractions in EDA (Luebcke-von Varel *et al.*, 2011) but at similar sites polycyclic aromatic hydrocarbons and legacy persistent organic pollutants mainly contributed to the activation of AhR (Otte *et al.*, 2013). These and other case

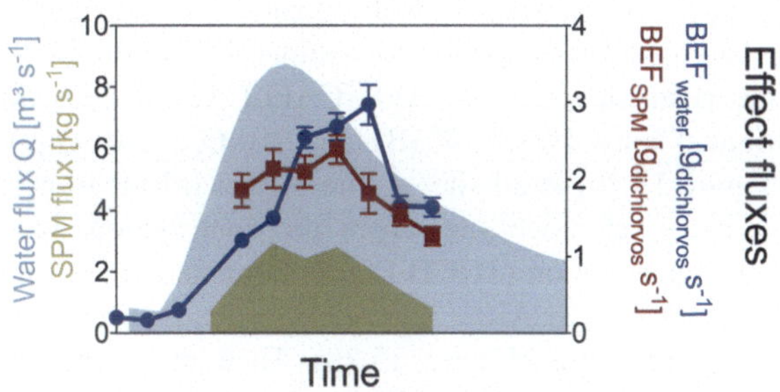

Figure 15.1 Equal role of the bioanalytical effect fluxes (BEF) in water and SPM expressed as dichlorvos BEF measured with AREc32 for oxidative stress response over the time course of a storm event despite the water flux *Q* and the SPM flux differed by more than a factor of 1000. The BEF is calculated by multiplying the water or SPM flux with the measured bioanalytical equivalent concentration (BEQ) with permission from Mueller *et al.* (2021). Storm event-driven occurrence and transport of dissolved and sorbed organic micropollutants and associated effects in the Ammer River, Southwestern Germany. *Environmental Toxicology and Chemistry*, **40**(1): 88–99. © 2020 The Authors. *Environmental Toxicology and Chemistry* published by Wiley Periodicals LLC on behalf of SETAC.

studies demonstrate the complexity and variability of contaminated sediments and how useful bioanalytical tools can be for monitoring sediment contamination.

Bioavailability is decisive in both soils and sediments to determine if there is an environmental risk (Reid *et al.*, 2000; Semple *et al.*, 2004; Mayer *et al.*, 2014). Therefore, biological assessment initially relies on direct contact testing with sediment-dwelling organisms (Tuikka *et al.*, 2011; de Baat *et al.*, 2019a, b). Application of bioanalytical tools is possible if the bioavailable fraction is extracted with passive sampling methods (Jonker *et al.*, 2020), and there is also the option of performing EDA after extractions of the bioavailable fraction (You and Li, 2017).

Numerous studies have performed bioanalytical assessment of total sediment extracts and *ex situ* passive sampling with silicone and a selection of results for three endpoints is given in Table 15.1. The concentrations per kg silicone can be almost converted one-to-one to concentrations per kg organic carbon (OC) (Li *et al.*, 2013; Niu *et al.*, 2020), hence it is possible to compare the total with the bioavailable effects expressed as bioanalytical equivalent concentrations (BEQ_{bio}, for definition see Chapter 7).

Li *et al.* (2013) compared the BEQ in total sediment extracts with the BEQ in extracts from PDMS *ex situ* passive sampling of sediment. They demonstrated that the 'bioavailable effects' matched quite well with the effects from total extracts for herbicidal activity and oxidative stress response but were substantially smaller for assays that are triggered by more hydrophobic chemicals such as the activation of AhR. Based upon these experiences and additionally accounting for binding to black carbon, Braunig *et al.* (2016) developed a bioanalytical effect balance model that allowed the differentiation between total effect burden and that associated with organic matter and black carbon (Figure 15.2a). The model was applied to sediments from different sources from urban to pristine river sites. In the AhR CAFLUX, the bioavailable fraction (BEQ_{OC}) contributed only 1%–11% ($BEQ_{OC}/(BEQ_{OC} + BEQ_{BC})$) to the total effect burden (Figure 15.2b) at all sites, some of which had strong urban impact, except at one pristine reference site where the contribution from bioavailable TCDD equivalent concentration (TCCD EQ_{OC}) dominated the overall effect. The situation was different for the AREc32, which responds to chemicals activating the oxidative stress response (Figure 15.2b). Here the contribution from the aqueous phase became important with 7%–26% of effect burden expressed as tBHQ EQ ($BEQ_w/(BEQ_w + BEQ_{OC} + BEQ_{BC})$). The contribution of OC to the total sorbed chemical mixtures ($BEQ_{OC}/(BEQ_{OC} + BEQ_{BC})$) was higher with 7%–24% compared to the 1%–11% for the AhR CAFLUX described above. This model underpins the bioassay-specific fate of effects in sediment–water systems, which is relevant for risk assessment.

Vethaak *et al.* (2017) later proposed a similar approach and applied total solvent and PDMS extracts to five *in vitro* bioassays (DR-Luc, ERLuc, AR-EcoScreen, a transthyretin (TTR) binding assay and *Aliivibrio fischeri* bioluminescence

Table 15.1 Global comparison of bioanalytical equivalent concentrations (BEQ_{bio}) of exhaustive sediments and PDMS extracts with AhR CALUX, AREc32 and PPARγ GeneBLAzer.

AhR Assay

Sampling Site	BEQ of Total Extract ($ng_{ref}/kg_{sed,dw}$)		BEQ of PDMS Extract (ng_{ref}/kg_{PDMS})	Ref.	Lit.
	Raw Sample	Clean-up Sample			
Sweden	83.5–143,640		755–90,000	TCDD	1
Coastal Svalbard and offshore deep sea	88.1–1877		316–1030	TCDD	1
Rivers/coastal areas in Queensland, Australia	21.4–506		222–5450	TCDD	1
French–German river catchment	499–18,668		535–8500	TCDD	1
Brisbane River, Australia	10–927		122–1186	TCDD	2
Gladstone Harbour, Australia	284–12,198 (wet basis)		1880–125,641	TCDD	3
Lake Tai, China		20–300	<110–6000	TCDD	4
Haihe River, China	330–930			TCDD	5
Dagu River, China	1200–13,900			TCDD	5
Wenyu River, China	8.5–336	2.7–63.8		TCDD	6
Yellow Sea region	0–28			TCDD	7
Lake Shihwa and Masan Bay, Korea	14–868			TCDD	8
River Elbe, Germany		15.5–322		TCDD	9
Pohang Area, Korea	0–800			TCDD	10

(Continued)

Table 15.1 Global comparison of bioanalytical equivalent concentrations (BEQ_{bio}) of exhaustive sediments and PDMS extracts with AhR CALUX, AREc32 and PPARγ GeneBLAzer (*Continued*).

AhR Assay

Sampling Site	BEQ of Total Extract ($ng_{ref}/kg_{sed,dw}$)		BEQ of PDMS Extract (ng_{ref}/kg_{PDMS})	Ref.	Lit.
	Raw Sample	Clean-up Sample			
UK estuaries	1100–177,000	1–106		TCDD	11
Tietê River, Brazil	160–24,170			TCDD	12
North Sea, south-western Baltic Sea and western Mediterranean		20–3493	97–7257	TCDD	13
West coast of South Korea	0–57,000			B[a]P	14
Beijing-Hangzhou Grand Canal	25.2–208		13.4–118	TCDD	15
Qiantang River	17.4–24.0		4.68–14.0	TCDD	15
Beijing-Hangzhou Grand Canal	66.3–545		35.3–311	B[a]P	15
Qiantang River	45.1–62.9		12.3–36.8	B[a]P	15

AREc32 Assay

Sampling Site	BEQ of Total Extract ($mg_{ref}/kg_{sed,dw}$)	BEQ of PDMS Extract (mg_{ref}/kg_{PDMS})	Ref.	Lit.
Sweden	6–1262	25–359	tBHQ	1
Coastal Svalbard and offshore deep sea	10–201	14.1–79.0	tBHQ	1
Rivers/coastal areas in Queensland, Australia	3.1–11.3	19.0–93.5	tBHQ	1

French–German river catchment	26–283	33.5–257	tBHQ	1
Brisbane River, Australia	79–564	1666–4591	tBHQ	2
Gladstone Harbour, Australia	5.7–31.3 (ww)	59.0–205	tBHQ	3
Beijing-Hangzhou Grand Canal	7.04–18.0	8.9–21.0	tBHQ	15
Qiantang River	3.85–6.25	10.2–22.0	tBHQ	15
PPARγ GeneBLAzer Assay				
Sampling Site	**BEQ of Total Extract ($\mu g_{ref}/kg_{sed,dw}$)**	**BEQ of PDMS Extract ($\mu g_{ref}/kg_{PDMS}$)**	**Ref.**	**Lit.**
Sweden	1.0–186	28.3–286	Rosiglitazone	1
Coastal Svalbard and offshore deep sea	2.3–7.1	9.9–91.0	Rosiglitazone	1
Rivers/coastal areas in Queensland, Australia	1.7–5.1	20.2–29.0	Rosiglitazone	1
French–German river catchment	3.3–29.6	31.5–298	Rosiglitazone	1
Beijing–Hangzhou Grand Canal	2.72–9.60	5.08–28.4	Rosiglitazone	15
Qiantang River	0.94–4.34	1.70–10.5	Rosiglitazone	15

Abbreviations: dw = dry weight; ww = wet weight; Ref. = reference chemical. Lit. = literature reference: [1]Jahnke *et al.* (2018); [2]Li *et al.* (2013); [3]Braunig *et al.* (2016); [4]Li *et al.* (2016); [5]Song *et al.* (2006); [6]Luo *et al.* (2009); [7]Hong *et al.* (2012); [8]Yoo *et al.* (2006); [9]Otte *et al.* (2013); [10]Hong *et al.* (2014); [11]Hurst *et al.* (2004); [12]Rocha *et al.* (2010); [13]Vethaak *et al.* (2017); [14]Jeon *et al.* (2017), [15]Niu *et al.* (2020).

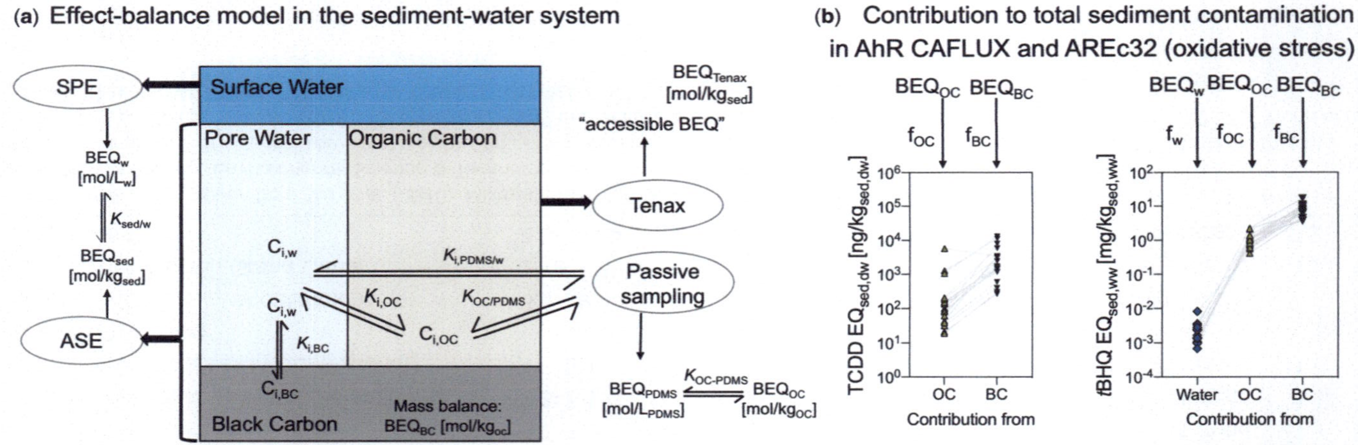

Figure 15.2 (a) Conceptual model describing the relationships and equilibria in a sediment system. Gray circles represent extracts used in the study. Bold arrows refer to exhaustive extractions, two-way arrows refer to equilibrium partitioning, while dashed arrows indicate limited exchange between phases. BEQ = bioanalytical equivalent concentration; SPE = solid-phase extraction; ASE = accelerated solvent extraction; K = partition constant; PDMS = polydimethylsiloxane (silicone) extraction; BC = black carbon; OC = organic carbon; IC = inorganic carbon, f = fraction. (b) Contribution of water, OC and BC to the total sediment contamination quantified with AhR CALUX and AREc32. TCDD EQ=TCDD equivalent concentration, tBHQ EQ = *t*-butylhydroquinone equivalent concentration. The lines connect data from the same sampling site. Figure (a) reprinted with permission from Braunig *et al.* (2016). Bioanalytical effect-balance model to determine the bioavailability of organic contaminants in sediments affected by black and natural carbon. *Chemosphere*, **156**: 181–190. © 2016 Elsevier.

inhibition) and one *in vivo* assay on the embryo toxicity of the sea urchin. They confirmed the earlier results qualitatively but did not convert the BEQ_{PDMS} in BEQ_{OC}, so a retrospect effect balance was not possible. They compared the bioassay results with chemical analysis and indicated that the detected chemicals could not entirely explain the effects detected with the bioassays.

Two more recent studies evaluated Chinese sediments (Niu *et al.*, 2020) and world-wide sampled sediments (Jahnke *et al.*, 2018; Muz *et al.*, 2020) using both total accelerated solvent extraction and PDMS and confirmed that typically less than 10% of the effects and BEQ were bioavailable (Table 15.1).

The responses in the AhR CALUX, PPARγ GeneBLAzer and AREc32 assay expressed as BEQ_{bio} and TU_{bio} were compared with the predicted effect BEQ_{chem} and TU_{chem} (for definition see Chapter 8) from chemical analysis of up to 650 organic chemicals with up to 420 chemicals detected (Niu *et al.*, 2020). Despite of this massive analytical effort only a small fraction of effects (BEQ_{chem}/BEQ_{bio} and TU_{chem}/TU_{bio}) could be explained by the detected chemicals: 0.1%–28% in whole sediment and 0.009%–3.3% in the bioavailable fraction of the Chinese sediments (Niu *et al.*, 2020). Similarly, less than 10% effect in the bioavailable fraction of the sediments sampled worldwide could be explained by the detected chemicals with the exception of only one site in Sweden, where benzo[a]pyrene and benzo[k]fluoranthene together explained more than 30% of the effect in AREc32 (Muz *et al.*, 2020).

These studies highlight how important it is to not only analyse chemicals in sediments but also assess the mixture effects with bioanalytical tools for a comprehensive understanding of the immediate risk (bioavailable effects from PDMS extracts) but also the total hazard potential (mixture effects from exhaustive extracts).

15.2.3 Soil

Soil has received far less attention with bioanalytical tools (Xiao *et al.*, 2006). The few studies available focus on soils in connection with the application of sewage sludge to agricultural soils (Liu *et al.*, 2014) or soil of e-waste recycling sites (Shen *et al.*, 2008) or other contaminated areas (Sidlova *et al.*, 2009; Lam *et al.*, 2018).

15.3 PARTICLES IN AIR AND DUST

Several studies have applied bioanalytical tools to test particle extracts from air and dust. Most studies initially focused on genotoxicity and mutagenicity of air particulate material (Lewtas *et al.*, 1990; Isidori *et al.*, 2003; Marvin and Hewitt, 2007) and the activation of the aryl hydrocarbon receptor (AhR, Skarek *et al.*, 2007; Khedidji *et al.*, 2017; Zhang *et al.*, 2018a, b; McDonough *et al.*, 2019) but other studies have also expanded the range of endpoints to endocrine activity (Clemons *et al.*, 1998; Kennedy *et al.*, 2009; Novak *et al.*, 2020), oxidative stress and inflammation (Al Hanai *et al.*, 2019; Chen *et al.*, 2020). Particles in air are usually collected using either passive or active air samplers, and chemicals bound to particulates (such as $PM_{2.5}$ and PM_{10}) are extracted by (accelerated) solvent extraction.

The studies have shown that chemicals bound to air particulate matter can induce significant mutagenicity, AhR activity, estrogenicity, oxidative stress and inflammation in cell-based bioassays, indicative of a risk to human health.

Dust is also a source of persistent organic pollutants, including flame retardants, and dust has been used as proxy for human exposure. Dioxin-like chemicals and the AhR activation were the focus of many studies on dust (Tue *et al.*, 2013; Suzuki *et al.*, 2017) but also inflammation (Allermann and Poulsen, 2000) and endocrine effects (Chou *et al.*, 2015).

Microplastic in water and air poses an environmental threat due to the plastic as such but also because plastic materials can carry micropollutants from production but also from sorption processes in the environment (Koelmans *et al.*, 2016). *In vitro* bioassays have been used to assess the micropollutant burden in plastic and microplastic (Rummel *et al.*, 2019; Zimmermann *et al.*, 2019, 2020).

15.4 BIOTA

Applications of *in vitro* bioassays in biota samples were reviewed by Jin *et al.* (2015b) in marine wildlife. Seminal studies have not only used commercially available cell lines, such as the ones typically used for water quality monitoring but have tested extracts in cell lines that were derived from tissues of those or similar test species.

15.4.1 Blood

Dogruer *et al.* (2018) applied a modified QuEChERS extraction method in a preliminary study to blood samples from marine turtles and were able to quantify cytotoxicity, activation of AhR and oxidative stress response and concluded that turtles foraging closer to agricultural areas were associated with higher burdens of mixture effects. Finlayson and co-workers developed several green turtle cell lines (Finlayson *et al.*, 2019a, 2019b) and measured cytotoxicity and oxidative stress in green turtle cells associated with extracts from QuEChERS of green turtle blood from those same sites (Finlayson *et al.*, 2020), concluding that turtles in coastal areas of the Great Barrier Reef were at risk from current concentrations of organic contaminants.

Green turtle blood has also been extracted with silicone and 69–98% of the detected response in the AhR CAFLUX assay could be explained by dioxin-like compounds such as polychlorinated dibenzodioxins, furans and PCBs (Jin *et al.*, 2015a). The samples also activated oxidative stress response in AREc32 but the causative chemicals could not be identified (Jin *et al.*, 2015a). Blood of polar bears also tested positively for AhR activation in the majority of tested samples while estrogenicity was agonistic or antagonistic in a smaller number of samples (Erdmann *et al.*, 2013).

Thyroid-disrupting compounds including various nonylphenol isomers and hydroxylated chlorinated biphenyls were identified in plasma samples of polar bears using EDA and a transthyretin (TTR)-binding assay (Simon *et al.*, 2013).

15.4.2 Tissue

Most studies that have applied bioanalysis of tissue samples in wildlife have worked with acid-treated tissue samples and some have applied passive sampling techniques. Similar test batteries as for water quality testing have been applied with more of a focus on the activation of the AhR and EROD activity because many of the earlier studies focused on persistent organic pollutants and more recently also hormone receptor activation (Jin *et al.*, 2015b).

Suzuki *et al.* (2011) extracted liver and blubber from diverse marine mammals and birds followed by acid-digest clean-up. The samples were run in a broad test battery including AhR, the estrogen receptor (ER), the androgen receptor (AR), the glucocorticoid receptor (GR) and the peroxisome proliferator-activated receptor (PPARγ) and found no activity in these assays for most tissues but they detected dioxin-like activity in AhR and AR antagonistic effects in blubber from Baikal seals. Desforges *et al.* (2017) extracted blubber of killer whales and polar bears and tested the extracts for several immunotoxicity endpoints, with cytotoxicity and strong effects on T cell proliferation and phagocytosis detected.

Other groups have applied silicone to extract organic compounds from the sample directly, without the need to purify the extract any further and thus ensure that the sample composition did not change during the extraction process. Jin *et al.* (2013) found a fairly good agreement between the BEQ_{bio} in the AhR CAFLUX assay from directly analysing blubber extracts of dugongs and the BEQ_{chem} predicted from the detected polychlorinated dibenzodioxins and furans (for definitions of BEQ_{bio} and BEQ_{chem}, see Chapters 7 and 8). Similar extracts tested in addition for their adaptive stress responses only showed activation of the oxidative stress response and the detected dibenzodioxins and furans together with some chlorinated pesticides could not explain the biological effect (Jin *et al.*, 2015c).

It must be noted, however, that even small fractions of coextracted lipids decrease the sensitivity of the *in vitro* bioassays (Reiter *et al.*, 2020). This is caused by decreased bioavailability of the contaminants, as they are retained in the co-dosed lipids in a similar manner as the serum-mediated passive dosing (SMPD) described in Chapter 9. Unlike the effect of SMPD, which is constant, the effect of lipids is diluted with dilution of the sample and eventually disappears. If the chemical burden is high enough, this effect can eventually become negligible with sufficiently large dilution, but caution must be applied for samples with low contamination.

15.5 HUMAN BIOMONITORING

Applications of bioanalytical tools in human biomonitoring are only slowly emerging. With the concept of the exposome (Wild, 2005), which is defined as the totality of exposure to humans over a lifetime, came the awareness of complex mixtures in our bodies and with it the idea to also apply bioanalytical tools to capture mixture effects (Escher *et al.*, 2017).

However, these tools are not widely applied yet. This can partly be explained out of practical necessity that biomonitoring often needs to rely on more easily accessible samples such as urine or hair. Metabolites of pollutants can be analysed in urine and the detected concentrations of metabolites can be translated to human exposure with toxicokinetic models but evidently mixture effects cannot reasonably be measured in those types of samples. The preferred matrices for quantification of mixture effects using bioanalytical tools are blood, as a proxy for the immediate past exposure to chemicals, and lipid tissues, as a measure of long-term storage of persistent organic pollutants.

Most studies applying *in vitro* bioassays have worked with blood or serum, often as part of cohort studies, thereby interrogating association with certain health outcomes. This is beyond the scope of the review in this chapter, for more details the reader is referred to the excellent review by Vinggaard *et al.* (2020), who collated and analysed 43 studies on human tissue that have applied *in vitro* bioassays.

By far the most evaluated endpoint remains AhR activity. This is because many persistent organic pollutants activate this receptor. It is also relevant for many contaminants of emerging concerns, for which metabolic activation and deactivation may also play a role. While AhR activation was mainly tested in blood samples, the activation of hormone receptors has often been probed with fat tissue or placenta extracts (as well as blood). Estrogenic effects and anti-androgenic effects were most prominent and could in some cases be related to adverse health outcomes. Applications with bioassays related to thyroid disruption are scarce but important additions.

Fractionation techniques are generally used to remove endogenous chemicals, in particular hormones, but some studies have also applied extensive EDA. Given the complex mixtures of endogenous compounds present at higher concentrations than the organic micropollutants, EDA seems most promising to ensure that neither false positives nor masking effects impact the bioassay results.

15.6 CONCLUSION

Bioanalytical tools have great potential to capture mixture effects of organic micropollutants in non-aqueous environmental and biota samples. The same bioassays can be applied and a lot of the developments with respect of mapping chemicals to modes of action and toxicity pathways and mixture effects modelling can be directly translated to these other matrices. Iceberg modelling, for instance, has already been applied to sediments (Muz *et al.*, 2020; Niu *et al.*, 2020) and to biota samples (Jin *et al.*, 2015a).

However, sample preparation and sample clean-up are much more challenging than for water samples, especially for biota samples and human specimen, where endogenous chemicals may contribute to the effect. EDA has great potential to not only identify causative chemicals but also help with separation and clean-up without altering the sample itself. Equilibrium passive sampling with polymers also seems very promising, however, samples with trace concentrations of pollutants might be below detection limits as enrichment is limited with this method.

Triggered by the exposome concept and increasing awareness of the relevance of mixture effects in the environment, we can expect tremendous growth in this field in the near future. We fully expect that the third edition of this book may not be called 'Bioanalytical tools in water quality assessment' but 'Bioanalytical tools for environmental monitoring and biomonitoring' (!)

Chapter 16

A promising future for bioanalytical tools

16.1 INTRODUCTION

In the almost 10 years since the publication of the first edition of this book, there have been major developments and innovations in this and related fields that have dramatically improved our ability to apply bioanalytical tools for water quality monitoring and interpret their results.

In this last chapter we provide a summary of the achievements in the field so far, followed by a knowledge gap analysis and an outlook into future research needs and opportunities. We also discuss what work still needs to be carried out for wider regulatory acceptance of bioanalytical tools in water quality assessment.

16.2 ACHIEVEMENTS SO FAR

In the previous chapters we have learnt about the scientific background of bioanalytical tools in the context of human and environmental health and how they can be applied to water quality monitoring. There is now a wide array of case studies in the literature where bioanalytical tools have delivered information on mixtures of organic micropollutants in diverse water types – from pristine water sources to contaminated sites and sewage. In addition to benchmarking water quality, the overall treatment efficacy of a specific water treatment technology or an entire treatment train can be assessed both in the validation and verification stages.

Authors: Beate Escher, Peta Neale and Frederic Leusch
doi: 10.2166/9781789061987_0339

The combination of bioassays with targeted chemical analysis and suspect screening is particularly powerful because they provide complementary information and mixture modelling techniques can make a quantitative link between the different techniques (Escher *et al.*, 2020).

16.2.1 A sound guidance for selection of bioassays based on the conceptual framework of toxicity pathways

The toxicity pathway and adverse outcome pathway concept presented in Chapter 4 and followed throughout the remainder of the book provides an excellent basis for the categorisation and rationalisation of the choice of bioanalytical tools. In addition, the toxicity pathway concept allows us to converge on the commonalities between human health and ecological risk assessment by focusing on cellular effects and responses.

With the diversity of toxicity pathways relevant for human and environmental health, it becomes evident that there is no ‘one size fits all’ bioassay, but rather that a test battery should include several bioassays with at least one from each mode of action category (*i.e.*, non-specific, reactive and specific toxicity). Activation of hormone receptors and endocrine pathways may have adverse effects on reproduction at sub-ng/L concentrations of pollutants that are difficult to quantify with routine chemical analysis due to the low concentrations, but reporter gene assays for hormone receptor activation are highly sensitive and detect even very small quantities of bioactive chemicals.

In addition, adaptive cellular response pathways are important as they constitute an aggregated response to the toxic insult and are more integrative than assessing the individual modes of action. Finally, cytotoxicity provides a sum parameter of all bioactive chemicals in complex mixtures and assures that ‘nothing has been overlooked’. These principles of test battery design are covered in extensive detail in Chapter 13, which relies on a tiered approach of indicator bioassays rather than attempting to measure every receptor and molecular initiating event we know of.

16.2.2 A more comprehensive measure of the realm of organic pollutants

Bioanalytical tools bridge the gap between individual chemical analysis and direct toxicity assessment of water samples. While direct toxicity assessment is very comprehensive in that all components including salts and macropollutants such as phosphate and nitrogen compounds (nitrate, nitrite) are evaluated together (Figure 16.1), it has drawbacks with respect to sensitivity, assessing the causative agents and avoiding artefacts (due to non-chemical factors, such as pH or temperature). Macropollutants and metals can be comprehensively analysed by chemical analysis, but organic micropollutants consist of such a wide range of compounds and their transformation products that chemical analysis alone cannot

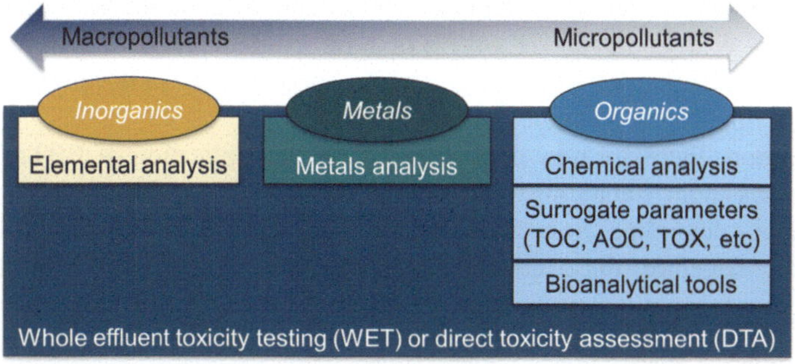

Figure 16.1 Bioanalytical tools as part of a comprehensive assessment of chemical pollution in water. TOC = total organic carbon; AOC = assimilable organic carbon; TOX = total organic halogens.

provide sufficient information and needs to be supplemented by alternative methods that provide sum and surrogate parameters (Figure 16.1). Bioanalytical tools provide a more refined sum parameter to the overall burden of organic micropollutants than chemical surrogates such as total organic carbon (TOC), assimilable organic carbon (AOC), or total organic halogen compounds (TOX).

Another advantage of bioanalytical tools is that bioassay responses are by nature risk-scaled, that is, more toxic chemicals are given more weight in a bioassay than less potent chemicals. A drawback is that the observed effect cannot be directly related to a causative agent, and bioanalytical tools should therefore always be used in conjunction with other techniques such as chemical analysis. 'Iceberg' mixture modelling provides a measure of how much of the measured effect is caused by the detected chemicals and effect-directed analysis (EDA) can be used to identify mixture risk drivers. These concepts are described in more detail in Chapter 13.

16.2.3 Effect-based trigger values

A major change since the publication of the first edition of this book has been the development of several approaches to derive effect-based trigger values (EBTs) for a wide range of both category 1 and 2 assays (see Chapter 13 for details). EBTs are specific for the water type, and might be different for drinking water, surface water and recycled water from non-potable reuse. They all share a common point of departure from safe concentrations for individual chemicals, such as acceptable daily intake (ADI) for humans or environmental quality standards (EQS) for the aquatic environment, but they go beyond single chemicals and include mixtures in their derivation.

The lack of a threshold or trigger level that could be used to determine whether a bioassay response was of concern or not has, understandably, previously been a major limitation to the wider adoption of bioanalytical tools for water quality monitoring. With a larger set of EBTs now available to assess risk to both humans and receiving ecosystems, and various frameworks to develop new EBTs as needed, this step is no longer a limiting issue.

In addition, there are now also frameworks for application of EBTs in water quality assessment (including our synthesised version in Figure 13.15) to help practitioners determine how to adequately respond to exceedances of EBT thresholds. Together, these new developments greatly enhance the usability of bioanalytical tools in water quality assessment.

16.3 CHALLENGES

Despite major advances in the field, there are still knowledge gaps that need to be addressed and require more research efforts. The most important ones in the context of water quality assessment relate to the validity of the sample preparation methods and experimental artefacts from the matrix itself. As cell-based bioassays stand somewhere between chemical analysis and *in vivo* toxicity testing, a quantitative connection between these different tools will help to better understand the performance and limits of bioanalytical tools. However, there is also a need for advances in basic science, with some endpoints and bioassays requiring further development.

16.3.1 Matrix effects and extraction methods

Among the practical limitations are issues of sample preparation as discussed in Chapter 12. This of course is not an inherent problem of the bioassays themselves, but it nevertheless is a problem that has prevented regulatory implementation in the past. Once a water sample is processed and not directly tested, it is unclear how much of the overall toxic potential is actually retained with common extraction and enrichment methods. Direct water testing is only possible with highly contaminated effluents and wastewater, while recycled and drinking water cannot be assessed without prior extraction and enrichment because (a) concentrations of micropollutants are too low to be detected and (b) matrix effects by salts and organic matter may interfere with the bioassays.

Quantifying the extraction yield is a problem that chemical analysis has overcome with the addition of labelled standards, which allow correction for loss during sample preparation. Internal standards should not be used with bioassays because of the difficulty of differentiating between added standard and chemicals originally present in the sample. Instead, bioassay studies have used spiked mixtures and calculated recovery of the mix ('effect recovery') using mixture modelling. While this approach cannot quantify individual sample recovery the way deuterated standards are used for in chemical analysis, it can at least establish the recovery efficiency for a particular sample extraction methodology.

While methods for aqueous sample extraction such as liquid–liquid extraction (LLE) or solid-phase extraction (SPE) have been used for decades, new developments in sorbent materials for SPE and new extraction methodologies need to be routinely evaluated to ensure optimal selection of the extraction technique and parameters for bioassay analysis of a particular matrix. Co-polymer SPE materials such as HLB or HRX can achieve remarkably good effect recoveries. This can be rationalised by the fact that these polymers are almost biomimetic – they efficiently extract chemicals of medium hydrophobicity and charged chemicals that are bioavailable to the cells. They perform more poorly for very water-soluble chemicals and those with multiple charges, but these often show low bioavailability and have a low contribution to mixture toxicity.

16.3.2 Dosing into cell-based bioassays

Dosing into a bioassay might seem straightforward at first sight but can also pose challenges with respect to loss of chemicals and dosing techniques, including the choice of solvent in which the water extract is dosed and if the solvent is removed prior to testing.

High-throughput screening (HTS) bioassays are only amenable to non-volatile chemicals and the typical cut-off of a medium–air partition constant of 10,000 (Escher *et al.*, 2019) comes as a surprise to many toxicologists unfamiliar with concepts of multimedia fate modelling. Loss from the well of a plate is not just related to the volatility of a chemical. One has to keep in mind that the medium components such as lipids and proteins retain semi-volatile chemicals in the assays to a certain extent and they also prevent too much loss due to binding to the polystyrene of the well plates.

Even sealing the plates may not prevent loss of volatile chemicals, as plate sealants must allow for gas exchange necessary for cellular respiration. Hence, different extraction techniques and bioassay set-up need to be used for volatile chemicals, for example, when testing the formation of disinfection by-products in chlorinated drinking water (Stalter *et al.*, 2016c).

Passive dosing techniques have been advocated in the past to overcome the issue of loss processes (Smith and Schäfer, 2016). While passive dosing might be very useful for testing of very hydrophobic chemicals to overcome solubility issues and to obtain effect concentrations expressed as freely dissolved or cellular concentrations, we do not recommend passive dosing for testing of water extracts that also contain many hydrophilic and charged chemicals. Furthermore, in cell-based assays that are run with protein-rich media, the medium itself serves as a passive dosing device that counteracts the depletion due to cellular uptake and loss processes (Fischer *et al.*, 2019).

There has also been much speculation about solvent toxicity and what solvent to use for dosing of water extracts. It has been widely reported that methanol and other volatile solvents have a lower toxicity than DMSO. This is true, but most of this difference is likely because volatile solvents are lost during the incubation period, especially for human cell assays that are incubated at 37°C. If solvents are lost anyway – and not in a predictable and reproducible fashion – why not remove the solvent right away prior to testing? Exchanging from a volatile solvent into medium is fairly simple for water samples but might be more challenging for extracts from tissues and sediments that contain more hydrophobic chemicals. The only difficulty with using volatile solvents is that they evaporate – even during storage at low temperatures. Therefore, vial weight needs to be continuously tracked during the experimental workflow and volatile solvents need to be topped up to initial weight, which seems laborious at first sight but can be automated and is worth the effort, given the subtle and overt toxic effects of DMSO as the most popular alternative. There are cases where it will be near impossible to replace DMSO, for example, if realistic mixtures of many chemicals are formulated or when hydrophobic chemicals are tested but it is important for the reader to keep in mind that there might be alternative approaches to 'what we have been doing the last twenty years'.

16.3.3 Linking bioanalysis with chemical analysis

All of our current risk/safety assessment and guidelines are based on single chemical testing, and all our regulations are written in this single chemical language. To move forward towards a more realistic mixture assessment without making our considerable single chemical toxicity information irrelevant, it is important to ensure that we understand how to correlate bioanalytical measures with chemical analysis. The scientific underpinning of mixture integration was discussed in Chapter 8, and there has been significant progress in linking chemical and bioassay results quantitatively over the past decade (see Chapter 13).

By definition, category 1 bioassays are activated by only a small number of highly potent chemicals and most water samples tested with these assays show good agreement between predicted and measured bioassay response, particularly when applied to relatively well-understood matrices. For these category 1 assays, EDA is also a way to identify causative chemicals. For those types of endpoints, bioassays and chemical analysis could be used interchangeably as both can in principle provide the same information but it remains beneficial to apply both techniques in parallel. Especially very potent chemicals might fall below individual detection limits but still contribute to the mixture effects and bioassays can give direct responses without the need for extensive mixture modelling. Further, applying both techniques gives more confidence that many chemicals present at low concentrations act indeed according to the easily predictable models of concentration addition and independent action (which actually yield the same results at very low effect levels, typically $<10\%$).

It is an urban myth that water samples might contain highly synergistically acting mixtures. Synergy is of academic interest and exploited for pesticide formulations –

mostly by inactivating detoxifying systems so that the synergy is rather a toxicokinetic effect than a true synergistic toxicodynamic interaction). Even if a small fraction of mixture components interacts synergistically, this will hardly affect the mixture effect of the entire water sample that contains thousands, potentially even hundreds of thousands of individual chemicals.

Category 2 bioassays are an entirely different story. There are many chemicals that activate these assays, and it is not possible to explain the mixture effects by the detected chemicals. EDA usually just results in every fraction being active or loss of activity in all fractions but no clear mixture effect drivers, unless the samples are from a contaminated site or an accident. While we do not know what the unknown active chemicals are, we can at least quantify how much of the effect is unknown: in a water sample the fraction of unexplained effect can be as high as 90–99% of the total effect quantified with a category 2 bioassay. This means that, for these endpoints, we need bioassays to complement the information of chemical analysis and that component-based prediction of the mixture effects from the identified chemicals will never close the 'effect' balance. This observation has actually been misused as an argument against these bioassays – some researchers and stakeholders reject category 2 bioassays because their effects cannot be explained by the detected chemicals. The opposite should be the case. If we can assure that a category 2 bioassay is describing a molecular initiating event or a key event that is linked to an adverse outcome, then the mixture effect is of biological relevance, whether we know what chemicals trigger it or not – provided of course that the bioassays are run to high QA/QC standards, that sample preparation is of high quality, and that there are no matrix artefacts. Then 'iceberg' modelling of effects observed in category 2 bioassays will provide a measure of chemical burden that chemical analysis hitherto overlooked entirely but may contribute significantly to the risk of mixtures.

Mathematical models have also been developed to meaningfully link category 2 bioassay responses to chemical guidelines. Additional studies will add to the weight-of-evidence needed to convince risk assessors and regulators that chemical and bioassay analysis complement each other.

16.3.4 Linking bioanalysis with whole-animal testing

Understanding how predictive *in vitro* bioassays are of whole organism effects is crucial to answer the 'so what?' question. The adverse outcome pathway provides the logical framework to connect the dots, but much research is still needed to connect *in vitro* bioassays with adverse outcomes in a quantitative manner.

Adverse outcome pathways are not linear, but they are actually networks, therefore there is no one-to-one relationship between one *in vitro* bioassay and the adverse outcome but often several endpoints need to be combined to predict the adverse outcome.

The development of alternative test methods for risk assessment of chemicals has led to a paradigm shift from exclusive reliance on *in vivo* toxicity testing to a process that relies on *in vitro* bioassays in conjunction with (quantitative) *in vitro* to *in vivo* extrapolation (QIVIVE) models (discussed in more detail in Chapter 9).

The validation of these alternative test methods can be used as the basis for application to water samples. Validated *in vitro* systems with a clear anchor and effect thresholds for adversity through inverse QIVIVE modelling can be applied for monitoring complex mixtures in the environment and for improving the derivation of EBTs. To adapt single chemical QIVIVE models to mixture QIVIVE models will most likely be possible in the future for defined mixture but might remain out of reach for complex environmental samples of unknown composition. Ultimately, it may be sufficient if we can make the connection for a large number of individual chemicals to extrapolate eventually to environmental mixtures.

16.3.5 Bioassays that require further development

As identified in Chapter 5 some endpoints are currently poorly modelled using *in vitro* methods, particularly if they rely on meta-cellular and integrated organism responses. Reproductive and developmental toxicity are two classical examples for which there is currently no comprehensive *in vitro* model. The development of pluripotent stem cells from non-embryonic tissue removes the ethical barrier of using embryonic stem cells, while advances in microfluidics leading to 'animal-on-a-chip' platforms (see Section 16.3.6) may enable future development in simulation of integrated organism responses at the *in vitro* level, although much work is still needed before this occurs.

While there are several good high-throughput bacterial assays to assess mutagenicity and genotoxicity, it is important to remember that there are significant differences in the organisation and protection of DNA between prokaryotes and eukaryotes. For example, DNA in eukaryotes is located inside the nuclear envelope while prokaryotic DNA is located in the cytoplasm, or eukaryotic DNA is bound to proteins and scaffolded into condensed chromatin while prokaryotic DNA is naked and unwound. There may therefore be differences in sensitivity of prokaryotes and eukaryotes to mutagenic and genotoxic chemicals. While there are options for eukaryotic DNA toxicology such as the micronucleus and Comet assays, they tend to be cumbersome and low throughput. With the advent of automated microscope imaging suitable for well plate formats and artificial intelligence-enabled image analysis, it is becoming possible to conduct these generally low-throughput methods in significantly higher throughput formats, which would make them suitable for application to water quality monitoring.

And there are a range of other endpoints where bioassays are currently limited, including activation of the constitutive androstane receptor (CAR) and assays for neurotoxicity, just to name a few. New developments would enable wider application of these assays to water quality testing.

We also need to keep in mind that most mammalian cell-based bioassay and reporter gene assays are derived from undifferentiated cancer cells. This is because primary cells survive only for a few cell cycles before they die off *in vitro*. Primary cells need to be constantly sourced from new tissue, resulting in ethical and reproducibility problems. In contrast, cancer cells are immortal and can be kept in a stable state for a very long time but are by definition mutants and can sometimes be very different from primary cells. Several innovations have enabled the creation of immortal non-cancerous cell lines, though these have not yet been widely used as platforms to develop reporter gene assays. Future work should be directed towards combining these emerging technologies of cell immortalisation with the development of novel bioassay.

16.4 FUTURE OPPORTUNITIES

High-throughput toxicogenomics, three-dimensional cell systems and organism-on-a-chip technology, and adaptation to online monitoring format offer future opportunities for adoption into water quality assessment once the immediate research gaps are closed.

16.4.1 The 'omics' revolution

The 'omics' revolution, the exciting developments in genomics, transcriptomics, proteomics and metabolomics have the potential to be brought into practical application. Technological advances have made it possible to run some of these in a high-throughput format using *in vitro* exposures in 96- and 384-well plate format, and they are increasingly used in chemical screening and screening-level risk assessment (Harrill *et al.*, 2019). The field is being developed from qualitative to quantitative assessment and from single compounds to mixtures (Spurgeon *et al.*, 2010), but some harmonisation and methodological refinements are still necessary to reduce assay variability (Sauer *et al.*, 2017). Once assay protocols are refined to reduce variability and standardised methods are worked out to assess and evaluate mixtures, it should be possible to also move on to complex and unresolved mixtures as they occur in water samples. There are still many questions to be solved – the most pertinent being 'what does it mean if a gene is transcribed x-fold more than in the control?' or 'how does a change in metabolite x affect the cell and the organism?', which relates to the issue of *in vitro* to *in vivo* extrapolation debated in Section 16.3.4.

16.4.2 Three-dimensional cell models and organ- and animal-on-a-chip systems to better model whole organism response

Undifferentiated cells grown artificially in a two-dimensional monolayer on plastic surfaces optimised for cell culture, as commonly used in *in vitro* toxicity testing, are clearly not fully representative of the *in vivo* environment. In particular, they lack interaction with the extracellular matrix and more complex inter-cellular communication, which is often vital in the function of organs. Therefore, three-dimensional cell models have been developed to simulate a variety of organs such as the small intestine, kidney tubules, bronchioles, liver, blood–brain barrier, lung alveoli and bone marrow (reviewed in Jensen and Teng, 2020).

The colon cancer cell line Caco-2 is a widely used *in vitro* model for intestinal uptake and is especially important in the pharmaceutical industry for testing the oral availability of pharmaceuticals. If grown on a microporous membrane, Caco-2 cells are capable of forming a three-dimensional epithelial barrier that expresses tight junctions and can support active and passive uptake processes (Cencic and Langerholc, 2010).

Tests for skin sensitisation by chemicals and permeation through skin are typically performed in three-dimensional (3D) reconstructed human epidermal models, which consist of several cell types and a dermal matrix. These 3D skin models can simulate many skin functions including barrier and immunological functions.

A disadvantage of many 3D cell models is that they are essentially not 'true' *in vitro* systems, but rather *ex vivo*, because they require primary cell lines or tissues from animals that are isolated from a living organism and cannot be maintained in culture for long periods. While they are becoming powerful tools in the earlier tiers of drug development pipelines (Langhans, 2018), application of 3D cell culture to HTS required for environmental monitoring applications remains a challenge.

By etching microscopic channels and microfluidic engineering, researchers have developed organ-on-a-chip platforms that mimic the function of various organs (*e.g.*, spleen, lung, liver, kidney, heart, gut, blood–brain barrier), going as far as developing body-on-a-chip (also known as animal-on-a-chip) platforms by including cells from multiple organ systems to integrate organ interactions (reviewed in Zheng *et al.*, 2016). Microfluidics platforms can also be combined with 3D cell models to produce advanced organ-on-a-chip models to investigate organ-level function (Shoemaker *et al.*, 2020). Chip-based methods are also available for ecotoxicological test systems (Campana and Wlodkowic, 2018).

Like 3D cell models, these new developments greatly enhance the ability of *in vitro* systems to mimic whole organism responses, but they are not yet compatible with HTS application.

16.4.3 Moving from offline to online monitoring

Surveillance monitoring requires changes in chemical pollutants or accidents to be detectable immediately, ideally using on site and online monitoring. Most bioanalytical tools are presently applied only offline, that is, water samples are taken and transported to the laboratory for processing and analysis. The bioassay then requires cells to be exposed for several minutes to several days, with 24 hours the most common exposure period. Since many commonly applied cell lines are genetically modified, they can only be maintained in certified facilities with secure quarantine and containment. The results are therefore obtained at best with a delay of 24 hours, usually more. Automation using online extractions and automatic HTS is technically possible as developments in analytical chemistry and pharmaceutical drug discovery as well as in the Tox21 programme have demonstrated. It should therefore be possible to shorten the interval between collecting a sample and receiving the analysis results and ensure that bioanalytical methods fulfil their potential in surveillance monitoring. There are already some online monitoring systems available to monitor toxicity to bacteria (*e.g.*, iTOXcontrol, Microtox CTM and NitriTox systems) and algae (*e.g.*, AquaSentinel, Algae Toximeter II) as well as whole-cell biochips for online monitoring (Elad and Belkin, 2012), now also for mobile phone applications (Lu *et al.*, 2019). Vertebrate cells require surrounding media with serum (including growth factors, amino acids and nutrients to feed and sustain the cells), and online monitoring with vertebrate cells is therefore significantly more challenging.

It is worth noting, however, that offline and online monitoring is not as binary as one might think. Indeed, in addition to 'offline' (sample taken manually, analysis in off-site laboratory) and 'online' monitoring (fully automated sampling, analysis on site), there is also 'at-line' monitoring (manual sampling, but analysis on site) and 'inline' monitoring (where a probe is placed directly in the process stream). Continuous monitoring of water quality requires at least at-line monitoring – and at-line bioanalytical monitoring is already possible. Using electric cell-substrate impendence sensing (ECIS), U.S. Army researchers have established and validated a field-portable drinking water toxicity sensor to monitor acute toxicity to rainbow trout gill epithelial cells as a way to monitor drinking water quality (Widder *et al.*, 2015). A variety of other sensors are also now becoming available, widening the field of the types of cells that can be used in at-line monitoring applications (Tan and Schirmer, 2017).

16.4.4 Towards ultra-high-throughput testing, multiplex assays and artificial intelligence-assisted bioinformatics

In their aim to test thousands of environmental chemicals in >1000 different assays, programmes such as ToxCast and Tox21 have deployed testing platforms that use 1536-well plates and low volume acoustic pipetting robots capable of accurately dispensing volumes as small as 2.5 nL. These tools create incredible opportunities for ultra-HTS of environmental samples.

Multiplex assays, where multiple effects are monitored simultaneously, would also offer a way to increase testing throughput by creating assays that can measure multiple responses at once. To a certain extent, most assays are already multiplexed: indeed, combining cell viability and cell vitality assessment for cytotoxicity with a reporter gene assay already provides two streams of toxicity information. But new developments in fluorophores, luciferase enzymes and emission spectra deconvolution open up avenues for highly multiplexed assays that retain the sensitivity of reporter gene assays.

Advancements in imaging technology have also made it possible to read out more from existing bioassays: phenotypic profiling, which is already popular for fish embryo toxicity assays can also be applied to cell-based assays. A study that applied multiparameter phenotypic profiling in MCF-7 showed how the size and structure of cells is related to biological processes like cell growth, death and communication and applied these tools to testing environmental waters (Wang *et al.*, 2018).

Ultra-HTS, more reliance on imaging techniques and assay multiplexing would dramatically increase our capacity to test the toxicity of chemicals, chemical mixtures and environmental samples, including water samples – but they also require increasingly complex bioinformatics workflows, often including a degree of artificial intelligence support. The risk is that bioassay analysis becomes a complex black box, which could lead to errors if improperly designed. The future of cell-based toxicity testing is bright, but we need to stride forwards with open eyes and critical minds.

16.5 THE ROAD TO REGULATORY ACCEPTANCE

At present, bioanalytical tools are popular research tools and while their development and validation are by no means fully accomplished, we believe it is time to move them forward to regulatory applications. Regulators and proponents of the water industry all agree that we need to overcome the limitations of a chemical-by-chemical approach. In their 'Chemicals Strategy for Sustainability towards a Toxic-free Environment', the European Commission has clearly stated that 'scientific consensus is emerging that the effect of chemical mixtures needs to be taken into account and integrated more generally into chemical risk assessments' (European Commission, 2020). What is preventing this crucial next step in the implementation of bioanalytical tools for water quality assessment?

We did not wait for the development of advanced chemical tools such as LC-MS/MS before we decided to define chemical-based water quality criteria. The definition of guideline values has actually boosted rapid advances in the field of chemical analysis. Likewise, we should not wait until we can produce the 'perfect' bioassay battery but instead consider its application today and start gaining more practical experience about applicability and limitations. We should trust in the fundamental principles of physical chemistry and biology that this field of science is based on, and not require everything to be demonstrated experimentally *ad nauseam*.

The most important benefit of bioanalytical tools is that they are risk-based measures, that is, a more toxic or more potent chemical contributes more to a bioassay response than a less potent chemical – just like it does in whole organisms. Bioanalysis therefore has huge potential as a prioritisation and monitoring tool. In a tiered approach bioanalytical tools could be used for initial screening and hazard identification, where only samples that exceed a given threshold such as an EBT go into more detailed evaluation (Figure 16.2). In a world of limited resources, this would allow a risk-based prioritisation of samples for complex chemical analysis.

The fundamental question then becomes 'where do we set that threshold?'. There has been significant innovation in this field in the past decade, and several approaches to derive EBTs are discussed in Chapter 13 (Section 13.5). Once a threshold has been set, the question becomes 'what to do if the threshold is exceeded?'. Again, the last decade has seen several approaches to respond to this question, and we summarised our current thinking in Chapter 13 (Figure 13.15). These two key questions have in the past been key to regulatory acceptance.

The State of California leads the world at the moment as the first to implement bioassays in the Water Quality Control Policy for Recycled Water (State Water Resources Control Board, 2019). The policy recommends a trigger level of 3.5 ng/L EEQ in an ERα reporter gene assay and 0.5 ng/L TCDD EQ in an AhR reporter gene assay. Other guidance documents recommend the application of bioassays for assessing the presence and possible risks associated with chemicals in water, although they do not go as far as recommending a trigger value. These include the Guidance for Producing Safe Drinking-Water (Section 3.3 in WHO, 2017c) and the Australian Guidelines for Water Recycling – Augmentation of Drinking Water Supplies (Section 4.5.1 in NRMMC/EPHC/NHMRC, 2008).

Water Safety Plans (WSPs) were developed by the World Health Organisation (WHO) and offer an internationally accepted approach to ensure the safety and acceptability of drinking water supplies, with a focus on hazard prevention

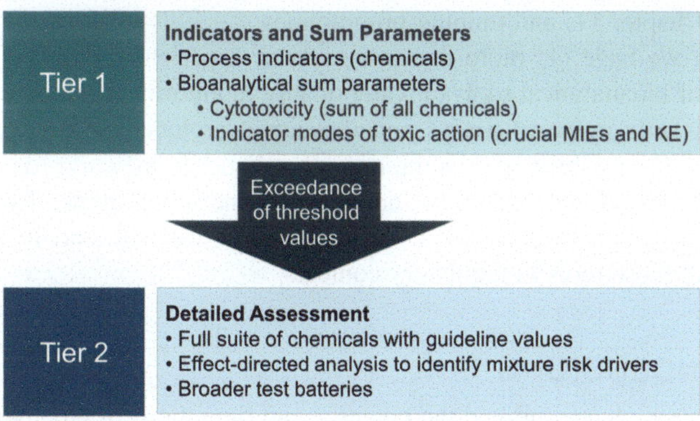

Figure 16.2 A tiered approach for water quality assessment with respect to chemicals. MIE = molecular initiating event; KE = key event.

Purpose	Monitoring within WSP	Application	WSP module
Benchmark water quality	System assessment monitoring	EBM used to characterize water quality. Compare with EBTs.	Module 2 Module 3 Module 10
Treatment process efficiency	Operational monitoring	EBM used to assess if treatment processes are working now. Continuous, rapid assessment needed. Compare with EBTs.	Module 6
	Verification monitoring	EBM used to verify routine monitoring. Compare with EBTs.	Module 7
Understanding processes (*e.g.*, critical steps)	Validation monitoring	EBM used to understand and validate critical processes.	Module 4 Module 5

Figure 16.3 Common purposes of effect-based monitoring (EBM) and how they fit into monitoring within Water Safety Plans (WSP) (Bartram *et al.*, 2009). EBT = effect-based trigger value.

(Bartram *et al.*, 2009). WSPs often provide the backbone, on which national drinking water guidelines are built, and understanding how to integrate bioanalytical tools in WSPs could help regulators appreciate how to incorporate bioanalytical tools in their water quality management frameworks. WSPs consist of 11 modules that systematically articulate a framework to characterise, monitor and manage drinking water quality. Bioanalytical tools would logically fit into seven of those modules (Figure 16.3), and a further two (modules 8 and 9) would benefit from the development of bioanalytical standard operating protocols (SOPs; see Chapter 11) and training programmes.

Now that we have (1) more than a decade of experience with the systematic application of bioanalytical tools to water quality monitoring, (2) established EBT values for a wide range of assays and developed protocols to derive EBTs for new assays, (3) established a simple and realistic framework to respond to exceedances of EBT values and (4) are seeing some initial uptake by regulatory bodies such as the California EPA, it is likely that we will start to see greater acceptance of these tools in regulatory contexts.

16.6 CONCLUSIONS

In this chapter we have outlined the potential and limitations of bioanalytical tools and have dreamt about possible future development. We hope that the review of the state-of-the-science and applications as presented in this book will help to encourage fellow researchers to persevere and progress bioanalytical approaches from research tools to practical application, give practitioners in (waste) water treatment recommendations on how to apply *in vitro* bioassays in their daily business and help regulators to implement them as regulatory tools.

Finally, we conclude on an optimistic notion, noting that we have seen tremendous progress in this field since the publication of the first edition of this book almost a decade ago, and we hope that the case studies that we are providing in this book will be outdated very soon by new advances in this field!

Glossary

AA-EQS: annual average EQS (→ EQS).

ABC transporters: ATP binding cassette transporters, family of drug transporters involved in active transport of organics (→ active transport).

Abiotic: a process that does not involve living organisms, for example, 'abiotic transformation reactions' take place via physical and chemical processes rather than via biological processes (→ biotic).

ACB: → activity concentration at baseline (benchmark value used by Tox21).

ACC: → activity concentration at cutoff (benchmark value used by Tox21).

Acceptable (or allowable) daily intake (ADI): the maximum amount of a chemical that can be ingested daily over a lifetime with no appreciable health risk (→ tolerable daily intake, → reference dose).

Acetylcholinesterase (AChE): an enzyme that catalyses the hydrolysis of the neurotransmitter acetylcholine.

Active transport: transport of molecules over cell membranes that requires energy (*e.g.*, ATP) (→ ABC transporters, → passive transport).

Activity concentration at baseline (ACB): concentration in a → concentration response curve that is associated with the three times the median absolute deviation of the effects of the two lowest concentration tested (benchmark value used by Tox21).

Activity concentration at cutoff (ACC): concentration in a → concentration response curve that is associated with a defined response cutoff, for example, 15%.

Acute exposure: exposure over a short period of time (hours or days) (→ chronic exposure).

Authors: Beate Escher, Peta Neale and Frederic Leusch
doi: 10.2166/9781789061987_0355

Adaptive stress response pathway: the cellular pathway for stress responses induced by chemicals and other stressors.

ADI: → acceptable (or allowable) daily intake.

ADME: absorption, excretion, distribution and metabolism (→ toxicokinetics).

Adverse outcome pathway (AOP): conceptual framework that leads from the initiating event of interaction between a toxicant and a receptor in an organism from cellular and organ response to an adverse outcome at organism- or population level (→ toxicity pathway).

ADWG: Australian Drinking Water Guidelines (NHMRC, 2011).

Aflatoxin: a type of mycotoxin (fungal toxin).

Agonist: chemical that mimics the action of a natural substrate, for example, binds to a receptor of a cell and triggers the natural response of that cell (→ antagonist).

AGWR: Australian Guidelines for Water Recycling (NRMMC/EPHC/AHMC 2006; NRMMC/EPHC/NHMRC 2008, 2009a, 2009b).

AhR: aryl hydrocarbon receptor (also termed dioxin receptor). The AhR is involved with the induction of cytochrome P450 monooxygenase and induced by dioxins and dioxin-like chemicals such as PAHs.

AhR CAFLUX: the AhR chemically activated fluorescence expression assay is a reporter gene assay for detection of dioxin-like activity in water samples.

***Aliivibrio fischeri*:** marine bioluminescent bacterium used in the Microtox assay (→ Microtox) formerly called *Vibrio fischeri*.

Ames test: assay for mutagenicity that measures the ability of toxicants to mutate a histidine-dependent strain of the bacterium *Salmonella typhimurium* to grow on a histidine-deficient substrate.

Anaemia: blood deficiency.

Androgen: natural or synthetic hormones including testosterone that regulate development and maintenance of masculine characteristics via the androgen receptor (→ AR).

Antagonist: chemical that blocks the action of an agonist (→ agonist).

Antibody: protein used by the immune system to detect and neutralise a foreign substance (*e.g.*, microorganism).

Antigen: foreign agent (*e.g.*, bacterium, virus) that triggers the production of an antibody (→ antibody).

Antimitotic drug: a drug used to fight cancer that inhibits cell division by interfering with mitosis.

AOC: assimilable organic carbon.

AOP: → adverse outcome pathway.

Apical endpoints: traditionally measured outcomes of toxicity in whole organisms, for example, lethality or reproductive failure.

Apoptosis: programmed cell death (as opposed to unplanned cell death, → necrosis).

AR: androgen receptor, important for male sexual development and reproduction and induced by natural and synthetic androgens (→ androgen).

ARE: antioxidant response element.

Aromatase: an enzyme in the → CYP family that is important for biosynthesis of estrogens and hence, for sexual development.

Assay: procedure in toxicology for testing the activity of a chemical on a biological system (cells, organisms, populations).

ATP: adenosine-5'-triphosphate, a multifunctional nucleotide that transports energy within cells.

Autoimmune disorder: a disease whereby the immune system starts attacking an organism's own tissues.

AWTP: advanced water treatment plant.

Axon: the projection of a neuron that conducts electrical impulses.

B[a]P: benzo[*a*]pyrene.

Basal toxicity: term for baseline toxicity in human toxicology (→ baseline toxicity).

Baseline toxicity: the minimal toxicity exhibited by any compound. Partitioning of chemicals into biological cell membranes causing non-specific disturbance of membrane integrity and functioning (→ narcosis, → basal toxicity).

BCF: bioconcentration factor.

BEC: → biologically effective concentration.

BED: → biologically effective dose.

Benchmark dose (BMD): dose associated with a specified level of response, the → benchmark response.

Benchmark response (BMR): certain response level from which → BMD are derived, typically 10%.

BEQ: → bioanalytical equivalent concentration.

Bioaccumulation: accumulation of chemicals within organisms exposed via the surrounding environment (*e.g.*, air, water, soil, sediment, food) (→ bioconcentration).

Bioactivation: biological activation of a chemical, that is, biological transformation that produces a metabolite that is more toxic than its precursor.

Bioanalytical equivalent concentration (BEQ): the concentration of a reference chemical that would elicit the same effect as the mixture of micropollutants in a water sample. BEQ differ from → TEQ in that effects on the level of a → MIE or → KE are assessed, which may or may not result in an → AO, while TEQ refer to toxicity, that is, an → AO.

Bioanalytical tool: cell-based or low-complexity *in vitro* bioassay indicative of a specified endpoint relevant for human and/or environmental health.

Bioavailability: refers to fraction of a chemical that can be taken up into cells.

Bioconcentration: accumulation of chemicals in aquatic organisms from the surrounding water via gills, skin and carapace. Bioconcentration does not include dietary accumulation and/or other non-water sources (→ bioaccumulation).

Biodegradation: biological degradation of chemicals by microorganism (*e.g.*, bacteria).

Biological pathway altering concentration (BPAC): effective concentration in an *in vitro* assay indicative of a specific toxicity pathway.

Biologically effective concentration (BEC) or biologically effective dose (BED): the BEC or BED is the amount of a toxicant that reaches cells, sites or membranes where adverse effects occur. The BEC or BED may represent only a fraction of the delivered concentration, but it is the most appropriate measure of exposure for predicting adverse effects.

Biomarker: biochemical, physiological or histological response of a cell or organism to a stressor that can be easily measured and used as a measure of exposure or effect.

Biotic: a process involving live organisms, for example, a 'biotic transformation reaction' takes place via biological processes as opposed to physical and chemical processes (→ abiotic).

Biotransformation: metabolic transformation of chemicals within cells and organisms.

BMD: → benchmark dose.

BMR: → benchmark response.

BPAC: → biological pathway altering concentration.

BTEX: benzene, toluene, ethylbenzene and xylene, a group of aromatic hydrocarbons.

CA/DA: concentration addition/dose addition (→ concentration addition).

CA: → concentration addition.

CALUX: the chemically activated luciferase gene-expression group of assays comprise cell lines for detection of receptor-mediated activity for several nuclear receptors, for example, → ER (*i.e.*, ER-CALUX), → AhR, → AR, → GR, → PR, → TR.

CAR: the constitutive androstane receptor is involved with protection of toxicity induced by bile acid and regulation of physiological functions.

Carbamate pesticides: class of insecticide related to organophosphate insecticides that likewise interfere with the enzyme → actylcholinesterase (AChE) at neuronal synapses and neuromuscular junctions.

Carcinogenesis: formation of cancer cells, that is, the initiation of cancer.

Carcinogenicity: the mode of toxic action for cancer, which can be caused by direct DNA damage (genotoxicity) or interference with gene regulation (epigenetic) (→ epigenetic carcinogen, → genotoxicity).

Carcinoma: malignant tissue growth (→ tumour).

Cardiomyocyte: cardiac muscle cell.

Cardiovascular toxicity: toxicity to the heart and vascular system (*i.e.*, blood vessels).

Catabolism: metabolic breakdown of endogenous and exogenous molecules.

Chemical-group motivated approach: test battery design approach that prioritises the detection of a specific group(s) of similarly acting chemicals rather than the protection goal (→ protection-goal motivated approach).

Chronic exposure: exposure over a longer time period compared to acute exposure (weeks to years) (→ acute exposure).

Comet assay: assay that detects DNA strand breaks in different cell types including human, other mammalian and fish cells. Also known as → single cell gel electrophoresis (SCGE) assay.

Concentration addition (CA): quantitative model for mixture toxicity of chemicals that act via the same target site and/or the same mode of toxic action, that is, the individual toxicities (expressed as effect concentrations) add up to the sum toxicity of the mixture. CA is applied in aquatic ecotoxicology, whereas dose addition (DA) is used in mammalian toxicology (→ independent action).

Concentration: the mass or molar amount of a chemical divided by the volume of the test system.

Concentration–response curve (CRC): a plot of the degree (%) of response observed in a test population against increasing exposure concentration (also known as dose–response curve for *in vivo* mammalian testing).

Confidence interval: the 95% confidence interval, for example, is the range of values between which 95% of the data will fall into.

Conjugate: adduct between a foreign chemical and a small hydrophilic biomolecule (*e.g.*, glucoronide or sulphate), catalysed by a metabolic enzyme, typically in the liver.

Conjugation: a phase II metabolic reaction that involves the addition of a hydrophilic biomolecule to a xenobiotic in order to form a (more) water-soluble product (conjugate) for excretion.

CRC: → concentration–response curve

CVr: repeatability coefficient of variation.

CVR: reproducibility coefficient of variation.

CWA: U.S. Clean Water Act.

Cyanotoxin: a type of toxin produced naturally by cyanobacteria (blue-green algae), which can form toxic blooms in both fresh- and marine water.

CYP: → cytochrome.

CYP450: → cytochrome P450.

Cytochrome: redox active proteins bonded to a heme, involved as electron transfer agents in many metabolic pathways, e.g., → cytochrome P450.

Cytochrome P450: a monooxygenase enzyme superfamily involved in the metabolism of endogenous and xenobiotic compounds.

Cytokine: cell-signalling molecule.

Cytoreductive drug: a drug than can reduce cell number (*e.g.*, cancer cells).

Cytotoxicity: toxicity of a chemical or other stressor to living cells.

DA: → dose addition.

DBP: → disinfection by-product.

DF: → dosing factor.

Dioxin: colloquial for → polychlorinated dibenzodioxin.

Direct toxicity assessment (DTA): term commonly used in Australia to describe whole effluent toxicity (WET) testing (→ whole effluent toxicity testing).

Disinfection by-product (DBP): chemical formed reactions between organic and inorganic matter during chemical disinfection of, for example, drinking water and pool water, including trihalomethanes (THM) and haloacetic acids (HAA).

Dithiocarbamate pesticides: a type of carbamate pesticide (→ carbamate pesticide).

Diuron: herbicide that acts on photosynthesis by inhibition of photosystem II (→ photosystem II).

DMSO: dimethyl sulphoxide, a solvent.

DNA: deoxyribonucleic acid, the macromolecule that encodes the genome and carries all the genetic information in living organisms.

DOC: dissolved organic carbon.

Dose addition (DA): same concept as → concentration addition, however, DA is used in mammalian toxicology.

Dose–response assessment: relationship between dose or level of exposure to a substance, and the incidence and severity of an effect (→ dose–response curve).

Dose: total quantity of a chemical delivered to a test animal or test system.

Dose–response curve: a plot of the degree (%) of toxicological response observed in a test population against increasing exposure dose (→ concentration–effect curve for *in vitro* testing).

Dosing factor (DF): dosing factor in a bioassay, the ratio of the volume of an extract dosed into a well of a microtitre plate divided by the total volume of the medium in the well.

DTA: → direct toxicity assessment.

DWEL: drinking water equivalent level, a lifetime exposure concentration protective of adverse, non-cancer health effects, assuming that all of the exposure to a contaminant is from drinking water (based on → acceptable (or allowable) daily intake).

DWTP: drinking water treatment plant.

EBT: → effect-based trigger value.

EC: effect concentration.

EC_{50}: effect concentration causing 50% of maximum effect in cell-based assays.

Ecological Risk Assessment (ERA): process of estimating potential impact of a chemical, biological or physical agent on a specific ecological species, population or ecosystem (→ human health risk assessment).

ECVAM: European Centre for the Validation of Alternative Methods.

EDA: → effect-directed analysis.

EEQ: 17β-estradiol equivalent concentration.

EF: enrichment factor or → extraction factor.

Effect unit (EU): Effect unit from chemical analysis (EU_{chem}) is the ratio of concentration of a chemical concentration divided by its effect concentration, and effect unit from bioassays (EU_{bio}) is the inverse of the effect concentration of a mixture or environmental samples (1/EC) (→ EC).

Effect-based trigger value (EBT): acceptable effect level or → BEQ in a specific water type (surface water, drinking water)

Effect-directed analysis (EDA): identification of mixture effect drivers by fractionation of a complex water sample and testing the effect in the fraction in several iterations until the causative agent is identified (→ toxicity identification evaluation).

ELISA: enzyme-linked immunosorbent assay, an immunological assay technique that relies on an enzyme bonded to a particular antibody (→ immunoassay).

Endocrine disrupting compound (EDC): a chemical capable of modifying natural hormone function (→ endocrine disruption).

Endocrine disruption: interference with the endocrine (hormone) system.

Endogenous: originating from within, that is, endogenous substances are produced within an organism or cell. Endogenous processes occur within an organism or cell and involve endogenous substances.

Endothelium: the inner cell layer of blood vessels or organs.

Endotoxin: bacterial toxin.

Endpoint: an observable or measurable biological event used as an indicator of effect.

Enrichment factor: → extraction factor.

Environmental Risk Assessment: → Ecological Risk Assessment.

Enzyme: an endogenous protein that catalyses (*i.e.*, increases the reaction rate of) chemical reactions.

Epidemiology: the study of disease clusters in human populations and the attempt to link these to human exposure to chemicals.

Epigenetic carcinogen: a chemical that induces cancer via interference with gene regulation mechanisms, as opposed to genotoxic carcinogens that induce cancer via direct damage to DNA structure (→ carcinogenicity, → genotoxicity).

Epithelial cell: skin cells or cells that form the outer layer of organs within an organism.

EQS: Environmental Quality Standard.

Equitoxic concentration: chemical concentrations of different chemicals that cause the same level of toxicity.

ER: estrogen receptor, important for female sexual development and reproduction. The ER is modulated by natural estrogens and xenoestrogens (→ estrogen, → xenoestrogen).

ERA: → Ecological Risk Assessment or → Environmental Risk Assessment.

ER-CALUX: assay for detection of estrogenicity in water samples (→ CALUX).

E-SCREEN: assay for detection of estrogenicity, which is based on cell proliferation in estrogen-dependent (MCF-7) cancer cells (→ MCF-7).

Estradiol (E2): a natural → estrogen (female sex hormone).

Estriol (E3): a natural → estrogen (female sex hormone).

Estrogen (or oestrogen): group of female sex hormones including three naturally occurring steroidal hormones (estrone (E1), estradiol (E2) and estriol (E3)) and synthetic compounds including 17α-ethinylestradiol (EE2) (→ ER, → estradiol, → estriol, → estrone, → 17α-ethinylestradiol).

Estrogenicity: toxic mode of action caused by estrogenic activity, for example, binding to the estrogen receptor (ER).

Estrone (E1): a natural → estrogen (female sex hormone).

17α-Ethinylestradiol (EE2): potent synthetic estrogen, active ingredient of birth control pills.

EU: → effect unit.

Eukaryote: an organism that has its DNA contained within the nucleus, that is, most living organisms except some bacteria (→ prokaryote).

Eutrophication: when a water body receives a surplus of nutrients, leading to excess plant growth and algal blooms.

***Ex vivo*:** experiments using tissue or cells isolated from a living organism and performed outside the organism.

Exogenous: originating from outside, that is, a process taking place within a cell or organism is referred to as 'exogenous' if it is caused by a (exogenous) substance with origin outside that organism or cell.

Exposure assessment: the determination of the emissions, pathways and rates of movement of a substance and its transformation or degradation in order to estimate the concentrations/doses to which human populations or environmental compartments are or may be exposed (→ risk assessment).

External exposure concentration: the concentration of a chemical in the exposure medium (*e.g.*, cell medium, water, sediment, food) as opposed to the BEC (→ biologically effective concentration).

Extraction factor (EF): the ratio between the mass or volume of a sample that is extracted to the volume of the final extract (synonym for → enrichment factor).

FCMN: flow cytometry micronucleus test for detection of micronucleus formation (DNA damage).

FDA: U.S. Food and Drug Administration.

FET: fish embryo toxicity test.

Frameshift mutation (DNA/RNA): a change in the genetic (three letter codon) reading frame caused by insertion or deletion of a number of nucleotides different to three from a DNA sequence (a type of direct genotoxicity).

FRET: Förster resonance energy transfer.

Furan: → PCDF.

GAC: granular activated carbon.

β-Galactosidase: hydrolysing enzyme often inserted in recombinant cell lines as a marker that can be measured by addition of a substrate that forms a coloured product upon hydrolysis.

GCB: graphitised carbon black used as solid material in → solid-phase extraction.

Gene activation: activation of gene expression, for example, by binding of a nuclear receptor–ligand complex to DNA.

Genetic polymorphism: the co-occurrence of two or more genetically different traits (phenotypes, morphs) within a population.

Genetically modified cell: a cell in which natural features have been over-expressed by genetic engineering to enable more sensitive detection and/or in which foreign features have been added for visualisation of effects (→ recombinant cell).

Genomics: the study of genomes (the total sum of genes in a cell or organism), that is, transcriptomics, proteomics and metabolomics (→ toxicogenomics).

Genotoxicity: the mode of action for DNA damage, for example, by direct reaction with chemicals and reactive oxygen species (→ epigenetic carcinogens, → carcinogenicity).

GFP: → green fluorescent protein.

GHS: Globally Harmonised System for the Classification and Labelling of Chemicals.

GI (tract): gastrointestinal (tract).

Glial cell: a cell type of the nervous system that is important for homeostasis, myelination and support of neurons (→ myelin, → neuron).

Glutathione (GSH): antioxidant tripeptide, important for cellular defence against ROS and conjugation of xenobiotics.

GR: glucorticoid receptor, important for regulation of development, metabolism and the immune system.

Granulosa cells: estrogen-secreting cells that form the lining around female oocytes (eggs).

Green fluorescent protein (GFP): a reporter gene often introduced to recombinant cell lines as an easily measurable marker (→ recombinant cell).

Grey water: water used for domestic purposes such as laundry, dishwashing and showering.

GSH: → glutathione.

GV: guideline value.

Haematopoiesis: production of blood cells.

Haematotoxicity: toxicity to the blood system.

Haloacetic acids (HAA): group of disinfection by-products formed from natural organic matter during chemical disinfection of drinking water and pool water (→ disinfection by-product).

Hazard: the inherent capacity of a chemical or mixture to cause adverse effects in humans or the environment under the conditions of exposure.

Hazard assessment: assessment of the possible adverse effects that could result from exposure to a hazard (→ risk assessment).

Hazard identification: a process that involves determining the nature and context of a risk management issue, sometimes referred to as issue identification (→ risk assessment).

Hazard quotient (HQ): → risk quotient.

Hepatocyte: liver cell.

Hepatotoxicity: liver toxicity.

HHRA: → Human Health Risk Assessment.

HLB: hydrophilic lipophilic balanced co-polymer used for → solid-phase extraction.

Homeostasis: maintenance of internal stability of conditions (redox and chemical steady state) within a cell or other system.

HQ: hazard quotient (→ risk quotient).

HT: high throughput.

HTS: high-throughput screening.

HTTK: high-throughput toxicokinetics.

HTTr: high-throughput transcriptomics.

Human Health Risk Assessment (HHRA): process of estimating potential impact of a chemical, biological or physical agent on a specific human population (→ Ecological Risk Assessment).

Hydrophilicity: affinity for water. A hydrophilic chemical is more likely to stay in the aqueous compartments than to be taken up by the lipids in cells.

Hydrophobicity: incompatibility with water. A hydrophobic chemical has high affinity for the lipids in cells and is more likely to be taken up in the lipid compartments than to stay in water.

Hyperplasia: excessive cell proliferation.

Hyperthermia: increased body temperature caused by dysfunctional heat regulation.

Hypoxia: oxygen deficiency.

IA: → independent action.

ICATM: International Cooperation on Alternative Test Methods.

ICCVAM: U.S. Interagency Coordinating Committee on the Validation of Alternative Methods.

Immortalised cell line: a cell line that has been either accidentally or deliberately mutated to proliferate indefinitely as opposed to a primary cell line with limited lifespan (→ primary cell line).

Immunoassay: a technique that allows detection of an antigen with affinity to bind to a specific antibody (*e.g.*, enzyme-linked immunosorbent assay ELISA, radioimmunoassay RIAs).

Immunotoxicity: toxicity to the immune system.

***In silico*:** refers to predictive computer models.

***In vitro*:** literally 'in glass', refers to tests conducted outside the organism, for example, using immortal cell lines or tissue/enzymes isolated and performed in a vial or dish (traditionally in glass but more recently rather in plastic well plates).

***In vivo*:** refers to tests performed with whole organisms.

Independent action (IA): quantitative model for mixture toxicity of chemicals that act via different target sites and different modes of toxic action, that is, the combined toxicity will be less than the sum of the individual effects (→ concentration addition).

Induction ratio (IR): ratio of signal to signal of unexposed cells in a reporter gene assays based on transcription factors.

Intercalation (DNA): the inclusion of a large planar molecule between two DNA bases.

Ionophoric shuttle mechanism: transport (shuttling) of ions across the cell membrane lipid bilayer by ionophores (lipid-soluble molecules, uncouplers).

IPCS: International Programme on Chemical Safety of the World Health Organisation (WHO).

ISO: International Organisation for Standardisation.

ITS: Integrated testing strategy, which incorporates multiple lines of investigation from predictive computer modelling (*in silico*) and *in vitro* testing to reduce, refine and replace whole animal testing.

KE: → key event.

Key event (KE): an observable effect, which is critical to the induction of a toxicological response following a molecular initiating event (→ molecular initiation event, → Adverse Outcome Pathway).

LC-MS/MS: liquid chromatography with tandem mass spectrometry.

LD_{50}: lethal dose for 50% of the test animals.

LDH: lactate dehydrogenase.

Leydig cells: testosterone-producing cells in the testis.

Ligand: a molecule with binding affinity for a specific biomolecule such as a receptor (→ receptor).

Lipid peroxidation: oxidative breakdown of lipids.

LLE: liquid–liquid extraction, a method used in the laboratory to extract chemicals from water using solvents (→ solid-phase extraction).

LOAEL: lowest observed adverse effect level.

LOD: limit of detection.

LOEC: lowest observed effect concentration.

LOEL: lowest observed effect level.

LOQ: limit of quantification.

Luciferase (Luc): a type of luminescent enzyme that is often utilised in recombinant cell lines as an easily measurable biomarker.

MAC-EQS: Maximum Acceptable Concentration – Environmental Quality Standard, a short-term EQS (→ EQS).

Macropollutant: toxicant, usually metal or salt, found in the mg/L to μg/L range (→ micropollutant).

MAF: → mixture assessment factor.

Margin of safety (MOS): ratio of the acceptable over anticipated exposure concentration. The larger the MOS, the lower the risk. Sometimes also referred to as margin of exposure (MOE).

Matrix effects: when components of the sample matrix (water) interfere with the bioassay.

MCF-7: human breast cancer cell line.

MCR: the maximum cumulative ratio is the ratio between the observed cumulative toxicity of a mixture and the maximum toxicity caused by an individual chemical in the mixture.

Mechanism of toxicity: crucial biochemical processes and/or xenobiotic–biological interactions underlying a given mode of action.

Metabolic activation: metabolic transformation of a chemical that, rather than the intended detoxification, produces a metabolite that is more toxic than its precursor. (Also termed bioactivation or biological activation).

Metabolic pathway: cellular pathway involved with metabolism.

Metabolism: the biological production (anabolism) and breakdown (catabolism) of organic molecules within living organisms. For xenobiotics, the primary role of metabolism is to catabolise chemicals for excretion. In the case of toxic chemicals, this can also be termed detoxification although in some cases, metabolism leads to metabolites more toxic than the precursor (→ metabolic activation, → phase I and II metabolism).

Metabolite: degradation product of metabolism (also called biotransformation product, breakdown product).

Metabolomics: global analysis of presence and abundance of low molecular weight metabolites in cells after exposure to a chemical stressor.

MF: membrane filtration.

Microcystin: a type of cyanotoxin that causes hepatotoxicity through inhibition of protein phosphatases.

Micropollutant: man-made organic chemicals including pesticides, industrial chemicals, consumer products and pharmaceuticals but also natural compounds such as hormones, usually occurring in the sub μg/L range (→ macropollutant).

Microtitre plate: also referred to as a microplate or microwell plate (*e.g.*, 96- or 384-well plate). A flat plate with multiple small wells applied in cell-based assays to hold dilution series of samples/standards.

Microtox: commercially available kit to measure bioluminescence inhibition in naturally luminescent bacteria, *Aliivibrio fischeri*.

MIE: → molecular initiating event.

Mixture assessment factor (MAF): extrapolation factor that is added to a single chemical's risk quotient to account for mixture effects.

Mode of (toxic) action (MOA): a common set of physiological and behavioural signs that characterise a type of (adverse) biological response; in a more recent definition related to AOP (→ adverse outcome pathway) is defined as a biologically plausible series of key events leading to an effect.

MOE: margin of exposure (→ margin of safety).

Molecular initiating event (MIE): the molecular interaction between a xenobiotic and a biomolecule that starts a cellular toxicity pathway (→ key event, → adverse outcome pathway).

Morphogenesis: development of shape.

MOS: → margin of safety.

msPAF: → multi-substance potentially affected fraction.

Multi-substance potentially affected fraction (msPAF): fraction of species exposed to a mixture concentration above their toxicity threshold.

Mutagenicity: mode of toxic action for toxicants causing mutations (→ genotoxicity).

Mutation: a change in the genomic sequence by, for example, insertion of incorrect bases following base excision or strand breaks from DNA damage.

Myelin sheet: electrically insulating layer that forms around the axon of neurons (nerve cells).

Myelinating cell: a cell forming the myelin sheet (→ myelin sheet).

NADPH: nicotinamide adenine dinucleotide phosphate, a metabolic coenzyme and electron donor (reducing agent).

NAM: → new approach method.

Narcosis (mode of action): physiological and behavioural responses elicited through exposure to baseline toxicants. Narcosis in this context refers to the minimum toxicity exhibited by any compound and is not related to narcosis/anaesthesia in clinical medicine (→ baseline toxicity).

Native cell: primary or immortalised cell line that has not been genetically modified (→ primary cell line, → immortalised cell line).

NCATS: National Center for Advancing Translational Sciences.

NCCT: National Center for Computational Toxicology.

Necrosis: unplanned cell death following irreversible damage (as opposed to programmed cell death, → apoptosis).

Negative control: a control sample of the test solvent or media to ensure these have no effect on the assay result.

NEL: no effect level.

Neonicotinoids: a group of neurotoxic insecticides.

Nephrotoxicity: toxicity to the kidneys.

Neuron: type of nerve cell that is important for generation and transfer of information (via neurotransmitters), neurons are supported by glial cells (→ glial cell, → neurotransmitter, → axon).

Neuronopathy: destruction of neurons.

Neurotoxicity: toxicity to the nervous system.

Neurotransmitter: endogenous chemical (*e.g.*, acetylcholine) that transmits information from neurons to target cells.

New approach method (NAM): any non-animal-based approaches that can be used to provide information in the context of chemical hazard and risk assessment.

Next-generation risk assessment (NGRA): risk assessment strategy that relies on → HTS methods.

NGRA: → next-generation risk assessment.

NIEHS: National Institute of Environmental Health Science.

Nitrosamines: a group of chemical compounds (with the structure $R^1N(-R^2)-N=O$) found in many common products such as rubbers, tobacco and foods (*e.g.*, formed in cured meats from amines in the meat and the additive sodium nitrite). Nitrosamines are also → DBPs formed during chloramination of drinking water. Many nitrosamines are carcinogenic.

NOAEL: no observed adverse effect level.

NOEC: no observed effect concentration.

NOEL: no observed effect level.

Non-specific (mode of action): physiological and behavioural responses elicited through exposure to baseline toxicants, often used synonymous to 'narcosis' (→ non-specific toxicity, → baseline toxicity).

Non-specific toxicity: → baseline toxicity.

Non-threshold chemical: a chemical for which it is assumed that there is no safe level of exposure at which there is no effect (*e.g.*, carcinogens) (→ threshold chemical).

NPDES: National Pollutant Discharge Elimination System (U.S.).

Nrf2: NF-E2-related factor 2, a transcription factor for defence against oxidative stress.

NRU: Neutral Red uptake assay using dye to assess cell viability.

NTP: National Toxicology Program of the → NIEHS (U.S.).

Nuclear receptor: a protein receptor that is capably of sensing hormones and of binding directly to DNA, thereby regulating gene expression (→ receptor).

Nuclear xenobiotic metabolism receptor: a protein receptor that senses xenobiotics that induce the gene expression of metabolic enzymes (→ nuclear receptor).

NWQMS: National Water Quality Management Strategy (Australia).

O_3: ozone (refers to water treatment by ozonation).

OECD: Organisation for Economic Co-operation and Development.

Organogenesis: the development of organs.

Organophosphate pesticide: a type of insecticide that causes neurotoxicity via inhibition of the enzyme AChE (→ acetylcholinesterase).

Oxidative stress: imbalance in the level of reactive oxygen species and the system's capacity for detoxification (→ reactive oxygen species).

P450 enzymes: → CYP450.

P53: family of transcription factors for an important adaptive stress response pathway for DNA damage (→ genotoxicity). (Also called 'tumour suppressor gene').

PAH: polycyclic aromatic hydrocarbon.

Passive dosing: a technique used in tests with hydrophobic chemicals, where the test compound is added via a solid phase to maintain a constant exposure concentration in the cell medium (also termed partition-controlled dosing).

Passive sampling: time-integrated sampling of water through deployment of passive sampling devices containing sorbent material with affinity for groups of chemicals with similar physicochemical properties.

Passive transport: passive diffusion of molecules across cell membranes via a concentration gradient from high to low concentration of the substance (→ active transport).

Pathogen: microorganism(s) causing disease in plants, animals and humans.

Pathway of toxicity: essentially synonymous to → toxicity pathway; cellular processes that mediate adverse outcomes of toxicants.

PBDE: polybrominated diphenyl ether. Group of structurally similar brominated compounds also referred to as brominated flame retardants.

PBT: persistent, bioaccumulative and toxic.

PBTK: physiologically based toxicokinetic model (→ toxicokinetics).

PCB: polychlorinated biphenyl. Group of 209 structurally similar industrial chemicals that were previously produced and used in large quantities in, for example, electrical appliances. Although these compounds have been banned as POPs/PBTs under the Stockholm Convention, traces of PCBs are still found in the environment.

PCDD: → polychlorinated dibenzodioxin.

PCDF: → polychlorinated dibenzofuran.

Phagocytes: white blood cells capable of eliminating many microorganisms by absorption.

Phase I metabolism: biotransformation of chemicals via oxidation, reduction and hydrolysis (→ metabolism).

Phase II metabolism: conjugation of the functional groups added in phase I metabolism with molecular entities such as sulphate and glucuronic acid to yield highly water-soluble metabolites, which are more easily excreted from the body (→ metabolism).

Phenotype: external characteristics of an organism that is the expression of its genotype, that is, its DNA make-up.

Photodegradation: the breakdown of organic chemicals through absorption of photons during sunlight exposure.

Photosynthesis: the conversion of carbon dioxide and water to sugars and oxygen by plants, algae and some bacteria, using sunlight energy.

Photosystem II: a protein complex that delivers electrons for photosynthesis to occur (→ photosynthesis).

Phytotoxicity: toxicity to plants.

Plasmid: circular DNA molecule, which carries a responsive element for a receptor of interest, followed by a reporter gene that encodes a measurable feature such as an enzyme or green fluorescence protein.

PNEC: predicted no effect concentration.

Point of departure (POD): point on a dose–response curve corresponding to an estimated low effect level or no effect level.

Point of inflexion: the point on a curve at which the slope changes.

Polychlorinated dibenzodioxin (PCDD): a group of structurally similar chlorinated compounds that are formed as by-products during the production of other chlorinated compounds such as some pesticides. The best-known example is 2,3,7,8-tetrachlorodibenzo-*p*-dioxin (TCDD), which was a major contaminant in Agent Orange, the herbicide used in the Vietnam war. Although these compounds have been banned as POPs/PBTs under the → Stockholm Convention, traces are still found in the environment.

Polychlorinated dibenzofuran (PCDF): a group of dioxin-like persistent organic pollutants (→ Stockholm Convention) structurally similar to → polychlorinated dibenzodioxins.

Polyhalogenated biphenyl: brominated, chlorinated and fluorinated biphenyls, for example, polychlorinated biphenyl (→ PCB).

POP: persistent organic pollutant (→ Stockholm Convention).

Positive control: a sample that contains a supra-maximal concentration of a test compound that is not the reference compound (→ supra-maximal concentration).

PPAR: peroxisome proliferator receptor, involved with metabolism of glucose, lipids and fatty acids.

PR: progesterone receptor, important for development and reproduction (fertility) and induced by progestogens and progestogen-like chemicals (→ progestogens).

Primary cell line: cell line isolated from living tissue. Most primary cell cultures have a limited lifespan (*i.e.*, have not been immortalised) (→ immortalised cell line).

Primary mechanism, primary effects: the type and degree of interaction of a toxicant with biomolecules at the target site.

Progestogens: steroid hormones such as progesterone that are important regulators of, for example, pregnancy and menstruation (→ PR).

Prokaryote: a single-celled organism that has no distinct nucleus (*e.g.*, bacteria, cyanobacteria) (→ eukaryote).

Promoter: a region in DNA that regulates transcription of a specific gene.

Protection-goal-motivated approach: a test battery design approach that targets the protection goal (*e.g.*, human health or ecosystem health) rather than specific chemical groups (→ chemical-group-motivated approach).

Proteolytic enzyme: an enzyme that splits proteins into their constitutive building blocks, that is, peptides or amino acids.

Proteomics: global analysis of presence and abundance of functional proteins in cells after exposure to a chemical stressor.

PRW: purified recycled water.

PXR: pregnane X receptor, involved in the induction of various phase I enzymes (→ phase I metabolism).

Pyrethroids: group of neurotoxic insecticides.

QA/QC: quality assurance and quality control.

qAOP: quantitative adverse outcome pathway (→adverse outcome pathway), (→ AOP).

QIVIVE: → quantitative *in vitro* to *in vivo* extrapolation.

QS: quality standard.

QSAR: quantitative structure–activity relationship.

Quantitative *in vitro* to *in vivo* extrapolation (QIVIVE): quantitative model that extrapolates from *in vitro* data to potential *in vivo* effects in humans.

Quinolones: group of synthetic antibiotics.

Radioimmunoassay (RIA): an immunoassay using radiolabelling to enable easy detection of an antigen.

RAR: retinoic acid receptor, important for regulation of, for example, development and homeostasis.

REACH: The European chemicals regulation titled Registration, Evaluation, Authorisation and Restriction of Chemical Substances.

Reactive oxygen species (ROS): endogenously produced oxygen containing reactive molecules including the superoxide radical ($O_2{\bullet}^-$), hydrogen peroxide (H_2O_2) and the hydroxyl radical (OH•). Chemical and other stress can increase ROS formation to levels that may cause lipid peroxidation, DNA damage and oxidation of proteins, followed by loss of enzymatic activity.

Reactive toxicity: mode of toxic action that is associated with chemical reactions where covalent bonds are formed. Direct reactivity is the direct reaction of electrophilic chemicals with biological nucleophiles such as DNA bases or proteins. Indirect reactivity is the formation of reactive oxygen species (ROS) that are potent oxidants.

Read-across: a technique for predicting the properties of one chemical by using data from another chemical.

Receptor binding assay (RBA): an assay that measures the competitive binding of a chemical or environmental sample and a native ligand molecule to a receptor (→ receptor, → ligand).

Receptor: a protein to which certain ligands (*e.g.*, hormones and hormone-mimicking chemicals) can bind. Each receptor is specific to binding of ligands with particular structure(s) (→ nuclear receptor).

Receptor-mediated toxicity: toxicity caused by induction of a specific receptor (*e.g.*, binding of xenoestrogen to the estrogen receptor) (→ receptor, → nuclear receptor).

Recombinant cell lines: cell lines that are created by insertion of a reporter plasmid, which carries a responsive element for a particular receptor followed by a reporter gene encoding a measurable marker (*e.g.*, green fluorescent protein).

REF: → relative enrichment factor, or relative extraction factor.

Reference dose (RfD): an estimate (with uncertainty factors) of the daily exposure (mg/kg/day) to the general population that is without harm (→ acceptable daily intake, → tolerable daily intake).

Relative enrichment factor (REF): is a measure of sample concentration that takes into account the enrichment of the sample that occurs during extraction and clean up, relative to the sample dilution that takes place during the assay. Product of → extraction factor (EF) and → dosing factor ($DF_{bioassay}$).

Relative extraction factor (REF): synonym for → relative enrichment factor.

Relative fluorescence units (RFU): fluorescence measurements with a plate reader that are not calibrated but are instrument-specific.

Relative light units (RLU): luminescence measurements that are specific to the luminometer and are therefore relative and can only be compared in the sample instrument.

REP: the relative effect potency is a measure of the relative toxicity of one chemical compared with another (often a reference compound) established through comparison of the individual concentration–effect curves.

Reporter gene: a gene with a particular characteristic that can be utilised through insertion into a gene that would not otherwise express this feature. An example of a reporter gene is the green fluorescent protein (GFP), which is introduced to cells in order to encode a measurable marker (→ recombinant cell).

Reporter plasmid: a transferable form of DNA that is separate to and independent of chromosomal DNA.

Responsive element: or hormone responsive element (HRE, or, for example, ERE for the estrogen receptor); a short sequence of DNA capable of binding to certain receptors (*e.g.*, ER), thereby regulating transcription.

Reverse osmosis (RO): a process that applies water pressure to force water movement over a membrane filter, which is impassable by many (but not all) micropollutants.

RfD: → reference dose.

RFU: → relative fluorescence units

rGTU: relative genotoxic unit (similar to → toxic unit, but specifically for genotoxicity).

RIA: → radioimmunoassay.

Risk assessment: a process, which entails the following elements: hazard identification, effects assessment, exposure assessment and risk characterisation.

Risk characterisation: estimate of the incidence and severity of the adverse effects likely to occur in a human population or environmental compartment due to actual or predicted exposure to a substance, and may include 'risk estimation', that is, the quantification of that likelihood (→ risk assessment).

Risk index (RI): cumulative risk measure, the sum of risk quotients (→ RQ). Sometimes also called hazard index (HI) (→ risk quotient).

Risk management: a decision-making process that entails weighing political, social, economic, and engineering information against risk-related information to develop, analyse and compare regulatory options and select the appropriate regulatory response to a potential health or environmental hazard (→ risk assessment).

Risk quotient (RQ): ratio of the exposure estimate over an acceptable effect level (also called hazard quotient HQ) (→ risk index, → hazard quotient).

Risk reduction: taking measures to protect humans and/or the environment from the risks identified.

Risk: the probability of an adverse effect on humans or the environment occurring as a result of a given exposure to a chemical or mixture.

RLU: → relative light units

RNA: ribonucleic acid, the macromolecule controlling, for example, gene expression and protein synthesis. DNA messenger.

RO: → reverse osmosis.

ROS: → reactive oxygen species.

RQ: → risk quotient.

rTU: relative toxic unit (similar to → toxic unit).

RXR: retinoid X receptor, forms heterodimers with many other nuclear receptors such as the RAR and thus has many regulatory functions (→ RAR).

S9: liver enzyme mix containing a wide range of metabolising enzymes including CYP450s. S9 mix is used in bioassays to study the effect of metabolism on xenobiotics, that is, some xenobiotics need metabolic activation (→ metabolic activation).

SCGE: → single cell gel electrophoresis

Sertoli cells: nurse cells within the testis.

Single cell gel electrophoresis (SCGE): an assay that detects DNA strand breaks in different cell types including human, other mammalian and fish cells. Also known as the Comet assay (→ Comet assay).

SMPD: serum-mediated passive dosing.

SOP: standard operating protocol (or procedure).

SOS Chromo: assay for measuring the SOS response in bacterial cells (→ SOS response).

SOS response: defence strategy of a cell in response to DNA damage. Cell assays targeting the SOS response include the *umu*C, SOS/*umu* and SOS Chromo.

SPE: solid-phase extraction, a laboratory method that uses cartridges packed with specialised sorbents to extract chemicals from water (→ LLE).

Specific mode of toxic action: a mode of toxic action caused by specific interaction with receptors or enzymes that result in higher toxicity than baseline toxicity.

Specificity ratio (SR): ratio between the inhibitory concentration predicted for baseline toxicity and experimental effect concentration. Chemicals with an SR close to 1 act as baseline toxicants, while chemicals with a high SR are specifically active in this particular endpoint.

Spermatogenesis: production and development of mature sperm cells.

SPM: → suspended particulate matter.

SPR: → suppression ratio.

SR: → specificity ratio.

Stable transfection: stable transfer of genetic material into a cell as opposed to transient transfection, which is unstable after reproduction.

Stockholm Convention: a global treaty to ban release and minimise exposure to persistent organic pollutants.

Super-minimal concentration: the highest concentration causing an effect that is not statistically different from the negative control (→ NOEC).

Suppression ratio (SPR): effect of an antagonist in a reporter gene assay run in antagonism mode, in some studies also called SR (to be avoided, might be confused with → suppression ratio).

Supra-maximal concentration: the lowest concentration causing 100% effect.

Suspended particulate matter (SPM): small particles composed of minerals and organic matter that are suspended in surface water

Synapse: junction between two nerve cells.

Synergy or synergism: when the combined toxicity of two or more toxicants is higher than the sum of the individual effects.

TCA: → tricarboxylic acid.

TCDD: 2,3,7,8-tetrachloro-dibenzodioxin.

tcpl: → ToxCast analysis pipeline.

TD: → toxicodynamics.

TDI: tolerable daily intake (→ acceptable daily intake, → reference dose).

TEF: toxic equivalency factor.

TEQ: → toxic equivalent concentration.

TEQ_{bio}: bioassay-derived TEQ (→ toxic equivalent concentration).

TEQ_{chem}: chemical analysis-derived TEQ (→ toxic equivalent concentration).

Teratogenesis: interference with embryonic or foetal development resulting in pre-natal or birth defects.

TF: → transcription factor.

THP1: human acute monocytic leukaemia cell line (monocyte cancer cell – the monocyte is a precursor to macrophages, a type of white blood cell).

THP1-CPA: THP1 cytokine production assay (→ THP1, → cytokine), a measure of immunotoxicity.

Threshold chemical: a chemical for which it is assumed that there is a safe dose or concentration, below which there is no appreciable risk to exposed organisms (→ non-threshold chemical).

Threshold of Toxicological Concern (TTC): level of human intake or exposure that is considered to be of negligible risk to human health.

TIE: → toxicity identification evaluation.

TIF2: human co-activator (→ co-activator).

TK: → toxicokinetics.

TMX: tamoxifen, an anti-estrogenic drug used in treatment of hormone-related cancer (→ estrogen).

TOC: total organic carbon.

TOX: total organic halogen compounds.

Tox21: high-throughput data initiative, collaboration between the National Center for Computational Toxicology (NCCT) of the U.S. EPA, the National Toxicology Program of the National Institute of Environmental Health Science (NIEHS) and the National Center for Advancing Translational Sciences (NCATS) and the U.S. Food and Drug Administration (FDA).

ToxCast analysis pipeline (tcpl): specific data evaluation pipeline in R for concentration–response modelling of high-throughput screening data developed by the → ToxCast project. (→ tcpl).

ToxCast: Toxicity ForeCaster project, launched by the National Center for Computational Toxicology (NCCT) of the U.S. EPA.

Toxic equivalent concentration (TEQ): the concentration of a reference chemical that would elicit the same effect toxicity as the mixture of micropollutants in a water sample.

Toxic unit (TU): the toxic unit from chemical analysis (TU_{chem}) is the ratio of concentration of a chemical concentration divided by its toxic concentration, e.g. → LC_{50}, and the toxic unit from bioassays (TU_{bio}) is the inverse of the toxic concentration of a mixture or environmental samples ($1/LC_{50}$) (→ effect unit).

Toxicity identification evaluation (TIE): a procedure that combines multiple fractionation and bioassay testing to isolate a toxic compound from a mixture (→ effect-directed analysis).

Toxicity pathway: a cellular response pathway that, when sufficiently perturbed, is expected to result in an adverse health outcome (→ pathway of toxicity, → adverse outcome pathway).

Toxicodynamics (TD): the actual toxicity pathways taking place inside the cell including the initial molecular interaction of the chemical and its biological target (→ toxicokinetics).

Toxicogenomics: the application of genomics in toxicological research.

Toxicokinetics (TK): the kinetic processes of uptake, distribution, metabolism and elimination that link external exposure (*e.g.*, via drinking water or eating food) to biologically effective concentration within a cell. These processes include absorption, excretion, internal distribution and metabolism (→ ADME) of a chemical within the whole body and within cells.

TR: thyroid receptor, important for regulation of heart rate, metabolism and development.

Transcription factor (TF): a protein responsible for transcription of an adaptive stress response pathway.

Transcription: generation of a RNA copy of DNA, the first step in gene expression.

Transcriptomics: expression profiling, that is, global analysis of RNA levels in cells after exposure to a chemical stressor.

Tricarboxylic acid (TCA) cycle: series of biochemical reaction used by all aerobic organisms to release stored chemical energy. Takes place in the mitochondria. Also known as the citric acid cycle.

Trihalomethanes (THM): group of disinfection by-products formed from natural organic matter during chemical disinfection of drinking water and pool water (→ disinfection by-product).

Trophic level: position that an organism occupies in a food chain. The lowest trophic level are primary producers (photosynthetic organisms), the highest trophic levels are organisms that feed on other carnivores.

TTC: → threshold of toxicological concern.

TU: → toxic unit.

Tumour: abnormal mass of tissue. Malignant tumours can lead to cancer.

U.S. EPA: U.S. Environmental Protection Agency.

***umu*C:** assay commonly used in water monitoring to measure SOS response in bacterial cells, also called SOS/*umu* (→ SOS response).

Uncoupler: chemical that disturbs mitochondrial energy transduction by a protonophoric shuttle mechanism.

Vasoactive agent: an agent capable of reducing or increasing blood pressure.

Vitellogenin (Vtg): a phospholipoprotein precursor of egg yolk, Vtg is stimulated by exposure to endogenous and exogenous estrogens (estrogenic compounds). Vtg is often measured *in vivo* but can be used in cell-based assays (→ endocrine disruption).

WET: → whole effluent toxicity.

WFD: European Union Water Framework Directive.

WHO: World Health Organisation.

Whole effluent toxicity (WET) testing: a procedure where the toxicity of water is tested exposing whole organisms to un-extracted water (→ direct toxicity assessment).

WWTP: wastewater treatment plant.

Xenobiotic receptor: a receptor that regulates metabolism of xenobiotics (*e.g.*, AhR).

Xenobiotic: foreign substance.

Xenoestrogen: exogenous estrogen (→ estrogen).

Yeast estrogen screen (YES): common assay used to test water samples for estrogenicity, which is measured via induction of human ER in recombinant yeast (→ ER, → recombinant cell lines).

Yeast two-hybrid assay: applies a recombinant yeast cell line, which has been transfected with two different reporter plasmids (→ reporter plasmid, → recombinant cell).

Zona radiata protein (Zrp): eggshell protein.

96- or 384-well plate: → microtitre plate.

References

Abbas A., Schneider I., Bollmann A., Funke J., Oehlmann J., Prasse C., Schulte-Oehlmann U., Seitz W., Ternes T., Weber M., Wesely H. and Wagner M. (2019). What you extract is what you see: Optimising the preparation of water and wastewater samples for *in vitro* bioassays. *Water Research*, **152**, 47–60.

Ademollo N., Patrolecco L., Polesello S., Valsecchi S., Wollgast J., Mariani G. and Hanke G. (2012). The analytical problem of measuring total concentrations of organic pollutants in whole water. *TrAC-Trends in Analytical Chemistry*, **36**, 71–81.

Aerni H. R., Kobler B., Rutishauser B. V., Wettstein F. E., Fischer R., Giger W., Hungerbuhler A., Marazuela M. D., Peter A., Schonenberger R., Vogeli A. C., Suter M. J. F. and Eggen R. I. L. (2004). Combined biological and chemical assessment of estrogenic activities in wastewater treatment plant effluents. *Analytical and Bioanalytical Chemistry*, **378**(3), 688–696.

Aguayo S., Muñoz M. J., de la Torre A., Roset J., de la Peña E. and Carballo M. (2004). Identification of organic compounds and ecotoxicological assessment of sewage treatment plants (STP) effluents. *Science of the Total Environment*, **328**(1–3), 69–81.

Ahn K. S. and Aggarwal B. B. (2005). Transcription factor NF-κB – A sensor for smoke and stress signals. In: Natural Products and Molecular Therapy G. J. Kotwal and D. K. Lahiri, Wiley, Hoboken, NJ, USA, pp. 218–233.

Al Hanai A. H., Antkiewicz D. S., Hemming J. D. C., Shafer M. M., Lai A. M., Arhami M., Hosseini V. and Schauer J. J. (2019). Seasonal variations in the oxidative stress and inflammatory potential of PM2.5 in Tehran using an alveolar macrophage model; The role of chemical composition and sources. *Environment International*, **123**, 417–427.

Albergamo V., Schollee J. E., Schymanski E. L., Helmus R., Timmer H., Hollender J. and de Voogt P. (2019). Nontarget screening reveals time trends of polar micropollutants in a riverbank filtration system. *Environmental Science & Technology*, **53**(13), 7584–7594.

Albergamo V., Escher B. I., Schymanski E. L., Helmus R., Dingemans M. M. L., Cornelissen E. R., Kraak M. H. S., Hollender J. and de Voogt P. (2020). Evaluation of reverse

Authors: Beate Escher, Peta Neale and Frederic Leusch
doi: 10.2166/9781789061987_0379

osmosis drinking water treatment of riverbank filtrate using bioanalytical tools and non-target screening. *Environmental Science-Water Research & Technology*, **6**(1), 103–116.

Aleem A. and Malik A. (2003). Genotoxicity of water extracts from the River Yamuna at Mathura, India. *Environmental Toxicology*, **18**(2), 69–77.

Aleem A. and Malik A. (2005). Genotoxicity of the Yamuna River water at Okhla (Delhi), India. *Ecotoxicology and Environmental Safety*, **61**(3), 404–412.

Allan H. L., van de Merwe J. P., Finlayson K. A., O'Brien J. W., Mueller J. F. and Leusch F. D. L. (2017). Analysis of sugarcane herbicides in marine turtle nesting areas and assessment of risk using *in vitro* toxicity assays. *Chemosphere*, **185**, 656–664.

Allermann L. and Poulsen O. M. (2000). Inflammatory potential of dust from waste handling facilities measured as IL-8 secretion from lung epithelial cells *in vitro*. *Annals of Occupational Hygiene*, **44**(4), 259–269.

Allinson G., Allinson M., Salzman S., Shiraishi F., Myers J., Theodoropoulos T., Hermon K. and Wightwick A. (2007). Hormones in recycled water. Final report. DPI (Department of Primary Industries), Queenscliff, Australia https://dro.deakin.edu.au/eserv/DU:30010499/salzman-hormonesintreated-2007.pdf. (Accessed 10 April 2021).

Altenburger R., Backhaus T., Boedeker W., Faust M., Scholze M. and Grimme L. H. (2000). Predictability of the toxicity of multiple chemical mixtures to *Vibrio fischeri*: mixtures composed of similarly acting chemicals. *Environmental Toxicology and Chemistry*, **19**(9), 2341–2347.

Altenburger R., Scholz S., Schmitt-Jansen M., Busch W. and Escher B. I. (2012). Mixture toxicity revisited from a toxicogenomic perspective. *Environmental Science & Technology*, **46**(5), 2508–2522.

Altenburger R., Scholze M., Busch W., Escher B. I., Jakobs G., Krauss M., Krüger J., Neale P. A., Aït-Aïssa S., Almeida A. C., Seiler T.-B., Brion F., Hilscherová K., Hollert H., Novák J., Schlichting R., Serra H., Shao Y., Tindall A. J., Tolefsen K. E., de Aragão Umbuzeiro G., Williams T. D. and Kortenkamp A. (2018). Mixture effects in samples of multiple contaminants – an inter-laboratory study with manifold bioassays. *Environment International*, **114**, 95–116.

Alygizakis N. A., Besselink H., Paulus G. K., Oswald P., Hornstra L. M., Oswaldova M., Medema G., Thomaidis N. S., Behnisch P. A. and Slobodnik J. (2019). Characterization of wastewater effluents in the Danube River Basin with chemical screening, *in vitro* bioassays and antibiotic resistant genes analysis. *Environment International*, **127**, 420–429.

Alyzakis N. A., Samanipour S., Hollender J., Ibanez M., Kaserzon S., Kokkali V., van Leerdam J. A., Mueller J. F., Pijnappels M., Reid M. J., Schymanski E. L., Slobodnik J., Thomaidis N. S. and Thomas K. V. (2018). Exploring the potential of a global emerging contaminant early warning network through the use of retrospective suspect screening with high-resolution mass spectrometry. *Environmental Science & Technology*, **52**(9), 5135–5144.

Ames B. N., McCann J. and Yamasaki E. (1975). Methods for detecting carcinogens and mutagens with the salmonella/mammalian-microsome mutagenicity test. *Mutation Research/Environmental Mutagenesis and Related Subjects*, **31**(6), 347–363.

Andersen M. E., Al-Zoughool M., Croteau M., Westphal M. and Krewski D. (2010). The future of toxicity testing. *Journal of Toxicology and Environmental Health-Part B-Critical Reviews*, **13**(2–4), 163–196.

Ankley G. T., Daston G. P., Degitz S. J., Denslow N. D., Hoke R. A., Kennedy S. W., Miracle A. L., Perkins E. J., Snape J., Tillitt D. E., Tyler C. R. and Versteeg D. (2006). Toxicogenomics in regulatory ecotoxicology. *Environmental Science & Technology*, **40**(13), 4055–4065.

Ankley G. T., Bennett R. S., Erickson R. J., Hoff D. J., Hornung M. W., Johnson R. D., Mount D. R., Nichols J. W., Russom C. L., Schmieder P. K., Serrrano J. A., Tietge J. E. and Villeneuve D. L. (2010). Adverse outcome pathways: a conceptual framework to support ecotoxicology research and risk assessment. *Environmental Toxicology and Chemistry*, **29**(3), 730–741.

ANZECC/ARMCANZ (2000). Australian and New Zealand Guidelines for Fresh and Marine Water Quality. Vol. **1**, Australian and New Zealand Environment Conservation Council and Agriculture and Resource Management Council of Australia and New Zealand, Canberra. www.waterquality.gov.au/anz-guidelines/resources/previous-guidelines/anzecc-armcanz-2000. (Accessed 18 February 2021).

Apel P., Kortenkamp A., Koch H. M., Vogel N., Ruther M., Kasper-Sonnenberg M., Conrad A., Bruning T. and Kolossa-Gehring M. (2020). Time course of phthalate cumulative risks to male developmental health over a 27-year period: Biomonitoring samples of the German Environmental Specimen Bank. *Environment International*, **137**, 105467.

Archer E., Wolfaardt G. M., van Wyk J. H. and van Blerk N. (2020). Investigating (anti) estrogenic activities within South African wastewater and receiving surface waters: Implication for reliable monitoring. *Environmental Pollution*, **263**, 114424.

Armitage J. M., Wania F. and Arnot J. A. (2014). Application of mass balance models and the chemical activity concept to facilitate the use of *in vitro* toxicity data for risk assessment. *Environmental Science & Technology*, **48**, 9770–9779.

Ashauer R., O'Connor I. and Escher B. I. (2017). Toxic mixtures in time – the sequence makes the poison. *Environmental Science & Technology*, **51**(5), 3084–3092.

Atkinson A. J., Colburn W. A., DeGruttola V. G., DeMets D. L., Downing G. J., Hoth D. F., Oates J. A., Peck C. C., Schooley R. T., Spilker B. A., Woodcock J. and Zeger S. L. and Biomarkers Definitions Working G. (2001). Biomarkers and surrogate endpoints: Preferred definitions and conceptual framework. *Clinical Pharmacology & Therapeutics*, **69**(3), 89–95.

Atterwill C. K., Bruinink A., Drejer J., Duarte E., Abdulla E. M., Meredith C., Nicotera P., Regan C., Rodriguezfarre E., Simpson M. G., Smith R., Veronesi B., Vijverberg H., Walum E. and Williams D. C. (1994). *In vitro* neurotoxicity testing – The report and recommendations of ECVAM workshop-3. *Atla-Alternatives to Laboratory Animals*, **22**(5), 350–362.

Auld D., Coassin P. A., Coussens N. P., Hensley P., Klumpp-Thomas C., Michael S., Sittampalam G. S., Trask O. J., Wagner B. K., Weidner J. R., Wildey M. J. and Dahlin J. L. (2020). Microplate selection and recommended practices in high-throughput screening and quantitative biology. In: Assay Guidance Manual [Internet], S. Markossian, G. Sittampalam and A. Grossman (eds.), Eli Lilly & Company and the National Center for Advancing Translational Sciences, Bethesda (MD). www.ncbi.nlm.nih.gov/books/NBK558077/ (Accessed on 2 January 2021).

Australian Government (2018a). Australian and New Zealand Guidelines for Fresh and Marine Water Quality. Water Quality Policy Sub Committee (WQPSC) and National

Water Reform Committee (NWRC), Australian Government, Canberra, ACT. www.waterquality.gov.au/anz-guidelines (Accessed 2 January 2021).

Australian Government (2018b). Charter: National Water Quality Management Strategy. Department of Agriculture and Water Resources, Canberra. www.waterquality.gov.au/about/charter (Accessed 2 January 2021).

Backhaus T. and Grimme L. H. (1999). The toxicity of antibiotic agents to the luminescent bacterium *Vibrio fischeri*. *Chemosphere*, **38**(14), 3291–3301.

Backhaus T. and Karlsson M. (2014). Screening level mixture risk assessment of pharmaceuticals in STP effluents. *Water Research*, **49**, 157–165.

Backhaus T., Altenburger R., Boedeker W., Faust M., Scholze M. and Grimme L. H. (2000). Predictability of the toxicity of multiple mixtures of dissimilarly acting chemicals to *Vibrio fischeri*. *Environmental Toxicology and Chemistry*, **19**(9), 2348–2356.

Baduel C., Mueller J. F., Tsai H. H. and Ramos M. J. G. (2015). Development of sample extraction and clean-up strategies for target and non-target analysis of environmental contaminants in biological matrices. *Journal of Chromatography A*, **1426**, 33–47.

Bahamonde P. A., Feswick A., Isaacs M. A., Munkittrick K. R. and Martyniuk C. J. (2016). Defining the role of omics in assessing ecosystem health: perspectives from the Canadian environmental monitoring program. *Environmental Toxicology and Chemistry*, **35**(1), 20–35.

Bailey H. C., Elphick J. R., Krassoi R., Mulhall A. M., Lovell A. J. and Slee D. J. (2005). Identification of chlorfenvinphos toxicity in a municipal effluent in Sydney, New South Wales, Australia. *Environmental Toxicology and Chemistry* **24**(7), 1773–1778.

Bain P. A., Williams M. and Kumar A. (2014). Assessment of multiple hormonal activities in wastewater at different stages of treatment. *Environmental Toxicology and Chemistry*, **33**(10), 2297–2307.

Baken K. A., Sjerps R. M. A., Schriks M. and van Wezel A. P. (2018). Toxicological risk assessment and prioritization of drinking water relevant contaminants of emerging concern. *Environment International*, **118**, 293–303.

Baker M. A., Cerniglia G. J. and Zaman A. (1990). Microtiter plate assay for the measurement of glutathione and glutathione disulfide in large numbers of biological samples. *Analytical Biochemistry*, **190**(2), 360–365.

Ballantyne B., Marrs T. and Syversen T. (1995). General and Applied Toxicology – Abridged edition. Macmillan Press, London, UK.

Bal-Price A. and Coecke S. (2011). Guidance on Good Cell Culture Practice (GCCP). In: Cell Culture Techniques, M. Aschner, C. Sunol and A. Bal-Price (eds.), Springer, Heidelberg, Germany, pp. 1–25.

Baltazar M. T., Cable S., Carmichael P. L., Cubberley R., Cull T., Delagrange M., Dent M. P., Hatherell S., Houghton J., Kukic P., Li H. Q., Lee M. Y., Malcomber S., Middleton A. M., Moxon T. E., Nathanail A. V., Nicol B., Pendlington R., Reynolds G., Reynolds J., White A. and Westmoreland C. (2020). A next-generation risk assessment case study for Coumarin in cosmetic products. *Toxicological Sciences*, **176**(1), 236–252.

Bambino K. and Chu J. (2017). Zebrafish in Toxicology and Environmental Health. In: Zebrafish at the Interface of Development and Disease Research, K. C. Sadler (ed.), Elsevier, pp. 331–367.

Bandelj E., Van den Heuvel M. R., Leusch F. D. L., Shannon N., Taylor S. and McCarthy L. H. (2006). Determination of the androgenic potency of whole effluents using mosquitofish and trout bioassays. *Aquatic Toxicology*, **80**(3), 237–248.

Barrueco C., Herrera A. and de la Pea E. (1991). Mutagenic evaluation of trichlorfon using different assay methods with *Salmonella typhimurium*. *Mutagenesis*, **6**(1), 71–76.

Bartram J., Corrales L., Davison A., Deere D., Drury D., Gordon B., Howard G., Rinehold A. and Stevens M. (2009). Water Safety Plan Manual: Step-by-Step Risk Management for Drinking-Water Suppliers. World Health Organization, Geneva. www.preventionweb.net/publications/view/8367 (Accessed 10 April 2021).

Baumer A., Bittermann K., Klüver N. and Escher B. (2017). Baseline toxicity and ion trapping models to describe the pH-dependence of bacterial toxicity of pharmaceuticals. *Environmental Science: Processes & Impacts*, **19**, 901–916.

Beckers L. M., Brack W., Dann J. P., Krauss M., Müller E. and Schulze T. (2020). Unraveling longitudinal pollution patterns of organic micropollutants in a river by non-target screening and cluster analysis. *Science of the Total Environment*, **727**, 138388.

Behnisch P. A., Hosoe K. and Sakai S.-i. (2001). Bioanalytical screening methods for dioxins and dioxin-like compounds – a review of bioassay/biomarker technology. *Environment International*, **27**(5), 413–439.

Bell S. M., Chang X. Q., Wambaugh J. F., Allen D. G., Bartels M., Brouwer K. L. R., Casey W. M., Choksi N., Ferguson S. S., Fraczkiewicz G., Jarabek A. M., Ke A., Lumen A., Lynn S. G., Paini A., Price P. S., Ring C., Simon T. W., Sipes N. S., Sprankle C. S., Strickland J., Troutman J., Wetmore B. A. and Kleinstreuer N. C. (2018). *In vitro* to *in vivo* extrapolation for high throughput prioritization and decision making. *Toxicology in Vitro*, **47**, 213–227.

Bellet V., Hernandez-Raquet G., Dagnino S., Seree L., Pardon P., Bancon-Montiny C., Fenet H., Creusot N., Ait-Aissa S., Cavailles V., Budzinski H., Antignac J.-P. and Balaguer P. (2012). Occurrence of androgens in sewage treatment plants influents is associated with antagonist activities on other steroid receptors. *Water Research*, **46**(6), 1912–1922.

Bengtson Nash S. M., Goddard J. and Muller J. F. (2006). Phytotoxicity of surface waters of the Thames and Brisbane River Estuaries: A combined chemical analysis and bioassay approach for the comparison of two systems. *Biosensors & Bioelectronics*, **21**(11), 2086–2093.

Bermudez D. S., Gray L. E. and Wilson V. S. (2012). Modelling defined mixtures of environmental oestrogens found in domestic animal and sewage treatment effluents using an *in vitro* oestrogen-mediated transcriptional activation assay (T47D-KBluc). *International Journal of Andrology*, **35**(3), 397–406.

Bernhard K., Stahl C., Martens R., Kohler H. R., Triebskorn R., Scheurer M. and Frey M. (2017). Two novel real time cell-based assays quantify beta-blocker and NSAID specific effects in effluents of municipal wastewater treatment plants. *Water Research*, **115**, 74–83.

Berninger J. P., DeMarini D. M., Warren S. H., Simmons J. E., Wilson V. S., Conley J. M., Armstrong M. D., Iwanowicz L. R., Kolpin D. W., Kuivila K. M., Reilly T. J., Romanok K. M., Villeneuve D. L. and Bradley P. M. (2019). Predictive analysis using chemical-gene interaction networks consistent with observed endocrine activity and mutagenicity of US streams. *Environmental Science & Technology*, **53**(15), 8611–8620.

Bicchi C., Schiliro T., Pignata C., Fea E., Cordero C., Canale F. and Gilli G. (2009). Analysis of environmental endocrine disrupting chemicals using the E-SCREEN method and stir bar sorptive extraction in wastewater treatment plant effluents. *Science of the Total Environment*, **407**(6), 1842–1851.

Birch H., Kramer N. I. and Mayer P. (2019). Time-resolved freely dissolved concentrations of semivolatile and hydrophobic test chemicals in *in vitro* assays. Measuring high losses and crossover by headspace solid-phase microextraction. *Chemical Research in Toxicology*, **32**(9), 1780–1790.

Bjorseth A., Carlberg G. E. and Moller M. (1979). Determination of halogenated organic-compounds and mutagenicity testing of spent bleach liquors. *Science of the Total Environment*, **11**(2), 197–211.

Blaauboer B. J. (2002). The applicability of *in vitro*-derived data in hazard identification and characterisation of chemicals. *Environmental Toxicology and Pharmacology*, **11**(3–4), 213–225.

Blaauboer B. J. (2008). The contribution of *in vitro* toxicity data in hazard and risk assessment: Current limitations and future perspectives. *Toxicology Letters*, **180**(2), 81–84.

Blaauboer B. J. (2010). Biokinetic modeling and *in vitro-in vivo* extrapolations. *Journal of Toxicology and Environmental Health, Part B: Critical Reviews*, **13**(2–4), 242–252.

Blackwell B. R., Ankley G. T., Bradley P. M., Houck K. A., Makarov S. S., Medvedev A. V., Swintek J. and Villeneuve D. L. (2019). Potential toxicity of complex mixtures in surface waters from a nationwide survey of United States streams: Identifying *in vitro* bioactivities and causative chemicals. *Environmental Science & Technology*, **53**(2), 973–983.

Boehler S., Strecker R., Heinrich P., Prochazka E., Northcott G. L., Ataria J. M., Leusch F. D. L., Braunbeck T. and Tremblay L. A. (2017). Assessment of urban stream sediment pollutants entering estuaries using chemical analysis and multiple bioassays to characterise biological activities. *Science of the Total Environment*, **593–594**, 498–507.

Boobis A., Budinsky R., Collie S., Crofton K., Embry M., Felter S., Hertzberg R., Kopp D., Mihlan G., Mumtaz M., Price P., Solomon K., Teuschler L., Yang R. and Zaleski R. (2011). Critical analysis of literature on low-dose synergy for use in screening chemical mixtures for risk assessment. *Critical Reviews in Toxicology*, **41**(5), 369–383.

Bopp S. K., Barouki R., Brack W., Dalla Costa S., Dorne J., Drakvik P. E., Faust M., Karjalainen T. K., Kephalopoulos S., van Klaveren J., Kolossa-Gehring M., Kortenkamp A., Lebret E., Lettieri T., Norager S., Ruegg J., Tarazona J. V., Trier X., van de Water B., van Gils J. and Bergman A. (2018). Current EU research activities on combined exposure to multiple chemicals. *Environment International*, **120**, 544–562.

Bopp S. K., Kienzler A., Richarz A. N., van der Linden S. C., Paini A., Parissis N. and Worth A. P. (2019). Regulatory assessment and risk management of chemical mixtures: challenges and ways forward. *Critical Reviews in Toxicology*, **49**(2), 174–189.

Borenfreund E. and Puerner J. A. (1985). Toxicity determined *in vitro* by morphological alterations and neutral red absorption. *Toxicology Letters*, **24**(2–3), 119–124.

Boxall A. B. A., Sinclair C. J., Fenner K., Kolpin D. and Maud S. J. (2004). When synthetic chemicals degrade in the environment. *Environmental Science & Technology*, **38**(19), 368A–375A.

Brack W., Altenburger R., Ensenbach U., Moder M., Segner H. and Schuurmann G. (1999). Bioassay-directed identification of organic toxicants in river sediment in the industrial region of Bitterfeld (Germany) – A contribution to hazard assessment. *Archives of Environmental Contamination and Toxicology*, **37**(2), 164–174.

Brack W., Schirmer K., Erdinger L. and Hollert H. (2005). Effect-directed analysis of mutagens and ethoxyresorufin-O-deethylase inducers in aquatic sediments. *Environmental Toxicology and Chemistry*, **24**(10), 2445–2458.

Brack W., Schmitt-Jansen M., Machala M., Brix R., Barcelo D., Schymanski E., Streck G. and Schulze T. (2008). How to confirm identified toxicants in effect-directed analysis. *Analytical and Bioanalytical Chemistry*, **390**(8), 1959–1973.

Brack W., Ait-Aissa S., Burgess R. M., Busch W., Creusot N., Di Paolo C., Escher B. I., Hewitt L. M., Hilscherova K., Hollender J., Hollert H., Jonker W., Kool J., Lamoree M., Muschket M., Neumann S., Rostkowski P., Ruttkies C., Schollee J., Schymanski E. L., Schulze T., Seiler T.-B., Tindall A. J., Umbuzeiro G. D. A., Vrana B. and Krauss M. (2016). Effect-directed analysis supporting monitoring of aquatic environments – An in-depth overview. *Science of the Total Environment*, **544**, 1073–1118.

Brack W., Ait Aissa S., Backhaus T., Dulio V., Escher B. I., Faust M., Hilscherova K., Hollender J., Hollert H., Muller C., Munthe J., Posthuma L., Seiler T. B., Slobodnik J., Teodorovic I., Tindall A. J., Umbuzeiro G. D., Zhang X. W. and Altenburger R. (2019). Effect-based methods are key. The European Collaborative Project SOLUTIONS recommends integrating effect-based methods for diagnosis and monitoring of water quality. *Environmental Sciences Europe*, **31**, 10.

Bradley P. M., Journey C. A., Romanok K. M., Barber L. B., Buxton H. T., Foreman W. T., Furlong E. T., Glassmeyer S. T., Hladik M. L., Iwanowicz L. R., Jones D. K., Kolpin D. W., Kuivila K. M., Loftin K. A., Mills M. A., Meyer M. T., Orlando J. L., Reilly T. J., Smalling K. L. and Villeneuve D. L. (2017). Expanded target-chemical analysis reveals extensive mixed organic-contaminant exposure in US streams. *Environmental Science & Technology*, **51**(9), 4792–4802.

Brand W., de Jongh C. M., van der Linden S. C., Mennes W., Puijker L. M., van Leeuwen C. J., van Wezel A. P., Schriks M. and Heringa M. B. (2013). Trigger values for investigation of hormonal activity in drinking water and its sources using CALUX bioassays. *Environment International*, **55**, 109–118.

Braunig J., Tang J. Y. M., Warne M. S. J. and Escher B. I. (2016). Bioanalytical effect-balance model to determine the bioavailability of organic contaminants in sediments affected by black and natural carbon. *Chemosphere*, **156**, 181–190.

Brettschneider D. J., Misovic A., Schulte-Oehlmann U., Oetken M. and Oehlmann J. (2019). Detection of chemically induced ecotoxicological effects in rivers of the Nidda catchment (Hessen, Germany) and development of an ecotoxicological, Water Framework Directive-compliant assessment system. *Environmental Sciences Europe*, **31**, 7.

Brian J. V., Harris C. A., Scholze M., Backhaus T., Booy P., Lamoree M., Pojana G., Jonkers N., Runnalls T., Bonfa A., Marcomini A. and Sumpter J. P. (2005). Accurate prediction of the response of freshwater fish to a mixture of estrogenic chemicals. *Environmental Health Perspectives*, **113**(6), 721–728.

Brion F., De Gussem V., Buchinger S., Hollert H., Carere M., Porcher J. M., Piccini B., Feray C., Dulio V., Konemann S., Simon E., Werner I., Kase R. and Ait-Aissa S. (2019). Monitoring estrogenic activities of waste and surface waters using a novel *in vivo* zebrafish embryonic (EASZY) assay: Comparison with *in vitro* cell-based assays and determination of effect-based trigger values. *Environment International*, **130**, 104896.

Brunner A. M., Dingemans M. M. L., Baken K. A. and van Wezel A. P. (2019). Prioritizing anthropogenic chemicals in drinking water and sources through combined use of mass spectrometry and ToxCast toxicity data. *Journal of Hazardous Materials*, **364**, 332–338.

Bunnell J. E., Tatu C. A., Lerch H. E., Orem W. H. and Pavlovic N. (2007). Evaluating nephrotoxicity of high-molecular-weight organic compounds in drinking water from lignite aquifers. *Journal of Toxicology and Environmental Health-Part A-Current Issues*, **70**(24), 2089–2091.

Burke M. D. and Mayer R. T. (1974). Ethoxyresorufin – direct fluorimetric assay of a microsomal o-dealkylation which is preferentially inducible by 3-methylcholanthrene. *Drug Metabolism and Disposition*, **2**(6), 583–588.

Burton G. A., Jr., Pitt R. and Clark S. (2000). The role of traditional and novel toxicity test methods in assessing stormwater and sediment contamination. *Critical Reviews in Environmental Science and Technology*, **30**(4), 413–447.

Buschini A., Carboni P., Frigerio S., Furlini M., Marabini L., Monarca S., Poli P., Radice S. and Rossi C. (2004). Genotoxicity and cytotoxicity assessment in lake drinking water produced in a treatment plant. *Mutagenesis*, **19**(5), 341–347.

Cahill P. A., Knight A. W., Billinton N., Barker M. G., Walsh L., Keenan P. O., Williams C. V., Tweats D. J. and Walmsley R. M. (2004). The GreenScreen((R)) genotoxicity assay: a screening validation programme. *Mutagenesis*, **19**(2), 105–119.

Campana O. and Wlodkowic D. (2018). Ecotoxicology goes on a chip: embracing miniaturized bioanalysis in aquatic risk assessment. *Environmental Science & Technology*, **52**(3), 932–946.

Campora C. E., Hokama Y., Tamaru C. S., Anderson B. and Vincent D. (2010). Evaluating the risk of ciguatera fish poisoning from reef fish grown at marine aquaculture facilities in Hawai'i. *Journal of the World Aquaculture Society*, **41**(1), 61–70.

Cantwell R. E. and Hofmann R. (2011). Ultraviolet absorption properties of suspended particulate matter in untreated surface waters. *Water Research*, **45**(3), 1322–1328.

Cao N., Yang M., Zhang Y., Hu J., Ike M., Hirotsuji J., Matsui H., Inoue D. and Sei K. (2009). Evaluation of wastewater reclamation technologies based on *in vitro* and *in vivo* bioassays. *Science of the Total Environment*, **407**(5), 1588–1597.

Cargouet M., Perdiz D., Mouatassim-Souali A., Tamisier-Karolak S. and Levi Y. (2004). Assessment of river contamination by estrogenic compounds in Paris area (France). *Science of the Total Environment*, **324**(1–3), 55–66.

Carlberg G. E., Gjos N., Moller M., Gustavsen K. O., Tveten G. and Renberg L. (1980). Chemical characterization and mutagenicity testing of chlorinated trihydroxybenzenes identified in spent bleach liquors from a sulfite plant. *Science of the Total Environment*, **15**(1), 3–15.

Carson R. (1962). Silent Spring. Houghton Mifflin, New York, NY, USA.

Castillo M., Alonso M. C., Riu J., Reinke M., Kloter G., Dizer H., Fischer B., Hansen P. D. and Barcelo D. (2001). Identification of cytotoxic compounds in European wastewaters during a field experiment. *Analytica Chimica Acta*, **426**(2), 265–277.

CDC (2019). Fourth National Report on Human Exposure to Environmental Chemicals. Updated Tables, January 2019. Centers for Disease Control and Prevention, US Department of Health and Human Services, Washington DC, USA. www.cdc.gov/exposurereport/index.html (Accessed 18 February 2021).

CDER/FDA (2001). Guidance for Industry: Bioanalytical Method Validation. Center for Drug Evaluation (CDER) and US Food and Drug Administration (FDA). www.fda.gov/files/drugs/published/Bioanalytical-Method-Validation-Guidance-for-Industry.pdf (Accessed 18 February 2021).

Cedergreen N. (2014). Quantifying synergy: A systematic review of mixture toxicity studies within environmental toxicology. *Plos One*, **9**(5), e96580.

Celander M. C., Goldstone J. V., Denslow N. D., Iguchi T., Kille P., Meyerhoff R. D., Smith B. A., Hutchinson T. H. and Wheeler J. R. (2011). Species extrapolation for the 21st century. *Environmental Toxicology and Chemistry*, **30**(1), 52–63.

Cencic A. and Langerholc T. (2010). Functional cell models of the gut and their applications in food microbiology: A review. *International Journal of Food Microbiology*, **141**, S4–S14.

Ceretti E., Zani C., Zerbini I., Guzzella L., Scaglia M., Berna V., Donato F., Monarca S. and Feretti D. (2010). Comparative assessment of genotoxicity of mineral water packed in polyethylene terephthalate (PET) and glass bottles. *Water Research*, **44**(5), 1462–1470.

Cetojevic-Simin D., Svircev Z. and Baltic V. V. (2009). *In vitro* cytotoxicity of cyanobacteria from water ecosystems of Serbia. *Journal of the Balkan Union of Oncology*, **14**(2), 289–294.

Chang J. C., Taylor P. B. and Leach F. R. (1981). Use of the Microtox® assay system for environmental samples. *Bulletin of Environmental Contamination and Toxicology*, **26**(2), 150–156.

Cheh A. M., Skochdopole J., Koski P. and Cole L. (1980). Non-volatile mutagens in drinking water – production by chlorination and destruction by sulfite. *Science*, **207**(4426), 90–92.

Chen Q. Y., Jia A., Snyder S. A., Gong Z. Y. and Lam S. H. (2016). Glucocorticoid activity detected by *in vivo* zebrafish assay and *in vitro* glucocorticoid receptor bioassay at environmental relevant concentrations. *Chemosphere*, **144**, 1162–1169.

Chen S., Li D. C., Wu X. N., Chen L. P., Zhang B., Tan Y. F., Yu D. K., Niu Y., Duan H. W., Li Q., Chen R., Aschner M., Zheng Y. X. and Chen W. (2020). Application of cell-based biological bioassays for health risk assessment of PM2.5 exposure in three megacities, China. *Environment International*, **139**, 105703.

Chinathamby K., Allinson M., Shiraishi F., Lopata A. L., Nugegoda D., Pettigrove V. and Allinson G. (2013). Screening for potential effects of endocrine-disrupting chemicals in peri-urban creeks and rivers in Melbourne, Australia using mosquitofish and recombinant receptor-reporter gene assays. *Environmental Science and Pollution Research*, **20**(3), 1831–1841.

Chou P. H., Lee C. H., Ko F. C., Lin Y. J., Kawanishi M., Yagi T. and Li I. C. (2015). Detection of hormone-like and genotoxic activities in indoor dust from Taiwan using a battery of *in vitro* bioassays. *Aerosol and Air Quality Research*, **15**(4), 1412–1421.

Clemons J. H., Allan L. M., Marvin C. H., Wu Z., McCarry B. E., Bryant D. W. and Zacharewski T. R. (1998). Evidence of estrogen- and TCDD-like activities in crude and fractionated extracts of PM10 air particulate material using *in vitro* gene expression assays. *Environmental Science & Technology*, **32**(12), 1853–1860.

Cloutier P. L., Fortin F., Groleau P. E., Brousseau P., Fournier M. and Desrosiers M. (2017). QuEChERS extraction for multi-residue analysis of PCBs, PAHs, PBDEs and PCDD/Fs in biological samples. *Talanta*, **165**, 332–338.

Coecke S., Goldberg A. M., Allen S., Buzanska L., Calamandrei G., Crofton K., Hareng L., Hartung T., Knaut H., Honegger P., Jacobs M., Lein P., Li A., Mundy W., Owen D., Schneider S., Silbergeld E., Reum T., Trnovec T., Monnet-Tschudi F. and Bal-Price A. (2007). Workgroup report: Incorporating *in vitro* alternative methods for developmental neurotoxicity into international hazard and risk assessment strategies. *Environmental Health Perspectives*, **115**(6), 924–931.

Collins F., Gray G. N. and Bucher J. R. (2008). Transforming environmental health protection. *Science*, **319**, 906–907.

Combes R., Balls M., Curren R., Fischbach M., Fusenig N., Kirkland D., Lasne C., Landolph J., LeBoeuf R., Marquardt H., McCormick J., Muller L., Rivedal E., Sabbioni E., Tanaka N., Vasseur P. and Yamasaki H. (1999). Cell transformation assays as predictors of human carcinogenicity – The report and recommendations of ECVAM Workshop 39. *ALTA-Alternatives to Laboratory Animals*, **27**(5), 745–767.

Conley J. M., Evans N., Cardon M. C., Rosenblum L., Iwanowicz L. R., Hartig P. C., Schenck K. M., Bradley P. M. and Wilson V. S. (2017a). Occurrence and *in vitro* bioactivity of estrogen, androgen, and glucocorticoid compounds in a nationwide screen of United States stream waters. *Environmental Science & Technology*, **51**(9), 4781–4791.

Conley J. M., Evans N., Mash H., Rosenblum L., Schenck K., Glassmeyer S., Furlong E. T., Kolpin D. W. and Wilson V. S. (2017b). Comparison of *in vitro* estrogenic activity and estrogen concentrations in source and treated waters from 25 US drinking water treatment plants. *Science of the Total Environment*, **579**, 1610–1617.

Conolly R. B., Ankley G. T., Cheng W. Y., Mayo M. L., Miller D. H., Perkins E. J., Villeneuve D. L. and Watanabe K. H. (2017). Quantitative adverse outcome pathways and their application to predictive toxicology. *Environmental Science & Technology*, **51**(8), 4661–4672.

Conroy O., Sáez A. E., Quanrud D., Ela W. and Arnold R. G. (2007). Changes in estrogen/anti-estrogen activities in ponded secondary effluent. *Science of the Total Environment*, **382**(2–3), 311–323.

Cortes C. and Marcos R. (2018). Genotoxicity of disinfection byproducts and disinfected waters: A review of recent literature. *Mutation Research-Genetic Toxicology and Environmental Mutagenesis*, **831**, 1–12.

Corvi R. and Madia F. (2017). *In vitro* genotoxicity testing – Can the performance be enhanced? *Food and Chemical Toxicology*, **106**, 600–608.

Costa L. G. (1998). Neurotoxicity testing: A discussion of *in vitro* alternatives. *Environmental Health Perspectives*, **106**, 505–510.

Costa L. D., Ottoni M. H. F., dos Santos M. G., Meireles A. B., de Almeida V. G., Pereira W. D., de Avelar-Freitas B. A. and Brito-Melo G. E. A. (2017). Dimethyl sulfoxide (DMSO) decreases cell proliferation and TNF-alpha, IFN-gamma, and IL-2 cytokines production in cultures of peripheral blood lymphocytes. *Molecules*, **22**(11), 1789.

Creusot N., Kinani S., Balaguer P., Tapie N., LeMenach K., Maillot-Marechal E., Porcher J. M., Budzinski H. and Ait-Aissa S. (2010). Evaluation of an hPXR reporter gene assay for the detection of aquatic emerging pollutants: screening of chemicals and application to water samples. *Analytical and Bioanalytical Chemistry*, **396**(2), 569–583.

Creusot N., Ait-Aissa S., Tapie N., Pardon P., Brion F., Sanchez W., Thybaud E., Porcher J. M. and Budzinski H. (2014). Identification of synthetic steroids in river water downstream from pharmaceutical manufacture discharges based on a bioanalytical approach and passive sampling. *Environmental Science & Technology*, **48**(7), 3649–3657.

Dagnino S., Gomez E., Picot B., Cavailles V., Casellas C., Balaguer P. and Fenet H. (2010). Estrogenic and AhR activities in dissolved phase and suspended solids from wastewater treatment plants. *Science of the Total Environment*, **408**(12), 2608–2615.

Daniels K. D., VanDervort D., Wu S. M., Leusch F. D. L., van de Merwe J. P., Jia A. and Snyder S. A. (2018). Downstream trends of *in vitro* bioassay responses in a wastewater effluent-dominated river. *Chemosphere*, **212**, 182–192.

Davey R. A. and Grossmann M. (2016). Androgen receptor structure, function and biology: From bench to bedside. *The Clinical Biochemist Reviews*, **37**(1), 3–15.

de Baat M. L., Bas D. A., van Beusekom S. A. M., Droge S. T. J., van der Meer F., de Vries M., Verdonschot P. F. M. and Kraak M. H. S. (2018). Nationwide screening of surface water toxicity to algae. *Science of the Total Environment*, **645**, 780–787.

de Baat M. L., Kraak M. H. S., Van der Oost R., De Voogt P. and Verdonschot P. F. M. (2019a). Effect-based nationwide surface water quality assessment to identify ecotoxicological risks. *Water Research*, **159**, 434–443.

de Baat M. L., Wieringa N., Droge S. T. J., van Hall B. G., van der Meer F. and Kraak M. H. S. (2019b). Smarter sediment screening: effect-based quality assessment, chemical profiling, and risk identification. *Environmental Science & Technology*, **53**(24), 14479–14488.

de Baat M. L., Van der Oost R., Van der Lee G. H., Wieringa N., Hamers T., Verdonschot P. F. M., De Voogt P. and Kraak M. H. S. (2020). Advancements in effect-based surface water quality assessment. *Water Research*, **183**, 116017.

Delgado L. F., Faucet-Marquis V., Pfohl-Leszkowicz A., Dorandeu C., Marion B., Schetrite S. and Albasi C. (2011). Cytotoxicity micropollutant removal in a crossflow membrane bioreactor. *Bioresource Technology*, **102**(6), 4395–4401.

DeMarini D. M. (2020). A review on the 40th anniversary of the first regulation of drinking water disinfection by-products. *Environmental and Molecular Mutagenesis*, **61**(6), 588–601.

Deneer J. W., Sinnige T., Seinen W. and Hermens J. (1988). The joint acute toxicity to *Daphnia magna* of industrial organic chemicals at low concentration. *Aquatic Toxicology*, **12**, 33–38.

Denison M. S., Soshilov A. A., He G. C., DeGroot D. E. and Zhao B. (2011). Exactly the same but different: promiscuity and diversity in the molecular mechanisms of action of the aryl hydrocarbon (dioxin) receptor. *Toxicological Sciences*, **124**(1), 1–22.

Denison M. S., Mehinto A., Olivieri A., Plumlee M., Schlenk D., Thompson S. and Waggoner C. (2020). Bioanalytical tools for detection and quantification of estrogenic and dioxin-like chemicals in water recycling and reuse. Guidance Document for Developing a Standard Operating Procedure. Prepared for WateReuse California and National Water Research Institute, www.cwea.org/news/nwri-releases-sop-guidance-document-for-bioassays/ (Accessed 7 June 2021).

Desforges J. P., Levin M., Jasperse L., De Guise S., Eulaers I., Letcher R. J., Acquarone M., Nordoy E., Folkow L. P., Jensen T. H., Grondahl C., Bertelsen M. F., Leger J. S., Almunia J., Sonne C. and Dietz R. (2017). Effects of polar bear and killer whale derived contaminant cocktails on marine mammal immunity. *Environmental Science & Technology*, **51**(19), 11431–11439.

Di Paolo C., Ottermanns R., Keiter S., Ait-Aissa S., Bluhm K., Brack W., Breitholtz M., Buchinger S., Carere M., Chalon C., Cousin X., Dulio V., Escher B. I., Hamers T., Hilscherova K., Jarque S., Jonas A., Maillot-Marechal E., Marneffe Y., Nguyen M. T., Pandard P., Schifferli A., Schulze T., Seidensticker S., Seiler T. B., Tang J., van der Oost R., Vermeirssen E., Zounkova R., Zwart N. and Hollert H. (2016). Bioassay battery interlaboratory investigation of emerging contaminants in spiked water extracts – Towards the implementation of bioanalytical monitoring tools in water quality assessment and monitoring. *Water Research*, **104**, 473–484.

Dingemans M. M. L., Baken K. A., van der Oost R., Schriks M. and van Wezel A. P. (2019). Risk-based approach in the revised European Union drinking water legislation: Opportunities for bioanalytical tools. *Integrated Environmental Assessment and Management*, **15**(1), 126–134.

Dizer H., Wittekindt E., Fischer B. and Hansen P. D. (2002). The cytotoxic and genotoxic potential of surface water and wastewater effluents as determined by bioluminescence, umu-assays and selected biomarkers. *Chemosphere*, **46**(2), 225–233.

Dogruer G., Weijs L., Tang J. Y.-M., Hollert H., Kock M., Bell I., Hof C. A. M. and Gaus C. (2018). Effect-based approach for screening of chemical mixtures in whole blood of green turtles from the Great Barrier Reef. *Science of the Total Environment*, **612**, 321–329.

Drakvik E., Altenburger R., Aoki Y., Backhaus T., Bahadori T., Barouki R., Brack W., Cronin M. T. D., Demeneix B., Bennekou S. H., van Klaveren J., Kneuer C., Kolossa-Gehring M., Lebret E., Posthuma L., Reiber L., Rider C., Ruegg J., Testa G., van der Burg B., van der Voet H., Warhurst A. M., van de Water B., Yamazaki K., Oberg M. and Bergman A. (2020). Statement on advancing the assessment of chemical mixtures and their risks for human health and the environment. *Environment International*, **134**, 105267.

Dyer S. D., White-Hull C. E. and Shephard B. K. (2000). Assessments of chemical mixtures via toxicity reference values overpredict hazard to Ohio fish communities. *Environmental Science & Technology*, **34**(12), 2518–2524.

Eggen R. and Segner H. (2003). The potential of mechanism-based bioanalytical tools in ecotoxicological exposure and effect assessment. *Analytical and Bioanalytical Chemistry*, **377**(3), 386–396.

Elad T. and Belkin S. (2012). Whole-cell biochips for online water monitoring. *Bioengineered*, **3**(2), 124–128.

Ellman G. L., Courtney K. D., Andres V. and Featherstone R. M. (1961). A new and rapid colorimetric determination of acetylcholinesterase activity. *Biochemical Pharmacology*, **7**(2), 88–95.

Embry M. R., Belanger S. E., Braunbeck T. A., Galay-Burgos M., Halder M., Hinton D. E., Léonard M. A., Lillicrap A., Norberg-King T. and Whale G. (2010). The fish embryo toxicity test as an animal alternative method in hazard and risk assessment and scientific research. *Aquatic Toxicology*, **97**(2), 79–87.

EMEA (2011). Guideline on validation of bioanalytical methods. EMEA/CHMP/EWP/192217/2009 Rev. 1 Corr. 2** Committee for Medicinal Products for Human Use (CHMP), European Medicines Agency, London, UK. www.ema.europa.eu/en/documents/scientific-guideline/guideline-bioanalytical-method-validation_en.pdf (Accessed 18 February 2021).

Emmanuel E., Perrodin Y., Keck G., Blanchard J. M. and Vermande P. (2005). Ecotoxicological risk assessment of hospital wastewater: A proposed framework for raw effluents discharging into urban sewer network. *Journal of Hazardous Materials*, **117**(1), 1–11.

enHealth (2012). Environmental Health Risk Assessment. Guidelines for Assessing Human Health Risks from Environmental Hazards. Department of Health and Aging and enHealth Council, Canberra, Australia. www1.health.gov.au/internet/main/publishing.nsf/Content/health-pubhlth-publicat-environ.htm (Accessed 3 November 2020).

Environment and Climate Change Canada (2020). Evaluating Existing Substances. www.ec.gc.ca/ese-ees/ (Accessed 15 October 2020).

EP&EC (1998). Council Directive 98/83/EC of 3 November 1998 on the quality of water intended for human consumption. European Parliament and European Council, Official Journal of the European Communities, L330/32.

EP&EC (2000). Directive 2000/60/EC of the European Parliament and the Council of 23 October 2000 establishing a framework for Community action in the field of water policy (short: Water Framework Directive). European Parliament and European Council, Official Journal of the European Communities, L327/1.

EP&EC (2006a). Regulation (EC) No 1907/2006 of the European Parliament and of the Council of 18 December 2006 concerning the Registration, Evaluation, Authorisation and Restriction of Chemicals (REACH), establishing a European Chemicals Agency, amending Directive 1999/45/EC and repealing Council Regulation (EEC) No 793/93 and Commission Regulation (EC) No 1488/94 as well as Council Directive 76/769/EEC and Commission Directives 91/155/EEC, 93/67/EEC, 93/105/EC and 2000/21/EC. European Parliament and European Council, Official Journal of the European Communities, 396/1.

EP&EC (2006b). Regulation (EC) no 1907/2006 REACH, Criteria for the identification of persistent, bioaccumulative and toxic substances, and very persistent and very bioaccumulative substances, Annex XIII. European Parliament and European Council, Official Journal of the European Communities, L69/7.

EP&EC (2010). Directive 2010/63/EU of 22 September 2010 on the protection of animals used for scientific purposes. European Parliament and European Council, Official Journal of the European Communities, L 276/233.

EP&EC (2013). Directive 2013/39/EU of the European Parliament and of the Council of of 12 August 2013 amending Directives 2000/60/EC and 2008/105/EC as regards priority substances in the field of water policy. European Parliament and European Council, Official Journal of the European Communities, L226/1.

EP&EC (2020). Directive of the European Parliament and of the Council of 16 December 2020 on the quality of water intended for human consumption (recast). European Parliament and European Council, Official Journal of the European Communities, L435/1.

Epler J. L., Larimer F. W., Rao T. K., Nix C. E. and Ho T. (1978). Energy-related pollutants in the environment – use of short-term tests for mutagenicity in the isolation and identification of biohazards. *Environmental Health Perspectives*, **27**, 11–20.

Erdmann S. E., Dietz R., Sonne C., Bechshoft T. O., Vorkamp K., Letcher R. J., Long M. H. and Bonefeld-Jorgensen E. C. (2013). Xenoestrogenic and dioxin-like activity in blood of East Greenland polar bears (*Ursus maritimus*). *Chemosphere*, **92** (5), 583–591.

Ermler S., Scholze M. and Kortenkamp A. (2014). Genotoxic mixtures and dissimilar action: concepts for prediction and assessment. *Archives of Toxicology*, **88**(3), 799–814.

Escher B. I. and Fenner K. (2011). Recent advances in the environmental risk assessment of transformation products. *Environmental Science & Technology*, **45**(9), 3835–3847.

Escher B. I. and Hermens J. L. M. (2002). Modes of action in ecotoxicology: their role in body burdens, species sensitivity, QSARs, and mixture effects. *Environmental Science & Technology*, **36**, 4201–4217.

Escher B. I. and Neale P. A. (2021). Effect-based trigger values for mixtures of chemicals in surface water detected with *in vitro* bioassays. *Environmental Toxicology and Chemistry*, **40**(2), 487–499.

Escher B. I., Eggen R. I. L., Schreiber U., Schreiber Z., Vye E., Wisner B. and Schwarzenbach R. P. (2002). Baseline toxicity (narcosis) of organic chemicals determined by *in vitro* membrane potential measurements in energy-transducing membranes. *Environmental Science & Technology*, **36**(9), 1971–1979.

Escher B. I., Bramaz N., Eggen R. I. L. and Richter M. (2005). *In vitro* assessment of modes of toxic action of pharmaceuticals in aquatic life. *Environmental Science & Technology*, **39**, 3090–3100.

Escher B. I., Bramaz N., Mueller J. F., Quayle P., Rutishauser S. and Vermeirssen E. L. M. (2008a). Toxic equivalent concentrations (TEQs) for baseline toxicity and specific modes of action as a tool to improve interpretation of ecotoxicity testing of environmental samples. *Journal of Environmental Monitoring*, **10**(5), 612–621.

Escher B. I., Bramaz N., Quayle P., Rutishauser S. and Vermeirssen E. L. M. (2008b). Monitoring of the ecotoxicological hazard potential by polar organic micropollutants in sewage treatment plants and surface waters using a mode-of-action based test battery. *Journal of Environmental Monitoring*, **10**(5), 622–631.

Escher B. I., Bramaz N. and Ort C. (2009). JEM Spotlight: Monitoring the treatment efficiency of a full scale ozonation on a sewage treatment plant with a mode-of-action based test battery. *Journal of Environmental Monitoring*, **11**(10), 1836–1846.

Escher B. I., Lawrence M., Macova M., Mueller J. F., Poussade Y., Robillot C., Roux A. and Gernjak W. (2011). Evaluation of contaminant removal of reverse osmosis and advanced oxidation in full-scale operation by combining passive sampling with chemical analysis and bioanalytical tools. *Environmental Science & Technology*, **45**, 5387–5394.

Escher B. I., Dutt M., Maylin E., Tang J. Y. M., Toze S., Wolf C. R. and Lang M. (2012). Water quality assessment using the AREc32 reporter gene assay indicative of the oxidative stress response pathway. *Journal of Environmental Monitoring*, **14**(11), 2877–2885.

Escher B. I., van Daele C., Dutt M., Tang J. Y. M. and Altenburger R. (2013). Most oxidative stress response in water samples comes from unknown chemicals: The need for effect-based water quality trigger values. *Environmental Science & Technology*, **47** (13), 7002–7011.

Escher B. I., Allinson M., Altenburger R., Bain P., Balaguer P., Busch W., Crago J., Humpage A., Denslow N. D., Dopp E., Hilscherova K., Kumar A., Grimaldi M., Jayasinghe B. S., Jarosova B., Jia A., Makarov S., Maruya K. A., Medvedev A., Mehinto A. C., Mendez J. E., Poulsen A., Prochazka E., Richard J., Schifferli A., Schlenk D., Scholz S., Shiraishi

F., Snyder S., Su G., Tang J., van der Burg B., van der Linden S., Werner I., Westerheide S. D., Wong C. K. C., Yang M., Yeung B., Zhang X. and Leusch F. D. L. (2014). Benchmarking organic micropollutants in wastewater, recycled water and drinking water with *in vitro* bioassays. *Environmental Science & Technology*, **48**, 1940–1956.

Escher B. I., Neale P. A. and Leusch F. D. L. (2015). Effect-based trigger values for *in vitro* bioassays: Reading across from existing water quality guideline values. *Water Research*, **81**, 137–148.

Escher B. I., Hackermüller J., Polte T., Scholz S., Aigner A., Altenburger R., Böhme A., Bopp S. K., Brack W., Busch W., Chadeau-Hyam M., Covaci A., Eisenträger A., Galligan J. J., Garcia-Reyero N., Hartung T., Hein M., Herberth G., Jahnke A., Kleinjans J., Klüver N., Krauss M., Lamoree M., Lehmann I., Luckenbach T., Miller G. W., Müller A., Phillips D. H., Reemtsma T., Rolle-Kampczyk U., Schüürmann G., Schwikowski B., Tan Y.-M., Trump S., Walter-Rohde S. and Wambaugh J. F. (2017). From the exposome to mechanistic understanding of chemical-induced adverse effects. *Environment International*, **99**, 97–106.

Escher B., Neale P. A. and Villeneuve D. (2018a). The advantages of linear concentration-response curves for *in vitro* bioassays with environmental samples. *Environmental Toxicology and Chemistry*, **37**(9), 2273–2280.

Escher B. I., Ait-Aissa S., Behnisch P. A., Brack W., Brion F., Brouwer A., Buchinger S., Crawford S. E., Du Pasquier D., Hamers T., Hettwer K., Hilscherova K., Hollert H., Kase R., Kienle C., Tindall A. J., Tuerk J., van der Oost R., Vermeirssen E. and Neale P. A. (2018b). Effect-based trigger values for *in vitro* and *in vivo* bioassays performed on surface water extracts supporting the environmental quality standards (EQS) of the European Water Framework Directive. *Science of the Total Environment*, **628–629**, 748–765.

Escher B. I., Glauch L., Konig M., Mayer P. and Schlichting R. (2019). Baseline toxicity and volatility cutoff in reporter gene assays used for high-throughput screening. *Chemical Research in Toxicology*, **32**(8), 1646–1655.

Escher B. I., Abagyan R., Embry M., Klüver N., Redman A. D., Zarfl C. and Parkerton T. F. (2020a). Recommendations for improving methods and models for aquatic hazard assessment of ionizable organic chemicals. *Environmental Toxicology and Chemistry*, **39**(1), 269–286.

Escher B. I., Braun G. and Zarfl C. (2020b). Exploring the concepts of concentration addition and independent action using a linear low-effect mixture model. *Environmental Toxicology and Chemistry*, **39**(12), 2552–2559.

Escher B. I., Henneberger L., Schlichting R. and Fischer F. C. (2020c). Cytotoxicity burst or baseline toxicity? Differentiating specific from nonspecific effects in reporter gene assays. *Environmental Health Perspectives*, **128**(7), 077007.

Escher B. I., Stapleton H. M. and Schymanski E. L. (2020d). Tracking complex mixtures of chemicals in our changing environment. *Science*, **367**, 388–392.

European Chemicals Agency (2011). Guidance on information requirements and chemical safety assessment. Part A: Introduction to the guidance document. echa.europa.eu/guidance-documents/guidance-on-information-requirements-and-chemical-safety-assessment (Accessed 15 October 2020).

European Chemicals Agency (2012). Guidance for the implementation of REACH, Guidance on information requirements and chemical safety assessment. Chapter R.19: Uncertainty analysis. echa.europa.eu/documents/10162/13632/information_requirements_r19_en.pdf (Accessed 23 October 2020).

European Chemicals Agency (2017). Read-Across Assessment Framework (RAAF) – considerations on multi-constituent substances and UVCBs. echa.europa.eu/documents/10162/13630/raaf_uvcb_report_en.pdf/3f79684d-07a5-e439-16c3-d2c8da96a316 (Accessed 20 December 2020).

European Commission (2009). Common Implementation Strategy for the WFD, Guidance Document No. 19, Guidance on Surface Water Chemical Monitoring. https://ec.europa.eu/environment/water/water-framework/facts_figures/guidance_docs_en.htm (Accessed 16 February 2021).

European Commission (2011). Common implementation strategy for the water framework directive (2000/60/EC). Guidance document No. 27. Technical guidance for deriving environmental quality standards, European Communities. Version 2018. https://ec.europa.eu/environment/water/water-framework/facts_figures/guidance_docs_en.htm (Accessed 16 February 2021).

European Commission (2020). Communication from the Commission to the European Parliament, the Council, the European Economic and Social Committee and the Committee of the Regions (COM (2020) 667 final). Chemicals Strategy for Sustainability towards a Toxic-free Environment, European Communities. eur-lex.europa.eu (Accessed 10 April 2021).

European Environment Agency (2013). Late lessons from early warnings: science, precaution, innovation. www.eea.europa.eu/publications/late-lessons-2 (Accessed 23 October 2020).

Fang Y.-X., Ying G.-G., Zhao J.-L., Chen F., Liu S., Zhang L.-J. and Yang B. (2012). Assessment of hormonal activities and genotoxicity of industrial effluents using *in vitro* bioassays combined with chemical analysis. *Environmental Toxicology and Chemistry*, **31**(6), 1273–1282.

Farmen E., Harman C., Hylland K. and Tollefsen K. E. (2010). Produced water extracts from North Sea oil production platforms result in cellular oxidative stress in a rainbow trout *in vitro* bioassay. *Marine Pollution Bulletin*, **60**(7), 1092–1098.

Farré M., Klöter G., Petrovic M., Alonso M. C., de Alda M. J. L. and Barceló D. (2002). Identification of toxic compounds in wastewater treatment plants during a field experiment. *Analytica Chimica Acta* **456**(1), 19–30.

Farre M. J., Day S., Neale P. A., Stalter D., Tang J. Y. M. and Escher B. I. (2013). Bioanalytical and chemical assessment of the disinfection by-product formation potential: Role of organic matter. *Water Research*, **47**(14), 5409–5421.

Fatima R. A. and Ahmad M. (2006). Genotoxicity of industrial wastewaters obtained from two different pollution sources in northern India: A comparison of three bioassays. *Mutation Research – Genetic Toxicology and Environmental Mutagenesis*, **609**(1), 81–91.

Faust M., Altenburger R. and Grimme L. H. (2001). Predicting the joint algal toxicity of multicomponent s-triazine mixtures at low-effect concentrations of individual toxicants. *Aquatic Toxicology*, **56**(1), 13–32.

Faust M., Altenburger R., Backhaus T., Blanck H., Boedeker W., Gramatica P., Hamer V., Scholze M., Vighi M. and Grimme L. H. (2003). Joint algal toxicity of 16 dissimilarly acting chemicals is predictable by the concept of independent action. *Aquatic Toxicology*, **63**(1), 43–63.

Fay K. A., Villeneuve D. L., Swintek J., Edwards S. W., Nelms M. D., Blackwell B. R. and Ankley G. T. (2018). Differentiating pathway-specific from nonspecific effects in high-throughput toxicity data: A foundation for prioritizing adverse outcome pathway development. *Toxicological Sciences*, **163**(2), 500–515.

Fedorenkova A., Vonk J. A., Lenders H. J. R., Ouborg N. J., Breure A. M. and Hendriks A. J. (2010). Ecotoxicogenomics: Bridging the gap between genes and populations. *Environmental Science & Technology*, **44**(11), 4328–4333.

Fenner K., Kooijman C., Scheringer M. and Hungerbuhler K. (2002). Including transformation products into the risk assessment for chemicals: The case of nonylphenol ethoxylate usage in Switzerland. *Environmental Science & Technology*, **36**(6), 1147–1154.

Ferk P. and Daris B. (2018). The influence of dimethyl sulfoxide (DMSO) on metabolic activity and morphology of melanoma cell line WM-266-4. *Cellular and Molecular Biology*, **64**(11), 41–43.

Filer D. L., Kothiya P., Setzer R. W., Judson R. S. and Martin M. T. (2016). tcpl: the ToxCast pipeline for high-throughput screening data. *Bioinformatics*, **33**(4), 618–620.

Finlayson K. A., Leusch F. D. L. and van de Merwe J. P. (2019a). Cytotoxicity of organic and inorganic compounds to primary cell cultures established from internal tissues of *Chelonia mydas*. *Science of the Total Environment*, **664**, 958–967.

Finlayson K. A., Leusch F. D. L. and van de Merwe J. P. (2019b). Primary green turtle (*Chelonia mydas*) skin fibroblasts as an *in vitro* model for assessing genotoxicity and oxidative stress. *Aquatic Toxicology*, **207**, 13–18.

Finlayson K. A., Madden Hof C. A. and van de Merwe J. P. (2020). Development and application of species-specific cell-based bioassays to assess toxicity in green sea turtles. *Science of the Total Environment*, **747**, 142095.

Fischer F., Henneberger L., König M., Bittermann K., Linden L., Goss K.-U. and Escher B. I. (2017). Modeling exposure in the Tox21 *in vitro* bioassays. *Chemical Research in Toxicology*, **30**, 1197–1208.

Fischer F., Abele C., Droge S. T. J., Henneberger L., König M., Schlichting R., Scholz S. and Escher B. I. (2018a). Cellular uptake kinetics of neutral and charged chemicals in *in vitro* assays measured by fluorescence microscopy. *Chemical Research in Toxicology*, **31**, 646–657.

Fischer F. C., Cirpka O., Goss K. U., Henneberger L. and Escher B. I. (2018b). Application of experimental polystyrene partition constants and diffusion coefficients to predict the sorption of organic chemicals to well plates in *in vitro* bioassays. *Environmental Science & Technology*, **52**, 13511–13522.

Fischer F. C., Henneberger L., Schlichting R. and Escher B. I. (2019). How to improve the dosing of chemicals in high-throughput *in vitro* mammalian cell assays. *Chemical Research in Toxicology*, **32**(8), 1462–1468.

Fischer F. C., Abele C., Henneberger L., Klüver N., König M., Mühlenbrink M., Schlichting R. and Escher B. I. (2020). Characterizing cellular metabolism in high-throughput *in vitro* reporter gene assays. *Chemical Research in Toxicology*, **33**(7), 1770–1779.

Flohe L. and Gunzler W. A. (1984). Assays of glutathione-peroxidase. *Methods in Enzymology*, **105**, 114–121.

Forsberg N. D., Wilson G. R. and Anderson K. A. (2011). Determination of parent and substituted polycyclic aromatic hydrocarbons in high-fat salmon using a modified QuEChERS extraction, dispersive SPE and GC-MS. *Journal of Agricultural and Food Chemistry*, **59**(15), 8108–8116.

Fox S. I. (1991). Perspectives on Human Biology. William C. Brown Publishers, Dubuque, IA, USA.

French V. A., King S. C., Kumar A., Northcott G., McGuinness K. and Parry D. (2015). Characterisation of microcontaminants in Darwin Harbour, a tropical estuary of northern Australia undergoing rapid development. *Science of the Total Environment*, **536**, 639–647.

Furlong E. T., Werner S. L., Anderson B. D. and Cahill J. D. (2008). Determination of human-health pharmaceuticals in filtered water by chemically modified styrene–divinylbenzene resin-based solid-phase extraction and high-performance liquid chromatography/mass spectrometry. In: US Geological Survey Techniques and Methods (ed.), USGS, pubs.usgs.gov/tm/tm5b5/pdf/tm-5-b5_508.pdf (Accessed 18 February 2021).

Furuichi T., Kannan K., Glesy J. P. and Masunaga S. (2004). Contribution of known endocrine disrupting substances to the estrogenic activity in Tama River water samples from Japan using instrumental analysis and *in vitro* reporter gene assay. *Water Research*, **38**(20), 4491–4501.

Gagné F. and Blaise C. (1998). Estrogenic properties of municipal and industrial wastewaters evaluated with a rapid and sensitive chemoluminescent *in situ* hybridization assay (CISH) in rainbow trout hepatocytes. *Aquatic Toxicology*, **44**(1–2), 83–91.

Gao X. Z., Huang C., Rao K. F., Xu Y. P., Huang Q. H., Wang F., Ma M. and Wang Z. J. (2018). Occurrences, sources, and transport of hydrophobic organic contaminants in the waters of Fildes Peninsula, Antarctica. *Environmental Pollution*, **241**, 950–958.

Garcia-Reyero N., Grau E., Castillo M., De Alda M. J. L., Barcelo D. and Pina B. (2001). Monitoring of endocrine disruptors in surface waters by the yeast recombinant assay. *Environmental Toxicology and Chemistry*, **20**(6), 1152–1158.

Gartiser S., Hafner C., Oeking S. and Paschke A. (2009). Results of a "Whole Effluent Assessment" study from different industrial sectors in Germany according to OSPAR's WEA strategy. *Journal of Environmental Monitoring*, **11**(2), 359–369.

Gartiser S., Hafner C., Hercher C., Kronenberger-Schafer K. and Paschke A. (2010). Whole effluent assessment of industrial wastewater for determination of bat compliance. Part I: paper manufacturing industry. *Environmental Science and Pollution Research*, **17**(4), 856–865.

Gehrmann L., Bielak H., Behr M., Itzel F., Lyko S., Simon A., Kunze G., Dopp E., Wagner M. and Tuerk J. (2018). (Anti-)estrogenic and (anti-)androgenic effects in wastewater during advanced treatment: Comparison of three *in vitro* bioassays. *Environmental Science and Pollution Research*, **25**, 4094–4104.

Genthe B., Steyn M., Aneck-Hahn N., Van Zijl C. and De Jager C. (2009). The feasibility of a health risk assessment framework to derive guidelines for oestrogen activity in treated drinking water. Water Research Commission, Pretoria, South Africa WRC. 1749, p. 02.

Gibb S. (2008). Toxicity testing in the 21st century: A vision and a strategy. *Reproductive Toxicology*, **25**(1), 136–138.

Gijsbers L., Man H. Y., Kloet S. K., de Haan L. H., Keijer J., Rietjens I. M., van der Burg B. and Aarts J. M. (2011). Stable reporter cell lines for peroxisome proliferator-activated receptor gamma (PPARgamma)-mediated modulation of gene expression. *Analytical Biochemistry*, **414**(1), 77–83.

Gilmore T. D. (2006). Introduction to NF-κB: players, pathways, perspectives. *Oncogene*, **25** (51), 6680–6684.

Glassmeyer S. T., Furlong E. T., Kolpin D. W., Batt A. L., Benson R., Boone J. S., Conerly O., Donohue M. J., King D. N., Kostich M. S., Mash H. E., Pfaller S. L., Schenck K. M., Simmons J. E., Varughese E. A., Vesper S. J., Villegas E. N. and Wilson V. S. (2017). Nationwide reconnaissance of contaminants of emerging concern in source and treated drinking waters of the United States. *Science of the Total Environment*, **581**, 909–922.

Glauch L. and Escher B. I. (2020). The combined algae test for the evaluation of mixture toxicity in environmental samples. *Environmental Toxicology and Chemistry*, **39**(12), 2496–2508.

Gohlke J. M. and Portier C. J. (2007). The forest for the trees: a systems approach to human health research. *Environmental Health Perspectives*, **115**, 1261–1263.

González A., Kroll K. J., Silva-Sanchez C., Carriquiriborde P., Fernandino J. I., Denslow N. D. and Somoza G. M. (2020). Steroid hormones and estrogenic activity in the wastewater outfall and receiving waters of the Chascomús chained shallow lakes system (Argentina). *Science of the Total Environment*, **743**, 140401.

Grimaldi M., Boulahtouf A., Delfosse V., Thouennon E., Bourguet W. and Balaguer P. (2015). Reporter cell lines for the characterization of the interactions between human nuclear receptors and endocrine disruptors. *Frontiers in Endocrinology*, **6**, 62.

Groothuis F. A., Heringa M. B., Nicol B., Hermens J. L. M., Blaauboer B. J. and Kramer N. I. (2015). Dose metric considerations in *in vitro* assays to improve quantitative *in vitro–in vivo* dose extrapolations. *Toxicology*, **332**, 30–40.

Grothe D. R., Dickson K. L. and Reed-Judkins D. K. (eds.) (1995). Whole Effluent Toxicity Testing. An Evaluation of the Methods and Prediction of Receiving System Impacts. SETAC Special Publication Series, SETAC Press, SETAC Press, Pensacola, FL, USA.

Gruener N. (1978). Mutagenicity of ozonated, recycled water. *Bulletin of Environmental Contamination and Toxicology*, **20**(4), 522–526.

Gruiz K., Fekete-Kertesz I., Kunglne-Nagy Z., Hajdu C., Feigl V., Vaszita E. and Molnar M. (2016). Direct toxicity assessment – Methods, evaluation, interpretation. *Science of the Total Environment*, **563**, 803–812.

Grung M., Lichtenthaler R., Ahel M., Tollefsen K.-E., Langford K. and Thomas K. V. (2007). Effects-directed analysis of organic toxicants in wastewater effluent from Zagreb, Croatia. *Chemosphere*, **67**(1), 108–120.

Guillouzo A., Corlu A., Aninat C., Glaise D., Morel F. and Guguen-Guillouzo C. (2007). The human hepatoma HepaRG cells: A highly differentiated model for studies of liver metabolism and toxicity of xenobiotics. *Chemico-Biological Interactions*, **168**(1), 66–73.

Gustavsson L., Hollert H., Jonsson S., van Bavel B. and Engwall M. (2007). Reed beds receiving industrial sludge containing nitroaromatic compounds – Effects of outgoing water and bed material extracts in the umu-C genotoxicity assay, DR-CALUX assay and on early life stage development in Zebrafish (*Danio rerio*). *Environmental Science and Pollution Research*, **14**(3), 202–211.

Gustavsson M., Kreuger J., Bundschuh M. and Backhaus T. (2017). Pesticide mixtures in the Swedish streams: Environmental risks, contributions of individual compounds and consequences of single-substance oriented risk mitigation. *Science of the Total Environment*, **598**, 973–983.

Guzzella L., Di Caterino F., Monarca S., Zani C., Feretti D., Zerbini I., Nardi G., Buschini A., Poli P. and Rossi C. (2006). Detection of mutagens in water-distribution systems after disinfection. *Mutation Research – Genetic Toxicology and Environmental Mutagenesis*, **608**(1), 72–81.

Guzzella L., Monarca S., Zani C., Feretti D., Zerbini I., Buschini A., Poli P., Rossi C. and Richardson S. D. (2004). *In vitro* potential genotoxic effects of surface drinking water treated with chlorine and alternative disinfectants. *Mutation Research – Genetic Toxicology and Environmental Mutagenesis*, **564**(2), 179–193.

Haber L. T., Dourson M. L., Allen B. C., Hertzberg R. C., Parker A., Vincent M. J., Maier A. and Boobis A. R. (2018). Benchmark dose (BMD) modeling: current practice, issues, and challenges. *Critical Reviews in Toxicology*, **48**(5), 387–415.

Hamers T., Kamstra J. H., van Gils J., Kotte M. C. and van Hattum A. G. M. (2015). The influence of extreme river discharge conditions on the quality of suspended particulate matter in Rivers Meuse and Rhine (The Netherlands). *Environmental Research*, **143**, 241–255.

Hamers T., Legradi J., Zwart N., Smedes F., de Weert J., van den Brandhof E. J., van de Meent D. and de Zwart D. (2018). Time-Integrative Passive sampling combined with TOxicity Profiling (TIPTOP): an effect-based strategy for cost-effective chemical water quality assessment. *Environmental Toxicology and Pharmacology*, **64**, 48–59.

Hamers T., Molin K. R. J., Koeman J. H. and Murk A. J. (2000). A small-volume bioassay for quantification of the esterase inhibiting potency of mixtures of organophosphate and carbamate insecticides in rainwater: Development and optimization. *Toxicological Sciences*, **58**(1), 60–67.

Hamilton L. A., Tremblay L. A., Northcott G. L., Boake M. and Lim R. P. (2016). The impact of variations of influent loading on the efficacy of an advanced tertiary sewage treatment plant to remove endocrine disrupting chemicals. *Science of the Total Environment*, **560**, 101–109.

Han Y. N., Li N., Oda Y., Ma M., Rao K. F., Wang Z. J., Jin W., Hong G., Li Z. G. and Luo Y. (2016). Evaluation of genotoxic effects of surface waters using a battery of bioassays indicating different mode of action. *Ecotoxicology and Environmental Safety*, **133**, 448–456.

Harder A., Escher B. I., Landini P., Tobler N. B. and Schwarzenbach R. P. (2003). Evaluation of bioanalytical tools for toxicity assessment and mode of toxic action classification of reactive chemicals. *Environmental Science & Technology*, **37**(21), 4962–4970.

Harrill J., Shah I., Setzer R. W., Haggard D., Auerbach S., Judson R. and Thomas R. S. (2019). Considerations for strategic use of high-throughput transcriptomics chemical screening data in regulatory decisions. *Current Opinion in Toxicology*, **15**, 64–75.

Hartung T. (2010). Lessons learned from alternative methods and their validation for a new toxicology in the 21st century. *Journal of Toxicology and Environmental Health, Part B: Critical Reviews*, **13**(2), 277–290.

Hashmi M. A. K., Escher B. I., Krauss M., Teodorovic I. and Brack W. (2018). Effect-directed analysis (EDA) of Danube River water sample receiving untreated municipal wastewater from Novi Sad, Serbia. *Science of the Total Environment*, **624**, 1072–1081.

Hashmi M. A. K., Krauss M., Escher B. I., Teodorovic I. and Brack W. (2020). Effect-directed analysis of progestogens and glucocorticoids at trace concentrations in river water. *Environmental Toxicology and Chemistry*, **39**, 189–199.

He G. C., Tsutsumi T., Zhao B., Baston D. S., Zhao J., Heath-Pagliuso S. and Denison M. S. (2011). Third-generation Ah receptor-responsive luciferase reporter plasmids: Amplification of dioxin-responsive elements dramatically increases CALUX bioassay sensitivity and responsiveness. *Toxicological Sciences*, **123**(2), 511–522.

Health Canada (2020). Guidelines for Canadian Drinking Water Quality – Summary Table. Water and Air Quality Bureau, Healthy Environments and Consumer Safety Branch, Health Canada, Ottawa, Ontario, Canada.

Hebert A., Feliers C., Lecarpentier C., Neale P. A., Schlichting R., Thibert S. and Escher B. I. (2018). Bioanalytical assessment of adaptive stress responses in drinking water as a tool to differentiate between micropollutants and disinfection by-products. *Water Research*, **132**, 340–349.

Heinrich P., Diehl U., Forster F. and Braunbeck T. (2014). Improving the *in vitro* ethoxyresorufin-O-deethylase (EROD) assay with RTL-W1 by metabolic normalization and use of beta-naphthofiavone as the reference substance. *Comparative Biochemistry and Physiology C-Toxicology & Pharmacology*, **164**, 27–34.

Henneberg A., Bender K., Blaha L., Giebner S., Kuch B., Kohler H. R., Maier D., Oehlmann J., Richter D., Scheurer M., Schulte-Oehlmann U., Sieratowicz A., Ziebart S. and Triebskorn R. (2014). Are *in vitro* methods for the detection of endocrine potentials in the aquatic environment predictive for *in vivo* effects? Outcomes of the projects SchussenAktiv and SchussenAktiv plus in the Lake Constance area, Germany. *Plos One*, **9**(6), e98307.

Henneberger L., Muhlenbrink M., Konig M., Schlichting R., Fischer F. C. and Escher B. I. (2019). Quantification of freely dissolved effect concentrations in *in vitro* cell-based bioassays. *Archives of Toxicology*, **93**(8), 2295–2305.

Heringa M. B. and Hermens J. L. (2003). Measurement of free concentration using negligible-depletion-solid phase microextraction (nd-SPME). *Trends in Analytical Chemistry*, **22**(1), 575–587.

Hermens J., Könemann H., Leeuwangh P. and Musch A. (1985). Quantitative structure-activity relationships in aquatic toxicity studies of chemicals and complex mixtures of chemicals. *Environmental Toxicology and Chemistry*, **4**, 273–279.

Hirota M., Motoyama A., Suzuki M., Yanagi M., Kitagaki M., Kouzuki H., Hagino S., Itagaki H., Sasa H., Kagatani S. and Aiba S. (2010). Changes of cell-surface thiols and intracellular signaling in human monocytic cell line THP-1 treated with diphenylcyclopropenone. *Journal of Toxicological Sciences*, **35**(6), 871–879.

Hirsch C. and Schildknecht S. (2019). *In vitro* research reproducibility: keeping up high standards. *Frontiers in Pharmacology*, **10**, 1484.

Hissin P. J. and Hilf R. (1976). A fluorometric method for determination of oxidized and reduced glutathione in tissues. *Analytical Biochemistry*, **74**(1), 214–226.

Hollender J., Schymanski E. L., Singer H. P. and Ferguson P. L. (2017). Nontarget screening with high resolution mass spectrometry in the environment: ready to go? *Environmental Science & Technology*, **51**(20), 11505–11512.

Holmes C. M., Brown C. D., Hamer M., Jones R., Maltby L., Posthuma L., Silberhorn E., Teeter J. S., Warne M. S. and Weltje L. (2018). Prospective aquatic risk assessment for chemical mixtures in agricultural landscapes. *Environmental Toxicology and Chemistry*, **37**(3), 674–689.

Honda G. S., Pearce R. G., Pham L. L., Setzer R. W., Wetmore B. A., Sipes N. S., Gilbert J., Franz B., Thomas R. S. and Wambaugh J. F. (2019). Using the concordance of *in vitro* and *in vivo* data to evaluate extrapolation assumptions. *Plos One*, **14**(5), e0217564.

Hong S., Khim J. S., Naile J. E., Park J., Kwon B.-O., Wang T., Lu Y., Shim W. J., Jones P. D. and Giesy J. R. (2012). AhR-mediated potency of sediments and soils in estuarine and coastal areas of the Yellow Sea region: A comparison between Korea and China. *Environmental Pollution*, **171**, 216–225.

Hong S., Khim J. S., Park J., Kim S., Lee S., Choi K., Kim C. S., Choi S. D., Park J., Ryu J., Jones P. D. and Giesy J. P. (2014). Instrumental and bioanalytical measures of dioxin-like compounds and activities in sediments of the Pohang Area, Korea. *Science of the Total Environment*, **470–471**, 1517–1525.

Houtman C. J., Legler J. and Thomas K. (2011). Effect-Directed Analysis of Endocrine Disruptors in Aquatic Ecosystems. In: Effect-Directed Analysis of Complex Environmental Contamination, W. Brack (ed.), Springer, Heidelberg, Germany, pp. 237–265.

Houtman C. J., ten Broek R. and Brouwer A. (2018). Steroid hormonal bioactivities, culprit natural and synthetic hormones and other emerging contaminants in waste water measured using bioassays and UPLC-tQ-MS. *Science of the Total Environment*, **630**, 1492–1501.

Houtman C. J., ten Broek R., van Oorschot Y., Kloes D., van der Oost R., Rosielle M. and Lamoree M. H. (2020). High resolution effect-directed analysis of steroid hormone (ant)agonists in surface and wastewater quality monitoring. *Environmental Toxicology and Pharmacology*, **80**, 103460.

Hrudey S. E. and Fawell J. (2015). 40 years on: what do we know about drinking water disinfection by-products (DBPs) and human health? *Water Science and Technology-Water Supply*, **15**(4), 667–674.

Hu J., Wang W., Zhu Z., Chang H., Pan F. and Lin B. (2007). Quantitative structure-activity relationship model for prediction of genotoxic potential for quinolone antibacterials. *Environmental Science & Technology*, **41**(13), 4806–4812.

Huang G. Y., Liu Y. S., Chen X. W., Liang Y. Q., Liu S. S., Yang Y. Y., Hu L. X., Shi W. J., Tian F., Zhao J. L., Chen J. and Ying G. G. (2016). Feminization and masculinization of western mosquitofish (*Gambusia affinis*) observed in rivers impacted by municipal wastewaters. *Scientific Reports*, **6**, 20884.

Huchthausen J., Muhlenbrink M., Konig M., Escher B. I. and Henneberger L. (2020). Experimental exposure assessment of ionizable organic chemicals in *in vitro* cell-based bioassays. *Chemical Research in Toxicology*, **33**(7), 1845–1854.

Huggett D. B., Foran C. M., Brooks B. W., Weston J., Peterson B., Marsh K. E., La Point T. W. and Schlenk D. (2003). Comparison of *in vitro* and *in vivo* bioassays for estrogenicity in effluent from North American municipal wastewater facilities. *Toxicological Sciences*, **72**(1), 77–83.

Hurst M. R., Balaam J., Chan-Man Y. L., Thain J. E. and Thomas K. V. (2004). Determination of dioxin and dioxin-like compounds in sediments from UK estuaries using a bio-analytical approach: chemical-activated luciferase expression (CALUX) assay. *Marine Pollution Bulletin*, **49**(7–8), 648–658.

Huuskonen S., Koponen K., Ritola O., Hahn M. and Lindstrom-Seppa P. (1998). Induction of CYP1A and porphyrin accumulation in fish hepatoma cells (PLHC-1) exposed to sediment or water from a PCB-contaminated lake (Lake Kernaala, Finland). *Marine Environmental Research*, **46**(1–5), 379–384.

Ihara M., Hanamoto S., Ihara M. O., Zhang H. and Tanaka H. (2020). Wastewater-derived antagonistic activities of G protein-coupled receptor-acting pharmaceuticals in river water. *Journal of Applied Toxicology*, **40**(7), 908–917.

Inoue D., Nakama K., Matsui H., Sei K. and Ike M. (2009). Detection of agonistic activities against five human nuclear receptors in river environments of Japan using a yeast two-hybrid assay. *Bulletin of Environmental Contamination and Toxicology*, **82**(4), 399–404.

IPCS (2009). Assessment of combined exposures to multiple chemicals: Report of a WHO/IPCS international workshop. W. H. O. International Programme on Chemical Safety, www.who.int/ipcs/methods/harmonization/areas/aggregate/en/index.html (Accessed 2 January 2021).

Isidori M., Ferrara M., Lavorgna M., Nardelli A. and Parrella A. (2003). *In situ* monitoring of urban air in Southern Italy with the tradescantia micronucleus bioassay and semipermeable membrane devices (SPMDs). *Chemosphere*, **52**(1), 121–126.

Isidori M., Lavorgna M., Palumbo M., Piccioli V. and Parrella A. (2007). Influence of alkylphenols and trace elements in toxic, genotoxic, and endocrine disruption activity of wastewater treatment plants. *Environmental Toxicology and Chemistry*, **26**(8), 1686–1694.

ISO6341 (1996). Water quality – Determination of the inhibition of the mobility of Daphnia magna Straus (Cladocera, Crustacea) – Acute toxicity test. International Organization for Standardization (ISO), Geneva, Switzerland.

ISO7346-3 (1996). Water Quality – Determination of the Acute Lethal Toxicity of Substances to a Freshwater Fish [Brachydanio rerio Hamilton-Buchanan (Teleostei, Cyprinidae)] – Part 3: Flow-through Method. International Organization for Standardization (ISO), Geneva, Switzerland.

ISO11369 (1997). Water quality – determination of selected plant treatment agents – method using high-performance liquid chromatography with UV detection after solid-liquid extraction. International Organization for Standardization (ISO), Geneva, Switzerland.

ISO11348-3 (1998). Water Quality – Determination of the Inhibitory Effect of Water Samples on the Light Emission of Vibrio fischeri (Luminescent Bacteria Test). International Organization for Standardization (ISO), Geneva, Switzerland.

ISO12890 (1999). Water Quality – Determination of Toxicity to Embryos and Larvae of Freshwater Fish – Semi-static Method. International Organization for Standardization (ISO), Geneva, Switzerland.

ISO10706 (2000). Water Quality – Determination of Long Term Toxicity of Substances to Daphnia magna Straus (Cladocera, Crustacea). International Organization for Standardization (ISO), Geneva, Switzerland.

ISO 13829 (2000). Water quality – Determination of the genotoxicity of water and waste water using the umu-test. International Organization for Standardization (ISO), Geneva, Switzerland.

ISO8692 (2004). Water quality – Freshwater Algal Growth Inhibition Test with Unicellular Green Algae. International Organization for Standardization (ISO), Geneva, Switzerland.

ISO15088 (2007). Water Quality – Determination of the Acute Toxicity of Waste Water to Zebrafish Eggs (Danio rerio). International Organization for Standardization (ISO), Geneva, Switzerland.

ISO 11350 (2012). Water Quality – Determination of the Genotoxicity of Water and Waste Water: Salmonella/Microsome Fluctuation Test (Ames Fluctuation Test). International Organization for Standardization (ISO), Geneva, Switzerland.

ISO/DIS19040-1 (2017). Water Quality – Determination of the Estrogenic Potential of Water and Waste Water – Part 1: Yeast Estrogen Screen (Saccharomyces cerevisiae). International Organization for Standardization (ISO), Geneva, Switzerland.

ISO/DIS19040-2 (2017). Water Quality – Determination of the Estrogenic Potential of Water and Waste Water – Part 2: Yeast Estrogen Screen (A-YES, Arxula adeninivorans). International Organization for Standardization (ISO), Geneva, Switzerland.

ISO/DIS19040-3 (2017). Water Quality – Determination of the Estrogenic Potential of Water and Waste Water – Part 3: In Vitro Human Cell-based Reporter Gene Assay. International Organization for Standardization (ISO), Geneva, Switzerland.

Jahnke A., Mayer P., Schaefer S., Witt G., Haase N. and Escher B. I. (2016). Strategies for transferring mixtures of organic contaminants from aquatic environments into bioassays. *Environmental Science & Technology*, **50**(11), 5424–5431.

Jahnke A., Sobek A., Bergmann M., Bräunig J., Landmann M., Schäfer S. and Escher B. I. (2018). Effect-based characterization of mixtures of environmental pollutants in diverse sediments. *Environmental Science: Processes & Impacts*, **20**, 1667.

Jakimska A., Huerta B., Barganska Z., Kot-Wasik A., Rodriguez-Mozaz S. and Barcelo D. (2013). Development of a liquid chromatography-tandem mass spectrometry procedure for determination of endocrine disrupting compounds in fish from Mediterranean rivers. *Journal of Chromatography A*, **1306**, 44–58.

Jalova V., Jarosova B., Blaha L., Giesy J. P., Ocelka T., Grabic R., Jurcikova J., Vrana B. and Hilscherova K. (2013). Estrogen-, androgen- and aryl hydrocarbon receptor mediated activities in passive and composite samples from municipal waste and surface waters. *Environment International*, **59**, 372–383.

Jarošová B., Blaha L., Vrana B., Randak T., Grabic R., Giesy J. P. and Hilscherova K. (2012). Changes in concentrations of hydrophilic organic contaminants and of endocrine-disrupting potential downstream of small communities located adjacent to headwaters. *Environment International*, **45**, 22–31.

Jarošová B., Blaha L., Giesy J. P. and Hilscherova K. (2014a). What level of estrogenic activity determined by *in vitro* assays in municipal wastewaters can be considered as safe? *Environment International*, **64**, 98–109.

Jarošová B., Erseková A., Hilscherová K., Loos R., Gawlik B. M., Giesy J. P. and Bláha L. (2014b). Europe-wide survey of estrogenicity in wastewater treatment plant effluents: the need for the effect-based monitoring. *Environmental Science and Pollution Research*, **21**(18), 10970–10982.

Javurek J., Sychrova E., Smutna M., Bittner M., Kohoutek J., Adamovsky O., Novakova K., Smetanova S. and Hilscherova K. (2015). Retinoid compounds associated with water blooms dominated by Microcystis species. *Harmful Algae*, **47**, 116–125.

Jellett J. F., Marks L. J., Stewart J. E., Dorey M. L., Watson-Wright W. and Lawrence J. F. (1992). Paralytic shellfish poison (saxitoxin family) bioassays: Automated endpoint determination and standardization of the *in vitro* tissue culture bioassay, and comparison with the standard mouse bioassay. *Toxicon*, **30**(10), 1143–1156.

Jensen C. and Teng Y. (2020). Is it time to start transitioning from 2D to 3D cell culture? *Frontiers in Molecular Biosciences*, **7**, 33.

Jeon J., Kretschmann A., Escher B. I. and Hollender J. (2013). Characterization of acetylcholinesterase inhibition and energy allocation in *Daphnia magna* exposed to carbaryl. *Ecotoxicology and Environmental Safety*, **98**, 28–35.

Jeon S., Hong S., Kwon B. O., Park J., Song S. J., Giesy J. P. and Khim J. S. (2017). Assessment of potential biological activities and distributions of endocrine-disrupting chemicals in sediments of the west coast of South Korea. *Chemosphere*, **168**, 441–449.

Jia A., Escher B. I., Leusch F. D. L., Tang J. Y. M., Prochazka E., Dong B., Snyder E. M. and Snyder S. A. (2015). *In vitro* bioassays to evaluate complex chemical mixtures in recycled water. *Water Research*, **80**, 1–11.

Jia A., Wu S., Daniels K. D. and Snyder S. A. (2016). Balancing the budget: Accounting for glucocorticoid bioactivity and fate during water treatment. *Environmental Science & Technology*, **50**(6), 2870–2880.

Jia Y. L., Hammers-Wirtz M., Crawford S. E., Chen Q. Q., Seiler T. B., Schaffer A. and Hollert H. (2019). Effect-based and chemical analyses of agonistic and antagonistic endocrine disruptors in multiple matrices of eutrophic freshwaters. *Science of the Total Environment*, **651**, 1096–1104.

Jin L., van Mourik L., Gaus C. and Escher B. (2013). Applicability of passive sampling to bioanalytical screening of bioaccumulative chemicals in marine wildlife. *Environmental Science & Technology*, **47**(14), 7982–7988.

Jin L., Escher B. I., Limpus C. and Gaus C. (2015a). Coupling passive sampling with *in vitro* bioassays and chemical analysis to understand combined effects of bioaccumulative chemicals in blood of marine turtles. *Chemosphere*, **138**, 292–299.

Jin L., Gaus C. and Escher B. (2015b). Bioanalytical Approaches to Understanding Toxicological Implications of Mixtures of Persistent Organic Pollutants in Marine Wildlife. In: Comprehensive Analytical Chemistry. Persistent Organic Pollutants (POPs): Analytical Techniques, Environmental Fate and Biological Effects, E. Zheng (ed.), Elsevier, pp. 58–84.

Jin L., Gaus C. and Escher B. I. (2015c). Adaptive stress response pathways induced by environmental mixtures of bioaccumulative chemicals in Dugongs. *Environmental Science & Technology*, **49**(11), 6963–6973.

Jobling S. and Tyler C. R. (2003). Endocrine disruption in wild freshwater fish. *Pure and Applied Chemistry*, **75**(11–12), 2219–2234.

Jobling S., Nolan M., Tyler C. R., Brighty G. and Sumpter J. P. (1998). Widespread sexual disruption in wild fish. *Environmental Science & Technology*, **32**(17), 2498–2506.

Jonker M. T. O., Burgess R. M., Ghosh U., Gschwend P. M., Hale S. E., Lohmann R., Lydy M. J., Maruya K. A., Reible D. and Smedes F. (2020). *Ex situ* determination of freely dissolved concentrations of hydrophobic organic chemicals in sediments and soils: Basis for interpreting toxicity and assessing bioavailability, risks and remediation necessity. *Nature Protocols*, **15**(5), 1800–1828.

Judson R., Richard A., Dix D. J., Houck K., Martin M., Kavlock R., Dellarco V., Henry T., Holderman T., Sayre P., Tan S., Carpenter T. and Smith E. (2009). The toxicity data landscape for environmental chemicals. *Environmental Health Perspectives*, **117**(5), 685–695.

Judson R. S., Kavlock R. J., Setzer R. W., Hubal E. A. C., Martin M. T., Knudsen T. B., Houck K. A., Thomas R. S., Wetmore B. A. and Dix D. J. (2011). Estimating toxicity-related biological pathway altering doses for high-throughput chemical risk assessment. *Chemical Research in Toxicology*, **24**(4), 451–462.

Judson R., Houck K., Martin M., Richard A. M., Knudsen T. B., Shah I., Little S., Wambaugh J., Setzer R. W., Kothya P., Phuong J., Filer D., Smith D., Reif D., Rotroff D., Kleinstreuer N., Sipes N., Xia M. H., Huang R. L., Crofton K. and Thomas R. S. (2016). Analysis of the effects of cell stress and cytotoxicity on *in vitro* assay activity across a diverse chemical and assay space. *Toxicological Sciences*, **152**(2), 323–339.

Jugan M. L., Oziol L., Bimbot M., Huteau V., Tamisier-Karolak S., Blondeau J. P. and Lévi Y. (2009). *In vitro* assessment of thyroid and estrogenic endocrine disruptors in wastewater treatment plants, rivers and drinking water supplies in the greater Paris area (France). *Science of the Total Environment*, **407**(11), 3579–3587.

Junghans M., Backhaus T., Faust M., Scholze M. and Grimme L. H. (2006). Application and validation of approaches for the predictive hazard assessment of realistic pesticide mixtures. *Aquatic Toxicology*, **76**(2), 93–110.

Kado N. Y., Guirguis G. N., Flessel C. P., Chan R. C., Chang K. I. and Wesolowski J. J. (1986). Mutagenicity of fine (<2.5 µm) airborne particles: Diurnal variation in community air determined by a *Salmonella* micro preincubation (microsuspension) procedure. *Environmental Mutagenesis*, **8**(1), 53–66.

Kamber M., Fluckiger-Isler S., Engelhardt G., Jaeckh R. and Zeiger E. (2009). Comparison of the Ames II and traditional Ames test responses with respect to mutagenicity, strain specificities, need for metabolism and correlation with rodent carcinogenicity. *Mutagenesis*, **24**(4), 359–366.

Kandie F. J., Krauss M., Beckers L. M., Massei R., Fillinger U., Becker J., Liess M., Torto B. and Brack W. (2020). Occurrence and risk assessment of organic micropollutants in freshwater systems within the Lake Victoria South Basin, Kenya. *Science of the Total Environment*, **714**, 136748.

Kargalioglu Y., McMillan B. J., Minear R. A. and Plewa M. J. (2002). Analysis of the cytotoxicity and mutagenicity of drinking water disinfection by-products in *Salmonella typhimurium*. *Teratogenesis, Carcinogenesis, and Mutagenesis*, **22**(2), 113–128.

Kase R., Javurkova B., Simon E., Swart K., Buchinger S., Konemann S., Escher B. I., Carere M., Dulio V., Ait-Aissa S., Hollert H., Valsecchi S., Polesello S., Behnisch P., di Paolo

C., Olbrich D., Sychrova E., Gundlach M., Schlichting R., Leborgne L., Clara M., Scheffknecht C., Marneffe Y., Chalon C., Tusil P., Soldan P., von Danwitz B., Schwaiger J., Palao A. M., Bersani F., Perceval O., Kienle C., Vermeirssen E., Hilscherova K., Reifferscheid G. and Werner I. (2018). Screening and risk management solutions for steroidal estrogens in surface and wastewater. *Trends in Analytical Chemistry*, **102**, 343–358.

Keenan P. O., Knight A. W., Billinton N., Cahill P. A., Dalrymple I. M., Hawkyard C. J., Stratton-Campbell D. and Walmsley R. M. (2007). Clear and present danger? The use of a yeast biosensor to monitor changes in the toxicity of industrial effluents subjected to oxidative colour removal treatments. *Journal of Environmental Monitoring*, **9**(12), 1394–1401.

Keiter S., Rastall A., Kosmehl T., Wurm K., Erdinger L., Braunbeck T. and Hollert H. (2006). Ecotoxicological assessment of sediment, suspended matter and water samples in the upper Danube River – A pilot study in search for the causes for the decline of fish catches. *Environmental Science and Pollution Research*, **13**(5), 308–319.

Kennedy K., Macova M., Leusch F., Bartkow M. E., Hawker D. W., Zhao B., Denison M. S. and Mueller J. F. (2009). Assessing indoor air exposures using passive sampling with bioanalytical methods for estrogenicity and aryl hydrocarbon receptor activity. *Analytical and Bioanalytical Chemistry*, **394**(5), 1413–1421.

Kerbrat A.-S., Darius H. T., Pauillac S., Chinain M. and Laurent D. (2010). Detection of ciguatoxin-like and paralysing toxins in *Trichodesmium* spp. from New Caledonia lagoon. *Marine Pollution Bulletin*, **61**(7–12), 360–366.

Kern S., Fenner K., Singer H. P., Schwarzenbach R. P. and Hollender J. (2009). Identification of transformation products of organic contaminants in natural waters by computer-aided prediction and high-resolution mass spectrometry. *Environmental Science & Technology* **43**, 7039–7046.

Khedidji S., Croes K., Yassaa N., Ladji R., Denison M. S., Baeyens W. and Elskens M. (2017). Assessment of dioxin-like activity in PM10 air samples from an industrial location in Algeria, using the DRE-CALUX bioassay. *Environmental Science and Pollution Research*, **24**(13), 11868–11877.

Kidd K. A., Blanchfield P. J., Mills K. H., Palace V. P., Evans R. E., Lazorchak J. M. and Flick R. W. (2007). Collapse of a fish population after exposure to a synthetic estrogen. *Proceedings of the National Academy of Sciences of the United States of America*, **104**(21), 8897–8901.

Kim H.-S., Yamada H. and Tsuno H. (2007). The removal of estrogenic activity and control of brominated by-products during ozonation of secondary effluents. *Water Research*, **41**(7), 1441–1446.

Kisitu J., Bennekou S. H. and Leist M. (2019). Chemical concentrations in cell culture compartments (C-5) concentration definitions. *ALTEX-Alternatives to Animal Experimentation*, **36**(1), 154–160.

Klaassen C. D. (2013). Casarett & Doull's Toxicology: The Basic Science of Poisons, 8th edn. McGraw-Hill.

Klee N., Gustavsson L., Kosmehl T., Engwall M., Erdinger L., Braunbeck T. and Hollert H. (2004). Changes in toxicity and genotoxicity of industrial sewage sludge samples containing nitro- and amino-aromatic compounds following treatment in bioreactors with different oxygen regimes. *Environmental Science and Pollution Research*, **11**(5), 313–320.

Klepper O., Bakker J., Traas T. P. and van de Meent D. (1998). Mapping the potentially affected fraction (PAF) of species as a basis for comparison of ecotoxicological risks between substances and regions. *Journal of Hazardous Materials*, **61**(1–3), 337–344.

Knight A. W., Little S., Houck K., Dix D., Judson R., Richard A., McCarroll N., Akerman G., Yang C. H., Birrell L. and Walmsley R. M. (2009). Evaluation of high-throughput genotoxicity assays used in profiling the US EPA ToxCast (TM) chemicals. *Regulatory Toxicology and Pharmacology*, **55**(2), 188–199.

Knudsen T. B., Houck K. A., Sipes N. S., Singh A. V., Judson R. S., Martin M. T., Weissman A., Kleinstreuer N. C., Mortensen H. M., Reif D. M., Rabinowitz J. R., Setzer R. W., Richard A. M., Dix D. J. and Kavlock R. (2011). Activity profiles of 309 ToxCast (TM) chemicals evaluated across 292 biochemical targets. *Toxicology*, **282** (1–2), 1–15.

Kobayashi M., Li L., Iwamoto N., Nakajima-Takagi Y., Kaneko H., Nakayama Y., Eguchi M., Wada Y., Kumagai Y. and Yamamoto M. (2009). The antioxidant defense system Keap1-Nrf2 comprises a multiple sensing mechanism for responding to a wide range of chemical compounds. *Molecular and Cellular Biology*, **29**(2), 493–502.

Koelmans A. A., Bakir A., Burton G. A. and Janssen C. R. (2016). Microplastic as a vector for chemicals in the aquatic environment: critical review and model-supported reinterpretation of empirical studies. *Environmental Science & Technology*, **50**(7), 3315–3326.

Kogure K., Tamplin M. L., Simidu U. and Colwell R. R. (1988). A tissue culture assay for tetrodotoxin, saxitoxin and related toxins. *Toxicon*, **26**(2), 191–197.

Kolkman A., Schriks M., Brand W., Bauerlein P. S., van der Kooi M. M. E., van Doorn R. H., Emke E., Reus A. A., van der Linden S. C., de Voogt P. and Heringa M. B. (2013). Sample preparation for combined chemical analysis and *in vitro* bioassay application in water quality assessment. *Environmental Toxicology and Pharmacology*, **36**(3), 1291–1303.

Könemann S., Kase R., Simon E., Swart K., Buchinger S., Schlüssener M., Hollert H., Escher B. I., Werner I., Ait-Assa S., Vermeirssen E., Dulio V., Valsecchi S., Polesello S., Behnisch P., Javurkova B., Perceval O., di Paolo C., Olbrich D., Sychrova E., Schlichting R., Leborgne L., Clara M., Scheffknecht C., Marneffe Y., Chalon C., Tusil P., Soldan P., von Danwitz B., Schwaiger J., San Martin Becares M. I., Bersani F., Hilscherova K., Reifferscheid G., Ternes T. and Carere M. (2018). Effect-based and chemical analytical methods to monitor estrogens under the European Water Framework Directive. *Trends in Analytical Chemistry*, **102**, 225–235.

König M., Escher B. I., Neale P. A., Krauss M., Hilscherová K., Novák J., Teodorović I., Schulze T., Seidensticker S., Kamal Hashmi M. A., Ahlheim J. and Brack W. (2017). Impact of untreated wastewater on a major European river evaluated with a combination of *in vitro* bioassays and chemical analysis. *Environmental Pollution*, **220**, 1220–1230.

Konsoula R. and Barile F. A. (2005). Correlation of *in vitro* cytotoxicity with paracellular permeability in Caco-2 cells. *Toxicology In Vitro*, **19**(5), 675–684.

Körner W., Spengler P., Bolz U., Schuller W., Hanf V. and Metzger J. W. (2001). Substances with estrogenic activity in effluents of sewage treatment plants in southwestern

Germany. 2. Biological analysis. *Environmental Toxicology and Chemistry*, **20**(10), 2142–2151.

Kortenkamp A. (2007). Ten years of mixing cocktails: a review of combination effects of endocrine-disrupting chemicals. *Environmental Health Perspectives*, **115**(Suppl. 1), 98–105.

Kortenkamp A. (2020). Which chemicals should be grouped together for mixture risk assessments of male reproductive disorders? *Molecular and Cellular Endocrinology*, **499**, 110581.

Kortenkamp A. and Faust M. (2018). Regulate to reduce chemical mixture risk. *Science*, **361**(6399), 224–226.

Kortenkamp A., Faust M., Scholze M. and Backhaus T. (2007). Low-level exposure to multiple chemicals: Reason for human health concerns? *Environmental Health Perspectives*, **115**, 106–114.

Kortenkamp A., Backhaus T. and Faust M. (2009). State of the art report on mixture toxicity. Report, European Commission 070307/2007/485103/ETU/D.1, ec.europa.eu/environment/chemicals/effects/pdf/report_mixture_toxicity.pdf (Accessed 14 February 2021).

Kortenkamp A., Faust M., Backhaus T., Altenburger R., Scholze M., Muller C., Ermler S., Posthuma L. and Brack W. (2019). Mixture risks threaten water quality: the European Collaborative Project SOLUTIONS recommends changes to the WFD and better coordination across all pieces of European chemicals legislation to improve protection from exposure of the aquatic environment to multiple pollutants. *Environmental Sciences Europe*, **31**(1), 69.

Kowalski L. A. (2001). *In vitro* carcinogenicity testing: Present and future perspectives in pharmaceutical development. *Current Opinion in Drug Discovery & Development* **4**, 29–35.

Kramer N. I., Busser F. J. M., Oosterwijk M. T. T., Schirmer K., Escher B. I. and Hermens J. L. M. (2010). Development of a partition-controlled dosing system for cell assays. *Chemical Research in Toxicology*, **23**(11), 1806–1814.

Kramer V. J., Etterson M. A., Hecker M., Murphy C. A., Roesijadi G., Spade D. J., Spromberg J. A., Wang M. and Ankley G. T. (2011). Adverse outcome pathways and ecological risk assessment bridging to population-level effects. *Environmental Toxicology and Chemistry*, **30**(1), 64–76.

Kramer N. I., Krismartina M., Rico-Rico A., Blaauboer B. J. and Hermens J. L. M. (2012). Quantifying processes determining the free concentration of phenanthrene in basal cytotoxicity assays. *Chemical Research in Toxicology*, **25**(2), 436–445.

Krauss M., Singer H. and Hollender J. (2010). LC-high resolution MS in environmental analysis: from target screening to the identification of unknowns. *Analytical and Bioanalytical Chemistry*, **397**(3), 943–951.

Kretschmann A., Ashauer R., Hitzfeld K., Spaak P., Hollender J. and Escher B. I. (2011). Mechanistic toxicodynamic model for receptor-mediated toxicity of diazoxon, the active metabolite of Diazinon, in *Daphnia magna*. *Environmental Science & Technology*, **45**(11), 4980–4987.

Kretschmann A., Ashauer R., Hollender J. and Escher B. I. (2012). Toxicokinetic and toxicodynamic model for diazinon toxicity-mechanistic explanation of differences in the sensitivity of *Daphnia magna* and *Gammarus pulex*. *Environmental Toxicology and Chemistry*, **31**(9), 2014–2022.

Krishnamurthi K., Devi S. S., Hengstler J. G., Hermes M., Kumar K., Dutta D., Vannan S. M., Subin T. S., Yadav R. R. and Chakrabarti T. (2008). Genotoxicity of sludges, wastewater and effluents from three different industries. *Archives of Toxicology*, **82** (12), 965–971.

Kunz P. Y., Kienle C., Carere M., Homazava N. and Kase R. (2015). *In vitro* bioassays to screen for endocrine active pharmaceuticals in surface and waste waters. *Journal of Pharmaceutical and Biomedical Analysis*, **106**, 107–115.

Kunz P. Y., Simon E., Creusot N., Jayasinghe B. S., Kienle C., Maletz S., Schifferli A., Schönlau C., Aït-Aïssa S., Denslow N. D., Hollert H., Werner I. and Vermeirssen E. L. M. (2017). Effect-based tools for monitoring estrogenic mixtures: Evaluation of five *in vitro* bioassays. *Water Research*, **110**, 378–388.

Kurelec B., Matijasevic Z., Rijavec M., Alacevic M., Britvic S., Muller W. E. G. and Zahn R. K. (1979). Induction of benzo(a)pyrene mono-oxygenase in fish and the *Salmonella* test as a tool for detecting mutagenic-carcinogenic xenobiotics in the aquatic environment. *Bulletin of Environmental Contamination and Toxicology*, **21** (6), 799–807.

Kusk K. O., Kruger T., Long M. H., Taxvig C., Lykkesfeldt A. E., Frederiksen H., Andersson A. M., Andersen H. R., Hansen K. M. S., Nellemann C. and Bonefeld-Jorgensen E. C. (2011). Endocrine potency of wastewater: contents of endocrine disrupting chemicals and effects measured by *in vivo* and *in vitro* assays. *Environmental Toxicology and Chemistry*, **30**(2), 413–426.

Lah B., Zinko B., Narat M. and Marinsek-Logar R. (2005). Monitoring of genotoxicity in drinking water using *in vitro* comet assay and Ames test. *Food Technology and Biotechnology*, **43**(2), 139–146.

Lahnsteiner F. (2008). The sensitivity and reproducibility of the zebrafish (*Danio rerio*) embryo test for the screening of waste water quality and for testing the toxicity of chemicals. *ATLA-Alternatives to Laboratory Animals*, **36**(3), 299–311.

Laingam S., Froscio S. M. and Humpage A. R. (2008). Flow-cytometric analysis of *in vitro* micronucleus formation: Comparative studies with WIL2-NS human lymphoblastoid and L5178Y mouse lymphoma cell lines. *Mutation Research – Genetic Toxicology and Environmental Mutagenesis*, **656**(1–2), 19–26.

Lam M. M., Engwall M., Denison M. S. and Larsson M. (2018). Methylated polycyclic aromatic hydrocarbons and/or their metabolites are important contributors to the overall estrogenic activity of polycyclic aromatic hydrocarbon-contaminated soils. *Environmental Toxicology and Chemistry*, **37**(2), 385–397.

Landis W. G. and Yu M.-H. (2004). Introduction to Environmental Toxicology: Impacts of Chemicals Upon Ecological Systems. 3rd edn. CRC Press.

Langevin R., Rasmussen J. B., Sloterduk H. and Blaise C. (1992). Genotoxicity in water and sediment extracts from the St Lawrence river system, using the SOS chromotest. *Water Research*, **26**(4), 419–429.

Langhans S. A. (2018). Three-dimensional *in vitro* cell culture models in drug discovery and drug repositioning. *Frontiers in Pharmacology*, **9**, 6.

Latif M. and Licek E. (2004). Toxicity assessment of wastewaters, river waters, and sediments in Austria using cost-effective microbiotests. *Environmental Toxicology*, **19**(4), 302–309.

Lazaro-Cote A., Sadoul B., Jackson L. J. and Vijayan M. M. (2018). Acute stress response of fathead minnows caged downstream of municipal wastewater treatment plants in the Bow River, Calgary. *Plos One*, **13**(6), e0198177.

le Maire A., Bourguet W. and Balaguer P. (2010). A structural view of nuclear hormone receptor: endocrine disruptor interactions. *Cellular and Molecular Life Sciences*, **67** (8), 1219–1237.

Lee D. G., Roehrdanz P. R., Feraud M., Ervin J., Anumol T., Jia A., Park M., Tamez C., Morelius E. W., Gardea-Torresdey J. L., Izbicki J., Means J. C., Snyder S. A. and Holden P. A. (2015). Wastewater compounds in urban shallow groundwater wells correspond to exfiltration probabilities of nearby sewers. *Water Research*, **85**, 467–475.

Legler J., Leonards P., Spenkelink A. and Murk A. J. (2003). *In vitro* biomonitoring in polar extracts of solid phase matrices reveals the presence of unknown compounds with estrogenic activity. *Ecotoxicology*, **12**(1–4), 239–249.

Legradi J. B., Di Paolo C., Kraak M. H. S., van der Geest H. G., Schymanski E. L., Williams A. J., Dingemans M. M. L., Massei R., Brack W., Cousin X., Begout M. L., van der Oost R., Carion A., Suarez-Ulloa V., Silvestre F., Escher B. I., Engwall M., Nilen G., Keiter S. H., Pollet D., Waldmann P., Kienle C., Werner I., Haigis A. C., Knapen D., Vergauwen L., Spehr M., Schulz W., Busch W., Leuthold D., Scholz S., vom Berg C. M., Basu N., Murphy C. A., Lampert A., Kuckelkorn J., Grummt T. and Hollert H. (2018). An ecotoxicological view on neurotoxicity assessment. *Environmental Sciences Europe*, **30**, 46.

Leusch F. D. L. and Snyder S. A. (2015). Bioanalytical tools: half a century of application for potable reuse. *Environmental Science-Water Research & Technology*, **1**(5), 606–621.

Leusch F. D. L., Chapman H. F., van den Heuvel M. R., Tan B. L. L., Gooneratne S. R. and Tremblay L. A. (2006). Bioassay-derived androgenic and estrogenic activity in municipal sewage in Australia and New Zealand. *Ecotoxicology and Environmental Safety*, **65**(3), 403–411.

Leusch F. D. L., De Jager C., Levi Y., Lim R., Puijker L., Sacher F., Tremblay L. A., Wilson V. S. and Chapman H. F. (2010). Comparison of five *in vitro* bioassays to measure estrogenic activity in environmental waters. *Environmental Science & Technology*, **44**(10), 3853–3860.

Leusch F. D. L., Khan S. J., Gagnon M. M., Quayle P., Trinh T., Coleman H., Rawson C., Chapman H. F., Blair P., Nice H. and Reitsema T. (2014a). Assessment of wastewater and recycled water quality: A comparison of lines of evidence from *in vitro*, *in vivo* and chemical analyses. *Water Research* **50**, 420–431.

Leusch F. D. L., Khan S. J., Laingam S., Prochazka E., Froscio S., Trinh T., Chapman H. F. and Humpage A. (2014b). Assessment of the application of bioanalytical tools as surrogate measure of chemical contaminants in recycled water. *Water Research*, **49**, 300–315.

Leusch F. D. L., Neale P. A., Hebert A., Scheurer M. and Schriks M. C. M. (2017). Analysis of the sensitivity of *in vitro* bioassays for androgenic, progestagenic, glucocorticoid, thyroid and estrogenic activity: Suitability for drinking and environmental waters. *Environment International*, **99**, 120–130.

Leusch F. D. L., Aneck-Hahn N. H., Cavanagh J.-A. E., Du Pasquier D., Hamers T., Hebert A., Neale P. A., Scheurer M., Simmons S. O. and Schriks M. (2018a). Comparison of *in vitro* and *in vivo* bioassays to measure thyroid hormone disrupting activity in water extracts. *Chemosphere*, **191**(Suppl. C), 868–875.

Leusch F. D. L., Neale P. A., Arnal C., Aneck-Hahn N. H., Balaguer P., Bruchet A., Escher B. I., Esperanza M., Grimaldi M., Leroy G., Scheurer M., Schlichting R., Schriks M. and Hebert A. (2018b). Analysis of endocrine activity in drinking water, surface water and treated wastewater from six countries. *Water Research*, **139**, 10–18.

Lewtas J., Chuang J., Nishioka M. and Petersen B. (1990). Bioassay-directed fractionation of the organic extract of SRM-1649 urban air particulate matter. *International Journal of Environmental Analytical Chemistry*, **39**(3), 245–256.

Li J., Ma M. and Wang Z. J. (2008). A two-hybrid yeast assay to quantify the effects of xenobiotics on thyroid hormone-mediated gene expression. *Environmental Toxicology and Chemistry*, **27**(1), 159–167.

Li J., Wang Z., Ma M. and Peng X. (2010). Analysis of environmental endocrine disrupting activities using recombinant yeast assay in wastewater treatment plant effluents. *Bulletin of Environmental Contamination and Toxicology*, **84**(5), 529–535.

Li N., Ma M., Rao K. and Wang Z. (2011). *In vitro* thyroid disrupting effects of organic extracts from WWTPs in Beijing. *Journal of Environmental Sciences*, **23**(4), 671–675.

Li J. Y., Tang J. Y. M., Jin L. and Escher B. I. (2013). Understanding bioavailability and toxicity of sediment-associated contaminants by combining passive sampling with *in vitro* bioassays in an urban river catchment. *Environmental Toxicology and Chemistry*, **32**(12), 2888–2896.

Li J. Y., Su L., Wei F. H., Yang J. H., Jin L. and Zhang X. W. (2016). Bioavailability-based assessment of aryl hydrocarbon receptor-mediated activity in Lake Tai Basin from Eastern China. *Science of the Total Environment*, **544**, 987–994.

Li H. Z., Zhang J. and You J. (2018). Diagnosis of complex mixture toxicity in sediments: Application of toxicity identification evaluation (TIE) and effect-directed analysis (EDA). *Environmental Pollution*, **237**, 944–954.

Li H. Z., Yi X. Y., Chen F., Tong Y. J., Mehler W. T. and You J. (2019). Identifying organic toxicants in sediment using effect-directed analysis: A combination of bioaccessibility-based extraction and high-throughput midge toxicity testing. *Environmental Science & Technology*, **53**(2), 996–1003.

Lienert J., Güdel K. and Escher B. I. (2007). Screening method for ecotoxicological hazard assessment of 42 pharmaceuticals considering human metabolism and excretory routes. *Environmental Science & Technology*, **41**(12), 4471–4478.

Lindim C., de Zwart D., Cousins I. T., Kutsarova S., Kuhne R. and Schuurmann G. (2019). Exposure and ecotoxicological risk assessment of mixtures of top prescribed pharmaceuticals in Swedish freshwaters. *Chemosphere*, **220**, 344–352.

Liu C., Xu Y., Ma M., Huang B., Wu J., Meng Q., Wang Z. and Gearheart R. A. (2014). Evaluation of endocrine disruption and dioxin-like effects of organic extracts from sewage sludge in autumn in Beijing, China. *Frontiers of Environmental Science & Engineering*, **8**(3), 433–440.

Liu Y. Y., Lin Y. S., Yen C. H., Miaw C. L., Chen T. C., Wu M. C. and Hsieh C. Y. (2018). Identification, contribution, and estrogenic activity of potential EDCs in a river receiving concentrated livestock effluent in Southern Taiwan. *Science of the Total Environment*, **636**, 464–476.

Liu J. C., Lu G. H., Yang H. H., Dang T. J. and Yan Z. H. (2020). Ecological impact assessment of 110 micropollutants in the Yarlung Tsangpo River on the Tibetan Plateau. *Journal of Environmental Management*, **262**, 110291.

Liviac D., Wagner E. D., Mitch W. A., Altonji M. J. and Plewa M. J. (2010). Genotoxicity of water concentrates from recreational pools after various disinfection methods. *Environmental Science & Technology*, **44**(9), 3527–3532.

Loos R., Carvalho R., Antonio D. C., Cornero S., Locoro G., Tavazzi S., Paracchini B., Ghiani M., Lettieri T., Blaha L., Jarosova B., Voorspoels S., Servaes K., Haglund P., Fick J., Lindberg R. H., Schwesig D. and Gawlik B. M. (2013). EU-wide monitoring survey on emerging polar organic contaminants in wastewater treatment plant effluents. *Water Research*, **47**(17), 6475–6487.

Louiz I., Kinani S., Gouze M. E., Ben-Attia M., Menif D., Bouchonnet S., Porcher J. M., Ben-Hassine O. K. and Ait-Aissa S. (2008). Monitoring of dioxin-like, estrogenic and anti-androgenic activities in sediments of the Bizerta lagoon (Tunisia) by means of *in vitro* cell-based bioassays: Contribution of low concentrations of polynuclear aromatic hydrocarbons (PAHs). *Science of the Total Environment*, **402** (2–3), 318–329.

Lu M. Y., Kao W. C., Belkin S. and Cheng J. Y. (2019). A smartphone-based whole-cell array sensor for detection of antibiotics in milk. *Sensors*, **19**(18), 3882.

Luebcke-von Varel U., Machala M., Ciganek M., Neca J., Pencikova K., Palkova L., Vondracek J., Loeffler I., Streck G., Reifferscheid G., Flueckiger-Isler S., Weiss J. M., Lamoree M. and Brack W. (2011). Polar compounds dominate *in vitro* effects of sediment extracts. *Environmental Science & Technology*, **45**(6), 2384–2390.

Lundholt B. K., Scudder K. M. and Pagliaro L. (2003). A simple technique for reducing edge effect in cell-based assays. *Journal of Biomolecular Screening*, **8**(5), 566–570.

Lundqvist J., Andersson A., Johannisson A., Lavonen E., Mandava G., Kylin H., Bastviken D. and Oskarsson A. (2019a). Innovative drinking water treatment techniques reduce the disinfection-induced oxidative stress and genotoxic activity. *Water Research*, **155**, 182–192.

Lundqvist J., Mandava G., Lungu-Mitea S., Lai F. Y. and Ahrens L. (2019b). *In vitro* bioanalytical evaluation of removal efficiency for bioactive chemicals in Swedish wastewater treatment plants. *Scientific Reports*, **9**, 1–9.

Luo J., Ma M., Zha J. and Wang Z. (2009). Characterization of aryl hydrocarbon receptor agonists in sediments of Wenyu River, Beijing, China. *Water Research*, **43**(9), 2441–2448.

Lv X. M., Xiao S. H., Zhang G., Jiang P. and Tang F. (2016). Occurrence and removal of phenolic endocrine disrupting chemicals in the water treatment processes. *Scientific Reports*, **6**, 1–10.

Ma M., Li J. and Wang Z. J. (2005). Assessing the detoxication efficiencies of wastewater treatment processes using a battery of bioassays/biomarkers. *Archives of Environmental Contamination and Toxicology*, **49**(4), 480–487.

Macova M., Escher B. I., Reungoat J., Carswell S., Chue K. L., Keller J. and Mueller J. F. (2010). Monitoring the biological activity of micropollutants during advanced wastewater treatment with ozonation and activated carbon filtration. *Water Research*, **44**(2), 477–492.

Macova M., Toze S., Hodgers L., Mueller J. F., Bartkow M. E. and Escher B. I. (2011). Bioanalytical tools for the evaluation of organic micropollutants during sewage treatment, water recycling and drinking water generation. *Water Research*, **45**(14), 4238–4247.

Maeder V., Escher B. I., Scheringer M. and Hungerbuhler K. (2004). Toxic ratio as an indicator of the intrinsic toxicity in the assessment of persistent, bioaccumulative, and toxic chemicals. *Environmental Science & Technology*, **38**(13), 3659–3666.

Maffei F., Carbone F., Forti G. C., Buschini A., Poli P., Rossi C., Marabini L., Radice S., Chiesara E. and Hrelia P. (2009). Drinking water quality: An *in vitro* approach for the assessment of cytotoxic and genotoxic load in water sampled along distribution system. *Environment International*, **35**(7), 1053–1061.

Mahjoub O., Leclercq M., Bachelot M., Casellas C., Escande A., Balaguer P., Bahri A., Gomez E. and Fenet H. (2009). Estrogen, aryl hydrocarbon and pregnane X receptors activities in reclaimed water and irrigated soils in Oued Souhil area (Nabeul, Tunisia). *Desalination*, **246**(1–3), 425–434.

Maier D., Benisek M., Blaha L., Dondero F., Giesy J. P., Kohler H. R., Richter D., Scheurer M. and Triebskorn R. (2016). Reduction of dioxin-like toxicity in effluents by additional wastewater treatment and related effects in fish. *Ecotoxicology and Environmental Safety*, **132**, 47–58.

Malaj E., von der Ohe P. C., Grote M., Kuehne R., Mondy C. P., Usseglio-Polatera P., Brack W. and Schaefer R. B. (2014). Organic chemicals jeopardize the health of freshwater ecosystems on the continental scale. *Proceedings of the National Academy of Sciences of the United States of America*, **111**(26), 9549–9554.

Manger R. L., Leja L. S., Lee S. Y., Hungerford J. M. and Wekell M. M. (1993). Tetrazolium-based cell bioassay for neurotoxins active on voltage-sensitive sodium-channels – semiautomated assay for saxitoxins, brevetoxins, and ciguatoxins. *Analytical Biochemistry*, **214**(1), 190–194.

Manger R. L., Leja L. S., Lee S. Y., Hungerford J. M., Hokama Y., Dickey R. W., Granade H. R., Lewis R., Yasumoto T. and Wekell M. M. (1995). Detection of sodium-channel toxins – directed cytotoxicity assays of purified ciguatoxins, brevetoxins, saxitoxins, and seafood extracts. *Journal of AOAC International*, **78**(2), 521–527.

Manusadzianas L., Balkelyte L., Sadauskas K., Blinova I., Põllumaa L. and Kahru A. (2003). Ecotoxicological study of Lithuanian and Estonian wastewaters: Selection of the biotests, and correspondence between toxicity and chemical-based indices. *Aquatic Toxicology*, **63**(1), 27–41.

Marabini L., Frigerio S., Chiesara E. and Radice S. (2006). Toxicity evaluation of surface water treated with different disinfectants in HepG2 cells. *Water Research*, **40**(2), 267–272.

Marabini L., Frigerio S., Chiesara E., Maffei F., Cantelli Forti G., Hrelia P., Buschini A., Martino A., Poli P., Rossi C. and Radice S. (2007). *In vitro* cytotoxicity and genotoxicity of chlorinated drinking waters sampled along the distribution system of two municipal networks. *Mutation Research – Genetic Toxicology and Environmental Mutagenesis*, **634**(1–2), 1–13.

Maron D. M. and Ames B. N. (1983). Revised methods for the *Salmonella* mutagenicity test. *Mutation Research*, **113**(3–4), 173–215.

Martin M. T., Dix D. J., Judson R. S., Kavlock R. J., Reif D. M., Richard A. M., Rotroff D. M., Romanov S., Medvedev A., Poltoratskaya N., Gambarian M., Moeser M.,

Makarov S. S. and Houck K. A. (2010). Impact of environmental chemicals on key transcription regulators and correlation to toxicity end points within EPA's ToxCast program. *Chemical Research in Toxicology*, **23**(3), 578–590.

Martinovic-Weigelt D., Mehinto A. C., Ankley G. T., Berninger J. P., Collette T. W., Davis J. M., Denslow N. D., Durhan E. J., Eid E., Ekman D. R., Jensen K. M., Kahl M. D., LaLone C. A., Teng Q. and Villeneuve D. L. (2017). Derivation and evaluation of putative adverse outcome pathways for the effects of cyclooxygenase inhibitors on reproductive processes in female fish. *Toxicological Sciences*, **156**(2), 344–361.

Martyniuk C. J., Mehinto A. C. and Denslow N. D. (2020). Organochlorine pesticides: Agrochemicals with potent endocrine-disrupting properties in fish. *Molecular and Cellular Endocrinology*, **507**, 110764.

Marvin C. H. and Hewitt L. M. (2007). Analytical methods in bioassay-directed investigations of mutagenicity of air particulate material. *Mutation Research-Reviews in Mutation Research*, **636**(1–3), 4–35.

Matsuoka S., Kikuchi M., Kimura S., Kurokawa Y. and Kawai S. (2005). Determination of estrogenic substances in the water of Muko River using *in vitro* assays, and the degradation of natural estrogens by aquatic bacteria. *Journal of Health Science*, **51**(2), 178–184.

Mayer P., Parkerton T. F., Adams R. G., Cargill J. G., Gan J., Gouin T., Gschwend P. M., Hawthorne S. B., Helm P., Witt G., You J. and Escher B. I. (2014). Passive sampling methods for contaminated sediments: Scientific rationale supporting use of freely dissolved concentrations. *Integrated Environmental Assessment and Management*, **10** (2), 197–209.

McDonough C. A., Franks D. G., Hahn M. E. and Lohmann R. (2019). Aryl hydrocarbon receptor-mediated activity of gas-phase ambient air derived from passive sampling and an *in vitro* bioassay. *Environmental Toxicology and Chemistry*, **38**(4), 748–759.

Medlock Kakaley E. K., Blackwell B. R., Cardon M. C., Conley J. M., Evans N., Feifarek D. J., Furlong E. T., Glassmeyer S. T., Gray L. E., Hartig P. C., Kolpin D. W., Mills M. A., Rosenblum L., Villeneuve D. L. and Wilson V. S. (2020). De facto water reuse: Bioassay suite approach delivers depth and breadth in endocrine active compound detection. *Science of the Total Environment*, **699**, 134297.

Meek M. E., Boobis A. R., Crofton K. M., Heinemeyer G., Van Raaij M. and Vickers C. (2011). Risk assessment of combined exposure to multiple chemicals: A WHO/IPCS framework. *Regulatory Toxicology and Pharmacology*, **60**(2), S1–S14.

Meek M. E., Boobis A., Cote I., Dellarco V., Fotakis G., Munn S., Seed J. and Vickers C. (2014). New developments in the evolution and application of the WHO/IPCS framework on mode of action/species concordance analysis. *Journal of Applied Toxicology*, **34**(1), 1–18.

Mehinto A. C., Jia A., Snyder S. A., Jayasinghe B. S., Denslow N. D., Crago J., Schlenk D., Menzie C., Westerheide S. D., Leusch F. D. L. and Maruya K. A. (2015). Interlaboratory comparison of *in vitro* bioassays for screening of endocrine active chemicals in recycled water. *Water Research*, **83**, 303–309.

Mehinto A. C., Jayasinghe B. S., Vandervort D. R., Denslow N. D. and Maruya K. A. (2016). Screening for endocrine activity in water using commercially available *in vitro* transactivation bioassays. *JOVE-Journal of Visualized Experiments*, **118**, 54725.

Mehinto A. C., VanDervort D. R., Lao W., He G., Denison M. S., Vliet S. M., Volz D. C., Mazor R. D. and Maruya K. A. (2017). High throughput *in vitro* and *in vivo* screening of inland waters of Southern California. *Environmental Science-Processes & Impacts*, **19**(9), 1142–1149.

Mendonca E., Picado A., Paixao S. M., Silva L., Cunha M. A., Leitao S., Moura I., Cortez C. and Brito F. (2009). Ecotoxicity tests in the environmental analysis of wastewater treatment plants: Case study in Portugal. *Journal of Hazardous Materials*, **163**(2–3), 665–670.

Michel B. (2005). After 30 years of study, the bacterial SOS response still surprises us. *Plos Biology*, **3**(7), 1174–1176.

Miege C., Karolak S., Gabet V., Jugan M. L., Oziol L., Chevreuil M., Levi Y. and Coquery M. (2009). Evaluation of estrogenic disrupting potency in aquatic environments and urban wastewaters by combining chemical and biological analysis. *TrAc-Trends in Analytical Chemistry*, **28**(2), 186–195.

Miloshev G., Mihaylov I. and Anachkova B. (2002). Application of the single cell gel electrophoresis on yeast cells. *Mutation Research – Genetic Toxicology and Environmental Mutagenesis*, **513**(1–2), 69–74.

Mnif W., Dagnino S., Escande A., Pillon A., Fenet H., Gomez E., Casellas C., Duchesne M.-J., Hernandez-Raquet G., Cavaillès V., Balaguer P. and Bartegi A. (2010). Biological analysis of endocrine-disrupting compounds in Tunisian sewage treatment plants. *Archives of Environmental Contamination and Toxicology*, **59**(1), 1–12.

Mnif W., Zidi I., Hassine A. I. H., Gomez E., Bartegi A., Roig B. and Balaguer P. (2012). Monitoring endocrine disrupter compounds in the Tunisian Hamdoun River using *in vitro* bioassays. *Soil & Sediment Contamination*, **21**(7), 815–830.

More S. J., Hardy A., Bampidis V., Benford D., Bennekou S. H., Bragard C., Boesten J., Halldorsson T. I., Hernandez-Jerez A. F., Jeger M. J., Knutsen H. K., Koutsoumanis K. P., Naegeli H., Noteborn H., Ockleford C., Ricci A., Rychen G., Schlatter J. R., Silano V., Nielsen S. S., Schrenk D., Solecki R., Turck D., Younes M., Benfenati E., Castle L., Cedergreen N., Laskowski R., Leblanc J. C., Kortenkamp A., Ragas A., Posthuma L., Svendsen C., Testai E., Dujardin B., Kass G. E. N., Manini P., Jeddi M. Z., Dorne J., Hogstrand C. and Comm E. S. (2019). Guidance on harmonised methodologies for human health, animal health and ecological risk assessment of combined exposure to multiple chemicals. *EFSA Journal*, **17**(3), 5634.

Moreland D. E. (1980). Mechanism of action of herbicides. *Annual Review Plant Physiology*, **31**, 597–638.

Mosmann T. (1983). Rapid colorimetric assay for cellular growth and survival: Application to proliferation and cytotoxicity assays. *Journal of Immunological Methods*, **65**(1–2), 55–63.

Moxon T. E., Li H. Q., Lee M. Y., Piechota P., Nicol B., Pickles J., Pendlington R., Sorrell I. and Baltazar M. T. (2020). Application of physiologically based kinetic (PBK) modelling in the next generation risk assessment of dermally applied consumer products. *Toxicology In Vitro*, **63**, 104746.

Mueller M. E., Werneburg M., Glaser C., Schwientek M., Zarfl C., Escher B. I. and Zwiener C. (2020). Influence of emission sources and tributaries on the spatial and temporal patterns of micropollutant mixtures and associated effects in a small river. *Environmental Toxicology and Chemistry*, **39**(7), 1382–1391.

Mueller M. E., Zwiener C. and Escher B. I. (2021). Storm event-driven occurrence and transport of dissolved and sorbed organic micropollutants and associated effects in the Ammer River, Southwestern Germany. *Environmental Toxicology and Chemistry*, **40**(1), 88–99.

Müller M. E., Escher B. I., Schwientek M., Werneburg M., Zarfl C. and Zwiener C. (2018). Combining *in vitro* reporter gene bioassays with chemical analysis to assess changes in the water quality along the Ammer River, Southwestern Germany. *Environmental Science Europe*, **30**, 20.

Murk A. J., Legler J., van Lipzig M. M. H., Meerman J. H. N., Belfroid A. C., Spenkelink A., van der Burg B., Rijs G. B. J. and Vethaak D. (2002). Detection of estrogenic potency in wastewater and surface water with three *in vitro* bioassays. *Environmental Toxicology and Chemistry*, **21**(1), 16–23.

Muz M., Escher B. I. and Jahnke A. (2020). Bioavailable environmental pollutant patterns in sediments from passive equilibrium sampling. *Environmental Science & Technology*, **54**(24), 15861–15871.

Nachlas M. M., Margulies S. I., Goldberg J. D. and Seligman A. M. (1960). The determination of lactic dehydrogenase with a tetrazolium salt. *Analytical Biochemistry*, **1**(4–5), 317–326.

Nagy S. R., Sanborn J. R., Hammock B. D. and Denison M. S. (2002). Development of a green fluorescent protein-based cell bioassay for the rapid and inexpensive detection and characterization of Ah receptor agonists. *Toxicological Sciences*, **65**(2), 200–210.

Narotsky M. G., Klinefelter G. R., Goldman J. M., Best D. S., McDonald A., Strader L. F., Suarez J. D., Murr A. S., Thillainadarajah I., Hunter E. S., Richardson S. D., Speth T. F., Miltner R. J., Pressman J. G., Teuschler L. K., Rice G. E., Moser V. C., Luebke R. W. and Simmons J. E. (2013). Comprehensive assessment of a chlorinated drinking water concentrate in a rat multigenerational reproductive toxicity study. *Environmental Science & Technology*, **47**(18), 10653–10659.

National Academies of Sciences, Engineering and Medicine (2017). Using 21st Century Science to Improve Risk-Related Evaluations. The National Academies Press, Washington, DC. www.nap.edu/catalog/24635/using-21st-century-science-to-improve-risk-related-evaluations (Accessed 18 February 2021).

Neale P. A. and Escher B. I. (2013). Coextracted dissolved organic carbon has a suppressive effect on the acetylcholinesterase inhibition assay. *Environmental Toxicology and Chemistry*, **32**(7), 1526–1534.

Neale P. A. and Escher B. I. (2014). Does co-extracted dissolved organic carbon cause artefacts in cell-based bioassays? *Chemosphere*, **108**, 281–288.

Neale P. A. and Leusch F. D. L. (2015). Considerations when assessing antagonism *in vitro*: Why standardizing the agonist concentration matters. *Chemosphere*, **135**, 20–30.

Neale P. A., Antony A., Bartkow M. E., Farre M. J., Heitz A., Kristiana I., Tang J. Y. M. and Escher B. I. (2012). Bioanalytical assessment of the formation of disinfection byproducts in a drinking water treatment plant. *Environmental Science & Technology*, **46**(18), 10317–10325.

Neale P.A., Escher B. I. and Leusch F. D. L. (2015a). Understanding the implications of dissolved organic carbon when assessing antagonism *in vitro:* An example with an estrogen receptor assay. *Chemosphere*, **135**, 341–346.

Neale P. A., Ait-Aissa S., Brack W., Creusot N., Denison M. S., Deutschmann B., Hilscherová K., Hollert H., Krauss M., Novák J., Schulze T., Seiler T.-B., Serra H., Shao Y. and Escher B. I. (2015b). Linking *in vitro* effects and detected organic micropollutants in surface water using mixture-toxicity modeling. *Environmental Science & Technology*, **49**(24), 14614–14624.

Neale P. A., Achard M., Escher B. I. and Leusch F. D. L. (2017a). Exploring the oxidative stress response mechanism triggered by environmental water samples. *Environmental Science: Processes & Impacts*, **19**, 1126–1133.

Neale P. A., Altenburger R., Ait-Aissa S., Brion F., Busch W., de Aragão Umbuzeiro G., Denison M. S., Du Pasquier D., Hilscherova K., Hollert H., Morales D. A., Novac J., Schlichting R., Seiler T.-B., Serra H., Shao Y., Tindall A. J., Tollefsen K. E., Williams T. D. and Escher B. I. (2017b). Development of a bioanalytical test battery for water quality monitoring: Fingerprinting identified micropollutants and their contribution to effects in surface water. *Water Research*, **123**, 734–750.

Neale P. A., Munz N. A., Ait-Aissa S., Altenburger R., Brion F., Busch W., Escher B. I., Hilscherova K., Kienle C., Novak J., Seiler T.-B., Shao Y., Stamm C. and Hollender J. (2017c). Integrating chemical analysis and bioanalysis to evaluate the contribution of wastewater effluent on the micropollutant burden in small streams. *Science of the Total Environment*, **576**, 785–795.

Neale P. A., Brack W., Ait-Aissa S., Busch W., Hollender J., Krauss M., Maillot-Maréchal E., Munz N. A., Schlichting R., Schulze T., Vogler B. and Escher B. I. (2018a). Solid-phase extraction as sample preparation of water samples for cell-based and other *in vitro* bioassays. *Environmental Science: Processes & Impacts*, **20**, 493–504.

Neale P. A., Leusch F. D. L. and Escher B. I. (2018b). What is driving the NF-kappa B response in environmental water extracts? *Chemosphere*, **210**, 645–652.

Neale P. A., Braun G., Brack W., Carmona E., Gunold R., König M., Krauss M., Liebmann L., Liess M., Link M., Schäfer R. B., Schlichting R., Schreiner V. C., Schulze T., Vormeier P., Weisner O. and Escher B. I. (2020a). Assessing the mixture effects in *in vitro* bioassays of chemicals occurring in small agricultural streams during rain events. *Environmental Science & Technology*, **54**(13), 8280–8290.

Neale P. A., Feliers C., Glauch L., König M., Lecarpentier C., Schlichting R., Thibert S. and Escher B. I. (2020b). Application of *in vitro* bioassays for water quality monitoring in three drinking water treatment plants using different treatment processes including biological treatment, nanofiltration and ozonation coupled with disinfection. *Environmental Science: Water Research & Technology*, **6**, 2444–2453.

Neale P. A., Grimaldi M., Boulahtouf A., Leusch F. D. L. and Balaguer P. (2020c). Assessing species-specific differences for nuclear receptor activation for environmental water extracts. *Water Research*, **185**, 116247.

Neale P. A., O'Brien J. W., Glauch L., König M., Krauss M., Mueller J. F., Tscharke B. and Escher B. I. (2020d). Wastewater treatment efficacy evaluated with *in vitro* bioassays. *Water Research X*, **9**, 100072.

Nelson E. D., Do H., Lewis R. S. and Carr S. A. (2011). Diurnal variability of pharmaceutical, personal care product, estrogen and alkylphenol concentrations in effluent from a tertiary wastewater treatment facility. *Environmental Science & Technology*, **45**(4), 1228–1234.

Nestmann E. R., Lebel G. L., Williams D. T. and Kowbel D. J. (1979). Mutagenicity of organic extracts from Canadian drinking water in the *Salmonella*/mammalian-microsome assay. *Environmental Mutagenesis*, **1**(4), 337–345.

Newman M. C. (2019). Fundamentals of Ecotoxicology. 5th edn. CRC Press, Boca Raton, FL, USA.

Nguyen T., Nioi P. and Pickett C. (2009). The Nrf2-antioxidant response element signaling pathway and its activation by oxidative stress. *Journal of Biological Chemistry*, **284**, 13291–13295.

NHMRC (2011). Australian drinking water guidelines 6. Version 3.5 Updated August 2018, National Health and Medical Research Council (NHMRC) and the Natural Resource Management Ministerial Council, Canberra, Australia. www.nhmrc.gov.au/about-us/publications/australian-drinking-water-guidelines. (Accessed 18 February 2021).

Nicholls D. G. (2013). Bioenergetics. 4th edn. Academic Press, New York, NY.

Nicoletti I., Migliorati G., Pagliacci M. C., Grignani F. and Riccardi C. (1991). A rapid and simple method for measuring thymocyte apoptosis by propidium iodide staining and flow-cytometry. *Journal of Immunological Methods*, **139**(2), 271–279.

Nisbet I. C. T. and Lagoy P. K. (1992). Toxic equivalency factors (TEFs) for polycyclic aromatic-hydrocarbons (PAHs). *Regulatory Toxicology and Pharmacology*, **16**(3), 290–300.

Niss F., Rosenmai A. K., Mandava G., Orn S., Oskarsson A. and Lundqvist J. (2018). Toxicity bioassays with concentrated cell culture media-A methodology to overcome the chemical loss by conventional preparation of water samples. *Environmental Science and Pollution Research*, **25**(12), 12183–12188.

Niu L. L., Carmona E., Konig M., Krauss M., Muz M., Xu C., Zou D. L. and Escher B. I. (2020). Mixture risk drivers in freshwater sediments and their bioavailability determined using passive equilibrium sampling. *Environmental Science & Technology*, **54**(20), 13197–13206.

Nivala J., Neale P. A., Haasis T., Kahl S., König M., Müller R. A. M., Reemtsma T., Schlichting R. and Escher B. I. (2018). Application of cell-based bioassays to evaluate treatment efficacy of conventional and intensified treatment wetlands. *Environmental Science: Water Research & Technology*, **4**, 206–217.

Norberg-King T. J., Embry M. R., Belanger S. E., Braunbeck T., Butler J. D., Dorn P. B., Farr B., Guiney P. D., Hughes S. A., Jeffries M., Journel R., Leonard M., McMaster M., Oris J. T., Ryder K., Segner H., Senac T., Van Der Kraak G., Whale G. and Wilson P. (2018). An international perspective on the tools and concepts for effluent toxicity assessments in the context of animal alternatives: reduction in vertebrate use. *Environmental Toxicology and Chemistry*, **37**(11), 2745–2757.

Norli H. R., Christiansen A. and Deribe E. (2011). Application of QuEChERS method for extraction of selected persistent organic pollutants in fish tissue and analysis by gas chromatography mass spectrometry. *Journal of Chromatography A*, **1218**(41), 7234–7241.

Novák J., Vrana B., Rusina T., Okonski K., Grabic R., Neale P. A., Escher B. I., Macova M., Ait-Aissa S., Creusot N., Allan I. and Hilscherová K. (2018). Effect-based monitoring of the Danube River using mobile passive sampling. *Science of the Total Environment*, **636**, 1608–1619.

Novak J., Vaculovic A., Klanova J., Giesy J. P. and Hilscherova K. (2020). Seasonal variation of endocrine disrupting potentials of pollutant mixtures associated with various size-fractions of inhalable air particulate matter. *Environmental Pollution*, **264**, 114654.

NRC (1983). Risk assessment in the federal government. Managing the process. National Research Council (NRC), National Academy Press, Washington, DC, USA.

NRC (2007). Toxicity Testing in the 21st Century: A Vision and a Strategy. National Research Council (NRC), National Academies Press, Washington, DC. www.nap.edu/catalog/11970/toxicity-testing-in-the-21st-century-a-vision-and-a (Accessed 18 February 2021).

NRC (2017). Incorporating 21st Century Science into Risk-Based Evaluations. National Research Council (NRC), National Academies Press, Washington, DC. www.nap.edu/catalog/24635/using-21st-century-science-to-improve-risk-related-evaluations (Accessed 18 February 2021).

NRMMC/EPHC/AHMC (2006). Australian Guidelines for Water Recycling: Managing Health and Environmental Risks (Phase 1). National Water Quality Management Strategy (NWQMS), Natural Resource Management Ministerial Council, Environment Protection and Heritage Council, and Australian Health Minister's Conference. Canberra, Australia.. www.waterquality.gov.au/media/87 (Accessed 2 January 2021).

NRMMC/EPHC/NHMRC (2008). Australian Guidelines for Water Recycling: Managing Health and Environmental Risks (Phase 2). Augmentation of drinking water supplies, National Water Quality Management Strategy (NWQMS), Natural Resource Management Ministerial Council, Environment Protection and Heritage Council, and National Health and Medical Research Council, Canberra, Australia. www.waterquality.gov.au/media/85 (Accessed 2 January 2021).

NRMMC/EPHC/NHMRC (2009a). Australian Guidelines for Water Recycling: Managing Health and Environmental Risks (Phase 2). Managed aquifer recharge, National Water Quality Management Strategy (NWQMS), Natural Resource Management Ministerial Council, Environment Protection and Heritage Council, and National Health and Medical Research Council, Canberra, Australia. www.waterquality.gov.au/media/88 (Accessed 2 January 2021).

NRMMC/EPHC/NHMRC (2009b). Australian Guidelines for Water Recycling: Managing Health and Environmental Risks (Phase 2). Stormwater harvesting and reuse, National Water Quality Management Strategy (NWQMS), Natural Resource Management Ministerial Council, Environment Protection and Heritage Council, and National Health and Medical Research Council, Canberra, Australia.. www.waterquality.gov.au/media/90 (Accessed 2 January 2021).

Nuwaysir E. F., Bittner M., Trent J., Barrett J. C. and Afshari C. A. (1999). Microarrays and toxicology: The advent of toxicogenomics. *Molecular Carcinogenesis*, **24**(3), 153–159.

NWC (2011). A national approach to health risk assessment, risk communication and management of chemical hazards from recycled water. Chapman HF, Leusch FDL, Prochazka E, Cumming J, Ross V, Humpage AR, Froscio S, Laingam S, Khan SJ,

Trinh T, McDonald JA. Waterlines report No 48. National Water Commission, Canberra, Australia. www.academia.edu/11644319/A_national_approach_to_health_risk_assessment_risk_communication_and_management_of_chemical_hazards_from_recycled_water (Accessed 18 February 2021).

Oberley L. W. and Spitz D. R. (1984). Assay of superoxide-dismutase activity in tumor-tissue. *Methods in Enzymology*, **105**, 457–464.

O'Connor S., McNamara L., Swerdin M. and Van Buskirk R. G. (1991). Multifluorescent assays reveal mechanisms underlying cytotoxicity – phase I-CTFA compounds. *In Vitro Toxicology*, **4**(3), 197–206.

Oda Y., Nakamura S.-I., Oki I., Kato T. and Shinagawa H. (1985). Evaluation of the new system (umu-test) for the detection of environmental mutagens and carcinogens. *Mutation Research*, **147**(5), 219–229.

OECD (1992). Test Guidelinek 203. Fish, Acute Toxicity Test. Organisation for Economic Co-operation and Development, Paris, France.

OECD (1998). OECD Series on Principles of GLP and Compliance Monitoring, Number 1: OECD Principles on Good Laboratory Practice. Environmental Directorate, Organisation for Economic Co-operation and Development, Paris, France.

OECD (2004). OECD Series on Principles of GLP and Compliance Monitoring, Number 14: The Application of the Principles of GLP to In Vitro Studies. Environmental Directorate, Organisation for Economic Co-operation and Development, Paris, France.

OECD (2006). Guidelines for the Testing of Chemicals. Organisation for Economic Co-operation and Development, Paris, France.

OECD (2011). Test Guideline 201. Freshwater Alga and Cyanobacteria, Growth Inhibition Test. Organisation for Economic Co-operation and Development, Paris, France.

OECD (2013). Test Guideline 236. Fish Embryo Acute Toxicity Test. Organisation for Economic Co-operation and Development, Paris, France.

OECD (2015). Test Guideline. 455. Performance-Based Test Guideline for Stably Transfected Transactivation In Vitro Assays to Detect Estrogen Receptor Agonists and Antagonist. Organisation for Economic Co-operation and Development, Paris, France.

OECD (2018). OECD Series on Testing and Assessment, Number 286: Guidance Document on Good In Vitro Method Practices (GIVIMP). Environmental Directorate, Organisation for Economic Co-operation and Development, Paris, France.

Oh S. M., Park K. and Chung K. H. (2006). Combination of *in vitro* bioassays encompassing different mechanisms to determine the endocrine-disrupting effects of river water. *Science of the Total Environment*, **354**(2–3), 252–264.

Ohe T., Suzuki A., Watanabe T., Hasei T., Nukaya H., Totsuka Y. and Wakabayashi K. (2009). Induction of SCEs in CHL cells by dichlorobiphenyl derivative water pollutants, 2-phenylbenzotriazole (PBTA) congeners and river water concentrates. *Mutation Research – Genetic Toxicology and Environmental Mutagenesis*, **678**(1), 38–42.

Omiecinski C. J., Heuvel J. P. V., Perdew G. H. and Peters J. M. (2011). Xenobiotic metabolism, disposition, and regulation by receptors: From biochemical phenomenon to predictors of major toxicities. *Toxicological Sciences*, **120**, S49–S75.

Orton F., Ermler S., Kugathas S., Rosivatz E., Scholze M. and Kortenkamp A. (2014). Mixture effects at very low doses with combinations of anti-androgenic pesticides, antioxidants, industrial pollutant and chemicals used in personal care products. *Toxicology and Applied Pharmacology*, **278**(3), 201–208.

Osorio V., Schriks M., Vughs D., de Voogt P. and Kolkman A. (2018). A novel sample preparation procedure for effect-directed analysis of micro-contaminants of emerging concern in surface waters. *Talanta*, **186**, 527–537.

Ostling O. and Johanson K. J. (1984). Microelectrophoretic study of radiation-induced DNA damages in individual mammalian cells. *Biochemical and Biophysical Research Communications*, **123**(1), 291–298.

Otte J. C., Keiter S., Fassbender C., Higley E. B., Rocha P. S., Brinkmann M., Wahrendorf D. S., Manz W., Wetzel M. A., Braunbeck T., Giesy J. P., Hecker M. and Hollert H. (2013). Contribution of priority PAHs and POPs to Ah receptor-mediated activities in sediment samples from the River Elbe Estuary, Germany. *Plos One*, **8**(10), e75596.

Owens C. W. I. and Belcher R. V. (1965). A colorimetric micro-method for determination of glutathione. *Biochemical Journal*, **94**(3), 705–711.

Page B., Page M. and Noel C. (1993). A new fluorometric assay for cytotoxicity measurements *in vitro*. *International Journal of Oncology*, **3**(3), 473–476.

Pawlowski S., Ternes T., Bonerz M., Kluczka T., van der Burg B., Nau H., Erdinger L. and Braunbeck T. (2003). Combined *in situ* and *in vitro* assessment of the estrogenic activity of sewage and surface water samples. *Toxicological Sciences*, **75**(1), 57–65.

Payne J., Scholze M. and Kortenkamp A. (2001). Mixtures of four organochlorines enhance human breast cancer cell proliferation. *Environmental Health Perspectives*, **109**(4), 391–397.

Pellacani C., Buschini A., Furlini M., Poli P. and Rossi C. (2006). A battery of *in vivo* and *in vitro* tests useful for genotoxic pollutant detection in surface waters. *Aquatic Toxicology*, **77**(1), 1–10.

Pelon W., Whitman B. F. and Beasley T. W. (1977). Reversion of histidine-dependent mutant strains of *Salmonella-typhimurium* by Mississippi River water samples. *Environmental Science & Technology*, **11**(6), 619–623.

Perry P. and Wolff S. (1974). New Giemsa method for differential staining of sister chromatids. *Nature*, **251**(5471), 156–158.

Pessala P., Schultz E., Nakari T., Joutti A. and Herve S. (2004). Evaluation of wastewater effluents by small-scale biotests and a fractionation procedure. *Ecotoxicology and Environmental Safety*, **59**(2), 263–272.

Petala M., Kokokiris L., Samaras P., Papadopoulos A. and Zouboulis A. (2009). Toxicological and ecotoxic impact of secondary and tertiary treated sewage effluents. *Water Research*, **43**(20), 5063–5074.

Petrie B., Proctor K., Youdan J., Barden R. and Kasprzyk-Hordern B. (2017). Critical evaluation of monitoring strategy for the multi-residue determination of 90 chiral and achiral micropollutants in effluent wastewater. *Science of the Total Environment*, **579**, 569–578.

Pieterse B., Felzel E., Winter R., van der Burg B. and Brouwer A. (2013). PAH-CALUX, an optimized bioassay for AhR-mediated hazard identification of polycyclic aromatic hydrocarbons (PAHs) as individual compounds and in complex mixtures. *Environmental Science & Technology*, **47**(20), 11651–11659.

Pillon A., Boussioux A.-M., Escande A., Aït-Aïssa S., Gomez E., Fenet H., Ruff M., Moras D., Vignon F., Duchesne M.-J., Casellas C., Nicolas J.-C. and Balaguer P. (2005).

Binding of estrogenic compounds to recombinant estrogen receptor-α: Application to environmental analysis. *Environmental Health Perspectives*, **113**(3), 278–284.

Plassmann M. M., Schmidt M., Brack W. and Krauss M. (2015). Detecting a wide range of environmental contaminants in human blood samples-combining QuEChERS with LC-MS and GC-MS methods. *Analytical and Bioanalytical Chemistry*, **407**(23), 7047–7054.

Plewa M. J., Kargalioglu Y., Vankerk D., Minear R. A. and Wagner E. D. (2000). Development of a quantitative comparative cytotoxicity and genotoxicity assays for environmental hazardous chemicals. *Water Science and Technology*, **42**(7–8), 109–116.

Plewa M. J., Kargalioglu Y., Vankerk D., Minear R. A. and Wagner E. D. (2002). Mammalian cell cytotoxicity and genotoxicity analysis of drinking water disinfection by-products. *Environmental and Molecular Mutagenesis*, **40**(2), 134–142.

Plewa M. J., Wagner E. D. and Mitch W. A. (2011). Comparative mammalian cell cytotoxicity of water concentrates from disinfected recreational pools. *Environmental Science & Technology*, **45**(9), 4159–4165.

Plewa M. J., Wagner E. D., Jazwierska P., Richardson S. D., Chen P. H. and McKague A. B. (2004a). Halonitromethane drinking water disinfection byproducts: Chemical characterization and mammalian cell cytotoxicity and genotoxicity. *Environmental Science & Technology*, **38**(1), 62–68.

Plewa M. J., Wagner E. D., Richardson S. D., Thruston A. D., Woo Y.-T. and McKague A. B. (2004b). Chemical and biological characterization of newly discovered iodoacid drinking water disinfection byproducts. *Environmental Science & Technology*, **38** (18), 4713–4722.

Poole C. F. (2003). New trends in solid-phase extraction. *TrAC-Trends in Analytical Chemistry*, **22**(6), 362–373.

Posthuma L., Suter G. W. and Traas T. P. (2002). Species Sensitivity Distributions in Ecotoxicology. Lewis, Boca Raton, FL, USA.

Posthuma L., Richards S., De Zwart D., Dyer S., Sibley P., Hickey C. and Altenberger R. (2008). Mixture extrapolation approaches. In: Extrapolation practice for ecotoxicological effect characterization of chemicals, K. Solomon, T. Brock, D. de Zwart *et al.* (eds.), Taylor & Francis, Boca Raton, FL, USA, pp. 137–185.

Posthuma L., van Gils J., Zijp M. C., van de Meent D. and de Zwart D. (2019). Species sensitivity distributions for use in environmental protection, assessment, and management of aquatic ecosystems for 12 386 chemicals. *Environmental Toxicology and Chemistry*, **38**(4), 905–917.

Powell D. J., Hertzberg R. P. and Macarron R. (2016). Design and implementation of high-throughput screening assays. In: High Throughput Screening: Methods and Protocols, 3rd edn. W. P. Janzen, Humana Press, Totowa, NJ, USA, pp. 1–32.

Prasse C., Stalter D., Schulte-Oehlmann U. and Oehlmann J. (2015). Spoilt for choice: A critical review on the chemical and biological assessment of current wastewater treatment technologies. *Water Research*, **87**, 237–270.

Price P. S. and Han X. (2011). Maximum cumulative ratio (MCR) as a tool for assessing the value of performing a cumulative risk assessment. *International Journal of Environmental Research and Public Health*, **8**(6), 2212–2225.

Prieto P., Blaauboer B. J., de Boer A. G., Boveri M., Cecchelli R., Clemedson C., Coecke S., Forsby A., Galla H. J., Garberg P., Greenwood J., Price A. and Tahti H. (2004). Blood-brain barrier *in vitro* models and their application in toxicology – The report and recommendations of ECVAM workshop 49. *ALTA-Alternatives to Laboratory Animals*, **32**(1), 37–50.

Prochazkova T., Sychrova E., Vecerkova J., Javurkova B., Otoupalikova A., Pernica M., Simek Z., Smutna M., Lepsova-Skacelova O. and Hilscherova K. (2018). Estrogenic activity and contributing compounds in stagnant water bodies with massive occurrence of phytoplankton. *Water Research*, **136**, 12–21.

Promega (2009). CytoTox 96® Non-radioactive cytotoxicity assay. Instructions for use of product G1780. www.promega.com/resources/protocols/technical-bulletins/0/ cytotox-96-non-radioactive-cytotoxicity-assay-protocol/ (Accessed 14 February 2021).

Purdom C. E., Hardiman P. A., Bye V. V. J., Eno N. C., Tyler C. R. and Sumpter J. P. (1994). Estrogenic effects of effluents from sewage treatment works. *Chemistry and Ecology*, **8**(4), 275–285.

Quillardet P., Huisman O., Dari R. and Hofnung M. (1982). SOS Chromotest, a direct assay of induction of an SOS function in *Escherichia coli* K-12 to measure genotoxicity. *Proceedings of the National Academy of Sciences of the United States of America-Biological Sciences*, **79**(19), 5971–5975.

Rajapakse N., Silva E. and Kortenkamp A. (2002). Combining xenoestrogens at levels below individual no-observed-effect concentrations dramatically enhances steroid hormone action. *Environmental Health Perspectives*, **110**(9), 917–921.

Ramalhosa M. J., Paiga P., Morais S., Delerue-Matos C. and Oliveira M. (2009). Analysis of polycyclic aromatic hydrocarbons in fish: Evaluation of a quick, easy, cheap, effective, rugged, and safe extraction method. *Journal of Separation Science*, **32**(20), 3529–3538.

Rand G. (1995). Fundamentals of Aquatic Toxicology: Effects, Environmental Fate and Risk Assessment. Taylor & Francis, Washington, DC.

Rappaport S. M., Richard M. G., Hollstein M. C. and Talcott R. E. (1979). Mutagenic activity in organic wastewater concentrates. *Environmental Science & Technology*, **13**(8), 957–961.

Reid B. J., Jones K. C. and Semple K. T. (2000). Bioavailability of persistent organic pollutants in soils and sediments – A perspective on mechanisms, consequences and assessment. *Environmental Pollution*, **108**(1), 103–112.

Reifferscheid G., Heil J., Oda Y. and Zahn R. K. (1991). A microplate version of the SOS/umu-test for rapid detection of genotoxins and genotoxic potentials of environmental samples. *Mutation Research – Environmental Mutagenesis and Related Subjects*, **253**(3), 215–222.

Reiter E. B., Jahnke A., Konig M., Siebert U. and Escher B. I. (2020). Influence of co-dosed lipids from biota extracts on the availability of chemicals in *in vitro* cell-based bioassays. *Environmental Science & Technology*, **54**(7), 4240–4247.

Reungoat J., Macova M., Escher B. I., Carswell S., Mueller J. F. and Keller J. (2010). Removal of micropollutants and reduction of biological activity in a full-scale reclamation plant using ozonation and activated carbon filtration. *Water Research*, **44**(2), 625–637.

Riccardi C. and Nicoletti I. (2006). Analysis of apoptosis by propidium iodide staining and flow cytometry. *Nature Protocols*, **1**(3), 1458–1461.

Richter M. and Escher B. I. (2005). Mixture toxicity of reactive chemicals by using two bacterial growth assays as indicators of protein and DNA damage. *Environmental Science & Technology*, **39**, 8753–8761.

Rider C. V., Dinse G. E., Umbach D. M., Simmons J. E. and Hertzberg R. C. (2018). Mixture toxicity with models of additivity. In: Chemical Mixtures and Combined Chemical and Nonchemical Stressors, C. Rider and J. Simmons (eds.), Springer, Cham.

Ritter L., Totman C., Krishnan K., Carrier R., Vezina A. and Morisset V. (2007). Deriving uncertainty factors for threshold chemical contaminants in drinking water. *Journal of Toxicology and Environmental Health-Part B-Critical Reviews*, **10**(7), 527–557.

Roberts J., Bain P. A., Kumar A., Hepplewhite C., Ellis D. J., Christy A. G. and Beavis S. G. (2015). Tracking multiple modes of endocrine activity in Australia's largest inland sewage treatment plant and effluent-receiving environment using a panel of *in vitro* bioassays. *Environmental Toxicology and Chemistry*, **34**(10), 2271–2281.

Rocha P. S., Azab E., Schmidt B., Storch V., Hollert H. and Braunbeck T. (2010). Changes in toxicity and dioxin-like activity of sediments from the Tiete River (Sao Paulo, Brazil). *Ecotoxicology and Environmental Safety*, **73**(4), 550–558.

Rodrigues F. P., Angeli J. P. F., Mantovani M. S., Guedes C. L. B. and Jordao B. Q. (2010). Genotoxic evaluation of an industrial effluent from an oil refinery using plant and animal bioassays. *Genetics and Molecular Biology*, **33**(1), 169–175.

Rosa R., Moreira-Santos M., Lopes I., Silva L., Rebola J., Mendonca E., Picado A. and Ribeiro R. (2010). Comparison of a test battery for assessing the toxicity of a bleached-kraft pulp mill effluent before and after secondary treatment implementation. *Environmental Monitoring and Assessment*, **161**(1–4), 439–451.

Rosenmai A. K., Lundqvist J., le Godec T., Ohisson A., Troger R., Hellman B. and Oskarsson A. (2018). *In vitro* bioanalysis of drinking water from source to tap. *Water Research*, **139**, 272–280.

Rotter S., Beronius A., Boobis A. R., Hanberg A., van Klaveren J., Luijten M., Machera K., Nikolopoulou D., van der Voet H., Zilliacus J. and Solecki R. (2018). Overview on legislation and scientific approaches for risk assessment of combined exposure to multiple chemicals: The potential EuroMix contribution. *Critical Reviews in Toxicology*, **48**(9), 796–814.

Roubicek D. A., Rech C. M. and Umbuzeiro G. A. (2020). Mutagenicity as a parameter in surface water monitoring programs-opportunity for water quality improvement. *Environmental and Molecular Mutagenesis*, **61**(1), 200–211.

Routledge E. J. and Sumpter J. P. (1996). Estrogenic activity of surfactants and some of their degradation products assessed using a recombinant yeast screen. *Environmental Toxicology and Chemistry*, **15**(3), 241–248.

Rummel C. D., Escher B. I., Sandblom O., Plassmann M. M., Arp H. P. H., MacLeod M. and Jahnke A. (2019). Effects of leachates from UV-weathered microplastic in cell-based bioassays. *Environmental Science & Technology*, **53**(15), 9214–9223.

Russom C. L., LaLone C. A., Villeneuve D. L. and Ankley G. T. (2014). Development of an adverse outcome pathway for acetylcholinesterase inhibition leading to acute mortality. *Environmental Toxicology and Chemistry*, **33**(10), 2157–2169.

Rutishauser B. V., Pesonen M., Escher B. I., Ackermann G. E., Aerni H. R., Suter M. J. F. and Eggen R. I. L. (2004). Comparative analysis of estrogenic activity in sewage treatment plant effluents involving three *in vitro* assays and chemical analysis of steroids. *Environmental Toxicology and Chemistry*, **23**(4), 857–864.

Rydberg B. and Johanson K. J. (1978). Estimation of DNA strand breaks in single mammalian cells. In: DNA repair mechanisms, P. C. Hanawalt, E. C. Friedberg and C. F. Fox (eds.), Academic Press, New York, NY, USA. pp. 465–468.

Sanchez P. S., Sato M. I. Z., Paschoal C., Alves M. N., Furlan E. V. and Martins M. T. (1988). Toxicity assessment of industrial effluents from S. Paulo State, Brazil, using short-term microbial assays. *Toxicity Assessment*, **3**(1), 55–80.

Sakamuru S, Zhu H, Xia M, Simeonov A and Huang R. 2020. Profiling the Tox21 Chemical Library for Environmental Hazards: Applications in Prioritisation, Predictive Modelling, and Mechanism of Toxicity Characterisation. In D Neagu and AN Richarz, eds, *Big Data in Predictive Toxicology*. -Issues in Toxicology, **41**, 242–263.

Sauer U. G., Deferme L., Gribaldo L., Hackermueller J., Tralau T., van Ravenzwaay B., Yauk C., Poole A., Tong W. and Gant T. W. (2017). The challenge of the application of 'omics technologies in chemicals risk assessment: Background and outlook. *Regulatory Toxicology and Pharmacology*, **91**, S14–S26.

Šauer P., Borik A., Golovko O., Grabic R., Stanova A. V., Valentova O., Stara A., Sandova M. and Kroupova H. K. (2018). Do progestins contribute to (anti-)androgenic activities in aquatic environments? *Environmental Pollution*, **242**, 417–425.

Saxena J. and Schwartz D. J. (1979). Mutagens in wastewaters renovated by advanced wastewater-treatment. *Bulletin of Environmental Contamination and Toxicology*, **22**(3), 319–326.

Scarsi M., Podvinec M., Roth A., Hug H., Kersten S., Albrecht H., Schwede T., Meyer U. A. and Rucker C. (2007). Sulfonylureas and glinides exhibit peroxisome proliferator-activated receptor gamma activity: A combined virtual screening and biological assay approach. *Molecular Pharmacology*, **71**(2), 398–406.

Schiliró T., Pignata C., Fea E. and Gilli G. (2004). Toxicity and estrogenic activity of a wastewater treatment plant in northern Italy. *Archives of Environmental Contamination and Toxicology*, **47**(4), 456–462.

Schirmer K. (2006). Proposal to improve vertebrate cell cultures to establish them as substitutes for the regulatory testing of chemicals and effluents using fish. *Toxicology*, **224**(3), 163–183.

Schirmer K., Tom D. J., Bols N. C. and Sherry J. P. (2001). Ability of fractionated petroleum refinery effluent to elicit cyto- and photocytotoxic responses and to induce 7-ethoxyresorufin-O-deethylase activity in fish cell lines. *Science of the Total Environment*, **271**(1–3), 61–78.

Schnurstein A. and Braunbeck T. (2001). Tail moment *versus* tail length – Application of an *in vitro* version of the comet assay in biomonitoring for genotoxicity in native surface waters using primary hepatocytes and gill cells from zebrafish (*Danio rerio*). *Ecotoxicology and Environmental Safety*, **49**(2), 187–196.

Scholze M., Boedeker W., Faust M., Backhaus T., Altenburger R. and Grimme L. H. (2001). A general best-fit method for concentration-response curves and the estimation of low-effect concentrations. *Environmental Toxicology and Chemistry*, **20**(2), 448–457.

Schreer A., Tinson C., Sherry J. P. and Schirmer K. (2005). Application of Alamar blue/5-carboxyfluorescein diacetate acetoxymethyl ester as a noninvasive cell viability assay in primary hepatocytes from rainbow trout. *Analytical Biochemistry*, **344** (1), 76–85.

Schriks M., Heringa M. B., van der Kooi M. M. E., de Voogt P. and van Wezel A. P. (2010a). Toxicological relevance of emerging contaminants for drinking water quality. *Water Research*, **44**(2), 461–476.

Schriks M., van Leerdam J. A., van der Linden S. C., van der Burg B., van Wezel A. P. and de Voogt P. (2010b). High-resolution mass spectrometric identification and quantification of glucocorticoid compounds in various wastewaters in the Netherlands. *Environmental Science & Technology*, **44**(12), 4766–4774.

Schriks M., van der Linden S. C., Stoks P. G. M., van der Burg B., Puijker L., de Voogt P. and Heringa M. B. (2013). Occurrence of glucocorticogenic activity in various surface waters in the Netherlands. *Chemosphere*, **93**(2), 450–454.

Schroeder A. L., Ankley G. T., Houck K. A. and Villeneuve D. L. (2016). Environmental surveillance and monitoring: The next frontiers for high-throughput toxicology. *Environmental Toxicology and Chemistry*, **35**(3), 513–525.

Schulze T., Ulrich M., Maier D., Maier M., Terytze K., Braunbeck T. and Hollert H. (2015). Evaluation of the hazard potentials of river suspended particulate matter and floodplain soils in the Rhine basin using chemical analysis and *in vitro* bioassays. *Environmental Science and Pollution Research*, **22**(19), 14606–14620.

Schulze T., Ahel M., Ahlheim J., Ait-Aissa S., Brion F., Di Paolo C., Froment J., Hidasi A. O., Hollender J., Hollert H., Hu M., Klolss A., Koprivica S., Krauss M., Muz M., Oswald P., Petre M., Schollee J. E., Seiler T. B., Shao Y., Slobodnik J., Sonavane M., Suter M. J. F., Tollefsen K. E., Tousova Z., Walz K. H. and Brack W. (2017). Assessment of a novel device for onsite integrative large-volume solid phase extraction of water samples to enable a comprehensive chemical and effect-based analysis. *Science of the Total Environment*, **581**, 350–358.

Schwarzenbach R. P., Escher B. I., Fenner K., Hofstetter T. B., Johnson C. A., von Gunten U. and Wehrli B. (2006). The challenge of micropollutants in aquatic systems. *Science*, **313**(5790), 1072–1077.

Schweigert N., Eggen R. I. L., Escher B. I., Burkhardt-Holm P. and Behra R. (2002). Ecotoxicological assessment of surface waters: A modular approach integrating *in vitro* methods. *ALTEX-Alternativen zu Tierexperimenten*, **19**(Suppl. 1), 30–37.

Schymanski E. L., Singer H. P., Longree P., Loos M., Ruff M., Stravs M. A., Vidal C. R. and Hollender J. (2014). Strategies to characterize polar organic contamination in wastewater: exploring the capability of high resolution mass spectrometry. *Environmental Science & Technology*, **48**(3), 1811–1818.

Scott P. D., Bartkow M., Blockwell S. J., Coleman H. M., Khan S. J., Lim R., McDonald J. A., Nice H., Nugegoda D., Pettigrove V., Tremblay L. A., Warne M. S. J. and Leusch F. D. L. (2014). An assessment of endocrine activity in Australian rivers using chemical and *in vitro* analyses. *Environmental Science and Pollution Research*, **21**(22), 12951–12967.

Scott P. D., Coleman H. M., Khan S., Limc R., McDonald J. A., Mondon J., Neale P. A., Prochazka E., Tremblay L. A., Warne M. S. and Leusch F. D. L. (2018). Histopathology, vitellogenin and chemical body burden in mosquitofish (*Gambusia holbrooki*) sampled from six river sites receiving a gradient of stressors. *Science of the Total Environment*, **616**, 1638–1648.

Seibert H., Balls M., Fentem J. H., Bianchi V., Clothier R. H., Dierickx P. J., Ekwall B., Garle M. J., GomezLechon M. J., Gribaldo L., Gulden M., Liebsch M., Rasmussen E., Roguet

R., Shrivastava R. and Walum E. (1996). Acute toxicity testing *in vitro* and the classification and labelling of chemicals – The report and recommendations of ECVAM workshop 16. *ATLA-Alternatives to Laboratory Animals*, **24**(4), 499–510.

Seiler T.-B., Best N., Fernqvist M. M., Hercht H., Smith K. E. C., Braunbeck T., Mayer P. and Hollert H. (2014). PAH toxicity at aqueous solubility in the fish embryo test with *Danio rerio* using passive dosing. *Chemosphere*, **112**, 77–84.

Seimandi M., Lemaire G., Pillon A., Perrin A., Carlavan I., Voegel J. J., Vignon F., Nicolas J. C. and Balaguer P. (2005). Differential responses of PPAR alpha, PPAR delta, and PPAR gamma reporter cell lines to selective PPAR synthetic ligands. *Analytical Biochemistry*, **344**(1), 8–15.

Semple K. T., Doick K. J., Jones K. C., Burauel P., Craven A. and Harms H. (2004). Defining bioavailability and bioaccessibility of contaminated soil and sediment is complicated. *Environmental Science and Technology*, **38**(12), 228A–231A.

Serra H., Brion F., Chardon C., Budzinski H., Schulze T., Brack W. and Ait-Aissa S. (2020). Estrogenic activity of surface waters using zebrafish- and human-based *in vitro* assays: The Danube as a case-study. *Environmental Toxicology and Pharmacology*, **78**, 103401.

Shen C., Huang S., Wang Z., Qiao M., Tang X., Yu C., Shi D., Zhu Y., Shi J., Chen X., Setty K. and Chen Y. (2008). Identification of Ah receptor agonists in soil of E-waste recycling sites from Taizhou area in China. *Environmental Science & Technology*, **42**(1), 49–55.

Shi W., Wang X. Y., Hu W., Sun H., Shen O. X., Liu H. L., Wang X. R., Giesy J. P., Cheng S. P. and Yu H. X. (2009a). Endocrine-disrupting equivalents in industrial effluents discharged into Yangtze River. *Ecotoxicology*, **18**(6), 685–692.

Shi Y., Cao X.-W., Tang F., Du H.-R., Wang Y.-Z., Qiu X.-D., Yu H.-P. and Lu B. (2009b). *In vitro* toxicity of surface water disinfected by different sequential treatments. *Water Research*, **43**(1), 218–228.

Shi P., Zhou S. C., Xiao H. X., Qiu J. F., Li A. M., Zhou Q., Pan Y. and Hollert H. (2018). Toxicological and chemical insights into representative source and drinking water in eastern China. *Environmental Pollution*, **233**, 35–44.

Shoemaker J. T., Zhang W., Atlas S. I., Bryan R. A., Inman S. W. and Vukasinovic J. (2020). A 3D cell culture organ-on-a-chip platform with a breathable hemoglobin analogue augments and extends primary human hepatocyte functions *in vitro*. *Frontiers in Molecular Biosciences*, **7**, 568777.

Shue M. F., Chen F. A., Kuo Y. T. and Chen T. C. (2009). Nonylphenol concentration and estrogenic activity in Kaoping River and its tributaries, Taiwan. *Water Science and Technology*, **59**(10), 2055–2063.

Shukla S. J., Huang R. L., Austin C. P. and Xia M. H. (2010). The future of toxicity testing: A focus on *in vitro* methods using a quantitative high-throughput screening platform. *Drug Discovery Today*, **15**(23–24), 997–1007.

Sidlova T., Novak J., Janosek J., Andel P., Giesy J. P. and Hilscherova K. (2009). Dioxin-like and endocrine disruptive activity of traffic-contaminated soil samples. *Archives of Environmental Contamination and Toxicology*, **57**(4), 639–650.

Silva E., Rajapakse N. and Kortenkamp A. (2002). Something from "nothing" – eight weak estrogenic chemicals combined at concentrations below NOECs produce significant mixture effects. *Environmental Science & Technology*, **36**, 1751–1756.

Simmon V. F. and Tardiff R. G. (1976). Mutagenic activity of drinking-water concentrates. *Mutation Research*, **38**(6), 389–390.

Simmons S. O., Fan C. Y. and Ramabhadran R. (2009). Cellular stress response pathway system as a sentinel ensemble in toxicological screening. *Toxicological Sciences*, **111** (2), 202–225.

Simon T., Britt J. K. and James R. C. (2007). Development of a neurotoxic equivalence scheme of relative potency for assessing the risk of PCB mixtures. *Regulatory Toxicology and Pharmacology*, **48**(2), 148–170.

Simon E., van Velzen M., Brandsma S. H., Lie E., Loken K., de Boer J., Bytingsvik J., Jenssen B. M., Aars J., Hamers T. and Lamoree M. H. (2013). Effect-directed analysis to explore the polar bear exposome: Identification of thyroid hormone disrupting compounds in plasma. *Environmental Science & Technology*, **47**(15), 8902–8912.

Simon E., Schifferli A., Bucher T. B., Olbrich D., Werner I. and Vermeirssen E. L. M. (2019). Solid-phase extraction of estrogens and herbicides from environmental waters for bioassay analysis – effects of sample volume on recoveries. *Analytical and Bioanalytical Chemistry*, **411**(10), 2057–2069.

Singh N. P., McCoy M. T., Tice R. R. and Schneider E. L. (1988). A simple technique for quantitation of low levels of DNA damage in individual cells. *Experimental Cell Research*, **175**(1), 184–191.

Sipes N. S., Wambaugh J. F., Pearce R., Auerbach S. S., Wetmore B. A., Hsieh J. H., Shapiro A. J., Svoboda D., DeVito M. J. and Ferguson S. S. (2017). An intuitive approach for predicting potential human health risk with the Tox21 10k Library. *Environmental Science & Technology*, **51**(18), 10786–10796.

Skarek M., Janosek J., Cupr P., Kohoutek J., Novotna-Rychetska A. and Holoubek I. (2007). Evaluation of genotoxic and non-genotoxic effects of organic air pollution using *in vitro* bioassays. *Environment International*, **33**(7), 859–866.

Skehan P., Storeng R., Scudiero D., Monks A., McMahon J., Vistica D., Warren J. T., Bokesch H., Kenney S. and Boyd M. R. (1990). New colorimetric cytotoxicity assay for anticancer-drug screening. *Journal of the National Cancer Institute*, **82**(13), 1107–1112.

Smith K. E. C. and Schäfer S. (2016). Defining and controlling exposure during *in vitro* toxicity testing and the potential of passive dosing. *Advances in Biochemical Engineering / Biotechnology*, **157**, 263–292.

Smith K. E. C., Oostingh G. J. and Mayer P. (2010). Passive dosing for producing defined and constant exposure of hydrophobic organic compounds during *in vitro* toxicity tests. *Chemical Research in Toxicology*, **23**(1), 55–65.

Smith K. E. C., Jeong Y. and Kim J. (2015). Passive dosing versus solvent spiking for controlling and maintaining hydrophobic organic compound exposure in the Microtox (R) assay. *Chemosphere*, **139**, 174–180.

Sohoni P. and Sumpter J. P. (1998). Several environmental oestrogens are also anti-androgens. *Journal of Endocrinology*, **158**(3), 327–339.

Song M., Jiang Q., Xu Y., Liu H., Lam P. K. S., O'Toole D. K., Zhang Q., Giesy J. P. and Jiang G. (2006). AhR-active compounds in sediments of the Haihe and Dagu Rivers, China. *Chemosphere*, **63**(7), 1222–1230.

Sonneveld E., Pieterse B., Schoonen W. G. and van der Burg B. (2011). Validation of *in vitro* screening models for progestagenic activities: Inter-assay comparison and correlation with *in vivo* activity in rabbits. *Toxicology In Vitro*, **25**(2), 545–554.

Soto A. M., Sonnenschein C., Chung K. L., Fernandez M. F., Olea N. and Serrano F. O. (1995). The E-SCREEN assay as a tool to identify estrogens: An update on estrogenic environmental pollutants. *Environmental Health Perspectives*, **103**(Suppl. 7), 113–122.

Souders C. L., Liang X. F., Wang X. H., Ector N., Zhao Y. H. and Martyniuk C. J. (2018). High-throughput assessment of oxidative respiration in fish embryos: Advancing adverse outcome pathways for mitochondrial dysfunction. *Aquatic Toxicology*, **199**, 162–173.

Spurgeon D. J., Jones O. A. H., Dorne J., Svendsen C., Swain S. and Sturzenbaum S. R. (2010). Systems toxicology approaches for understanding the joint effects of environmental chemical mixtures. *Science of the Total Environment*, **408**(18), 3725–3734.

Spycher S., Netzeva T. I., Worth A. and Escher B. I. (2008). Mode of action-based classification and prediction of activity of uncouplers for the screening of chemical inventories. *SAR and QSAR in Environmental Research*, **19**, 433–463.

Stadnicka-Michalak J., Tanneberger K., Schirmer K. and Ashauer R. (2014). Measured and modeled toxicokinetics in cultured fish cells and application to *in vitro-in vivo* toxicity extrapolation. *Plos One*, **9**(3), e92303.

Stalter D., Magdeburg A. and Oehlmann J. (2010a). Comparative toxicity assessment of ozone and activated carbon treated sewage effluents using an *in vivo* test battery. *Water Research*, **44**(8), 2610–2620.

Stalter D., Magdeburg A., Weil M., Knacker T. and Oehlmann J. (2010b). Toxication or detoxication? *In vivo* toxicity assessment of ozonation as advanced wastewater treatment with the rainbow trout. *Water Research*, **44**(2), 439–448.

Stalter D., Magdeburg A., Wagner M. and Oehlmann J. (2011). Ozonation and activated carbon treatment of sewage effluents: Removal of endocrine activity and cytotoxicity. *Water Research*, **45**, 1015–1024.

Stalter D., Dutt M. and Escher B. I. (2013). Headspace-free setup of *in vitro* bioassays for the evaluation of volatile disinfection by-products. *Chemical Research in Toxicology*, **26**, 1605–1614.

Stalter D., O'Malley E., von Gunten U. and Escher B. I. (2016a). Fingerprinting the reactive toxicity pathways of 50 drinking water disinfection by-products. *Water Research*, **91**, 19–30.

Stalter D., O'Malley E., von Gunten U. and Escher B. I. (2016b). Point-of-use water filters can effectively remove disinfection by-products and toxicity from chlorinated and chloraminated tap water. *Environmental Science: Water Research & Technology*, **2** (5), 875–883.

Stalter D., Peters L. I., O'Malley E., Tang J. Y. M., Revalor M., Farre M. J., Watson K., von Gunten U. and Escher B. I. (2016c). Sample enrichment for bioanalytical assessment of disinfected drinking water: Concentrating the polar, the volatiles, and the unknowns. *Environmental Science & Technology*, **50**, 6495–6505.

State Water Resources Control Board (2019). Water quality control policy for recycled water. California Environmental Protection Agency, USA. www.waterboards.ca.gov/

board_decisions/adopted_orders/resolutions/2018/121118_7_final_amendment_oal.pdf (Accessed 16 November 2020).

Sumida K., Igarashi Y., Toritsuka N., Matsushita T., Abe-Tomizawa K., Aoki M., Urushidani T., Yamada H. and Ohno Y. (2011). Effects of DMSO on gene expression in human and rat hepatocytes. *Human & Experimental Toxicology*, **30**(10), 1701–1709.

Sumpter J. P. (2002). Endocrine disruption and feminization in fish. *Toxicology*, **178**(1), 39–40.

Sumpter J. P. and Jobling S. (1995). Vitellogenesis as a biomarker for estrogenic contamination of the aquatic environment. *Environmental Health Perspectives*, **103**, 173–178.

Sun J., Zhang R., Qin L., Zhu H. X., Huang Y., Xue Y. G., An S. Q., Xie X. C. and Li A. M. (2017). Genotoxicity and cytotoxicity reduction of the polluted urban river after ecological restoration: A field-scale study of Jialu River in northern China. *Environmental Science and Pollution Research*, **24**(7), 6715–6723.

Suzuki G., Tue N. M., van der Linden S., Brouwer A., van der Burg B., van Velzen M., Lamoree M., Someya M., Takahashi S., Isobe T., Tajima Y., Yamada T. K., Takigami H. and Tanabe S. (2011). Identification of major dioxin-like compounds and androgen receptor antagonist in acid-treated tissue extracts of high trophic-level animals. *Environmental Science and Technology*, **45**(23), 10203–10211.

Suzuki G., Nakamura M., Michinaka C., Nguyen Minh T., Handa H. and Takigami H. (2017). Separate screening of brominated and chlorinated dioxins in field samples using *in vitro* reporter gene assays with rat and mouse hepatoma cell lines. *Analytica Chimica Acta*, **975**, 86–95.

Swart J. C., Pool E. J. and van Wyk J. H. (2011). The implementation of a battery of *in vivo* and *in vitro* bioassays to assess river water for estrogenic endocrine disrupting chemicals. *Ecotoxicology and Environmental Safety*, **74**(1), 138–143.

Tan L. and Schirmer K. (2017). Cell culture-based biosensing techniques for detecting toxicity in water. *Current Opinion in Biotechnology*, **45**, 59–68.

Tang J. Y. M. and Escher B. I. (2014). Realistic environmental mixtures of micropollutants in wastewater, recycled water and surface water: herbicides dominate the mixture toxicity towards algae. *Environmental Toxicology and Chemistry*, **33**(6), 1427–1436.

Tang J. Y. M., Glenn E., Aryal R., Gernjak W. and Escher B. I. (2013a). Toxicity characterization of urban stormwater with bioanalytical tools. *Water Research* **47**(15), 5594–5606.

Tang J. Y. M., McCarty S., Glenn E., Neale P. A., Warne M. S. and Escher B. I. (2013b). Mixture effects of organic micropollutants present in water: Towards the development of effect-based water quality trigger values for baseline toxicity. *Water Research* **47** (10), 3300–3314.

Tang J. Y. M., Busetti F., Charrois J. W. A. and Escher B. I. (2014). Which chemicals drive biological effects in wastewater and recycled water? *Water Research*, **60**, 289–299.

Tanneberger K., Knöbel M., Busser F. J. M., Sinnige T. L., Hermens J. L. M. and Schirmer K. (2013). Predicting fish acute toxicity using a fish gill cell line-based toxicity assay. *Environmental Science & Technology*, **47**(2), 1110–1119.

Terada H. (1990). Uncouplers of oxidative phosphorylation. *Environmental Health Perspectives*, **87**, 213–218.

Teuschler L. K. (2007). Deciding which chemical mixtures risk assessment methods work best for what mixtures. *Toxicology and Applied Pharmacology*, **223**(2), 139–147.

Thomas R. S., Paules R. S., Simeonov A., Fitzpatrick S. C., Crofton K. M., Casey W. M. and Mendrick D. L. (2018). The US Federal Tox21 program: A strategic and operational plan for continued leadership. *ALTEX-Alternatives to Animal Experimentation*, **35**(2), 163–168.

Thomas R. S., Bahadori T., Buckley T. J., Cowden J., Deisenroth C., Dionisio K. L., Frithsen J. B., Grulke C. M., Gwinn M. R., Harrill J. A., Higuchi M., Houck K. A., Hughes M. F., Hunter E. S., Isaacs K. K., Judson R. S., Knudsen T. B., Lambert J. C., Linnenbrink M., Martin T. M., Newton S. R., Padilla S., Patlewicz G., Paul-Friedman K., Phillips K. A., Richard A. M., Sams R., Shafer T. J., Setzer R. W., Shah I., Simmons J. E., Simmons S. O., Singh A., Sobus J. R., Strynar M., Swank A., Tornero-Valez R., Ulrich E. M., Villeneuve D. L., Wambaugh J. F., Wetmore B. A. and Williams A. J. (2019). The next generation blueprint of computational toxicology at the US environmental protection agency. *Toxicological Sciences*, **169**(2), 317–332.

Thorpe K. L., Gross-Sorokin M., Johnson I., Brighty G. and Tyler C. R. (2006). An assessment of the model of concentration addition for predicting the estrogenic activity of chemical mixtures in wastewater treatment works effluents. *Environmental Health Perspectives*, **114**, 90–97.

Tiffany-Castiglioni E., Hong S., Qian Y., Tang Y. and Donnelly K. C. (2006). *In vitro* models for assessing neurotoxicity of mixtures. *Neurotoxicology*, **27**(5), 835–839.

Timbrell J. (2009). Principles of Biochemical Toxicology, 4th edn. Informa Healthcare, New York, US.

Timm M., Saaby L., Moesby L. and Hansen E. W. (2013). Considerations regarding use of solvents in *in vitro* cell based assays. *Cytotechnology*, **65**(5), 887–894.

Timourian H., Felton J. S., Stuermer D. H., Healy S., Berry P., Tompkins M., Battaglia G., Hatch F. T., Thompson L. H., Carrano A. V., Minkler J. and Salazar E. (1982). Mutagenic and toxic activity of environmental effluents from underground coal gasification experiments. *Journal of Toxicology and Environmental Health*, **9**(5–6), 975–994.

Toušová Z., Oswald P., Slobodnik J., Blaha L., Muz M., Hu M., Brack W., Krauss M., Di Paolo C., Tarcai Z., Seiler T. B., Hollert H., Koprivica S., Ahel M., Schollee J. E., Hollender J., Suter M. J. F., Hidasi A. O., Schirmer K., Sonavane M., Ait-Aissa S., Creusot N., Brion F., Froment J., Almeida A. C., Thomas K., Tollefsen K. E., Tufi S., Ouyang X. Y., Leonards P., Lamoree M., Torrens V. O., Kolkman A., Schriks M., Spirhanzlova P., Tindall A. and Schulze T. (2017). European demonstration program on the effect-based and chemical identification and monitoring of organic pollutants in European surface waters. *Science of the Total Environment*, **601**, 1849–1868.

Toušová Z., Vrana B., Smutná M., Novák J., Klučárová V., Grabic R., Slobodník J., Giesy J. P. and Hilscherová K. (2019). Analytical and bioanalytical assessments of organic micropollutants in the Bosna River using a combination of passive sampling, bioassays and multi-residue analysis. *Science of the Total Environment*, **650**, 1599–1612.

Troger R., Klockner P., Ahrens L. and Wiberg K. (2018). Micropollutants in drinking water from source to tap – Method development and application of a multiresidue screening method. *Science of the Total Environment*, **627**, 1404–1432.

Tsuno H., Arakawa K., Kato Y. and Nagare H. (2008). Advanced sewage treatment with ozone under excess sludge reduction, disinfection and removal of EDCs. *Ozone-Science & Engineering*, **30**(3), 238–245.

Tue N. M., Suzuki G., Takahashi S., Kannan K., Takigami H. and Tanabe S. (2013). Dioxin-related compounds in house dust from New York State: Occurrence, *in vitro* toxic evaluation and implications for indoor exposure. *Environmental Pollution*, **181**, 75–80.

Tuikka A. I., Schmitt C., Hoss S., Bandow N., von der Ohe P. C., de Zwart D., de Deckere E., Streck G., Mothes S., van Hattum B., Kocan A., Brix R., Brack W., Barcelo D., Sormunen A. J. and Kukkonen J. V. (2011). Toxicity assessment of sediments from three European river basins using a sediment contact test battery. *Ecotoxicology and Environmental Safety*, **74**(1), 123–131.

Tuncer S., Gurbanov R., Sheraj I., Solel E., Esenturk O. and Banerjee S. (2018). Low dose dimethyl sulfoxide driven gross molecular changes have the potential to interfere with various cellular processes. *Scientific Reports*, **8**, 14828.

U.S. EPA (1976). Toxic Substances Control Act. Washington, DC, USA.

U.S. EPA (1986). Guidelines for the health risk assessment of chemical mixtures. *Federal Register*, **51**, 34014–34025.

U.S. EPA (2002a). Guidance on cumulative risk assessment of pesticide chemicals that have a common mechanism of toxicity. Washington, DC. www.epa.gov/pesticide-science-and-assessing-pesticide-risks/guidance-cumulative-risk-assessment-pesticide (Accesssed on 18 February 2021).

U.S. EPA (2002b). Methods for Measuring the Acute Toxicity of Effluents and Receiving Waters to Freshwater and Marine Organisms. 5th edn. EPA-821-R-02-012. www.epa.gov/sites/production/files/2015-08/documents/acute-freshwater-and-marine-wet-manual_2002.pdf (Accesssed on 18 February 2021).

U.S. EPA (2002c). Short-Term Methods for Estimating the Chronic Toxicity of Effluents and Receiving Water to Marine and Estuarine Organisms. www.epa.gov/sites/production/files/2015-08/documents/short-term-chronic-marine-and-estuarine-wet-manual_2002.pdf (Accesssed on 18 February 2021).

U.S. EPA (2007). Method: 1694, Pharmaceuticals and personal care products in water, soil, sediment, and biosolids by HPLC/MS/MS. In: EPA-821-R-08-002, Washington DC. www.epa.gov/sites/production/files/2015-10/documents/method_1694_2007.pdf (Accesssed on 18 February 2021).

U.S. EPA (2018). 2018 Edition of the drinking water standards and health advisories. www.epa.gov/sites/production/files/2018-03/documents/dwtable2018.pdf (Accesssed on 18 February 2021).

U.S. EPA (2020). CompTox Chemistry Dashboard. comptox.epa.gov/dashboard/ (Accessed 20 December 2020).

Ulitzur S., Weiser I. and Yannai S. (1980). A new, sensitive and simple bioluminescence test for mutagenic compounds. *Mutation Research – Environmental Mutagenesis and Related Subjects*, **74**(2), 113–124.

United Nations (1992). Report of the United Nations Conference on Environment and Development. www.un.org/en/development/desa/population/migration/generalassembly/docs/globalcompact/A_CONF.151_26_Vol.I_Declaration.pdf (Accessed 23 October 2020).

United Nations (2009). Stockholm Convention on Persistent Organic Pollutants. Geneva, Switzerland, and New York, NY, USA.

United Nations (2019). Globally Harmonized System of Classification and Labelling of Chemicals (GHS), 8th revised edition. Geneva, Switzerland, and New York, NY, USA.

Valcarcel Y., Valdehita A., Becerra E., de Alda M. L., Gil A., Gorga M., Petrovic M., Barcelo D. and Navas J. M. (2018). Determining the presence of chemicals with suspected endocrine activity in drinking water from the Madrid region (Spain) and assessment of their estrogenic, androgenic and thyroidal activities. *Chemosphere*, **201**, 388–398.

Välitalo P., Massei R., Heiskanen I., Behnisch P., Brack W., Tindall A. J., Du Pasquier D., Küster E., Mikola A., Schulze T. and Sillanpää M. (2017). Effect-based assessment of toxicity removal during wastewater treatment. *Water Research*, **126**, 153–163.

Vallotton N. and Price P. S. (2016). Use of the maximum cumulative ratio as an approach for prioritizing aquatic coexposure to plant protection products: A case study of a large surface water monitoring database. *Environmental Science & Technology*, **50**(10), 5286–5293.

van Broekhuizen F. L. P. and Traas T. (2016). Addressing combined effects of chemicals in environmental safety assessment under REACH – a thought starter. RIVM Letter report 2016-0162, National Institute for Public Health and the Environment, Bilthoven, The Netherlands. www.rivm.nl/bibliotheek/rapporten/2016-0162.pdf (Accessed 10 April 2021).

van Dam R. A. and Chapman J. C. (2001). Direct toxicity assessment (DTA) for water quality guidelines in Australia and New Zealand. *Australasian Journal of Ecotoxicology*, **7**, 175–198.

van de Merwe J. P. and Leusch F. D. L. (2015). A sensitive and high throughput bacterial luminescence assay for assessing aquatic toxicity – the BLT-screen. *Environmental Science-Processes & Impacts*, **17**(5), 947–955.

Van den Berg M., Birnbaum L. S., Denison M., De Vito M., Farland W., Feeley M., Fiedler H., Hakansson H., Hanberg A., Haws L., Rose M., Safe S., Schrenk D., Tohyama C., Tritscher A., Tuomisto J., Tysklind M., Walker N. and Peterson R. E. (2006). The 2005 World Health Organization reevaluation of human and mammalian toxic equivalency factors for dioxins and dioxin-like compounds. *Toxicological Sciences*, **93**(2), 223–241.

van der Lelie D., Regniers L., Borremans B., Provoost A. and Verschaeve L. (1997). The VITOTOX(R) test, an SOS bioluminescence *Salmonella typhimurium* test to measure genotoxicity kinetics. *Mutation Research-Genetic Toxicology and Environmental Mutagenesis*, **389**(2–3), 279–290.

van der Linden S. C., Heringa M. B., Man H. Y., Sonneveld E., Puijker L. M., Brouwer A. and Van der Burg B. (2008). Detection of multiple hormonal activities in wastewater effluents and surface water, using a panel of steroid receptor CALUX bioassays. *Environmental Science & Technology*, **42**(15), 5814–5820.

van der Linden S. C., von Bergh A. R. M., van Vught-Lussenburg B. M. A., Jonker L. R. A., Teunis M., Krul C. A. M. and van der Burg B. (2014). Development of a panel of high-throughput reporter-gene assays to detect genotoxicity and oxidative stress. *Mutation Research-Genetic Toxicology and Environmental Mutagenesis*, **760**, 23–32.

van der Oost R., Sileno G., Janse T., Nguyen M. T., Besselink H. and Brouwer A. (2017a). SIMONI (Smart Integrated Monitoring) as a novel bioanalytical strategy for water quality assessment: Part II–field feasibility survey. *Environmental Toxicology and Chemistry*, **36**(9), 2400–2416.

van der Oost R., Sileno G., Suarez-Munoz M., Nguyen M. T., Besselink H. and Brouwer A. (2017b). SIMONI (Smart Integrated Monitoring) as a novel bioanalytical strategy for water quality assessment: Part I–model design and effect-based trigger values. *Environmental Toxicology and Chemistry*, **36**(9), 2385–2399.

van Leeuwen C. J. and Vermeire T. G. (2007). Risk Assessment of Chemicals: An Introduction. 2nd edn. Springer, Dordrecht, The Netherlands.

van Leeuwen K., Schultz T. W., Henry T., Diderich B. and Veith G. D. (2009). Using chemical categories to fill data gaps in hazard assessment. *SAR and QSAR in Environmental Research*, **20**(3–4), 207–220.

van Wezel A. P. and Opperhuizen A. (1995). Narcosis due to environmental pollutants in aquatic organisms: Residue-based toxicity, mechanisms, and membrane burdens. *Critical Reviews in Toxicology*, **25**, 255–279.

Van Zijl M. C., Aneck-Hahn N. H., Swart P., Hayward S., Genthe B. and De Jager C. (2017). Estrogenic activity, chemical levels and health risk assessment of municipal distribution point water from Pretoria and Cape Town, South Africa. *Chemosphere*, **186**, 305–313.

Vankreijl C. F., Kool H. J., Devries M., Vankranen H. J. and Degreef E. (1980). Mutagenic activity in the rivers Rhine and Meuse in the Netherlands. *Science of the Total Environment*, **15**(2), 137–147.

Vergauwen L., Schmidt S. N., Stinckens E., Maho W., Blust R., Mayer P., Covaci A. and Knapen D. (2015). A high throughput passive dosing format for the fish embryo acute toxicity test. *Chemosphere*, **139**, 9–17.

Verheijen M., Lienhard M., Schrooders Y., Clayton O., Nudischer R., Boerno S., Timmermann B., Selevsek N., Schlapbach R., Gmuender H., Gotta S., Geraedts J., Herwig R., Kleinjans J. and Caiment F. (2019). DMSO induces drastic changes in human cellular processes and epigenetic landscape *in vitro*. *Scientific Reports*, **9**, 4641.

Vermeirssen E. L. M., Korner O., Schonenberger R., Suter M. J. F. and Burkhardt-Holm P. (2005). Characterization of environmental estrogens in river water using a three pronged approach: Active and passive water sampling and the analysis of accumulated estrogens in the bile of caged fish. *Environmental Science & Technology*, **39**(21), 8191–8198.

Verschaeve L., Van Gompel J., Thilemans L., Regniers L., Vanparys P. and van der Lelie D. (1999). VITOTOX (R) bacterial genotoxicity and toxicity test for the rapid screening of chemicals. *Environmental and Molecular Mutagenesis*, **33**(3), 240–248.

Vethaak A. D., Hamers T., Martinez-Gomez C., Kamstra J. H., de Weert J., Leonards P. E. G. and Smedes F. (2017). Toxicity profiling of marine surface sediments: A case study using rapid screening bioassays of exhaustive total extracts, elutriates and passive sampler extracts. *Marine Environmental Research*, **124**, 81–91.

Villeneuve D. L., Blankenship A. L. and Giesy J. P. (2000). Derivation and application of relative potency estimates based on *in vitro* bioassay results. *Environmental Toxicology and Chemistry*, **19**(11), 2835–2843.

Villeneuve D. L., Coady K., Escher B. I., Mihaich E., Murphy C., Schlekat T. and Garcia-Reyero N. (2019). High throughput screening and environmental risk assessment – state of the science and emerging applications. *Environmental Toxicology and Chemistry*, **38**(1), 12–26.

Vinggaard A. M., Bonefeld-Jorgensen E. C., Jensen T. K., Fernandez M. F., Rosenmai A. K., Taxvig C., Rodriguez-Carrillo A., Wielsoe M., Long M., Olea N., Antignac J.-P., Hamers T. and Lamoree M. (2020). Receptor-based *in vitro* activities to assess human exposure to chemical mixtures and related health impacts. *Environment International*, **146**, 106191–106191.

Vinken M. and Blaauboer B. J. (2017). *In vitro* testing of basal cytotoxicity: Establishment of an adverse outcome pathway from chemical insult to cell death. *Toxicology In Vitro*, **39**, 104–110.

Vogs C. and Altenburger R. (2016). Time-dependent effects in algae for chemicals with different adverse outcome pathways: A novel approach. *Environmental Science & Technology*, **50**(14), 7770–7780.

Volker J., Vogt T., Castronovo S., Wick A., Ternes T. A., Joss A., Oehlmann J. and Wagner M. (2017). Extended anaerobic conditions in the biological wastewater treatment: Higher reduction of toxicity compared to target organic micropollutants. *Water Research*, **116**, 220–230.

Volker J., Stapf M., Miehe U. and Wagner M. (2019). Systematic review of toxicity removal by advanced wastewater treatment technologies via ozonation and activated carbon. *Environmental Science & Technology*, **53**(13), 7215–7233.

von der Ohe P. C., Dulio V., Slobodnik J., De Deckere E., Kuhne R., Ebert R. U., Ginebreda A., De Cooman W., Schuurmann G. and Brack W. (2011). A new risk assessment approach for the prioritization of 500 classical and emerging organic microcontaminants as potential river basin specific pollutants under the European Water Framework Directive. *Science of the Total Environment*, **409**(11), 2064–2077.

Wagner M. and Oehlmann J. (2009). Endocrine disruptors in bottled mineral water: Total estrogenic burden and migration from plastic bottles. *Environmental Science and Pollution Research*, **16**(3), 278–286.

Wagner E. D. and Plewa M. J. (2008). Chapter 3 Microplate-based comet assay. In: The Comet Assay in Toxicology, The Royal Society of Chemistry, pp. 79–97.

Walker C. W. and Watson J. E. (2010). Adsorption of estrogens on laboratory materials and filters during sample preparation. *Journal of Environmental Quality*, **39**(2), 744–748.

Walker C. H., Hopkin S. P., Sibly R. M. and Peakall D. B. (2012). Principles of Ecotoxicology. 4th edn. Taylor & Francis, Milton Park, UK.

Walter H., Consolaro F., Gramatica P., Scholze M. and Altenburger R. (2002). Mixture toxicity of priority pollutants at no observed effect concentrations (NOECs). *Ecotoxicology*, **11**, 299–310.

Wambaugh J. F., Hughes M. F., Ring C. L., MacMillan D. K., Ford J., Fennell T. R., Black S. R., Snyder R. W., Sipes N. S., Wetmore B. A., Westerhout J., Setzer R. W., Pearce R. G., Simmons J. E. and Thomas R. S. (2018). Evaluating *in vitro-in vivo* extrapolation of toxicokinetics. *Toxicological Sciences*, **163**(1), 152–169.

Wang H. and Joseph J. A. (1999). Quantifying cellular oxidative stress by dichlorofluorescein assay using microplate reader. *Free Radical Biology and Medicine*, **27**(5–6), 612–616.

Wang W. L., Tada M., Nakajima D., Sakai M., Yoneda M. and Sone H. (2018). Multiparameter phenotypic profiling in MCF-7 cells for assessing the toxicity and estrogenic activity of whole environmental water. *Environmental Science & Technology*, **52**(16), 9277–9284.

Wang Z., Walker G., Muir D. and Nagatani-Yoshida K. (2020). Toward a global understanding of chemical pollution: A first comprehensive analysis of national and regional chemical inventories. *Environmental Science & Technology*, **54**(5), 2575–2584.

Wania F. (2003). Assessing the potential of persistent organic chemicals for long-range transport and accumulation in polar regions. *Environmental Science & Technology*, **37**(7), 1344–1351.

Warne M. S. J. and Hawker D. W. (1995). The number of components in a mixture determines whether synergistic and antagonistic or additive toxicity predominate – the funnel hypothesis. *Ecotoxicology and Environmental Safety*, **31**(1), 23–28.

Wernersson A.-S., Carere M., Maggi C., Tusil P., Soldan P., James A., Sanchez W., Dulio V., Broeg K., Reifferscheid G., Buchinger S., Maas H., Van Der Grinten E., O'Toole S., Ausili A., Manfra L., Marziali L., Polesello S., Lacchetti I., Mancini L., Lilja K., Linderoth M., Lundeberg T., Fjallborg B., Porsbring T., Larsson D. G. J., Bengtsson-Palme J., Forlin L., Kienle C., Kunz P., Vermeirssen E., Werner I., Robinson C. D., Lyons B., Katsiadaki I., Whalley C., den Haan K., Messiaen M., Clayton H., Lettieri T., Carvalho R. N., Gawlik B. M., Hollert H., Di Paolo C., Brack W., Kammann U. and Kase R. (2015). The European technical report on aquatic effect-based monitoring tools under the water framework directive. *Environmental Sciences Europe*, **27**, 1–11.

Wetmore B. A. (2015). Quantitative *in vitro-in vivo* extrapolation in a high-throughput environment. *Toxicology*, **332**, 94–101.

White P. A., Rasmussen J. B. and Blaise C. (1996). A semi-automated, microplate version of the SOS Chromotest for the analysis of complex environmental extracts. *Mutation Research – Environmental Mutagenesis and Related Subjects*, **360**(1), 51–74.

WHO (2017a). Chemical mixtures in source water and drinking-water. World Health Organization, www.who.int/water_sanitation_health/publications/chemical-mixtures-in-water/en/ (Accessed 14 February 2021).

WHO (2017b). Guidelines for Drinking-Water Quality. 4th edn, incorporating the 1st addendum. Geneva, Switzerland. www.who.int/water_sanitation_health/publications/2011/dwq_guidelines/en/index.html. (Accessed 18 February 2021).

WHO (2017c). Potable Reuse: Guidance for Producing Safe Drinking-Water. World Health Organization. www.who.int/water_sanitation_health/publications/potable-reuse-guidelines/en/ (Accessed 18 February 2021).

Widder M. W., Brennan L. M., Hanft E. A., Schrock M. E., James R. R. and van der Schalie W. H. (2015). Evaluation and refinement of a field-portable drinking water toxicity sensor utilizing electric cell-substrate impedance sensing and a fluidic biochip. *Journal of Applied Toxicology*, **35**(7), 701–708.

Wild C. P. (2005). Complementing the genome with an "exposome": The outstanding challenge of environmental exposure measurement in molecular epidemiology. *Cancer Epidemiology Biomarkers & Prevention*, **14**(8), 1847–1850.

Wilkinson M. D., Dumontier M., Aalbersberg I. J., Appleton G., Axton M., Baak A., Blomberg N., Boiten J. W., Santos L. B. D., Bourne P. E., Bouwman J., Brookes A. J., Clark T., Crosas M., Dillo I., Dumon O., Edmunds S., Evelo C. T., Finkers R., Gonzalez-Beltran A., Gray A. J. G., Groth P., Goble C., Grethe J. S., Heringa J., Hoen P. A. C., Hooft R., Kuhn T., Kok R., Kok J., Lusher S. J., Martone M. E.,

Mons A., Packer A. L., Persson B., Rocca-Serra P., Roos M., van Schaik R., Sansone S. A., Schultes E., Sengstag T., Slater T., Strawn G., Swertz M. A., Thompson M., van der Lei J., van Mulligen E., Velterop J., Waagmeester A., Wittenburg P., Wolstencroft K., Zhao J. and Mons B. (2016). The FAIR Guiding Principles for scientific data management and stewardship. *Scientific Data*, **6**, 160018.

Wilson V. S., Bobseine K. and Gray L. E. , Jr. (2004). Development and characterization of a cell line that stably expresses an estrogen-responsive luciferase reporter for the detection of estrogen receptor agonist and antagonists. *Toxicological Sciences*, **81**(1), 69–77.

Wölz J., Engwall M., Maletz S., Takner H. O., van Bavel B., Kammann U., Klempt M., Weber R., Braunbeck T. and Hollert H. (2008). Changes in toxicity and Ah receptor agonist activity of suspended particulate matter during flood events at the rivers Neckar and Rhine – a mass balance approach using *in vitro* methods and chemical analysis. *Environmental Science and Pollution Research*, **15**(7), 536–553.

Wood S. A., Holland P. T., Stirling D. J., Briggs L. R., Sprosen J., Ruck J. G. and Wear R. G. (2006). Survey of cyanotoxins in New Zealand water bodies between 2001 and 2004. *New Zealand Journal of Marine and Freshwater Research*, **40**(4), 585–597.

Woutersen M., Belkin S., Brouwer B., van Wezel A. P. and Heringa M. B. (2011). Are luminescent bacteria suitable for online detection and monitoring of toxic compounds in drinking water and its sources? *Analytical and Bioanalytical Chemistry*, **400**(4), 915–929.

Wu Q.-Y., Hu H.-Y., Zhao X. and Sun Y.-X. (2009). Effect of chlorination on the estrogenic/antiestrogenic activities of biologically treated wastewater. *Environmental Science & Technology*, **43**(13), 4940–4945.

Xiao R., Wang Z., Wang C. and Yu G. (2006). Soil screening for identifying ecological risk stressors using a battery of *in vitro* cell bioassays. *Chemosphere*, **64**(1), 71–78.

Xiao S. H., Lv X. M., Lu Y., Yang X. M., Dong X. R., Ma K. P., Zeng Y. F., Jin T. and Tang F. (2016). Occurrence and change of estrogenic activity in the process of drinking water treatment and distribution. *Environmental Science and Pollution Research*, **23**(17), 16977–16986.

Xiao S. H., Lv X. M., Zeng Y. F., Jin T., Luo L., Zhang B. B., Zhang G., Wang Y. H., Feng L., Zhu Y. and Tang F. (2017). Mutagenicity and estrogenicity of raw water and drinking water in an industrialized city in the Yangtze River Delta. *Chemosphere*, **185**, 647–655.

Xie S. H., Liu A. L., Chen Y. Y., Zhang L., Zhang H. J., Jin B. X., Lu W. H., Li X. Y. and Lu W. Q. (2010). DNA damage and oxidative stress in human liver cell L-02 caused by surface water extracts during drinking water treatment in a waterworks in China. *Environmental and Molecular Mutagenesis*, **51**(3), 229–235.

Yagi K. (1998). Simple assay for the level of total lipid peroxides in serum or plasma. *Methods in Molecular Biology*, **108**(1), 101–106.

Yang C., Zhou J. Y., Zhong H. J., Wang H. Y., Yan J., Liu Q., Huang S. N. and Jiang J. X. (2011). Exogenous norepinephrine correlates with macrophage endoplasmic reticulum stress response in association with XBP-1. *Journal of Surgical Research*, **168**(2), 262–271.

Yeh R. Y. L., Farré M. J., Stalter D., Tang J. Y. M., Molendijk J. and Escher B. I. (2014). Bioanalytical and chemical evaluation of disinfection by-products in swimming pool water. *Water Research*, **59**, 172–184.

Yoo H., Khim J. S. and Giesy J. P. (2006). Receptor-mediated *in vitro* bioassay for characterization of Ah-R-active compounds and activities in sediment from Korea. *Chemosphere*, **62**(8), 1261–1271.

Yoon M., Blaauboer B. J. and Clewell H. J. (2015). Quantitative *in vitro* to *in vivo* extrapolation (QIVIVE): An essential element for *in vitro*-based risk assessment. *Toxicology*, **332**, 1–3.

You J. and Li H. Z. (2017). Improving the accuracy of effect-directed analysis: The role of bioavailability. *Environmental Science-Processes & Impacts*, **19**(12), 1484–1498.

Zani C., Feretti D., Buschini A., Poli P., Rossi C., Guzzella L., Caterino F. D. and Monarca S. (2005). Toxicity and genotoxicity of surface water before and after various potabilization steps. *Mutation Research – Genetic Toxicology and Environmental Mutagenesis*, **587**(1–2), 26–37.

Zegura B., Heath E., Cernosa A. and Filipic M. (2009). Combination of *in vitro* bioassays for the determination of cytotoxic and genotoxic potential of wastewater, surface water and drinking water samples. *Chemosphere*, **75**(11), 1453–1460.

Zeng S. Y., Huang Y. Q., Sun F., Li D. and He M. (2016). Probabilistic ecological risk assessment of effluent toxicity of a wastewater reclamation plant based on process modeling. *Water Research*, **100**, 367–376.

Zhang D. D. (2006). Mechanistic studies of the Nrf2-Keap1 signaling pathway. *Drug Metabolism Reviews*, **38**(4), 769–789.

Zhang J. H., Chung T. D. Y. and Oldenburg K. R. (1999). A simple statistical parameter for use in evaluation and validation of high throughput screening assays. *Journal of Biomolecular Screening*, **4**(2), 67–73.

Zhang Z. H., Feng Y. J., Gao P., Wang C. and Ren N. Q. (2011). Occurrence and removal efficiencies of eight EDCs and estrogenicity in a STP. *Journal of Environmental Monitoring*, **13**(5), 1366–1373.

Zhang H., Ihara M., Hanamoto S., Nakada N., Jurgens M. D., Johnson A. C. and Tanaka H. (2018a). Quantification of pharmaceutical related biological activity in effluents from wastewater treatment plants in UK and Japan. *Environmental Science & Technology*, **52**(20), 11848–11856.

Zhang S. Y., Li S. Z., Zhou Z. G., Fu H. L., Xu L., Xie H. and Zhao B. (2018b). Development and application of a novel bioassay system for dioxin determination and aryl hydrocarbon receptor activation evaluation in ambient-air samples. *Environmental Science & Technology*, **52**(5), 2926–2933.

Zhang J. Y., Yang Y., Liu W. P., Schlenk D. and Liu J. (2019). Glucocorticoid and mineralocorticoid receptors and corticosteroid homeostasis are potential targets for endocrine-disrupting chemicals. *Environment International*, **133**, 105133.

Zhao J. L., Ying G. G., Yang B., Liu S., Zhou L. J., Chen Z. F. and Lai H. J. (2011). Screening of multiple hormonal activities in surface water and sediment from the Pearl River system, South China, using effect-directed *in vitro* bioassays. *Environmental Toxicology and Chemistry*, **30**(10), 2208–2215.

Zheng F., Fu F., Cheng Y., Wang C., Zhao Y. and Gu Z. (2016). Organ-on-a-chip systems: Microengineering to biomimic living systems. *Small*, **12**(17), 2253–2282.

Zhong Y., Feng S. L., Luo Y., Zhang G. D. and Kong Z. M. (2001). Evaluating the genotoxicity of surface water of Yangzhong City using the *Vicia faba* micronucleus test and the comet assay. *Bulletin of Environmental Contamination and Toxicology*, **67** (2), 217–224.

Zimmermann F. K., Kern R. and Rasenberger H. (1975). A yeast strain for simultaneous detection of induced mitotic crossing over, mitotic gene conversion and reverse mutation. *Mutation Research – Fundamental and Molecular Mechanisms of Mutagenesis*, **28**(3), 381–388.

Zimmermann L., Dierkes G., Ternes T. A., Volker C. and Wagner M. (2019). Benchmarking the *in vitro* toxicity and chemical composition of plastic consumer products. *Environmental Science & Technology*, **53**(19), 11467–11477.

Zimmermann L., Dombrowski A., Volker Ç. and Wagner M. (2020). Are bioplastics and plant-based materials safer than conventional plastics? *In vitro* toxicity and chemical composition. *Environment International*, **145**, 106066.

Zounkova R., Jalova V., Janisova M., Ocelka T., Jurcikova J., Halirova J., Giesy J. P. and Hilscherova K. (2014). *In situ* effects of urban river pollution on the mudsnail *Potamopyrgus antipodarum* as part of an integrated assessment. *Aquatic Toxicology*, **150**, 83–92.

Zucker R. M., Elstein K. H., Easterling R. E. and Massaro E. J. (1988). Flow cytometric analysis of the cellular toxicity of tributyltin. *Toxicology Letters*, **43**(1–3), 201–218.

Zurita J., del Peso A., Rojas R., Maisanaba S. and Repetto G. (2019). Integration of fish cell cultures in the toxicological assessment of effluents. *Ecotoxicology and Environmental Safety*, **176**, 309–320.

Zwart N., Nio S. L., Houtman C. J., de Boer J., Kool J., Hamers T. and Lamoree M. H. (2018). High-throughput effect-directed analysis using downscaled *in vitro* reporter gene assays to identify endocrine disruptors in surface water. *Environmental Science & Technology*, **52**(7), 4367–4377.

Zwart N., Jonker W., ten Broek R., de Boer J., Somsen G., Kool J., Hamers T., Houtman C. J. and Lamoree M. H. (2020). Identification of mutagenic and endocrine disrupting compounds in surface water and wastewater treatment plant effluents using high-resolution effect-directed analysis. *Water Research*, **168**, 15204.

Index

Note: t = Table, f = Figure

A

AA-EQS (annual average environmental quality standard), 41, 284, 285, 355
ABC transporters, 53, 355. *See also* Active transport
Abiotic, 4, 5t. *See also* Biotic
Absorption, 51, 53f, 73, 74, 75, 77, 81, 158
ACC (activity concentration at cutoff), 108, 109f, 150, 319, 355
Acceptable (or allowable) daily intake (ADI), 29, 37, 40, 281, 284, 341, 355, 356
Accuracy, 46, 225–227, 227f, 243
Acetylcholinesterase (AChE), 60t, 62–63, 86, 96–97, 96f, 118, 124, 200–201, 201t, 229, 283t, 309t, 311t, 312t, 355, 358, 368
AChE. *See* Acetylcholinesterase
ACR. *See* acute-to-chronic ratio
Active transport, 52–53, 61, 74, 77, 79, 158, 171, 355, 369. *See also* ABC transporters, Passive transport
Acute exposure, 355, 359. *See also* Chronic exposure
Acute toxicity, 6, 15, 30, 41, 44–48, 92, 215, 268, 349
Acute-to-chronic ratio (ACR), 291, 294
Adaptive stress response pathway, 67–70, 68f, 212, 356, 369, 376
Adenosine-5'-triphosphate (ATP), 52, 60t, 61–63, 77, 79–80, 95, 97, 355, 357
ADI. *See* Acceptable (or allowable) daily intake
Advanced water treatment plant (AWTP), 115f, 357
Adverse outcome pathway (AOP), 56, 56f, 58, 61, 73, 75f, 91–99, 94f–98f, 125, 125f, 126, 152, 154, 268f, 340, 345, 356. *See also* Toxicity pathway
ADWG (Australian drinking water guideline), 39, 285, 287, 307, 356
Aflatoxin, 78, 356
Agonist, 57, 63–64, 65f, 89, 112–113, 112f, 124, 195, 242, 276, 277f, 304t, 306f, 309t, 311t, 312t, 335, 356. *See also* Antagonist
AGWR (Australian guidelines for water recycling), 39, 40, 285, 287, 351, 356
AhR. *See* Aryl hydrocarbon receptor
AhR CAFLUX, 172–173, 174t, 176t, 328, 334–335, 356
Algae, 2, 12, 30, 44, 44f, 45, 47, 48, 59, 92–96, 94f, 95f, 121, 127t, 128t, 136, 137f, 138, 141, 199, 200t, 221, 222t, 275, 279, 279f, 284, 289t, 290, 317f, 349
Aliivibrio fischeri, 14, 106f, 107f, 115f, 121, 127t, 136, 137f, 208t, 218, 242, 248, 287, 328, 356. *See also Vibrio fischeri*
American Society for Testing and Materials (ASTM), 45

Ames test, 14–15, 202, 208t, 242, 271, 356
Anaemia, 83, 356
Androgen, 14, 78, 87–89, 118, 127t, 179, 182–188, 185t, 186t, 187t, 249, 268, 275–277, 277f, 282t, 306, 309t–313t, 335, 356
Androgen receptor (AR), 89, 98, 127t, 179, 182–188, 306, 311t–313t, 316, 335, 357
Annual average EQS (AA-EQS), 41. *See also* EQS
Antagonism, 64, 112, 113, 120, 120t, 123–125, 127t, 182, 186, 190, 192, 195, 228
Antagonist, 57, 63–64, 65f, 89, 98, 112–113, 113f, 123–125, 194–195, 242, 276–278, 277f, 304t, 311t, 312t, 335, 356. *See also* Agonist
Antibody, 84, 356
Antigen, 83–84, 356. *See also* Antibody
Antimitotic drug, 83, 356
Antioxidant response element (ARE), 32, 69, 136, 137, 137f, 212, 213t, 214t, 238t, 240, 241f, 259f, 274f, 275, 288t, 293f, 313t, 320, 321, 322f, 327f, 328, 329t, 330t, 332f, 333, 334, 357
AOC. *See* Assimilable organic carbon
AOP. *See* Adverse Outcome Pathway
Apical endpoints, 20, 92–93, 139, 267, 290, 299, 356
Apoptosis, 58, 61, 65, 69–71, 77, 82, 166, 202, 213, 356
AR. *See* Androgen receptor
ARE. *See* Antioxidant response element
AREc32, 212, 213t, 214t, 274f, 275, 288, 293f
Aromatase, 17, 63–64, 154–155, 357
Aryl hydrocarbon receptor (AhR), 13, 17, 54–55, 54f, 55t, 60t, 129, 137, 172–175, 174t, 176t, 222, 236, 251, 254, 269–271, 274f, 275, 278–280, 279f, 288t, 289t, 293f, 296–297, 304t, 312, 313t, 314–318, 317f, 320–324, 322f, 323f, 327–328, 329t, 330t, 332f, 333–336, 356
Assay, 109–110, 114–117, 126–127, 136, 144–250, 357
Assimilable organic carbon (AOC), 341, 341f, 356
ASTM. *See* American Society for Testing and Materials
ATP. *See* Adenosine-5'-triphosphate
Attagene, 150, 154, 173, 269, 270f, 318, 319f, 320, 320f
Australian and New Zealand Environment and Conservation Council (ANZECC), 42
Australian Drinking Water Guidelines (ADWG), 39, 285, 287, 357
Australian Guidelines for Water Recycling (AGWR), 39–40, 285, 287, 357
Autoimmune disease, 85
AWTP. *See* Advanced water treatment plant
Axon, 86, 357

B

Basal toxicity, 59–60, 61f, 357. *See also* Baseline toxicity
Baseline toxicity, 13, 59, 60t, 61, 77, 94–97, 135, 137, 163–168, 218–219, 278f, 290, 357
BEC. *See* Biologically effective concentration
BED. *See* Biologically effective dose
Benchmark dose (BMD), 28, 101, 103, 103f, 357
Benchmark response (BMR), 357
BEQ. *See* Bioanalytical equivalent concentration
BEQ_{bio}, 117–118, 258, 260, 272–275, 290, 295–296, 316–317, 321–323, 329–330
BEQ_{chem}, 258–260, 259f, 272–275, 274f, 290, 295–296, 317, 321, 322f–323f, 333, 335
Bioaccumulation, 28, 75, 91, 357
Bioactivation, 63, 178, 357
Bioanalytical effect flux (BEF), 327, 327f
Bioanalytical equivalent concentration (BEQ), 102, 117–118, 137, 137f, 208, 230, 236, 240, 258, 258f, 259f, 272, 274f, 284, 285, 290, 291, 294, 316, 317f, 321, 322, 322f, 327f, 328, 329t, 330t, 333, 357

Bioanalytical tool, 1–2, 6, 7f, 9–14, 15f, 18–22, 21t, 23, 33, 47–48, 51, 99, 120, 126, 142, 169–223, 225–226, 235, 245, 271, 272f, 299, 307–308, 324, 325–337, 339–353, 357
Bioavailability, 145, 284, 328, 331f, 332f, 335, 343, 357
Bioconcentration, 28, 154, 155f, 357. *See also* Bioaccumulation
Biodegradation, 4, 5t, 21, 249, 358
Biologically effective concentration (BEC), 51, 94, 105, 158, 358
Biologically effective dose (BED), 51, 94, 105, 158, 159t, 168, 357, 358
Biological pathway altering concentration (BPAC), 149, 358
Biomarker, 9–10, 33, 46–47, 149, 170–172, 268, 358
Biota, 7, 41, 52, 98, 239, 245, 325–337
Biotic, 4, 5t, 358. *See also* Abiotic
Biotransformation, 4, 53–54, 75, 78, 173, 249, 358
Blank, 114, 115, 231, 236–239, 245, 246, 254, 256, 257, 264
BMD. *See* Benchmark dose
BMDL (benchmark dose level), 103
BMR. *See* Benchmark response
BPAC. *See* Biological pathway altering concentration
BPADL (lower limit of biological pathways altering dose), 149
BTEX (benzene, toluene, ethylbenzene, xylene), 3t, 358

C

CA. *See* Concentration addition
CA/DA. *See* Concentration addition/dose addition
CALUX, 108f, 108, 137, 173, 174f, 176–199, 212, 213, 213f, 215, 215f, 216f, 216f, 236, 236f, 238t, 248, 257, 274f, 275, 277f, 278, 282t, 283f, 284, 288, 291, 294, 308t, 313t, 320, 321, 322t, 323f, 329t, 330t, 333
CAR. *See* Constitutive androstane receptor
Carbamate pesticides, 86, 358. *See also* Acetylcholinesterase
Carcinogenesis, 29, 53, 81, 203, 358
Carcinogenicity, 14, 28, 55, 73, 77, 81–82, 173, 203, 358. *See also* Epigenetic carcinogen, genotoxicity
Carcinoma, 163, 220t, 358. *See also* Tumour
Cardiomyocyte, 80, 358
Cardiovascular toxicity, 77, 79–81, 358
Catabolism, 77, 89, 358
Category 1 bioassay, 266, 267, 274f, 275, 278–286, 278f, 279f, 281f, 282t, 283t, 287f, 290, 294–296, 344
Category 2 bioassay, 267, 274f, 275, 278–280, 279f, 285, 286–291, 287f, 288t–289t, 293f, 294–296, 321, 345
Chemical-group motivated approach, 359. *See also* Protection-goal motivated approach
Chronic exposure, 78, 359
Chronic toxicity, 41, 44, 45
Clean Water Act (CWA), 32, 40, 41, 359
C_{max} (maximum plasma concentration), 151, 151f, 153
Coefficient of variation for repeatability (CVr), 227–228, 359
Coefficient of variation for reproducibility (CVR), 227–228
Combined algae test (CAT), 135, 199, 200t, 221, 222t, 248, 283t
Comet assay, 14, 202–203, 204t, 206, 346, 359. *See also* Single Cell Gel Electrophoresis
Complex mixture, 21t, 43, 49, 102, 119, 120, 138, 142, 252, 265, 266, 270f, 275, 326, 336, 340, 346
CompTox Chemistry Dashboard, 147, 167, 101, 359
Concentration addition (CA), 119–133, 136–140, 321f, 358, 359
Concentration addition/dose addition (CA/DA), 120–133, 127t, 135, 358
Concentration-effect curve, 45, 110–111, 116f, 229, 306f
Concentration–response curve (CRC), 101, 105–108, 106f, 108f, 110–112, 111f, 112f, 113–116, 116f, 120, 139, 174t, 181t, 193t, 196t, 229f, 230f, 231f, 235, 236, 272, 276, 306, 359

Confidence interval, 163, 359
Conjugate, 53, 75, 359
Conjugation, 75, 89, 359
Constitutive androstane receptor (CAR), 55t, 173, 346, 358
Control chart, 232, 236, 240–242, 241f
CRC. *See* Concentration–response curve
CVr. *See* Coefficient of variation for repeatability
CVR. *See* Coefficient of variation for reproducibility
CWA. *See* Clean Water Act
Cyanotoxin, 202, 359
Cytochrome (CYP), 17, 53, 55t, 67, 69, 75, 77, 124, 154, 173, 359
Cytokine, 69, 78, 80, 84, 310t, 359
Cytoreductive drug, 83, 359
Cytotoxicity, 9, 12, 14, 17, 18, 20, 23, 52, 58, 61, 68, 77, 78, 101, 102, 104, 106–111, 109f, 111f, 113, 127t, 137–139, 152, 165, 165f, 166, 170, 171, 182, 190, 193t, 194, 204t, 206, 208, 213, 215–221, 220t, 221t, 230, 231, 236, 238t, 239, 243, 245, 256, 257, 266–268, 271–273, 276, 278, 287, 287f, 290, 292f, 296, 299, 306, 306f, 310t–312t, 313, 316, 334, 335, 340, 350, 359. *See also* Basal toxicity

D

DA (dose addition), 119, 120t, 122–126, 123f, 127t, 129–131, 133, 135, 358, 360
Daphnia magna, 22, 45, 95, 96f, 97, 127t, 138, 139, 290
DBP. *See* Disinfection by-product
Denaturing, 68
DF (dosing factor), 47, 114, 248, 262, 360
Dimethyl sulfoxide (DMSO), 229, 237–238, 238t, 242, 257, 262, 264, 343–344, 360
Dioxin, 55t, 129, 360
Direct toxicity assessment (DTA), 42, 215, 340, 360. *See also* WET
Disinfection by-product (DBP), 4–6, 62, 202, 203, 204t–205t, 206, 207, 208, 208t, 210, 212, 218, 249, 255, 266, 299, 304, 305, 307, 308, 311t, 343, 360
Dissolved organic carbon (DOC), 201, 228, 252, 360
Dithiocarbamate pesticides, 66, 360
Diuron, 63, 95, 118, 135–137, 199–200, 200t, 283, 289t, 317, 322, 360. *See also* Photosystem II
DMSO. *See* Dimethyl sulfoxide
DNA, 13, 14, 23, 53–55, 57–59, 59t, 60t, 64–68, 66f, 69t, 70, 81, 82, 110, 157, 166, 173, 202, 203, 204t–205t, 206, 210, 213, 219, 221t, 238, 310t, 346, 360
DOC. *See* Dissolved organic carbon
Dose, 101, 360
Dose addition (DA), 119, 120–133, 120t, 127t, 135, 358
Dose-metric, 101, 105, 158–161, 159t
Dose–response assessment, 28, 101–118, 360. *See also* Dose–response curve
Dose–response curve, 101, 102–103, 105, 130, 360
Dosing factor (DF), 47, 114, 248, 262, 360
Drinking water, 4, 5, 11, 15–17, 20, 22, 35, 36, 37f, 38–40, 42, 51, 62, 73, 74, 76f, 77, 83, 110, 114, 133, 136, 169, 173, 175, 176t, 177, 177t, 178, 179t, 180, 182, 184t, 185, 185t, 186, 187t, 189t, 190, 191t–194t, 195, 197t–201t, 198–203, 206–212, 207t, 209t, 214t, 215, 216t, 217t, 217–219, 218t, 221–223, 222t, 239, 240, 245, 247, 247f, 248, 251, 253, 255, 256, 262, 267–269, 271, 280, 281, 281f, 284–286, 286f, 287f, 289–290, 296, 297, 299, 304–308, 341–343, 349, 351, 352
Drinking water equivalent level (DWEL), 38, 360
Drinking water treatment plant (DWTP), 115f, 206, 269, 305–306, 360
DTA. *See* Direct toxicity assessment
DWEL. *See* Drinking water equivalent level
DWTP. *See* Drinking water treatment plant

E

17α-Ethinylestradio (EE2), 3t, 182f, 268, 273, 275, 278, 284, 285, 362
EBT. *See* Effect-based trigger value

EC_{10}, 108, 108f, 110, 111, 111f, 118, 165, 173–175, 174t, 177, 177t, 178, 178t, 179t, 180, 181t, 184t, 186, 186t, 187t, 188, 190, 190t, 191t, 193t, 196t, 197t, 200, 201t, 215t, 222t, 229, 230, 230f, 291, 306, 307f
EC_{50}, 45, 48, 108, 108f, 111, 115, 122, 125, 126, 127t, 140, 174, 181t, 195, 196t, 198t, 199, 201, 210, 218, 218t, 222, 230, 236, 240–241, 289–291, 360
$EC_{IR1.5}$, 111, 111f, 165, 203, 207t, 212, 213, 213t–215t, 215, 216t, 217t, 229, 240, 241f, 290, 307, 307f
Ecological risk assessment (ERA), 8–9, 8f, 26, 29f, 30, 43, 56, 125, 149f, 154–157, 155f, 360, 361, 362. *See also* Human health risk assessment
EC_{SPR20}, 112, 113, 182, 185t, 186, 188t, 189t, 190, 192, 192t, 194t
ECVAM, 145, 360
EDA. *See* Effect-directed analysis
EDC. *See* Endocrine disrupting compound
EE2. *See* 17α-Ethinylestradiol
EF *See* Extraction factor or enrichment factor
Effect-based trigger value (EBT), 166, 246, 267, 278f, 280–296, 281f, 282t–283t, 287f, 288t–289t, 292f, 293f, 304, 341–342, 360, 361
Effect concentration (EC), 30, 31f, 45, 61, 102, 111f, 112, 117, 154, 165, 165f, 178, 182, 208, 209t, 229, 272, 278f, 290, 291, 307f, 315, 315f, 343, 360
Effect-directed analysis (EDA), 20, 138, 267, 275–278, 276f, 277f, 294, 326–328, 335–337, 341, 344, 345, 360, 361
Effect unit (EU), 117–118, 122, 239, 273, 361, 362
Enrichment factor (EF), 114, 215, 248, 263f, 361
Environmental risk assessment, 8–9, 8f, 26, 43, 56, 125, 149f, 154–157, 155f, 361, 362
Enzyme-linked immunosorbent assay (ELISA), 310t, 361. *See also* Immunoassay
EQS (environmental quality standard), 22, 36, 43f, 284, 341, 361
ERA. *See* Ecological or environmental risk assessment
ER (estrogen receptor), 17, 57, 89, 98, 99, 112, 112f, 118, 124, 150, 179–182, 242, 266, 278, 306, 313t, 335, 361
ER-CALUX, 108, 108f, 179–180, 181t, 182, 183t–185t, 362
EROD. *See* Ethoxyresorufin-O-deethylase
E-SCREEN, 128t, 181t, 182f, 183t–184t, 282t, 284, 362
Estriol (E3), 266, 284, 362
Ethoxyresorufin-O-deethylase (EROD), 174, 175, 251, 335
EU (effect unit), 117–118, 122, 239, 273, 361, 362
Eukaryote, 171, 362
Exposure activity ratio (EAR), 305, 318, 320f
Extraction factor (EF), 114, 361, 362

F

FCMN (flow cytometry micronucleus), 205t, 310t, 362
FET. *See* Fish embryo toxicity
Filtration, 47, 48, 212, 242, 249–252, 260, 305, 314, 327
Fish, 1, 9, 10, 12, 15, 16, 27, 30, 41, 44–48, 44f, 64, 92, 93, 97–99, 97f, 98f, 126, 127t, 132, 135, 139–141, 154, 155f, 171, 175, 204t, 219–221, 221t, 266–268, 326
Fish embryo, 2, 9, 45, 48, 97, 160, 161, 162
Fish embryo toxicity (FET), 10, 45, 47–48, 139, 160f, 162, 215, 222, 242, 268, 271, 289t, 290, 318, 350, 362
Flow cytometry, 205t, 206, 221t, 310t
Food chain, 44, 44f, 92, 93, 325
Frameshift mutation (DNA/RNA), 208, 362
Freely dissolved concentration, 151, 154, 158, 159, 159t, 161, 167, 167t, 168
Furans (PCDF), 129, 370

G

β-Galactosidase, 13, 172, 204t, 363
GAC (Granular activated carbon), 305, 363
Gastrointestinal tract (GI tract), 75, 76f, 77
Gene activation, 57, 61, 139, 299, 363
GeneBLAzer, 137, 154, 177, 177t, 179, 180, 181t, 182, 183t–194t, 188, 190, 192, 194, 195, 196t–198t, 198, 199, 212, 213, 213t–217t 215, 228, 230, 230f, 231f, 240, 241f, 259f, 274f, 275, 282t, 288t, 293f, 306f, 313t, 320, 321, 322f, 329t–331t, 333
Genetic polymorphism, 57, 363
Genetically modified cell, 1, 10, 13, 363
Genomics, 347, 363
Genotoxicity, 6, 14, 22, 60t, 65, 126, 153, 166, 202–207, 204t–205t, 207t, 210, 212, 265, 266, 271, 294, 296, 299, 310t–312t, 333, 346, 363
GFP. *See* Green fluorescent protein
GHS (Globally Harmonised System), 8, 27, 133, 363
Glial cells, 86, 87, 201, 363. *See also* Myelin sheet, Neuron
Glucocorticoid receptor (GR), 179, 188–190, 190t, 191t, 192t, 276, 306, 313, 313t, 335
Glutathione (GSH), 60t, 66, 67, 69, 70, 210, 210t, 211, 211t, 363
GR. *See* Glucocorticoid receptor
Granulosa cells, 363
Green fluorescent protein (GFP), 205t, 206, 363. *See also* Recombinant cell
Grey water, 39, 363
Guideline, 36–40, 42, 43, 45, 49, 221, 242–243, 285, 287, 294, 296, 344, 345, 350
Guideline value, 22, 35–37, 37f, 40, 42, 48, 267, 281, 281f, 284–287, 289, 291, 294, 296, 350, 363
GV. *See* Guideline value

H

Haematopoiesis, 83, 363
Haematotoxicity, 83, 84f, 363
Haloacetic acids (HAAs), 5t, 6, 62, 204t, 205t, 208t, 363. *See also* Disinfection by-product
Hazard quotient (HQ), 30, 131, 364. *See also* Risk quotient
Hazard assessment, 9, 11, 27, 151, 364. *See also* Risk assessment
Hazard identification, 8, 11, 17, 26, 27–28, 33, 294, 351, 364. *See also* Risk assessment
Hazard, 4, 25, 27–28, 30, 132, 133, 145, 152f, 308, 327, 333, 364
Hepatocytes, 54, 77, 78, 170, 210t, 211t
Hepatotoxicity, 77–78, 78f, 309t, 311t, 312t, 364
Herbicide, 17, 18, 59, 63, 93, 95, 95f, 124, 127t, 135, 136f, 138, 199, 200, 266, 290, 317, 318, 322
HHRA. *See* Human health risk assessment
High throughput, 1, 2, 143, 148, 149f, 150f, 151, 153, 169, 225, 276, 346, 347, 349, 364
HLB (hydrophilic-lipophilic balance), 253, 255–257, 261f, 343, 364
Homeostasis, 55, 56, 67, 71, 78, 80, 83, 87, 88, 179, 212, 364
HTS (high throughput screening), 10, 108, 147, 149f, 155f, 169, 203, 243, 343, 364
HTTK (high-throughput toxicokinetics), 151, 153, 364
HTTr (high-throughput transcriptomics), 147, 152, 153, 168, 364
Human Health Risk Assessment (HHRA), 26, 28–31, 29f, 131, 144, 146, 146f, 148–149, 154, 157, 215, 364. *See also* Ecological risk assessment
Hydrophilicity, 364
Hydrophobicity, 160, 161, 164, 165f, 254, 326, 364
Hyperplasia, 78, 88, 364
Hyperthermia, 68, 364
Hypoxia, 69t, 83, 87, 364
Hypoxia-inducible factor (HIF), 173

I

IA. *See* Independent action

ICATM (International Cooperation on Alternative Test Methods), 145, 364
ICCVAM (US Interagency Coordinating Committee on the Validation of Alternative Methods), 145, 364
Iceberg modelling, 147, 154, 266, 271–278, 273f, 280, 290, 296, 304, 305, 314, 317, 321f, 324, 337, 341, 345
Immortalised cell line, 170, 364. *See also* Primary cell line
Immunoassays, 2, 9
Immunotoxicity, 83, 84–85, 310t, 311f, 312t, 335, 364
Independent Action (IA), 119, 120, 121–122, 124, 126, 128t, 130f, 136, 139, 344, 365. *See also* Concentration addition
Inflammation, 69t, 84, 87, 212, 213, 334
Insecticide, 46, 63, 93, 96–97, 124, 138, 201, 290
Integrated (or intelligent) testing strategy (ITS), 143, 145–146, 365
Intercalation (DNA), 57, 365
Internal concentration, 95, 221, 284
International Cooperation on Alternative Test Methods (ICATM), 145, 364
International Organization for Standardization (ISO), 45, 47, 242, 282t, 365
In silico, 144, 146, 151, 168, 365
Invertebrate, 1, 44, 45, 63, 92, 93, 95–97, 162, 268, 289t
In vitro, 1, 9, 10, 11, 15, 33, 45, 47, , 52, 56, 57f, 74f, 90, 92, 98–99, 101, 104, 105, 113, 114, 116f, 117, 120, 122, 123, 126, 129, 138, 143–168, 69–71, 199t, 200, 201, 202, 204t, 205, 208t, 210t, 211t, 221t, 225, 226, 232, 242, 252, 267, 278f, 280, 284, 285, 291, 303t, 304t, 305f, 322f, 326, 328, 334, 348, 365
In vivo, 1, 9–12, 33, 37f, 45, 48, 49, 52, 74f, 104, 116, 123, 126, 129, 146, 150f, 151f, 155f, 203, 233f, 284, 285, 289, 291–4, 303–303t, 333, 342, 365
Ionophoric shuttle mechanism, 60t, 365
IPCS (International Programme on Chemical Safety), 130, 130f, 131, 133, 365
ISO, 45, 47, 242, 282t, 365
ITS, 143–146, 308, 365>

K

Key event (KE), 55–56, 58, 61, 92, 101, 120, 147, 345, 365

L

Lactate dehydrogenase (LDH), 219, 220t, 365
LC_{50} (lethal concentration for 50% of the test species), 30, 45, 46, 122, 140
LC-MS/MS (liquid chromatography with tandem mass spectrometry), 350, 365
LD_{50}, 101, 103f, 125, 365
Legacy, 147, 327
Leydig cells, 365
Ligand, 54, 56, 175, 178, 280, 365. *See also* Receptor
Limit of detection (LOD), 21t, 110, 119, 177, 180, 184t, 195, 197t, 202, 229, 365
Limit of quantification (LOQ), 22, 229, 365
Lipid peroxidation, 66, 67, 211, 211t, 365
Liquid-liquid extraction (LLE), 47, 114, 169, 245, 252f, 254–255, 266, 343, 365. *See also* Solid phase extraction
LLE. *See* Liquid-liquid extraction
LOAEL (lowest observed adverse effect level), 103, 103f, 107, 365
LOEC. *See* Lowest observed effect concentration
Lowest Observed Effect Concentration (LOEC), 107, 107f, 108, 212, 291, 365
Luciferase (Luc), 13, 108, 172, 241, 309–310t, 350, 365

M

MAC-EQS, 41, 366. *See also* EQS
Macrocosm, 46, 92. *See also* Mesocosm
Macropollutant, 3, 340, 366. *See also* Micropollutant
MAF (mixture assessment factor), 134, 135, 366, 367

Margin of exposure (MOE), 31, 366. *See also* Margin of safety
Margin of safety (MOS), 31, 31f, 40, 153, 366
Matrix, 47, 114, 169, 176t, 177t, 179t, 183t, 184t, 185t, 187t, 189t, 192t, 193t, 194t, 197t, 198t, 199t, 200t, 201t, 207t, 209t, 214t, 216t, 217t, 218t, 222t, 228–229, 252f, 253, 325, 342–343, 366
Matrix effects, 325, 342–343, 366
Maximum culculative ratio (MCR), 131–132, 134, 366
MCF-7, 181t, 213t, 220t, 350, 366
MCR. *See* Maximum cumulative ratio
Mechanism of toxicity, 125, 148f, 366
Membrane filtration (MF), 212, 366
Mesocosm, 46, 92
Metabolic activation, 5t, 82, 96, 97, 124, 158, 203, 206, 207t, 213, 306, 336, 366
Metabolic pathway, 54, 366
Metabolism precursor, 5t, 53, 78, 153, 175, 218, 258. *See also* Metabolic activation, Phase I and II metabolism
Metabolite, 4, 48, 53, 62, 70, 78, 89, 90, 158, 163, 336, 347, 366
Metabolomics, 11, 347, 366
Microcystin, 78, 366
Micropollutant, 42, 64, 66, 70, 110, 139, 169, 170, 215, 246, 315f, 324, 366. *See also* Macropollutant
Microtiter plate, 9, 105, 366
Microtox, 14, 47, 115, 127t, 218, 271, 289t, 366
MIE. *See* Molecular initiating event
Mineralocorticoid receptor (MR), 79, 195–197, 304t
Mixture factor (MF), 283t, 287, 287f, 289t
Mixture assessment factor (MAF), 134, 135, 366, 367
Mode of (toxic) action (MOA), 12, 20, 58–67, 93, 120, 125, 127t, 133, 172, 284, 309t, 310t, 313t,340, 367
Molecular initiating event (MIE), 55, 58, 59t, 95, 96, 97, 101, 120, 126, 147, 150f, 152f, 154, 156f, 157, 165, 340, 345, 351f, 367
Monitoring, 1, 18–22, 90, 154, 169, 268f, 269–271, 348–349
Morphogenesis, 83, 367
Multiplex, 152, 154, 159, 269, 270, 296, 304, 349–350
Multi-substance potentially affected fraction (msPAF), 141, 367
Mutagenicity, 14, 15, 28, 207–209, 242, 271, 276, 296, 306, 346, 367. *See also* Genotoxicity
Mutation, 65, 81, 367
Myelinating cells, 86, 367. *See also* Myelin sheet
Myelin sheet, 87, 367

N

NADPH (reduced nicotinamide adenine dinucleotide phosphate), 62, 66, 67f, 69, 173, 367
NADPH-quinoneoxidoreductase (NQO1), 69, 173
NAM. *See* New approach method
Narcosis (mode of action), 59, 61, 290, 367. *See also* Baseline toxicity
National Center for Advancing Translational Sciences (NCATS), 147, 148f, 367, 375
National Center for Computational Toxicology (NCCT), 147, 367, 375
National Institute of Environmental Health Science (NIEHS), 147, 368
National Pollutant Discharge Elimination System (NPDES), (USA), 41, 42, 368
National Research Council (NRC), 26, 143, 147
National Water Quality Management Strategy (NWQMS), (Australia), 39, 42, 368
Native cell, 12–13, 23, 171, 367. *See also* Primary cell lines, Immortalised cell lines
Necrosis, 58, 61, 77, 367
Negative control, 103, 110–111, 228–230, 231f, 232, 236, 237, 290, 367
Neonicotinoid, 63, 367
Nephelometric turbidity units (NTU), 251

Nephrotoxicity, 78–79, 79f, 367
Neuron, 80, 85–87, 367. *See also* Glial cell, Neurotransmitter, Axon
Neuronopathy, 86, 368
Neurotoxicity, 60t, 63, 81, 85–87, 86f, 200–202, 309t, 311t, 312t, 346, 368
Neurotransmitter, 63, 86, 96, 201, 202, 368
Neutral red uptake (NRU), 219, 220t, 221t, 368
New approach method (NAM), 9, 143, 144–148, 154–157, 368
Next-generation risk assessment (NGRA), 148, 151–153, 368
Nicotinamide adenine dinucleotide phosphate (NADPH), 62, 66, 67f, 69, 173, 367
Nitrosamines, 6, 208, 368. *See also* Disinfection by-product
Nominal concentration, 151, 153, 159t, 161, 162, 258
Non-specific (mode of action), 368. *See also* Non-specific toxicity, Baseline toxicity
Non-specific toxicity, 12–14, 58–62, 93, 127t, 136, 137f, 218, 251, 294, 296, 310t, 368. *See also* Baseline toxicity
Non-threshold chemicals, 26, 368. *See also* Threshold chemicals
No Effect Level (NEL), 29t, 149
No Observable Adverse Effect Level (NOAEL), 28, 37, 103, 121, 122, 368
No Observable Effect Concentration (NOEC), 28, 31, 37, 44, 45, 92, 101, 107, 121, 122, 291, 368
No Observable Effect Level (NOEL), 368
Nrf2, 69, 69t, 173, 212, 213t, 214t, 288t, 368
NRU. *See* Neutral red uptake
Nuclear factor kappa B (NF-κB), 69
Nuclear receptor, 14, 54, 55t, 59, 68, 99, 105, 147, 153, 166, 175, 178, 179, 269, 278, 368. *See also* Receptor

O

Organisation for Economic Co-operation and Development (OECD), 145, 146, 221, 232, 242, 368
Organogenesis, 82, 368
Organophosphate pesticide, 63, 368. *See also* Acetylcholinesterase
Oxidation, 4, 5t, 21, 53, 178, 211, 238
Oxidative stress, 14, 57, 66–67, 69t, 136, 137f, 203, 211–212, 333, 334, 369. *See also* ROS
Ozone (O_3), 22, 368

P

p53, 69t, 70, 166, 213, 215t, 216t, 306, 316, 369. *See also* Genotoxicity, Tumour suppressor gene
Partition constant, 159, 160, 161, 162f, 165f, 167, 237, 332f, 343
Passive dosing, 162, 163, 343, 369
Passive sampling, 169, 252f, 254, 255, 294, 326, 328, 335, 337, 369
Passive transport, 369. *See also* Active transport
Pathogen, 4, 6, 35, 84, 308, 369
Pathway of toxicity, 369
PBTK (physiologically based toxicokinetic model), 153, 369
PBT. *See* persistent, bioaccumulative and toxic
PCB. *See* Polychlorinated biphenyls
PCDD (polychlorinated dibenzodioxin), 55t, 129, 334, 335, 370
PCDF (polychlorinated dibenzofuran), 129, 370
PDMS. *See* polydimethylsiloxane
persistent, bioaccumulative and toxic (PCT), 6, 8, 28, 369
Persistent organic pollutants (POP), 28, 133, 327, 334–336, 370. *See also* Stockholm Convention
Pesticide, 5t, 20, 23, 39, 42, 62, 66, 85, 124, 132, 136, 141, 179, 209, 248, 290, 308, 315
Phagocytes, 369
Phase I metabolism, 53, 70, 78, 178, 369. *See also* Metabolic activation
Phase II metabolism, 53, 369. *See also* Metabolic activation
Phenotype, 98, 369
Photodegradation, 4, 5t, 238, 247, 369

Photosynthesis, 2, 12, 59, 60t, 63, 94, 95, 118, 135–138, 137f, 266, 275, 278, 283t, 284, 317, 370
Photosystem II, 63, 95, 199, 200t, 279, 290, 316, 370. *See also* Photosynthesis
Phytotoxicity, 17, 199–200, 271, 370
Plasmid, 13, 172, 172f, 370
Point of departure POD), 103, 152f, 281, 287, 341, 370
Point of inflexion, 370
Polycyclic aromatic hydrocarbon (PAH), 38, 53, 54f, 85, 130, 173, 327, 369
Polydimethylsiloxane (PDMS), 326, 332f
Polychlorinated biphenyls, 75, 370. *See also* (PCBs)
POPs. *See* persistent organic pollutants
Positive control, 108, 228, 237, 239, 370
PPAR, 55t, 173, 175, 370
PR. *See* Progesterone receptor
Precautionary principle, 32
Precision, 226–227, 235
Predicted No Effect Concentration (PNEC), 8, 28, 29f, 30, 42, 284, 291, 370
Primary cell lines, 348, 370
Primary mechanism, primary effects, 370
Primary producer, 44, 93
Progesterone receptor (PR), 89, 179, 190–194, 268f, 276, 304t, 306, 311t, 312t, 313t, 370. *See* Progestogens
Progestogens, 88, 89, 370. *See also* Progesterone receptor
Prokaryote, 346, 370. *See also* Eukaryote
Promoter, 64, 69, 172, 172f, 370
Protection-goal motivated approach, 16–17, 371. *See also* Chemical-group motivated approach
Proteolytic enzymes, 84, 371
Proteomics, 11, 347, 371
Purified recycled water (PRW), 16, 371
PXR, 55t, 139, 173, 178, 178t, 179t, 254, 269, 318, 371. *See also* Phase I metabolism
Pyrethroids, 63, 371

Q

qAOP (quantitative adverse outcome pathway), 326, 334
QIVIVE (quantitative *in vitro* to *in vivo* extrapolation), 144, 150–151, 233, 284, 346, 371
Quality assurance and quality control (QA/QC), 92, 161, 225–243, 246, 371
Quality standards (QS), 41, 371
Quantitative structure-activity relationship (QSAR), 146, 278f, 371
QuEChERS (quick, easy, cheap, effective, rugged and safe), 326, 334
Quinolones, 127t, 371

R

Radioimmunoassay (RIA), 371
RAR. *See* retinoic acid receptor
REACH (Registration, Evaluation, Authorisation and Restriction of Chemicals), 9, 20, 26–28, 32, 41, 133, 134, 371
Reactive oxygen species (ROS), 60t, 65–67, 67f, 69t, 81, 84, 166, 211–212, 371, 373
Reactive toxicity, 13, 14, 64–67, 70, 165, 202–212, 309t, 371
Read-across, 22, 93, 133, 146, 281, 291–294, 371
Receptor, 13f, 41, 55, 57–59, 63, 139, 172, 336, 371. *See also* Nuclear receptor
Receptor binding assay (RBA), 2, 9, 168, 371. *See also* Receptor, Ligand
Receptor-mediated toxicity, 14, 130, 372. *See also* Receptor, Nuclear receptor
Recombinant cell lines, 12, 372
Recovery, 77, 249, 257–260, 342
REF. *See* Relative enrichment factor or relative extraction factor
Reference dose (RfD), 29, 37, 281, 372
Relative effect potency (REP), 116, 117, 161, 180, 182f, 284, 285, 291, 372
Relative enrichment factor (REF), 102, 114, 115f, 117, 118, 155f, 186, 230, 231f, 248, 262, 289, 294, 306, 307f, 315f, 372
Relative extraction factor (REF), 114, 186, 248, 372
Relative fluorescence unit (RFU), 236, 372
Relative light units (RLU), 372

Relative genotoxic unit (rGTU), 309t–310t, 372
Relative toxic unit (rTU), 310t, 373
REP. *See* Relative effect potency
Replication, 57, 60t, 65, 233–235
Reporter gene, 9, 13, 61, 98, 99, 108–110, 126, 139, 163, 166–168, 172–174, 172f, 178, 195, 236, 242, 269, 276, 318, 320f, 347, 350, 372
Reporter plasmid, 13, 372
Responsive element, 13, 372
Retinoic acid receptor (RAR), 22, 179, 197–199, 268f, 304, 371
Retinoic X receptor (RXR), 179, 198, 268f, 304f, 373
Reverse dosimetry, 151
Reverse osmosis (RO), 4, 22, 48, 178, 308, 372
Reverse toxicokinetics, 153
Revertant ratio (RR), 208, 209
RIA. *See* Radioimmunoassay
Risk, 2, 8, 25, 350, 373
Risk assessment, 9, 25–33, 130–135, 142–168, 373
Risk characterisation, 26, 30–32, 40, 142, 373. *See also* Risk assessment
Risk index (RI), 131, 140, 373. *See also* Risk quotient
Risk management, 25, 26, 32–33, 33f, 373
Risk quotient (RQ), 30, 131, 134f, 373. *See also* Risk index, Hazard quotient
Risk reduction, 31, 33, 373
RNA, 60t, 373
Robustness, 47, 169, 227–228
ROS. *See* reactive oxygen species
RQ. *See* Risk Quotient
RXR. *See* Retinoic X receptor

S

S9, 158, 163, 203, 206, 208, 213, 373. *See also* Metabolic activation
SCGE. *See* Single cell gel electrophoresis
Sediment, 30, 41, 114, 238, 325–337
Selectivity, 166, 226
Sensitivity, 57, 82, 180, 227, 229–231, 335, 350
Sertoli cells, 90, 373
Serum-mediated passive dosing (SMPD), 161–163, 335, 373
Shewart chart, 240
Single cell gel electrophoresis (SCGE), 203, 204t, 373. *See also* Comet assay
SMPD. *See* Serum-mediated passive dosing
Soil, 91, 114, 239, 251, 326–333, 357
Solid phase extraction (SPE), 48, 114, 169, 237, 245, 250f, 252, 253, 255–260, 266, 306, 343, 374. *See also* Liquid-liquid extraction
Solid phase microextraction (SPME), 158, 159t, 168
Solvent, 229, 237–239, 253, 254, 257, 261, 264, 333, 343, 344
SOS Chromo, 203, 204t, 373. *See also* SOS response
SOS response, 203, 204t, 374
SPE. *See* Solid phase extraction
Specificity, 138, 139, 163–166, 226, 287
Specificity ratio (SR), 112, 165f, 165, 278, 278f, 279f, 287f, 292f, 374
Specific mode of toxic action, 374
Spermatogenesis, 90, 374
Species sensitivity distribution (SSD), 30, 37, 92, 141
SPM. *See* Suspended particulate matter
SPME. *See* Solid phase microextraction
SPR. *See* Suppression ratio
SR. *See* Specificity ratio
SSD. *See also* species sensitivity distributions
Stable transfection, 374
Standard operating protocol (or procedure) (SOP), 226, 242, 352, 373
Steady-state, 151f
Stockholm Convention, 28, 133, 374. *See also* POP
Substances of very high concern (SVHC), 8, 28
Super-minimal concentration, 374. *See also* NOEC
Superoxide dismutase (SOD), 211t

Suppression ratio (SPR), 112, 112f, 182, 185t, 374
Supra-maximal concentration, 374
Surface water, 4, 5t, 15, 22, 36, 37f, 38, 41, 176t, 177t, 183t, 189t, 191t, 192t, 193t, 194t, 199t, 200t, 201t, 208, 214, 222, 248, 269, 275, 282t, 283t, 286f, 292f, 293f, 304, 314–318, 320–324
Suspended particulate matter (SPM), 250, 326–333, 374
SVHC. *See* substance of very high concern, 8, 28
Synapse, 63, 96, 374. *See also* Axon
Synergism, 120, 124, 125, 344–345, 374

T

Tamoxifen (TMX), 113f, 182, 185t, 309t, 313t, 375
Target concentration, 61, 158
Target site, 2, 53, 58, 59t, 62, 73, 87, 122, 124, 158, 159t
TCDD, 129, 173–175, 174t, 236, 329t–330t, 374
Tcpl (ToxCast Analysis Pipeline), 108, 109f, 147, 242, 375
TD. *See* Toxicodynamics
TDI, 29, 37, 374. *See also* Acceptable daily intake, Reference dose
Test battery, 2, 14, 16–17, 149, 215, 267–268, 270, 296, 304, 305, 309t, 310t, 314, 315f, 340
TEQ. *See* Toxic equivalent concentration
TEQ_{bio}, 117, 374. *See also* Toxic equivalent concentration
TEQ_{chem}, 374. *See also* Toxic equivalent concentration
Teratogenesis, 83, 374
TF. *See* Transcription factor
The International Programme on Chemical Safety (IPCS) of the World Health Organization (WHO), 130, 365
THP1, 310t, 375
THP1-CPA, 310t, 375. *See also* THP1, Cytokine
Threshold chemical, 38, 375. *See also* Non-threshold chemical
Threshold of toxicological concern (TTC), 40, 375
Thyroid receptor (TR), 179, 194–195, 196f, 197f, 268t, 304t, 311t, 312t, 376
TIE. *See* Toxicity identification evalutation
TIF2, 375. *See also* Co-activator
Tissue, 59t, 77, 78, 104, 175, 238, 335
TMX. *See* Tamoxifen
TOC (total organic carbon), 341
TOX (total organic halogens), 341
Tox21, 12, 108, 144, 147–148, 150f, 154, 157, 159, 161, 243, 349, 375
ToxCast, 108, 109f, 144, 147, 148, 154, 168, 174, 242, 317, 319, 349, 375
Toxic equivalency factor (TEF), 116, 129, 133, 374
Toxic equivalent concentration (TEQ), 116–117, 129–131, 375
Toxicity identification and evaluation (TIE), 46, 275, 375
Toxicity pathway, 7, 11, 51–71, 73–90, 98–99, 119, 124–126, 144, 154, 169, 170f, 340, 375. *See also* Adverse outcome pathway
Toxicity Testing in the 21st Century, 144, 147–148
Toxicodynamics (TD), 97, 124, 375. *See also* Toxicokinetics
Toxicogenomics, 11, 376
Toxicokinetics, 51–55, 96, 153, 163, 281, 376. *See also* Toxicodynamics
Toxic ratio (TR), 164
Toxic unit (TU), 116, 117, 122, 130, 139, 273, 290, 310t, 375
Toxtracker, 153
TR. *See* Thyroid receptor
Transcription, 54, 57, 64, 68, 69, 82, 178
Transcription factor (TF), 68, 69t, 175, 376
Transcriptomics, 11, 153, 347, 376
Transthyretin (TTR), 283t, 328, 335
Trihalomethanes (THM), 5–6, 255, 376. *See also* Disinfection by-product
Trophic level, 30, 44, 92, 376
TTC. *See* Threshold of toxicological concern
TU. *See* Toxic unit
Tumour, 70, 82, 84, 376

Tumour suppressor gene (p53), 65, 69t, 70, 166, 212, 213–215, 216t, 249, 294, 306, 316, 369

U

*umu*C, 203, 204t, 206, 207t, 271, 294, 376. *See* SOS response
Uncertainty analysis, 32
Uncoupler, 62, 376

V

Validation, 145, 147, 168, 226–231, 239, 246, 346, 350
Vasoactive agent, 376
Vibrio fischeri, 14, 218. *See also Aliivibrio fischeri*
Vitellogenin (Vtg), 10, 46, 47, 98, 126, 154, 171, 376. *See also* Endocrine disruption

W

96-Well plate: microtiter plate, 157, 158, 169, 218, 225, 248, 262, 347
Water Framework Directive (WFD), 9, 36, 43f, 142, 250, 305, 376
Water Safety Plan (WSP), 351, 352f
Wastewater treatment plant (WWTP), 46, 48, 112, 114, 115f, 132, 135, 136f, 137f, 141, 175, 231f, 247f, 266, 273, 299, 314f, 314–318, 376
Well, 1, 47, 147, 157, 158, 159, 160, 160f, 161, 167, 169, 170, 215, 218, 222, 225, 232, 237, 243, 246f, 248, 262, 263f, 276, 283t, 343, 346, 347, 349, 377
WET, 41–49, 252, 376
WFD. *See* Water Framework Directive
WHO, 29, 36–39, 42, 133, 281, 351, 376
WWTP. *See* Wastewater treatment plant

X

Xenobiotic, 53, 54–57, 69, 75, 76f, 175, 202, 210, 377
Xenobiotic metabolism, 53, 53f, 68, 170, 173–179, 222, 258, 267, 270, 288t, 290, 291, 297, 304t, 309t, 311, 313t, 318
Xenobiotic receptor, 54, 55, 377
Xenoestrogen, 97, 123, 126, 267, 377. *See also* Estrogen

Y

Yeast estrogen screen (YES), 99, 126, 180, 242, 377. *See also* Estrogen receptor, Recombinant cell lines
Yeast two-hybrid assay, 195, 199

Z

Z-factor, 228, 229f, 241
Zona radiata protein (Zrp), 377

IWA Publishing's authorised EU representative for General Product Safety Regulations is Diane D'Arras, 15 rue Duret, 75116 Paris, France, e-mail: safety@iwap.co.uk.

Printed and bound by CPI Group (UK) Ltd, Croydon, CR0 4YY
12/05/2026
02108877-0003